S0-AWV-341

Frontiers in Crystal Engineering

Frontiers in Crystal Engineering

Edited by

Edward R.T. Tiekink

Department of Chemistry, The University of Texas at San Antonio

Jagadese J. Vittal

Department of Chemistry, National University of Singapore

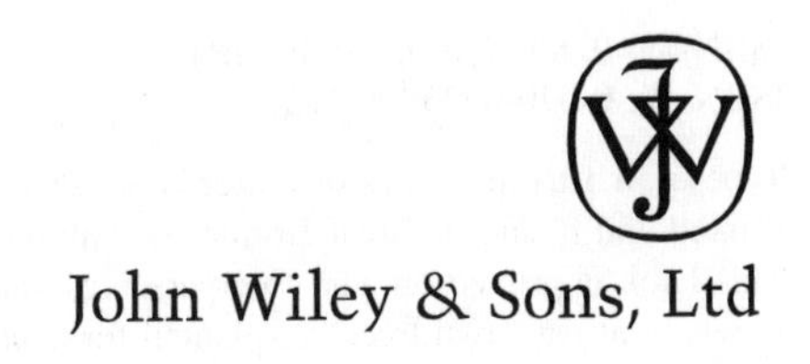

John Wiley & Sons, Ltd

Telephone (+44) 1243 779777

Email (for orders and customer service enquiries): cs-books@wiley.co.uk
Visit our Home Page on www.wiley.com

Other Wiley Editorial Offices

John Wiley & Sons Inc., 111 River Street, Hoboken, NJ 07030, USA

Jossey-Bass, 989 Market Street, San Francisco, CA 94103-1741, USA

Wiley-VCH Verlag GmbH, Boschstr. 12, D-69469 Weinheim, Germany

John Wiley & Sons Australia Ltd, 42 McDougall Street, Milton, Queensland 4064, Australia

John Wiley & Sons (Asia) Pte Ltd, 2 Clementi Loop #02-01, Jin Xing Distripark, Singapore 129809

John Wiley & Sons Canada Ltd, 22 Worcester Road, Etobicoke, Ontario, Canada M9W 1L1

Wiley also publishes its books in a variety of electronic formats. Some content that appears in print may not be available in electronic books.

Library of Congress Cataloging-in-Publication Data:

Frontiers in crystal engineering / Edward R.T. Tiekink & Jagadese J. Vittal (editors).
p. cm.
Includes bibliographical references and index.
ISBN-13: 978-0-470-02258-0 (HB) (acid-free paper)
ISBN-10: 0-470-02258-2 (HB) (acid-free paper)
1. Molecular crystals – Research. I. Tiekink, Edward R. T. II. Vittal, Jagadese J.
QD921.F76 2006
548 – dc22

2005016323

British Library Cataloguing in Publication Data

A catalogue record for this book is available from the British Library

ISBN-13: 978-0-470-02258-0 (HB)
ISBN-10: 0-470-02258-2 (HB)

Typeset in 10/12pt Times by Laserwords Private Limited, Chennai, India
Printed and bound in Great Britain by Antony Rowe Ltd, Chippenham, Wiltshire
This book is printed on acid-free paper responsibly manufactured from sustainable forestry in which at least two trees are planted for each one used for paper production.

Contents

List of Contributors

Brendan F. Abrahams, School of Chemistry, University of Melbourne, Victoria, Australia

Stuart R. Batten, School of Chemistry, Monash University, Clayton, Victoria, Australia

Roger Bishop, School of Chemistry, The University of New South Wales, Sydney, New South Wales, Australia

Dario Braga, Dipartimento di Chimica G. Ciamician, University of Bologna, Bologna, Italy

Xiao-Ming Chen, School of Chemistry and Chemical Engineering, Sun Yat-Sen University, Guangzhou, China

Daniela D'Addario, Dipartimento di Chimica G. Ciamician, University of Bologna, Bologna, Italy

Gautam R. Desiraju, School of Chemistry, University of Hyderabad, Hyderabad, India

Tomislav Friščić, Department of Chemistry, University of Iowa, Iowa City, Iowa, USA

Shuhei Furukawa, Department of Synthetic Chemistry and Biological Chemistry, Kyoto University, Katsura, Japan

Stefano Giaffreda, Dipartimento di Chimica G. Ciamician, University of Bologna, Bologna, Italy

Fabrizia Grepioni, Dipartimento di Chimica G. Ciamician, University of Bologna, Bologna, Italy

Michaele J. Hardie, School of Chemistry, University of Leeds, Leeds, UK

Susumu Kitagawa, Department of Synthetic Chemistry and Biological Chemistry, Kyoto University, Katsura, Japan

Leonard R. MacGillivray, Department of Chemistry, University of Iowa, Iowa City, Iowa, USA

Lucia Maini, Dipartimento di Chimica G. Ciamician, University of Bologna, Bologna, Italy

Jennifer A. McMahon, Department of Chemistry, University of South Florida, Tampa, Florida, USA

Marco Polito, Dipartimento di Chimica G. Ciamician, University of Bologna, Bologna, Italy

Katia Rubini, Dipartimento di Chimica G. Ciamician, University of Bologna, Bologna, Italy

Jonathan W. Steed, Department of Chemistry, University of Durham, Durham, UK

Edward R. T. Tiekink, Department of Chemistry, The University of Texas at San Antonio, San Antonio, Texas, USA

Ming-Liang Tong, School of Chemistry and Chemical Engineering, Sun Yat-Sen University, Guangzhou, China

Peddy Vishweshwar, Department of Chemistry, University of South Florida, Tampa, Florida, USA

Jagadese J. Vittal, Department of Chemistry, National University of Singapore, Singapore

Michael J. Zaworotko, Department of Chemistry, University of South Florida, Tampa, Florida, USA

Foreword

Crystal engineering, the design of functional molecular solids and coordination polymers, has emerged as one of the most appealing areas of chemical research in recent years. The subject attracts attention for both fundamental and applied reasons. Given a molecular structure, what is the crystal structure? Or alternatively, given a ligand structure and specified metal ions, what is the network structure that will form? These questions appear to be simple, but the answers are far from trivial. A molecule, or tecton, may be rigid or flexible. Assembly of tectons in the crystal occurs through the intermediacy of particular patterns of intermolecular interactions, the all-important supramolecular synthons. However, prediction of crystal structures is still difficult. Even rigid molecules employ a variety of alternative synthon patterns, giving rise to polymorphism, and our ability to predict which synthons would actually appear in crystal structures depends acutely on a corresponding ability to understand and compare the nature of the interactions that constitute these synthons. For flexible molecules, the problems of crystal engineering quickly reach formidable levels of difficulty, because the intra- and intermolecular interactions are comparable; accordingly, molecular structure and crystal structure affect each other in mutually unpredictable ways. Crystal engineering becomes even more lively with the large palette of interactions now available, and only the imagination of the chemist can limit the ways in which these interactions – strong or weak, familiar or exotic – may be employed in crystal design strategy.

Crystal engineering of coordination polymers is a popular area today. Yet, it is difficult not only to predict what network will be formed for a given ligand–metal combination but also to anticipate how these networks will interact with one another in the crystal. Understanding the relationships between molecular and supramolecular stereochemistry is one of the major challenges in this area. Interpenetration, intercalation and host–guest properties are all manifestations of these challenges. Unless the host–host interactions are extremely strong and specific, the formation of a large host network that is filled with plenty of guest molecules is probably the result of guest-induced host assembly. How then can the chemist ensure that evacuation of the guest will result in an empty host, which can then be "refilled" with another guest? These issues also bring into focus the fact that an engineered structure should also be a useful structure in terms of its physical or chemical properties. In the early days of crystal engineering, one could afford to design crystal structures simply for their aesthetic value or because, in doing so, a better understanding of the relevant intermolecular interactions emerged. Today, when the subject has matured, it is more than ever likely that good uses for these crystals will be sought, be it in catalysis, pharmaceutical chemistry, synthesis or materials science.

The editors of this volume have assembled a collection of chapters that highlight recent progress in these important areas of crystal engineering. The book is likely to be useful to both entrants to the field as well as to established practitioners, and demonstrates the variety of approaches that are being used to tackle the many difficult problems associated with complexity, design and functionality of crystalline molecular solids.

Gautam R. Desiraju
Hyderabad

1

Applications of Crystal Engineering Strategies in Solvent-free Reactions: Toward a Supramolecular Green Chemistry

DARIO BRAGA, DANIELA D'ADDARIO, LUCIA MAINI, MARCO POLITO, STEFANO GIAFFREDA, KATIA RUBINI and FABRIZIA GREPIONI
Dipartimento di Chimica G. Ciamician, University of Bologna, Via Selmi 2, 40126 Bologna, Italy.

1. INTRODUCTION

Making crystals by design is the paradigm of crystal engineering [1]. The goal of this field of research is that of assembling functionalized molecular and ionic components into a target network of supramolecular interactions [2]. This "bottom-up" process generates *collective* supramolecular properties from the convolution of the physical and chemical properties of the individual building blocks with the periodicity and symmetry operators of the crystal (Figure 1) [3].

One can envisage two main subareas of crystal engineering, namely, those of coordination networks [4] and of molecular materials [5], even though all possible intermediate situations are possible. The preparation of coordination networks or polymers can be appropriately described as *periodical coordination chemistry* and exploits the possibility of *divergent ligand-metal coordination*, as opposed to the more traditional *convergent coordination* chemistry operated by chelating polydentate ligands [6] (Figure 2).

The possibility of exploiting engineered coordination networks for practical applications (such as absorption of molecules, reactions in cavities, etc.) very much depends on whether the networks contain large empty spaces (channels, cavities, etc.) [7] or whether the network is close packed because of interpenetration and self-entanglement [8]. The

Frontiers in Crystal Engineering. Edited by Edward R.T. Tiekink and Jagadese J. Vittal

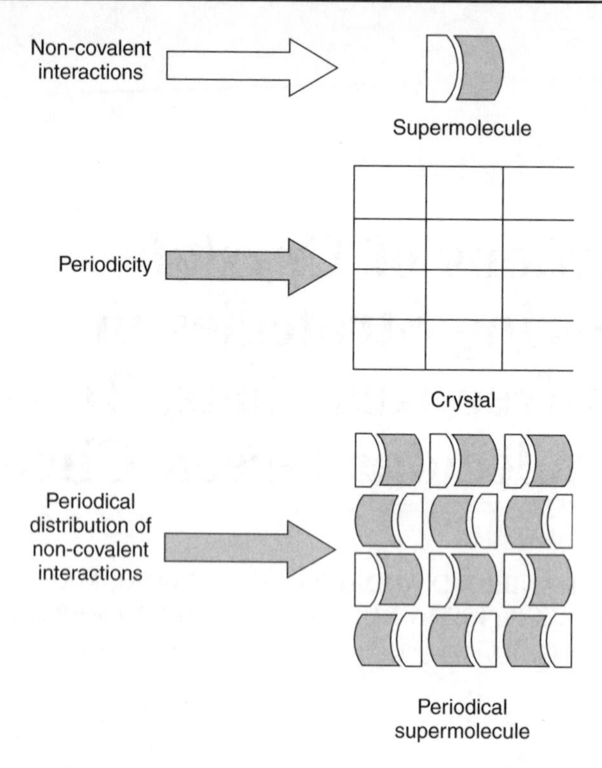

Figure 1. From molecules to periodical supermolecules: the collective properties of molecular crystals result from the convolution of the properties of the individual molecular/ionic building blocks with the periodical distribution of intermolecular non-covalent bonding of the crystal. Reproduced from Ref. 6 by permission of The Royal Society of Chemistry.

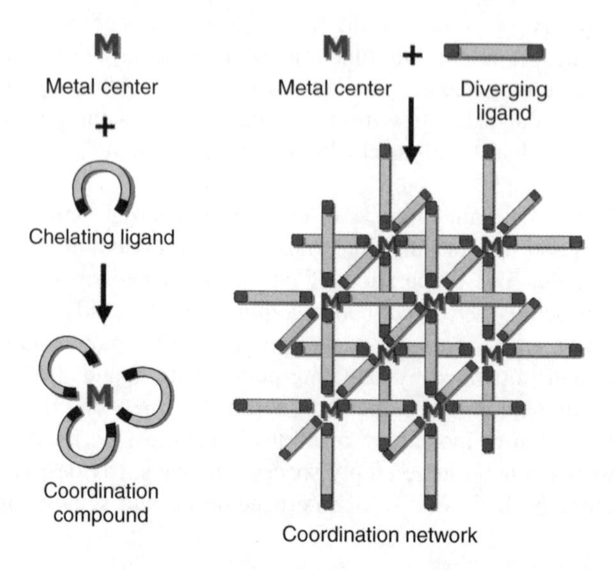

Figure 2. The relationship between molecular (left) and periodical (right) coordination chemistry: the use of bidentate ligand spacers allows construction of *periodical* coordination complexes. Reproduced from Ref. 6 by permission of The Royal Society of Chemistry.

possibility of sponge-like behavior by which the network can change, that is, swell/shrink to accommodate/release guest molecules, should also be taken into account [9].

While in periodical coordination chemistry it is useful to focus on the *knots* and *spacers* in order to describe the topology of the network, when dealing with molecular materials what matters most are the characteristics of the component molecules or ions and the type of interactions holding these building blocks together. These interactions are mainly of the non-covalent type (van der Waals, hydrogen bonds, π-stacking, ionic interactions, ion pairs, etc.) [10–12]. The intermolecular links will be weaker than the covalent chemical bonds within the individual components, which, in general, will retain their chemical and physical identity once evaporated or dissolved. This is not so for coordination networks that usually cannot be reversibly assembled and disassembled.

1.1. Making Crystals by *Smashing* Crystals?

Since the focus of crystal engineering is *making* crystals with a purpose, crystal *makers* invariably (and inevitably) end up facing the problem of obtaining crystals, possibly single crystals of reasonable size, in order to benefit from the speed and accuracy of single-crystal x-ray diffraction experiments. Even though amorphous materials can be extremely interesting, and certainly are so in the biological world and are providing inspiration to scientists [13], in crystal engineering studies, the desired materials need to be *by definition* in the crystalline form and will be obtained by a crystallization method, whether from solution, melt or vapor or from more forceful hydrothermal syntheses.

In this chapter, we will provide evidence that reactions between solids and between solids and vapors offer alternative ways to prepare crystals, both of the coordination network and of the molecular crystal type [14]. It may be useful to stress that, since reactions involving solid reactants or occurring between solids and gases do not generally require recovery, storage and disposal of solvents, they are of interest in the field of "green chemistry", where environmentally friendly processes are actively sought [15]. Furthermore, solvent-less reactions often lead to very pure products and reduce the formation of solvate species [16].

In the following, we will discuss two types of solvent-free processes: those involving gas uptake by a molecular crystal to form a new crystalline solid and those involving reactions between molecular crystals or between a molecular and an ionic crystal to yield new crystalline materials [17]. Since the "Bolognese" crystal engineering laboratory has been traditionally interested in using organometallic building blocks, the vast majority of cases discussed throughout this chapter will show the utilization of organometallic building blocks [18].

In previous papers, we have argued that reactions of molecular crystals with gases or other crystals ought to be regarded as *supramolecular reactions* whereby non-covalent interactions (including coordination bonds) between guest and the host are broken and formed. The two types of processes are depicted in Figure 3.

The absence of solvent requires that other means be used to bring molecules into contact for the formation of supramolecular bonds. Since our reactants are, in general, molecular crystals, the utilization of finely ground powders favors reactions with vapors because of the large surface area. On the other hand, the reactions between two molecular crystals (as shown in Figure 3) often require co-grinding to obtain intercrystal reactions. Hence, in both types of reactions, crystals need to be ground, a condition that may appear

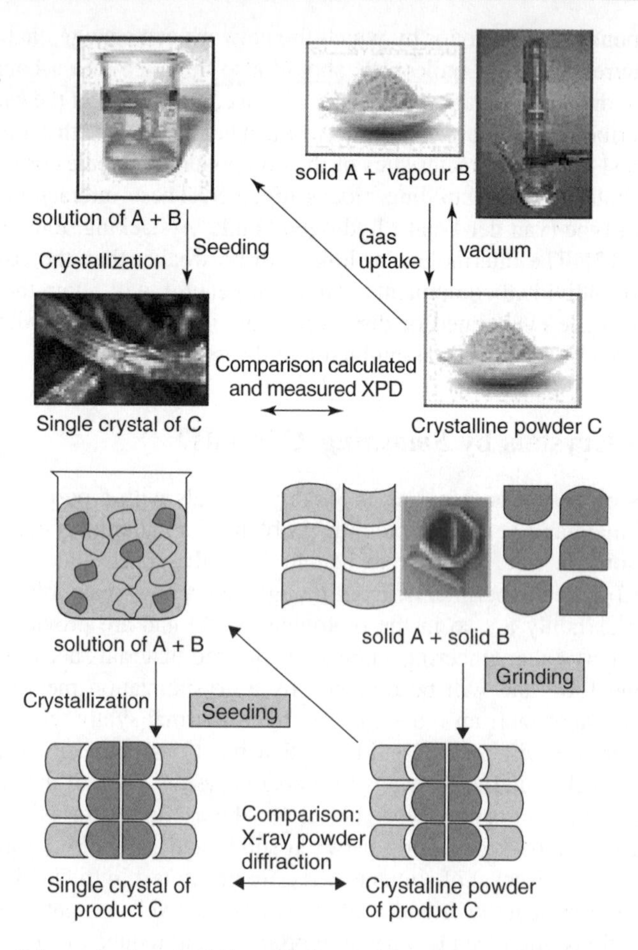

Figure 3. The solid–gas (top) and solid–solid (bottom) processes and the strategy to obtain single crystals by recrystallization of the solid reaction product in the presence of *seeds* of the desired crystals.

to contradict the *single-crystal dogma*, as the vast majority of crystal engineering studies are based on the type of structural knowledge provided by single-crystal x-ray diffraction experiments. This is also true for the cases discussed herein. The characterization of the products of both solid–gas and solid–solid strategies rely on the possibility of obtaining single crystals of the desired product, either by direct crystallization from solution or via *seeding*, in order to be able to compare the powder diffractograms measured on the product of the solvent-free process with that calculated on the basis of the single crystal structures. This aspect of the work will be briefly expanded upon in the next section.

1.2. Milling, Grinding, Kneading and Seeding

As mentioned in the previous section, the size of the crystals will dictate the experimental method of choice. Unless one resorts to high-intensity synchrotron radiation, microcrystals will allow only powder diffraction experiments, which only rarely can be used for *ab initio* structure determination in order to get those precise structural information that are so essential to the crystal engineer. Clearly, the solvent-less reactions of a crystalline powder with a vapor or that between two crystalline powder are, generally speaking, not compatible with the formation of large single crystals, since it will generally produce a material in the form of a powder.

Nonsolution methods to obtain new products require the chemist, or crystal engineer, to explore/exploit methods that are not routinely used in chemical laboratories. Reactions of the type described herein require broadening the view of typical chemical processes. Beside the conventional "academic" chemical reaction procedures (typically Schlenk techniques, if air-sensitive organometallic molecules are synthesized as building blocks), one has to resolve to methods, such as *grinding* and *milling*, that are less popular – when they are not dismissed as nonchemical – in research laboratories. These methods are related to the mechanochemical activation of reactions occurring between solids and to the control exerted on the crystallization process [19, 20].

Grinding and milling

Typical mechanochemical reactions are those activated by co-grinding or milling of powder materials, usually carried out either manually, in an agate mortar, or electromechanically, as in ball milling. In both cases, the main difficulty is in controlling reaction conditions: grinding time, temperature, pressure exerted by the operator, and so on. Furthermore, the heat generated in the course of the mechanochemical process can induce local melting of crystals or melting at the interface between the different crystals, so that the reaction takes place in the liquid phase even though solid products are ultimately recovered. One should also keep in mind that mechanical stress, by fracturing the crystals, increases surface area and facilitates interpenetration and reaction depending on the ability of molecules to diffuse through the crystal surfaces. With this viewpoint, intersolid reactions between molecular crystals can be conceptually related to the uptake of a vapor from a molecular solid to form a new crystalline solid.

Mechanochemical processes, and more generally solid-state reactions, though little exploited at the level of academic research, are commonly used at industrial level, mainly with inorganic solids and materials [21].

Kneading

Even though the discussion of the role of solvent in a chapter devoted to solvent-free reactions may seem contradictory, it is useful to remind the reader that, in some cases, the use of a small quantity of solvent can accelerate solid-state reactions carried out by grinding or milling [22]. The method of the grinding of powdered reactants in the presence of a small amount of solvent, also known as *kneading*, is commonly exploited, for instance, in the preparation of cyclodextrin inclusion compounds. Studies of *kneading* and development of laboratory/industry *kneaders* (mainly of pharmaceutical powders) have been carried out [23].

As an example of a laboratory scale preparation, one could mention the preparation by *kneading* of binary β-cyclodextrin-bifonazole [24], and of β-cyclodextrin inclusion compounds of ketoprofen [25a], ketoconazole [25b], and carbaryl [25c].

Clearly, the objection whether a *kneaded* reaction between two solid phases can be regarded as a *bona fide* solid-state process is justified. However, in the context of this work, our interest lies more in the methods used to make new crystalline materials rather than in the mechanisms. *Kneading* has been described as a sort of "solvent catalysis" of the solid-state process, whereby the small amount of solvent provides a *lubricant* for molecular diffusion.

Seeding

Another apparent contradiction of the utilization of "noncrystallization" methods to prepare new crystalline materials arises from the fact that the products of grinding, milling and kneading processes are usually in the form of a powdered material, while single crystals would be desirable or indispensable for the characterization of the reaction product. Crystallization by *seeding*, that is, crystallization in the presence of microcrystals of the desired materials, is one way to control the growth of a given polycrystalline powder to a size adequate for single-crystal x-ray diffraction experiments.

Seeding procedures are commonly employed in pharmaceutical industries to make sure that the desired crystal form is always obtained from a preparative process, a relevant problem when different polymorphic modifications can be obtained [26]. It is also important to appreciate that *seeding* often prevents formation of kinetically favored products and allows those thermodynamically favored.

Seeds of isostructural or *quasi*-isostructural species that crystallize well can also be employed to induce crystallization of unyielding materials, a process that may be termed *heteromolecular seeding* [27, 28]. For instance, chiral co-crystals of tryptamine and hydrocinnamic acid have also been prepared by crystallization in the presence of seeds of different chiral crystals [29]. Of course, unintentional *seeding* may also alter the crystallization process in an undesired manner [30].

The use of these methods will now be discussed mainly on the basis of examples taken from our recent work.

2. MECHANOCHEMICAL PREPARATION OF HYDROGEN-BONDED ADDUCTS

Manual grinding of the ferrocenyl dicarboxylic acid complex $[Fe(\eta^5\text{-}C_5H_4COOH)_2]$ with solid nitrogen-containing bases, namely, 1,4-diazabicyclo[2.2.2]octane, 1,4-phenylenediamine, piperazine, *trans*-1,4-cyclohexanediamine and guanidinium carbonate, generates quantitatively the corresponding organic-organometallic adducts [31] (Figure 4a). The case of the adduct $[HC_6N_2H_{12}][Fe(\eta^5\text{-}C_5H_4COOH)(\eta^5\text{-}C_5H_4COO)]$ (Figure 4b) is particularly noteworthy because the same product can be obtained in three different ways: (i) by reaction of solid $[Fe(\eta^5\text{-}C_5H_4COOH)_2]$ with vapors of 1,4-diazabicyclo[2.2.2]octane (which possesses a small but significant vapor pressure), (ii) by reaction of solid $[Fe(\eta^5\text{-}C_5H_4COOH)_2]$ with solid 1,4-diazabicyclo[2.2.2]octane, that is, by co-grinding of the two crystalline powders, and by reaction of the two reactants in MeOH solution. Clearly, the fastest process is the solid–solid reaction. It is also interesting to note that the base can be removed by mild treatment regenerating the structure of the starting dicarboxylic acid. The processes imply breaking and reassembling of hydrogen-bonded networks, conformational change from cis to trans of the –COO/–COOH groups on the ferrocene diacid, and

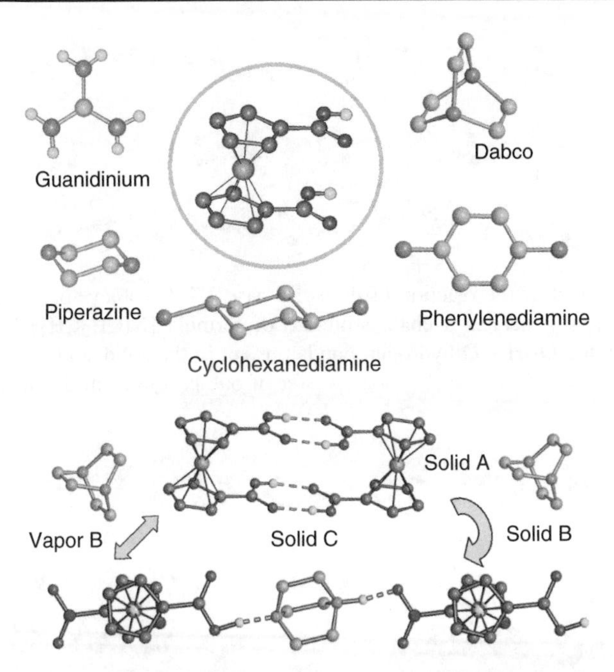

Figure 4. (a) Grinding of the organometallic complex $[Fe(\eta^5\text{-}C_5H_4COOH)_2]$ (top center) as a solid polycrystalline material with the solid bases 1,4-diazabicyclo[2.2.2]octane, $C_6H_{12}N_2$ (top right), guanidinium carbonate, $[C(NH_2)_3]_2[CO_3]$ (top left – only cation is shown), 1,4-phenylenediamine, $p\text{-}(NH_2)_2C_6H_4$, (bottom right), piperazine, $HN(C_2H_4)_2NH$, (bottom left) and *trans*-1,4-cyclohexanediamine, $p\text{-}(NH_2)_2C_6H_{10}$, (bottom center) generates quantitatively the corresponding adducts $[HC_6H_{12}N_2][Fe(\eta^5\text{-}C_5H_4COOH)(\eta^5\text{-}C_5H_4COO)]$, $[C(NH_2)_3]_2[Fe(\eta^5\text{-}C_5H_4COO)_2]\cdot 2H_2O$, $[HC_6H_8N_2][Fe(\eta^5\text{-}C_5H_4COOH)(\eta^5\text{-}C_5H_4COO)]$, $[H_2C_4H_{10}N_2][Fe(\eta^5\text{-}C_5H_4COO)_2]$, and $[H_2C_6H_{14}N_2][Fe(\eta^5\text{-}C_5H_4COO)_2]\cdot 2H_2O$, and (b) the solid–gas and solid–solid reactions involving 1,4-diazabicyclo[2.2.2]octane with formation of the linear chain.

proton transfer from acid to base. As mentioned above, in some cases, it was necessary to resolve to *seeding*, that is, to the use of a tiny amount of power of the desired compound, to grow crystals suitable for single-crystal x-ray experiments.

The effect of mechanical mixing of solid dicarboxylic acids $HOOC(CH_2)_nCOOH$ (n = 1–7) of variable chain length together with the solid base 1,4-diazabicyclo[2.2.2]octane, $C_6H_{12}N_2$, to generate the corresponding salts or co-crystals of formula $[N(CH_2CH_2)_3N]\text{-}H\text{-}[OOC(CH_2)_nCOOH]$ (n = 1–7) has also been investigated [32]. The reactions implied transformation of interacid O–H---O bonds into hydrogen bonds of the O–H---N type between acid and base, an example is shown in Figure 5. The nature (whether neutral O–H---N or charged $^{(-)}$O---H–N$^{(+)}$) of the hydrogen bond was established by means of solid-state NMR measurement, the chemical shift tensors of the compounds obtained with chain length from 3 to 7 [32].

The mechanochemical formation of hydrogen-bonded co-crystals between sulfonamide (4-amino-N-(4,6-dimethylpyrimidin-2-yl)benzenesulfonamide) and aromatic carboxylic acids has been investigated by Caira *et al.* [33].

In a related study [34], it has been shown that the reaction of $[N(CH_2CH_2)_3N]$ with malonic acid $[HOOC(CH_2)COOH]$ in the molar 1:2 ratio yields two different crystal forms

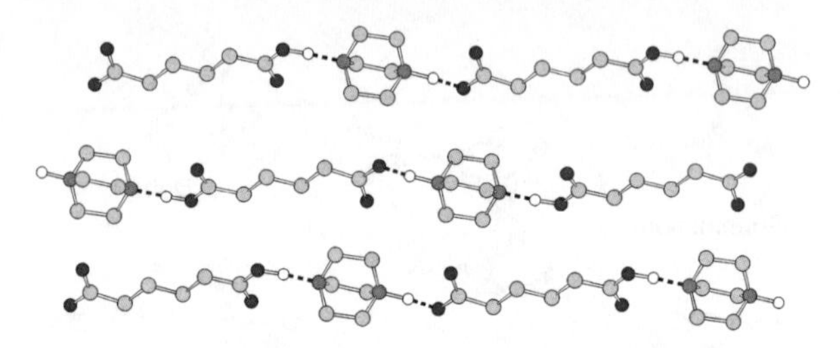

Figure 5. The product of the reaction of the solid base 1,4-diazabicyclo[2.2.2]octane, $C_6H_{12}N_2$, with solid adipic acid generates a chain structure of formula $[N(CH_2CH_2)_3N]$-H-$[OOC(CH_2)_4$-COOH]. Note how the O–H – O hydrogen-bonds present in the solid acid are replaced by neutral O–H---N and charged$^{(-)}$ O–H---N$^{(+)}$ upon transfer of one proton from the acid to the base.

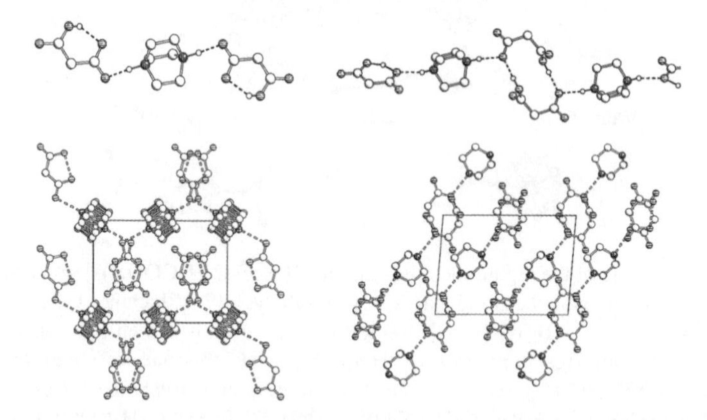

Figure 6. Form I (right) and II (left) of $[HN(CH_2CH_2)_3NH][OOC(CH_2)COOH]_2$ and their respective crystal packings. Form I is obtained by solid-state co-grinding or by rapid crystallization while form II is obtained by slow crystallization.

of the salt $[HN(CH_2CH_2)_3NH][OOC(CH_2)COOH]_2$ (Figure 6) depending on the preparation technique and crystallization speed: the less dense form I, containing mono-hydrogen malonate anions forming conventional intramolecular hydrogen bonds between hydrogen malonate anions, is obtained by solid-state co-grinding or by rapid crystallization, while a denser form II, containing intermolecular hydrogen bonds is obtained by slow crystallization. Forms I and II do not interconvert, while form I undergoes an order-disorder phase transition on cooling. These observations led the authors to wonder whether the two forms could be treated as *bona fide* polymorphs or should be regarded more appropriately as hydrogen-bond isomers of the same *solid supermolecule*.

3. MECHANICALLY INDUCED FORMATION OF COVALENT BONDS

While the reactions described in the previous section can be regarded as supramolecular reactions since only hydrogen-bond breaking and forming are implied, in this section we will discuss examples of mechanochemical formation of covalent bonds for the preparation of building blocks. Bis-substituted pyridine/pyrimidine ferrocenyl complexes have been obtained by mechanically induced Suzuki-coupling reaction [35] in the solid state starting from the complex ferrocene-1,1′-diboronic acid, $[Fe(\eta^5\text{-}C_5H_4\text{-}B(OH)_2)_2]$. It is worth recalling that boronic esters and acids are thermally stable, relatively un-reactive to both oxygen and water, and thus easily handled without special precautions. It has been reported that the use of an alumina/potassium fluoride mixture without solvent is very effective in palladium-catalyzed reactions, in particular in the Suzuki coupling of phenyl iodides with phenylboronic acids [36] and in the synthesis of thiophene oligomers *via* Suzuki coupling [36d]. The use of KF/alumina as a solid-phase support for solvent-less Suzuki reactions offers a convenient, environmentally friendly, route to the synthesis of mono- and bis-substituted pyridine and pyrimidine ferrocenyl derivatives (Figure 8), as an alternative to the preparation in solution. In the case of $[Fe(\eta^5\text{-}C_5H_4\text{-}1\text{-}C_5H_4N)_2]$, the solvent-less process is much faster, and more selective than the same reaction carried out in solution. However, the reactions in Figure 7 appear to be facilitated by the addition of tiny amounts of MeOH and, as discussed above, cannot be considered entirely solvent-less.

All reactions depicted in Figure 8 were carried out in air at room temperature. In the Suzuki reaction, the yield depends critically on having a good dispersion of the palladium catalyst on the $KF-Al_2O_3$. This dispersion was obtained by grinding the palladium catalyst with $KF-Al_2O_3$ before the reaction and, later, by adding to the mixture of Al_2O_3/reagents/catalyst a few drops (0.1–0.2 ml) of methanol, which was subsequently evaporated under reduced pressure.

Beside shorter reaction times, less workup, higher yield and the absence of solvents, the solid-state reaction affords the possibility of combining different synthetic steps in order to obtain homo- and hetero-ligand ferrocenyl complexes.

The di-substituted ferrocenyl derivatives can be utilized to prepare "complexes of complexes" [37]. In particular, novel mixed-metal macrocyclic complexes were obtained by reacting $[Fe(\eta^5\text{-}C_5H_4\text{-}1\text{-}C_5H_4N)_2]$ with metal salts, such as $AgNO_3$, $Cd(NO_3)_2$, $Cu(CH_3COO)_2$, $Zn(CH_3COO)_2$, and $ZnCl_2$ [38]. A family of hetero-bimetallic metalla-macrocycles was obtained and characterized: $[Fe(\eta^5\text{-}C_5H_4\text{-}1\text{-}C_5H_4N)_2]_2Ag_2(NO_3)_2\cdot 1.5H_2O$, $[Fe(\eta^5\text{-}C_5H_4\text{-}1\text{-}C_5H_4N)_2]_2\ Cu_2(CH_3COO)_4\cdot 3H_2O$, $[Fe(\eta^5\text{-}C_5H_4\text{-}1\text{-}C_5H_4N)_2]_2Cd_2(NO_3)_4\cdot CH_3OH\cdot 0.5C_6H_6$, $[Fe(\eta^5\text{-}C_5H_4\text{-}1\text{-}C_5H_4N)_2]_2Zn_2(CH_3COO)_4$ and $[Fe(\eta^5\text{-}C_5H_4\text{-}1\text{-}C_5H_4N)_2]_2Zn_2Cl_4$ (Figure 8). Beside the metalla-macrocycles, the reaction of mechanochemically prepared $[Fe(\eta^5\text{-}C_5H_4\text{-}1\text{-}C_5H_4N)_2]$ with the ferrocenyl dicarboxylic acid complex $[Fe(\eta^5\text{-}C_5H_4COOH)_2]$ has led to the supramolecular adduct $[Fe(\eta^5\text{-}C_5H_4\text{-}1\text{-}C_5H_4N)_2]$ $[Fe(\eta^5\text{-}C_5H_4COOH)_2]$. However, in these cases, mechanochemical mixing leads to formation of an amorphous material.

As a matter of fact, there are not yet many examples of the utilization of mechanochemical procedures in coordination chemistry. Balema *et al.* have shown, for instance, that the cis-platinum complexes *cis*-$(Ph_3P)_2PtCl_2$ and *cis*-$(Ph_3P)_2PtCO_3$ can be prepared mechanochemically from solid reactants in the absence of solvent [39]. Orita *et al.*, on the other hand, have reported that the reaction of (ethylenediamine) $Pt(NO_3)_2$ with 4,4′-bipyridine, which takes as long as 4 weeks at 100 °C to form metalla-macrocycles molecular squares, is brought to completion within 10 min at room temperature by mixing reactants without solvents [40]. Similar reaction acceleration has also been observed with triazine-based ligands. Double helix formation under solvent-free conditions has also been achieved by reacting chiral oligo(bipyridine) copper complexes with $[(CH_3CN)_4Cu]PF_6$. The progress

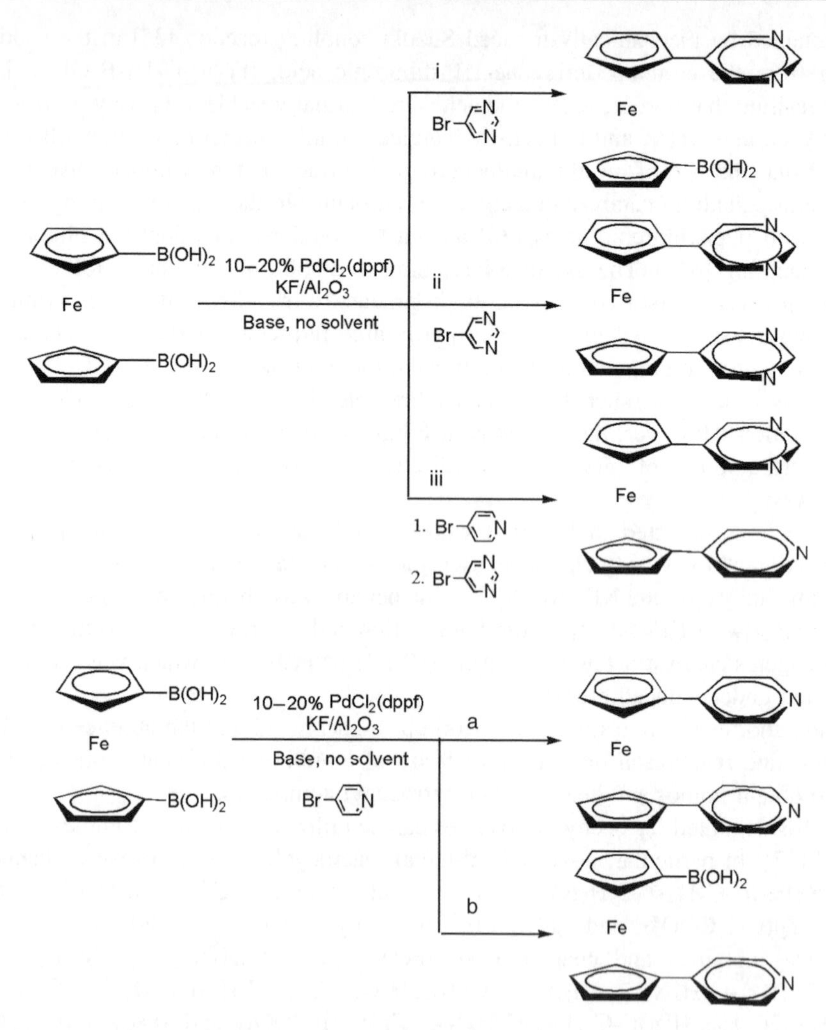

Figure 7. The solid-state synthesis of mono- and bis-substituted pyridine and pyrimidine ferrocenyl derivatives (i: stoichiometry (1:1); ii: stoichiometry (1:2); iii: stoichiometry (1:1:1); a: stoichiometry (1:2); b: stoichiometry (1:1)).

of the reaction was monitored by measuring solid-state CD-spectra showing that after grinding for 5 min the desired helicate had been obtained.

3.1. Mechanochemical Preparation of Coordination Networks

Even though the mechanochemical preparation of the metalla-macrocycles described in the previous section was not possible, coordination polymers with bidentate nitrogen bases can be prepared mechanochemically [41].

The coordination polymer $Ag[N(CH_2CH_2)_3N]_2[CH_3COO]\cdot 5H_2O$ has been obtained by co-grinding in the solid state, and in air, of silver acetate and $[N(CH_2CH_2)_3N]$ in a 1:2 ratio (Figure 9). Single crystals suitable for x-ray diffraction were obtained from a

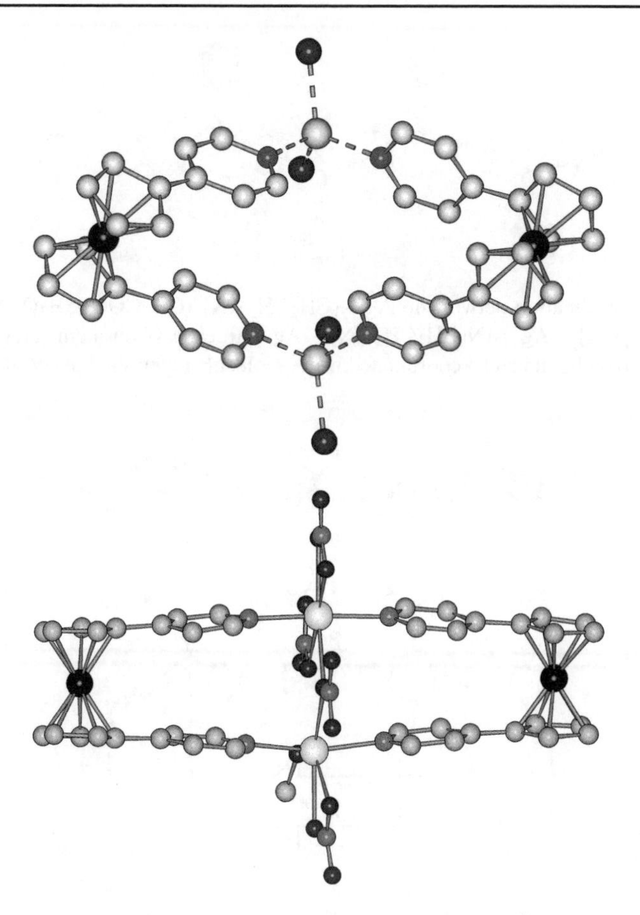

Figure 8. The metalla-macrocycles produced by reaction of $[Fe(\eta^5\text{-}C_5H_4\text{-}1\text{-}C_5H_4N)_2]$ and the salts $ZnCl_2$ (a) and $Cd(NO_3)_2$ (b). The starting material has been obtained by Suzuki-coupling reaction in the solid-state starting from the complex ferrocene-1,1′-diboronic acid $[Fe(\eta^5\text{-}C_5H_4\text{-}B(OH)_2)_2]$.

water–methanol solution and used to compare calculated and experimental x-ray powder diffractograms. When $ZnCl_2$ is used instead of $AgCH_3COO$ in the equimolar reaction with $[N(CH_2CH_2)_3N]$, different products are obtained from solution and solid-state reactions, respectively. The preparation of single crystals of $Ag[N(CH_2CH_2)_3N]_2[CH_3COO]\cdot 5H_2O$ was obviously indispensable for the determination of the exact nature of the co-grinding product. In order to do so, the powder diffraction pattern computed on the basis of the single-crystal structure was compared with the one measured on the product of the solid-state preparation. Figure 10 shows that the structure of $Zn[N(CH_2CH_2)_3N]Cl_2$ is based on a one-dimensional coordination network comprising alternating $[N(CH_2CH_2)_3N]$ and $ZnCl_2$ units, joined by Zn–N bonds. As mentioned above, upon co-grinding of the solid reactants, a new zinc compound of unknown stoichiometry was obtained as a powder material. Even though attempts to obtain single crystals of this latter compound have failed, there is a relationship between the compound obtained initially by co-grinding and the one obtained from solution. In fact, the co-grind phase can be partially transformed

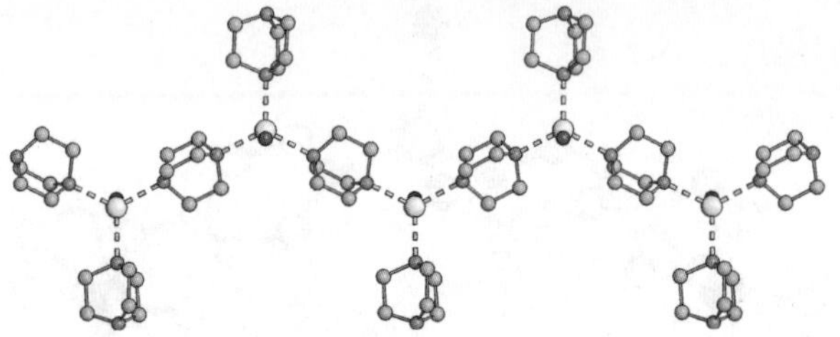

Figure 9. The coordination network in $Ag[N(CH_2CH_2)_3N]_2[CH_3COO]\cdot 5H_2O$. Note the chain of Ag---$[N(CH_2CH_2)_3N]$---Ag---$[N(CH_2CH_2)_3N]$---Ag with each silver atom carrying an extra pendant $[N(CH_2CH_2)_3N]$ ligand and a coordinated water molecule in tetrahedral coordination geometry.

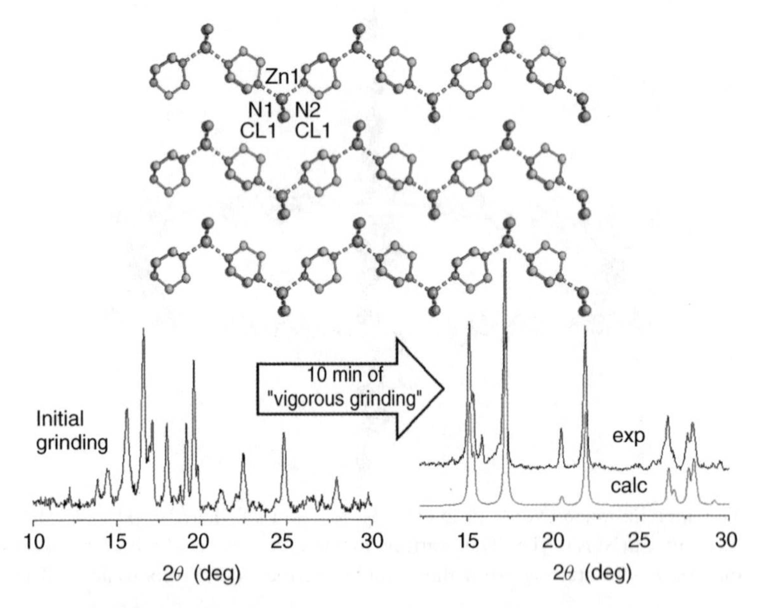

Figure 10. The one-dimensional coordination network present in crystals of $Zn[N(CH_2CH_2)_3N]Cl_2$ and a comparison of the powder diffraction pattern measured on the product of initial co-grinding and that obtained after prolonged grinding. Note how the latter coincides with the diffractogram computed on the basis of the single-crystal structure depicted on top.

by prolonged grinding into the known anhydrous phase $Zn[N(CH_2CH_2)_3N]Cl_2$, shown in Figure 10.

Steed and Raston *et al.* have explored the use of mechanochemistry in the synthesis of extended supramolecular arrays [42]. Grinding of $Ni(NO_3)_2$ with 1,10-phenanthroline (phen) resulted in the facile preparation of $[Ni(phen)_3]^{2+}$ accompanied by a dramatic and rapid color change. Addition of the solid sodium salt of tetrasulfonatocalix[4]arene (tsc) gives two porous π-stacked supramolecular arrays $[Ni(phen)_3]_2[tsc^{4-}]\cdot nH_2O$ and the related $[Na(H_2O)_4(phen)][Ni(phen)_3]_4$ $[tsc^{4-}][tsc^{5-}]\cdot nH_2O$ depending on stoichiometry. It has also been reported that the co-grinding of copper(II) acetate hydrate with 1,3-di(4-pyridyl)propane (dpp) gives a gradual color change from blue to blue-green over approximately 15 min. The resulting material was shown by solid-state NMR spectroscopy to comprise a 1D coordination polymer with water-filled pores. The same host structure, $[\{Cu(OAc)_2\}_2(\mu\text{-dpp})]_n$, could be obtained from solution containing methanol, acetic acid or ethylene glycol guest species [43].

4. THE SOLVENT-FREE CHEMISTRY OF THE ZWITTERION $[CO^{III}(\eta^5\text{-}C_5H_4COOH)(\eta^5\text{-}C_5H_4COO)]$

In this section we summarize the solvent-free chemistry of the zwitterionic sandwich complex $[Co^{III}(\eta^5\text{-}C_5H_4COOH)(\eta^5\text{-}C_5H_4COO)]$ [44]. Thanks to its amphoteric behavior, the complex undergoes reversible gas–solid reactions with the hydrated vapors of a variety of acids (e.g. HCl, CF_3COOH, CCl_3COOH, CHF_2COOH, HBF_4 and HCOOH [45–48], and bases (e.g. NH_3, NMe_3 and NH_2Me [45]) as well as solid–solid reactions with crystalline salts MX ($M = K^+$, Rb^+, Cs^+ and NH_4^+; $X = Cl^-$, Br^-, I^- and PF_6^-, though not in all permutations of cations and anions). The reactions with crystalline alkali salts carried out by manual co-grinding of the powdered materials yields crystalline solids of formula $[Co^{III}(\eta^5\text{-}C_5H_4COOH)(\eta^5\text{-}C_5H_4COO)]_2{\cdot}M^+X^-$ [49]. The gas–solid and solid–solid reactivity of $[Co^{III}(\eta^5\text{-}C_5H_4COOH)(\eta^5\text{-}C_5H_4COO)]$ is summarized in Table 1.

The zwitterion $[Co^{III}(\eta^5\text{-}C_5H_4COOH)(\eta^5\text{-}C_5H_4COO)]$ can be quantitatively prepared from the corresponding dicarboxylic cationic acid $[Co^{III}(\eta^5\text{-}C_5H_4COOH)_2]^+$. The amphoteric behavior of the zwitterion depends on the presence of one –COOH group, which can react with bases, and one $\text{–COO}^{(-)}$ group, which can react with acids. Incidentally, the organometallic zwitterion of $[Co^{III}(\eta^5\text{-}C_5H_4COOH)(\eta^5\text{-}C_5H_4COO)]$ is easy to handle and it is thermally stable up to a temperature of 506 K. $[Co^{III}(\eta^5\text{-}C_5H_4COOH)(\eta^5\text{-}C_5H_4COO)]$ undergoes fully reversible heterogeneous reactions with the hydrated vapors of a variety of acids (e.g. HCl, CF_3COOH, CCl_3COOH, CHF_2COOH, HBF_4 and HCOOH) and bases (e.g. NH_3, NMe_3 and NH_2Me), with formation of the corresponding salts.

Table 1. Summary of the gas–solid (top) and solid–solid (bottom) reactions of the zwitterionic sandwich complex $[Co^{III}(\eta^5\text{-}C_5H_4COOH)(\eta^5\text{-}C_5H_4COO)]$

Solid–gas reactivity of $[Co^{III}(\eta^5\text{-}C_5H_4COOH)(\eta^5\text{-}C_5H_4COO)]$	
Acid vapor	Product
HCl	$[Co^{III}(\eta^5\text{-}C_5H_4COOH)_2]Cl{*}H_2O$
CF_3COOH	$[Co^{III}(\eta^5\text{-}C_5H_4COOH)_2][CF_3COO]$
$CH_2ClCOOH$	$[Co^{III}(\eta^5\text{-}C_5H_4COOH)_2][CH_2ClCOO]{*}H_2O$
CHF_2COOH	$[Co^{III}(\eta^5\text{-}C_5H_4COOH)_2][CHF_2COO]$
HBF_4	$[Co^{III}(\eta^5\text{-}C_5H_4COOH)_2][BF_4]$
HCOOH	$[Co^{III}(\eta^5\text{-}C_5H_4COOH)(\eta^5\text{-}C_5H_4COO)][HCOOH]$
Base vapor	Product
NH_3	$[Co^{III}(\eta^5\text{-}C_5H_4COOH)_2][NH_4]{*}3H_2O$
$(CH_3)\ NH_2$, $(CH_3)_3N$	Novel, unidentified products
Solid–solid reactivity of $[Co^{III}(\eta^5\text{-}C_5H_4COOH)(\eta^5\text{-}C_5H_4COO)]$	
Solid salt of formula MX	Product
LiCl, LiBr, LiI, NaBr	Novel, unidentified products
NaCl, KCl, RbCl, CsCl, KI, RbI	No reaction
KBr, RbBr, CsBr, NaI, CsI	$[Co^{III}(\eta^5\text{-}C_5H_4COOH)(\eta^5\text{-}C_5H_4COO)]_2{*}M^+X^-$

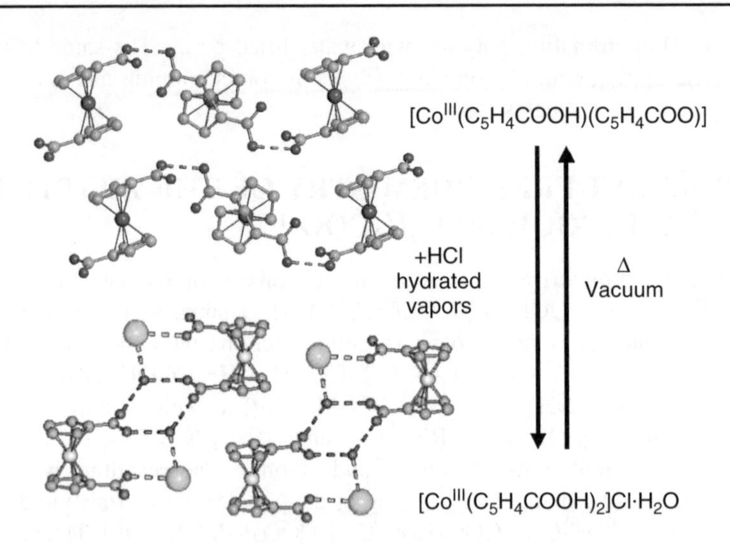

Figure 11. The reversible reaction between anhydrous [Co^{III}(η^5-C$_5$H$_4$COOH)(η^5-C$_5$H$_4$COO)] and HCl leading to formation of [Co^{III}(η^5-C$_5$H$_4$COOH)$_2$]Cl·H$_2$O (bottom). The solid-state structure of [Co^{III}(η^5-C$_5$H$_4$COOH)$_2$]Cl·H$_2$O has been obtained from single-crystal x-ray diffraction experiment on crystals obtained from solution.

For instance, complete conversion of the neutral crystalline zwitterion into the corresponding crystalline chloride salt [Co^{III}(η^5-C$_5$H$_4$COOH)$_2$]Cl·H$_2$O is attained in 5 min of exposure to vapors of aqueous HCl 36% (Figure 11). Formation of the salt in the heterogeneous reaction is easily assessed by comparing the observed x-ray powder diffraction pattern with that calculated on the basis of the single-crystal structure. Crystals can be grown to adequate size by seeding the water solution of the salt obtained from the heterogeneous reactions.

Crystalline [Co^{III}(η^5-C$_5$H$_4$COOH)$_2$]Cl·H$_2$O (Figure 11), can be converted back to neutral [Co^{III}(η^5-C$_5$H$_4$COOH)(η^5-C$_5$H$_4$COO)] by heating the sample for 1 h at 440 K under low pressure (10^{-2} mbar). A thermogravimetric analysis demonstrates that the solid product releases, stepwise, one water molecule and one HCl molecule per molecular unit at 394 and 498 K, respectively. The powder diffractogram of the product after thermal treatment corresponds precisely to that of anhydrous [Co^{III}(η^5-C$_5$H$_4$COOH)(η^5-C$_5$H$_4$COO)]. The formation of [Co^{III}(η^5-C$_5$H$_4$COOH)$_2$]Cl·H$_2$O from the zwitterions implies a substantial rearrangement of the hydrogen-bonding interactions: O–H---O bonds between zwitterionic molecules are broken and new links of the types O–H---Cl$^-$ and O–H---O(water) are established.

The behavior of the zwitterion toward NH$_3$ is similar to that toward HCl but, obviously, opposite in terms of proton exchange. Single crystals of the ammonium salt for x-ray structure determination can be obtained if the reaction of the zwitterion with ammonia is carried out in aqueous solution. Experiments showed that 1–10 mg of the neutral system quantitatively transforms into the hydrated ammonium salt [Co^{III}(η^5-C$_5$H$_4$COO)$_2$][NH$_4$]·3H$_2$O upon 5 min exposure to vapors of aqueous ammonia 30% (Figure 12). The salt is characterized by the presence of charge-assisted $^{(+)}$N–H---O$^{(-)}$ interactions between the ammonium cations and the deprotonated –COO$^{(-)}$ groups on the organometallic anion.

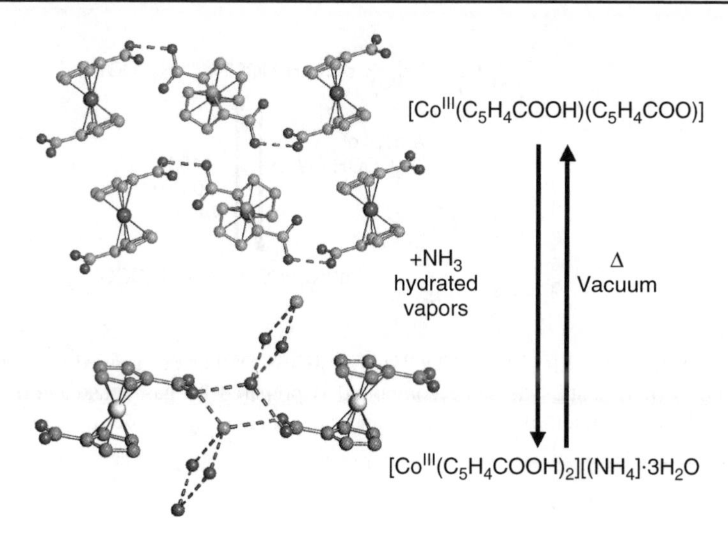

Figure 12. The reversible reaction between anhydrous $[Co^{III}(\eta^5\text{-}C_5H_4COOH)(\eta^5\text{-}C_5H_4COO)]$ and NH_3 leading to formation of $[Co^{III}(\eta^5\text{-}C_5H_4COO)_2][NH_4]\cdot 3H_2O$ as obtained from single-crystal x-ray diffraction experiments.

As in the case of the chloride salt, formation of $[Co^{III}(\eta^5\text{-}C_5H_4COO)_2][NH_4]\cdot 3H_2O$ in the heterogeneous reaction is assessed *via* comparison of the observed and calculated x-ray powder patterns. Absorption of ammonia is also fully reversible: upon thermal treatment (1 h at 373 K, ambient pressure) the salts converts quantitatively into the starting material.

It is worth stressing that the two crystalline powders $[Co^{III}(\eta^5\text{-}C_5H_4COOH)_2]Cl\cdot H_2O$ and $[Co^{III}(\eta^5\text{-}C_5H_4COO)_2][NH_4]\cdot 3H_2O$ can be cycled through several absorption and release processes of HCl or ammonia without decomposition or detectable formation of amorphous material.

Similar behavior is shown toward other volatile acids. Exposure of the zwitterion to vapors of CF_3COOH and HBF_4, for instance, quantitatively produces the corresponding salts of the cation $[Co^{III}(\eta^5\text{-}C_5H_4COOH)_2]^+$, namely, $[Co^{III}(\eta^5\text{-}C_5H_4COOH)_2][CF_3COO]$ (Figure 13), and $[Co^{III}(\eta^5\text{-}C_5H_4COOH)_2][BF_4]$ (Figure 14). As in the previous cases, all heterogeneous reactions are fully reversible and the acids can be removed by thermal treatment, which quantitatively regenerates the starting material. In terms of crystal structure organization, formation of $[Co^{III}(\eta^5\text{-}C_5H_4COOH)_2][CF_3COO]$ and of $[Co^{III}(\eta^5\text{-}C_5H_4COOH)_2][BF_4]$, besides leading from a formally neutral system to molecular salts, implies profound molecular rearrangements and breaking and forming of non-covalent interactions. From the analogy between gas–solid and solution reactions, one may be brought to suppose that the gas–solid reactions occur via a process of dissolution and recrystallization as the vapors are adsorbed by the crystalline powder. The reverse process, that is, reconstruction of the zwitterionic crystals, is more difficult to explain as it implies proton removal from the cationic acid. Moreover, the TGA experiments show that water of hydration is always released first while the acid and the base come off only subsequently. Hence, the participation in the reverse process of an intermediate liquid-phase is unlikely. Probably, removal of HCl or ammonia causes phase reconstruction as the gas leaves the crystals.

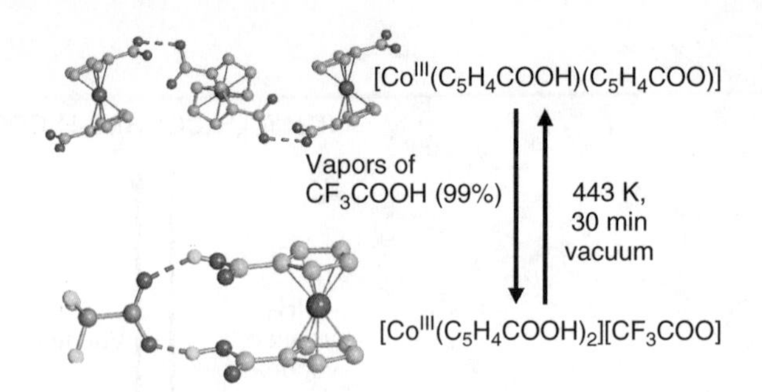

Figure 13. The structure of $[(\eta^5\text{-}C_5H_4COOH)_2Co^{III}][CF_3COO]$ as determined from single-crystal x-ray diffraction experiments. The same compound is prepared by gas uptake. Formation of the salt is assessed by comparison of the observed and calculated x-ray powder patterns.

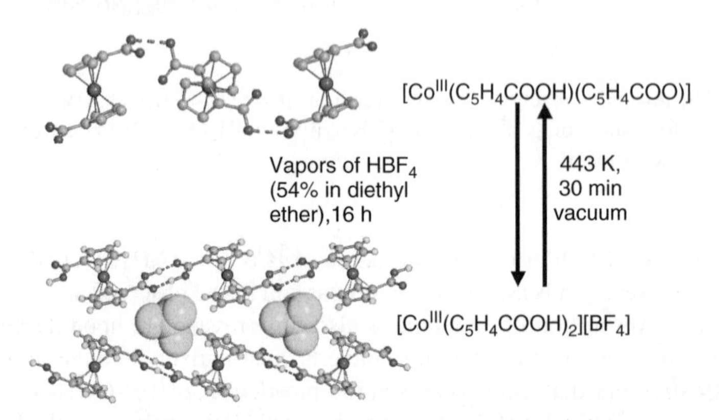

Figure 14. The structure of $[(\eta^5\text{-}C_5H_4COOH)_2Co^{III}][BF_4]$ as determined from single-crystal x-ray diffraction experiments. The same compound is prepared by gas uptake. Formation of the salt is assessed by comparison of the observed and calculated x-ray powder patterns.

Exposure of the solid zwitterion to vapors of CHF_2COOH quantitatively produces the corresponding salt of the cation, $[Co^{III}(\eta^5\text{-}C_5H_4COOH)_2][CHF_2COO]$ (Figure 15). The solid–gas reaction implies a profound rearrangement of the hydrogen-bonding patterns with formation of ionic pairs between organometallic and organic moieties. The reaction with hydrated vapors of $CH_2ClCOOH$ produces the hydrated salt $[Co^{III}(\eta^5\text{-}C_5H_4COOH)_2][CH_2ClCOO]\cdot H_2O$.

Finally, we should report that the zwitterion also reversibly absorbs formic acid from humid vapors forming selectively a 1:1 co-crystal, $[Co^{III}(\eta^5\text{-}C_5H_4COOH)(\eta^5\text{-}C_5H_4COO)]$ $[HCOOH]$, from which the starting material can be fully recovered by mild thermal treatment (Figure 16). Contrary to the other compounds of this class, no proton transfer from the adsorbed acid to the organometallic moiety is observed. Hence, the reaction between $[Co^{III}(\eta^5\text{-}C_5H_4COOH)(\eta^5\text{-}C_5H_4COO)]$(solid) and HCOOH(vapor) would be more appropriately described as a special kind of solvation rather than as a heterogeneous acid–base reaction.

As shown in Figure 16, crystalline $[Co^{III}(\eta^5\text{-}C_5H_4COOH)(\eta^5\text{-}C_5H_4COO)][HCOOH]$ comprises pairs of zwitterion molecules linked by O–H---O bonds between the protonated

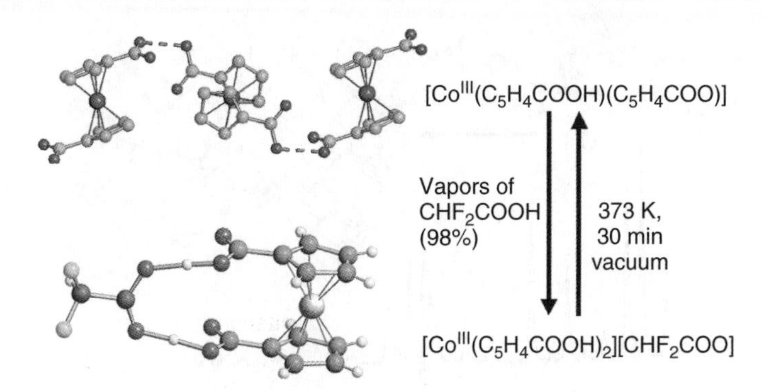

Figure 15. The packing in $[Co^{III}(\eta^5\text{-}C_5H_4COOH)_2][CHF_2COO]$ as determined from single-crystal x-ray diffraction experiments. The same compound is prepared by gas uptake. Formation of the salt is assessed by comparison of the observed and calculated x-ray powder patterns.

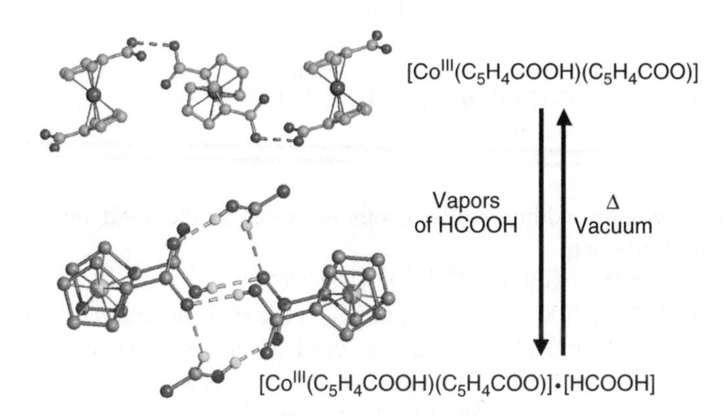

Figure 16. The structure of $[Co^{III}(\eta^5\text{-}C_5H_4COOH)(\eta^5\text{-}C_5H_4COO)][HCOOH]$ (bottom) as determined from single-crystal x-ray diffraction experiments.

–COOH and the deprotonated $–COO^-$ groups [O---O separation 2.526(4) Å]. On the other hand, the C–O distances within the HCOOH moiety [1.305(5) and 1.199(5) Å] indicate that the formic acid molecule retains its acidic hydrogen atom. This is also confirmed by ^{13}C CPMAS NMR spectrometry. Conversion to the starting material is attained by leaving the sample at room temperature in the air for a few days or by mild heating.

We have also exploited the great versatility of the organometallic zwitterion in the preparation of hybrid organometallic-inorganic salts by reacting $[Co^{III}(\eta^5\text{-}C_5H_4COOH)(\eta^5\text{-}C_5H_4COO)]$ with a number of MX salts ($M = K^+$, Rb^+, Cs^+ and $NH_4{}^+$; $X = Cl^-$, Br^-, I^- and $PF_6{}^-$ though not in all permutations of cations and anions) obtaining compounds of general formula $[Co^{III}(\eta^5\text{-}C_5H_4COOH)(\eta^5\text{-}C_5H_4COO)]_2{\cdot}M^+X^-$. As in the cases discussed above, exact information about the solid-state structures of the reaction products were obtained by single-crystal x-ray diffraction experiments carried out on crystals obtained from the reaction powders. Information on the hydrogen-bonding nature and on the relationship between structures in solution and those obtained in the solid state by mechanical grinding were obtained by a combination of solution and solid-state NMR methods. In some cases ($M = Rb^+$, Cs^+, $X = Cl^-$, Br^- and I^-), it was necessary

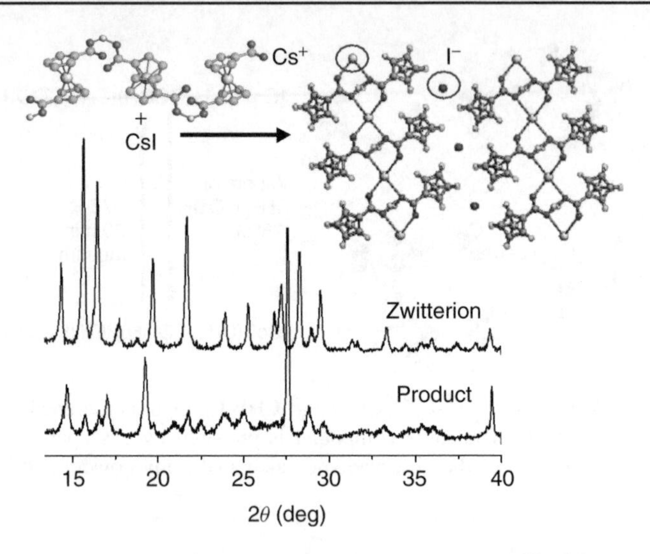

Figure 17. A pictorial representation of the process leading from $[Co^{III}(\eta^5\text{-}C_5H_4COOH)(\eta^5\text{-}C_5H_4\text{-}COO)]$ and CsI to $[Co^{III}(\eta^5\text{-}C_5H_4COOH)(\eta^5\text{-}C_5H_4COO)]_2{\cdot}Cs^+I^-$ and a comparison of calculated and observed powder diffractograms.

to resort to *kneading* by adding a few drops of water to the solid mixture in order to obtain the desired product.

Since all compounds of formula $[Co^{III}(\eta^5\text{-}C_5H_4COOH)(\eta^5\text{-}C_5H_4COO)]_2{\cdot}M^+X^-$, ($M = K^+$, Rb^+, Cs^+ and $[NH_4]^+$ $X = Br^-$, I^- and $PF_6{}^-$) are isostructural, compound $[Co^{III}(\eta^5\text{-}C_5H_4COOH)(\eta^5\text{-}C_5H_4COO)]_2{\cdot}Cs^+I^-$ can be used as an example to describe structure and supramolecular architecture. This class of compounds is characterized by the presence of a supramolecular cage formed by four zwitterionic molecules encapsulating the alkali or ammonium cations. The cage is sustained by O–H---O hydrogen bonds between carboxylic –COOH and carboxylate $\text{–COO}^{(-)}$ groups, and by C–H---O bonds between –CH_{Cp} and –CO groups, while the anions are layered in between the cationic complexes, as shown in Figure 17. The process leading to formation of the cage can be seen as a sophisticated solvation operated by the organometallic complex. The zwitterion is capable of "extracting" via $O\text{---}X^-$ interactions the alkali cations from their lattice, while the anions are "extruded" and left to interact with the peripheral C–H groups via numerous C–H---X interactions. The solid–solid process can thus be seen as the dissolution of one solid (e.g. the alkali salt) into a *solid solvent.*

5. CONCLUDING REMARKS

One of the core paradigms of crystal engineering is that of being able to assemble molecular or ionic components into a target functional structure by controlling the periodical distribution of supramolecular interactions responsible for molecular recognition, aggregation, nucleation and growth of the crystalline material [1–6]. In other words, crystal engineering amounts to the deliberate construction of a molecular solid (whether a molecular complex, an adduct or a co-crystal [50]) that can perform desired functions. Hence, it is conceptually related to the construction of a supermolecule [51]. In both molecular crystals and supermolecules, the *collective properties* depend on the aggregation *via* intermolecular bonds of two or more component units. These supramolecular interactions can be coordination bonds between ligands and metal centers and non-covalent bonds between neutral molecules or ions or, of course, any of their combinations. The process that leads, via breaking and forming of such non-covalent bonds, from reactants to products is, therefore, a supramolecular reaction.

In this chapter, we have shown that non-covalent bonds can be broken and formed in a controlled way by reactions that do not imply the use of solvent but that can be carried out directly between two crystalline solids or between a crystalline solid and a vapor. Reactions of this type have been the subject of investigation for decades in the fields of organic and inorganic chemistry. Although solid–gas and solid–solid reactions are the basis for a number of industrial processes that range from preparation of pharmaceutical compounds [52] to inorganic alloying [53], they still enjoy little popularity in the field or organometallic and coordination chemistry [54]. This is probably due, on the one hand, to the fact that crystals are depicted (even at the level of crystallography courses) as rigid, stiff, fragile materials that are good for little else beside structural analysis, and, on the other hand, to the belief that molecular crystals, being held together by non-covalent interactions, cannot compete with covalent or ionic inorganic solids in terms of cohesion and stability and are not the best materials for gas uptake and/or mechanical treatment.

Our experience is that adequately chosen organometallic crystalline materials can withstand reversible gas–solid reactions with vapors of both acidic and basic substances as well as mechanically activated reactions with other molecular crystals and inorganic salts.

In this Chapter, we have confined ourselves to essentially four classes of reactions involving organometallic molecular crystals as reactants: (i) reactions between a hydrogen-bonded molecular crystal and a vapor with formation of hydrogen-bonded supramolecular adducts, (ii) reactions leading to formation of covalent bonds for the preparation of crystal engineering building blocks, (iii) reactions between hydrogen-bonded molecular crystals to produce new molecular crystals based on hydrogen bonds, and (iv) reactions between molecular and ionic crystals via "solid-state solvation". All these reactions involve molecular crystals and lead from a solid reactant (or a mixture of solid reactants) to a molecular crystal product. In such processes, hydrogen bonds, π-stacking, van der Waals, ion pairing interactions, and so on, are broken and formed through the reaction process leading to formation of supramolecular compounds or hybrid molecular crystals. Clearly, all these reactions (perhaps with the exception of those of type (ii)) are diffusion controlled and are not necessarily reactions in the solid state as mechanical stress may cause local melting; co-grinding may form an intermediate eutectic phase, and *kneading* probably generates locally *hyper*saturated solutions wherefrom crystals of the new phase nucleate. In all these cases, the crystal lattice is destroyed and reformed through recrystallization. It should also be stressed, that in terms of chemistry, gas–solid and solid–solid processes of the types discussed herein, are not conceptually different. In both reactions, the starting material is a molecular crystal and both reactions occur in the absence of solvent. Since the reaction products are new molecular crystals, these reactions are solvent-free means to prepare new crystalline materials. This is a useful notion for crystal engineers.

6. ACKNOWLEDGMENTS

We thank MIUR (COFIN and FIRB), the Universities of Bologna and Sassari for financial support.

REFERENCES

1. (a) G. R. Desiraju, *Crystal Engineering: The Design of Organic Solids*, Elsevier, Amsterdam, 1989; (b) D. Braga, F. Grepioni and G. R. Desiraju, *Chem. Rev.*, **98**, 1375 (1998); (c) A. J. Blake, N. R. Champness, P. Hubberstey, W. S. Li, M. A. Withersby and M. Schroder, *Coord. Chem. Rev.*, **183**, 117 (1999); (d) Proceedings of the Dalton discussion on inorganic crystal engineering, the whole issue, *J. Chem. Soc., Dalton Trans.*, 3705 (2000); (e) B. Moulton and M. J. Zaworotko, *Chem. Rev.*, **101**, 1629 (2001); (f) D. Braga, G. R. Desiraju, J. Miller, A. G. Orpen and S. Price, *CrystEngComm*, **4**, 500 (2002); (g) M. D. Hollingsworth, *Science*, **295**, 2410 (2002).
2. (a) G. R. Desiraju, *Angew. Chem., Int. Ed. Engl.*, **34**, 2311 (1995); (b) D. Braga and F. Grepioni, *Acc. Chem. Res.*, **33**, 601 (2000).
3. D. Braga, F. Grepioni and A. G. Orpen (Eds.), *Crystal Engineering: From Molecules and Crystals to Materials*, Kluwer Academic Publishers, Dordrecht, 1999.
4. (a) S. R. Batten, B. F. Hoskins and R. Robson, *Chem. Eur. J.*, **6**, 156 (2000); (b) S. A. Bourne, J. Lu, B. Moulton and M. J. Zaworotko, *Chem. Commun.*, 861 (2001); (c) B. Rather and M. J. Zaworotko, *Chem. Commun.*, 830 (2003); (d) B. Moulton, H. Abourahma, M. W. Bradner, J. Lu, G. J. McManus and M. J. Zaworotko, *Chem. Commun.*, 1342 (2003); (e) M. Fujita, *Chem. Soc. Rev.*, **27**, 417 (1998); (f) B. Olenyuk, A. Fechtenkötter and P. J. Stang, *J. Chem. Soc., Dalton Trans.*, 1707 (1998); (g) M. D. Ward, *Science*, **300**, 1124 (2003); (h) L. Pan, M. B. Sander, X. Huang, J. Li, M. Smith, E. Bittner, B. Bockrath and J. K. Johnson, *J. Am. Chem. Soc.*, **126**, 1309 (2004); (i) G. Férey, M. Latroche, C. Serre, F. Millange, T. Loiseau and A. Percheron-Guégan, *Chem. Commun.*, 2976 (2003); (j) F. A. Cotton, C. Lin and C. A. Murillo, *J. Chem. Soc., Dalton Trans.*, 499 (2001); (k) F. A. Cotton, C. Lin and C. A. Murillo, *Chem. Commun.*, 11 (2001); (l) W. Mori and S. Takamizawa, *J. Solid State Chem.*, **152**, 120 (2000); (m) L. Carlucci, G. Ciani, D. M. Proserpio and S. Rizzato, *CrystEngComm*, **4**, 121 (2002).
5. (a) D. W. Bruce, D. O'Hare (Eds.), *Inorganic Materials*, Wiley, Chichester, 1992; (b) S. R. Marder, *Inorg. Mater.*, **115** (1992); (c) N. J. Long, *Angew. Chem., Int. Ed. Engl.*, **34**, 21 (1995); (d) T. J. Marks and M. A. Ratner, *Angew. Chem., Int. Ed. Engl.*, **34**, 155 (1995); (e) D. R. Kanis, M. A. Ratner and T. J. Marks, *Chem. Rev.*, **94**, 195 (1994); (f) O. Khan, *Molecular Magnetism*, VCH, New York, 1993; (g) D. Gatteschi, *Adv. Mater.*, **6**, 635 (1994); (h) J. S. Miller and A. J. Epstein, *Chem. Eng. News*, **73**, 30 (1995); (i) J. S. Miller, *Angew. Chem., Int. Ed. Engl.*, **42**, 27 (2003).
6. D. Braga, *Chem. Commun.*, 2751 (2003).
7. (a) N. L. Rosi, M. Eddaouddi, J. Kim, M. O'Keeffe and O. M. Yaghi, *Angew. Chem., Int. Ed. Engl.*, **41**, 284 (2001); (b) O. M. Yaghi, H. L. Li, C. Davis, D. Richardson and T. L. Groy, *Acc. Chem. Res.*, **31**, 474 (1998); (c) H. Li, M. Eddaouddi, M. O'Keeffe and O. M. Yaghi, *Nature*, **402**, 276 (1999); (d) M. Eddaouddi, J. Kim, N. Rosi, D. Vodak, J. Wachter, M. O'Keeffe and O. M. Yaghi, *Science*, **295**, 469 (2002). (e) N. L. Rosi, M. Eddaouddi, J. Kim, M. O'Keeffe and O. M. Yaghi, *CrystEngComm*, **4**, 401 (2002); (f) N. L. Rosi, J. Eckert, M. Eddaouddi, D. T. Vodak, J. Kim, M. O'Keeffe and O. M. Yaghi, *Science*, **300**, 1127 (2003); (g) M. E. Braun, C. D. Steffek, J. Kim, P. G. Rasmussen and O. M. Yaghi, *Chem. Commun.*, 2532 (2001).
8. (a) S. R. Batten and R. Robson, *Angew. Chem., Int. Ed. Engl.*, **37**, 1461 (1998); (b) B. Moulton and M. J. Zaworotko, *Chem. Rev.*, **101**, 1629 (2001); (c) L. Carlucci,

G. Ciani and D. M. Proserpio, *CrystEngComm*, **5**, 269 (2003); (d) L. Carlucci, G. Ciani and D. M. Proserpio, *Coord. Chem. Rev.*, **246**, 247 (2003); (e) S. A. Barnett and N. R. Champness, *Coord. Chem. Rev.*, **246**, 145 (2003).

9. (a) S. Takamizawa, E. Nakata, T. Saito, T. Akatsuka and K. Kojima, *CrystEngComm*, **6**, 197 (2004); (b) S. Takamizawa, E. Nakata and T. Saito, *Angew. Chem., Int. Ed. Engl.*, **43**, 1368 (2004); (c) S. Takamizawa, E. Nakata and T. Saito, *CrystEngComm*, **6**, 39 (2004); (d) S. Takamizawa, E. Nakata, H. Yokoyama, K. Mochizuki and W. Mori, *Angew. Chem., Int. Ed. Engl.*, **42**, 4331 (2003); (e) E. Le Fur, E. Demers, T. Maris and J. D. Wuest, *Chem. Commun.*, 2966 (2003); (f) D. Laliberte, T. Maris and J. D. Wuest, *J. Org. Chem.*, **69**, 1776 (2004); (g) O. Saied, T. Maris and J. D. Wuest, *J. Am. Chem. Soc.*, **125**, 14956 (2003).
10. (a) M. W. Hosseini and A. De Cian, *Chem. Commun.*, 727 (1998); (b) D. Braga and F. Grepioni, *Chem. Commun.*, 571 (1996).
11. (a) C. B. Aakeröy and K. R. Seddon, *Chem. Soc. Rev.*, 397 (1993); (b) A. M. Beatty, *CrystEngComm*, 51 (2001); (c) K. Biradha, *CrystEngComm*, **5**, 374 (2003); (d) H. W. Roesky and M. Andruh, *Coord. Chem. Rev.*, **236**, 91 (2003); (e) A. M. Beatty, *Coord. Chem. Rev.*, **246**, 131 (2003); (f) E. A. Bruton, L. Brammer, F. C. Pigge, C. B. Aakeröy and D. S. Leinen, *New J. Chem.*, **27**, 1084 (2003).
12. (a) G. R. Desiraju and T. Steiner (Eds.), *The Weak Hydrogen Bond in Structural Chemistry and Biology*, Oxford University Press, Oxford, 1999; (b) D. Braga, F. Grepioni, K. Biradha, V. R. Pedireddi and G. R. Desiraju, *J. Am. Chem. Soc.*, **117**, 3156 (1995); (c) M. J. Zaworotko, *Chem. Soc. Rev.*, **23**, 283 (1994); (d) T. Steiner, *Angew. Chem., Int. Ed. Engl.*, 41 (2002); (e) M. J. Calhorda, *Chem. Commun.*, 801 (2000); (f) D. Braga, L. Maini, M. Polito and F. Grepioni, *Struct. Bonding*, **111**, 1 (2004); (g) C. Janiak, *Angew. Chem. Int. Ed. Engl.*, **36**, 1431 (1997); (h) S. George, A. Nangia, M. Bagieu-Beucher, R. Masse and J. F. Nicoud, *New J. Chem.*, **27**, 568 (2003); (i) M. Nishio, *CrystEngComm*, **6**, 130 (2004); (j) L. Brammer, M. D. Burgard, M. D. Eddleston, C. S. Rodger, N. P. Rath and H. Adam, *CrystEngComm*, 239 (2002); (k) J. Hulliger, *Chem. Eur. J.*, **8**, 4579 (2002); (l) L. Brammer, *Dalton Trans.*, **16**, 3145 (2003); (m) A. D. Burrows, *Struct. Bonding*, **108**, 55 (2004).
13. (a) L. Addadi and M. Geva, *CrystEngComm*, **5**, 140 (2003); (b) J. Aizenberg, G. Lambert, S. Weiner and L. Addadi, *J. Am. Chem. Soc.*, **124**, 32 (2002); (c) E. Beniash, J. Aizenberg, L. Addadi and S. Weiner, *Proc. R. Soc. Lond., B, Biol. Sci.*, **264**, 461 (1997); (d) L. Addadi, S. Raz and S. Weiner, *Adv. Mater.*, **15**, 959 (2003); (e) J. Johnston, H. E. Merwin and E. D. Williamson, *Am. J. Sci.*, **41**, 473 (1916); (f) S. Raz, S. Weiner and L. Addadi, *Adv. Mater.*, **12**, 38 (2000); (g) E. Loste and F. C. Meldrum, *Chem. Commun.*, 901 (2001); (h) J. Aizenberg, G. Lambert, L. Addadi and S. Weiner, *Adv. Mater.*, **8**, 222 (1996); (i) D. Braga, *Angew. Chem., Int. Ed. Engl.*, **42**, 5544 (2003).
14. (a) D. Braga and F. Grepioni, *Angew. Chem., Int. Ed. Engl.*, **43**, 2 (2004); (b) D. Braga, D. D'Addario, S. L. Giaffreda, L. Maini, M. Polito and F. Grepioni, *Topics in Current Chemistry* **254**, 71–94 (2005) (Ed. F. Toda).
15. P. T. Anastas and J. C. Warner, *Green Chemistry: Theory and Practice*, Oxford University Press, New York, 1998.
16. (a) K. Tanaka and F. Toda, *Chem. Rev.*, **100**, 1025 (2000); (b) K. Tanaka, *Solvent-free Organic Synthesis*, Wiley-VCH, 2003; (c) D. Bradley, *Chem. Br.*, 42 (2002); (d) G. W. V. Cave, C. L. Raston and J. L. Scott, *Chem. Commun.*, 2159 (2001); (e) G. Rothenberg, A. P. Downie, C. L. Raston and J. L. Scott, *J. Am. Chem. Soc.*, **123**, 8701 (2001); (f) F. Toda, *CrystEngComm*, **4**, 215 (2002); (g) L. R. Nassimbeni, *Acc. Chem. Res.*, **36**, 631 (2003).
17. (a) D. Braga and F. Grepioni, *Chem. Soc. Rev.*, **4**, 229 (2000); (b) D. Braga and F. Grepioni, in *Crystal Design, Structure and Function. Perspectives in Supramolecular Chemistry*, Vol. 7 (Ed. G. R. Desiraju), John Wiley, Chichester, 2003.
18. (a) D. Braga and F. Grepioni, *J. Chem. Soc., Dalton Trans.*, 1 (1999); (b) D. Braga and F. Grepioni, *Coord. Chem. Rev.*, **183**, 19 (1999); (c) D. Braga, G. Cojazzi, L. Maini, M. Polito,

L. Scaccianoce and F. Grepioni, *Coord. Chem. Rev.*, **216**, 225 (2001); (d) D. Braga, L. Maini, M. Polito, E. Tagliavini and F. Grepioni, *Coord. Chem. Rev.*, **246**, 53 (2003).
19. (a) V. V. Boldyrev and K. Tkacova, *J. Mater. Synth. Process.*, **8**, 121 (2000); (b) J. F. Fernandez-Bertran, *Pure Appl. Chem.*, **71**, 581 (1999).
20. (a) G. Kaupp, in *Comprehensive Supramolecular Chemistry*, Vol. 8, (Ed. J. E. D. Davies), Elsevier, Oxford, 381 (1996); (b) G. Kaupp, *CrystEngComm*, **5**, 117 (2003).
21. (a) J. M. Cairney, S. G. Harris, L. W. Ma, P. R. Munroe and E. D. Doyle, *J. Mater. Sci.*, **39**, 3569 (2004); (b) G. Bettinetti, M. R. Caira, A. Callegari, M. Merli, M. Sorrenti and C. Tadini, *J. Pharm. Sci.*, **89**, 478 (2000); (c) Y. A. Kim, T. Hayashi, Y. Fukai, M. Endo, T. Yanagisawa and M. S. Dresselhaus, *Chem. Phys. Lett.*, **355**, 279 (2002); (d) K. Wieczorek-Ciurowa, K. Gamrat and K. Fela, *Solid State Ionics*, **164**, 193 (2003).
22. N. Shan, F. Toda and W. Jones, *Chem. Commun.*, 2372 (2002).
23. S. Watano, T. Okamoto, M. Tsuhari, I. Koizumi and Y. Osako, *Chem. Pharm. Bull.*, **50**, 341 (2002); S. Watano, J. Furukawa, K. Miyanami and Y. Osako, *Adv. Powder. Technol.*, **12**, 427 (2001).
24. N. Morin, A. Chilouet, J. Millet and J. C. Rouland, *J. Therm. Anal. Calorim.*, **62**, 187 (2000).
25. (a) G. Bruni, A. Marini, V. Berbenni, R. Riccardi and M. Villa, *J. Incol. Phenom. Macro. Chem.*, **35**, 517 (1999); (b) F. Taneri, T. Guneri, Z. Aigner and M. Kata, *J. Incol. Phenom. Macro. Chem.*, **44**, 257 (2002); (c) R. Saikosin, T. Limpaseni and P. Pongsawadsi, *J. Incol. Phenom. Macro. Chem.*, **44**, 191 (2002).
26. (a) T. L. Threlfall, *Analyst*, **120**, 2435 (1995); (b) N. Kubota, N. Doki, M. Yokota and D. Jagadesh, *J. Chem. Eng. Jpn.*, **35**, 1063 (2002).
27. P. Seiler and J. D. Dunitz, *Acta Crystallogr., Sect. B* **38**, 1741 (1982); D. Braga, G. Cojazzi, D. Paolucci and F. Grepioni, *CrystEngComm*, **1** (2001).
28. R. J. Davey, N. Blagden, G. D. Potts and R. Docherty, *J. Am. Chem. Soc.*, **119**, 1767 (1997).
29. H. Koshima and M. Miyauchi, *Cryst. Growth Des.*, **1**, 355 (2001).
30. J. Dunitz and J. Bernstein, *Acc. Chem. Res.*, **28**, 193 (1995).
31. (a) D. Braga, L. Maini, M. Polito, L. Mirolo and F. Grepioni, *Chem. Commun.*, **24**, 2960 (2002); (b) D. Braga, L. Maini, M. Polito, L. Mirolo and F. Grepioni, *Chem. Eur. J.*, **9**, 4362 (2003).
32. D. Braga, L. Maini, G. de Sanctis, K. Rubini, F. Grepioni, M. R. Chierotti and R. Gobetto, *Chem. Eur. J.*, **9**, 5538 (2003).
33. M. R. Caira, L. R. Nassimbeni and A. F. Wildervanck, *J. Chem. Soc., Perkin Trans.*, **2**, 2213 (1995).
34. D. Braga and L. Maini, *Chem. Commun.*, 976 (2004).
35. D. Braga, D. D'Addario, M. Polito and F. Grepioni, *Organometallics*, **23**, 2810 (2004).
36. (a) A. Suzuki and N. Miyaura, *Chem. Rev.*, **95**, 2457 (1995); (b) A. Suzuki and N. Miyaura, *J. Org. Chem.*, **63**, 4726 (1998); (c) D. Villemin and F. Caillot, *Tetrahedron Lett.*, **42**, 639 (2001); (d) G. W. Kabalka, G. R. M. Pagni and C. M. Hair, *Org. Lett.*, **1**, 1423 (1999); (e) M. Melucci, G. Barbarella and G. Sotgiu, *J. Org. Chem.*, **67**, 8877 (2002).
37. M. Bracaccini, D. Braga, D. D'Addario, F. Grepioni, M. Polito, L. Sturba and E. Tagliavini, *Organometallics*, **22**, 214 (2003).
38. D. Braga, D. D'Addario, F. Grepioni, M. Polito, D. M. Proserpio, J. W. Steed and E. Tagliavini, *Organometallics*, **22**, 4532 (2003).
39. V. P. Balema, J. W. Wiench, M. Pruski and V. K. Pecharsky, *Chem. Commun.*, 1606 (2002).
40. A. Orita, L. S. Jiang, T. Nakano, N. C. Ma and J. Otera, *Chem. Commun.*, 1362 (2002).
41. D. Braga, S. Giaffreda, F. Grepioni and M. Polito, *CrystEngComm*, **6**, 458 (2004).
42. P. J. Nichols, C. L. Raston and J. W. Steed, *Chem. Commun.*, 1062 (2001).
43. W. J. Belcher, C. A. Longstaff, M. R. Neckenig and J. W. Steed, *Chem. Commun.*, 1602 (2002).
44. D. Braga, L. Maini, M. Polito and F. Grepioni, *Organometallics*, **18**, 2577 (1999).

45. D. Braga, G. Cojazzi, D. Emiliani, L. Maini and F. Grepioni, *Chem. Commun.*, **21**, 2272 (2001).
46. D. Braga, G. Cojazzi, D. Emiliani, L. Maini and F. Grepioni, *Organometallics*, **21**, 1315 (2002).
47. D. Braga, L. Maini, M. Mazzotti, K. Rubini and F. Grepioni, *CrystEngComm*, **29**, 154 (2003).
48. D. Braga, L. Maini, M. Mazzotti, K. Rubini, A. Masic, R. Gobetto and F. Grepioni, *Chem. Commun.*, **20**, 2296 (2002).
49. D. Braga, L. Maini, S. Giaffreda, F. Grepioni, M. R. Chierotti and R. Gobetto, *Chem. Eur. J.*, **10**, 3261 (2004).
50. G. R. Desiraju, *CrystEngComm*, **5**, 466 (2003); J. D. Dunitz, *CrystEngComm*, **5**, 506 (2003).
51. (a) M. Lehn, *Supramolecular Chemistry: Concepts and Perspectives*. VCH: Weinheim, 1995; (b) W. Steed and J. L. Atwood, *Supramolecular Chemistry*, John Wiley & Sons, 2000.
52. (a) A. Burger, in *Topics in Pharmaceutical Sciences*, (Eds. D. Breimer and P. Speiser), Elsevier, Amsterdam, 1983, p. 347; (b) S. R. Byrn, *Solid State Chemistry of Drugs*, Academic Press, New York, 1982, p. 79.
53. V. V. Boldyrev and K. Tkacova, *J. Mater. Synth. Process.*, **8**, 121 (2000).
54. (a) M. Albrecht, M. Lutz, A. L. Spek and G. van Koten, *Nature*, **406**, 970 (2000); (b) M. Albrecht, R. A. Gossage, M. Lutz, A. L. Spek and G. van Koten, *Chem. Eur. J.*, **6**, 1431 (2000).

2

Crystal Engineering of Pharmaceutical Co-crystals

PEDDY VISHWESHWAR, JENNIFER A. McMAHON
and MICHAEL J. ZAWOROTKO
Department of Chemistry, University of South Florida, Tampa, Florida, USA.

1. INTRODUCTION

The concept of crystal engineering was introduced by Pepinsky [1] in 1955 and further implemented and elaborated by Schmidt in the context of organic solid-state photochemical reactions [2]. Desiraju subsequently defined crystal engineering as "*the understanding of intermolecular interactions in the context of crystal packing and in the utilization of such understanding in the design of new solids with desired physical and chemical properties*" [3]. Crystal engineering has subsequently matured into a paradigm for the preparation or supramolecular synthesis of new compounds. A salient feature of crystal-engineered structures is that they are designed from first principles using the principles of supramolecular chemistry [4] and can therefore consist of a diverse range of chemical components as exemplified by coordination polymers (i.e. metals and organic ligands) [5–10], polymers sustained by organometallic linkages [11] and networks sustained by non-covalent bonds [5, 12–14]. Active Pharmaceutical Ingredients, APIs, are extremely valuable materials but they have traditionally been limited to salts, polymorphs, hydrates and solvates [15]. Therefore, it is perhaps surprising that crystal engineering has only recently addressed APIs via systematic approaches to the development of a new broad class of API, pharmaceutical co-crystals [16–27]. Pharmaceutical co-crystals, that is, co-crystals that are formed between an API and a co-crystal former that is a solid under ambient conditions [22], represent a new paradigm in API formulation that might address important intellectual and physical property issues in the context of drug development and delivery. Co-crystals are long known as *addition compounds* [28] or *organic molecular compounds* [29]. However, the 322, 421 crystal structures deposited in the July 2004

Frontiers in Crystal Engineering. Edited by Edward R.T. Tiekink and Jagadese J. Vittal

release of the Cambridge Structural Database, CSD [30], indicate that co-crystals from two or more solid molecular components remain relatively unexplored, with few entries prior to 1960. Indeed, even now there are only about 1350 co-crystals in which the components are hydrogen bonded to one another as against over 32,000 hydrates. Interest in co-crystals appears to be increasing, since potential applications of co-crystals include the generation of novel Nonlinear optical materials [31], solvent-free organic synthesis [32, 33], host–guest chemistry [34–37] and modification of photographic films [38]. However, the relevance of co-crystals in the context of formulation of APIs is the focus of this contribution.

Analysis and design are the two principal components of crystal engineering and they are the keys to successful preparation of pharmaceutical co-crystals. Analysis involves examination of the intermolecular interactions that govern crystal packing. This is typically executed via analysis of structural data archived in the CSD. The CSD facilitates a statistical approach to the study of packing motifs and thereby provides empirical information concerning common functional groups and how they engage in molecular association. The potential of the CSD in the context of design was envisaged at least 21 years ago when Allen and Kennard noted [39] that "*the systematic analysis of large numbers of related structures is a powerful research technique, capable of yielding results that could not be obtained by any other method.*" Two decades later, it has become clear that crystal engineering is indeed synonymous with supramolecular synthesis and herein we address how analysis of the CSD facilitates the design and preparation of pharmaceutical co-crystals from first principles.

1.1. What Are Co-crystals?

The issue of how one defines a co-crystal is a matter of recent debate [40, 41]. Herein, we shall use the following operating definition: a co-crystal is a multiple-component crystal in which two or more molecules that are solid under ambient conditions coexist through a hydrogen bond [22]. This definition means that co-crystals represent a long known but relatively unexplored class of compounds. Further, they are attractive to the pharmaceutical industry because they offer opportunities to modify the chemical and/or physical properties of an API without the need to make or break covalent bonds. Indeed, such an approach, for which the term *non-covalent derivatization* [42, 43] was coined, has already been used in the context of modifying the stability of Polaroid film [38, 43]. Structural information on co-crystals was largely absent until the 1960s, when the term *complexes* was coined, primarily in the context of molecular recognition between nucleic bases [44, 45]. It is evident that many of the reported co-crystals appear to be the result of serendipity, although several groups have successfully exploited crystal engineering principles for design and synthesis of co-crystals [46–54].

1.2. How Are Co-crystals Prepared?

Synthesis of a co-crystal from solution might be thought of as counterintuitive since crystallization is an efficient and effective method of purification. However, if different molecules with complementary functional groups result in hydrogen bonds that are energetically more favorable than those between like molecules of either component, then co-crystals are likely to be thermodynamically (although not necessarily kinetically) favored.

Complementary supramolecular synthons [55] that seem to favor formation of co-crystals are exemplified by carboxylic acid-pyridine [56–58], carboxylic acid-amide [59–62] and alcohol-pyridine [63–65]. Co-crystals involving these supramolecular synthons are usually synthesized by slow evaporation from a solution containing stoichiometric amounts of the components (co-crystal formers); however, sublimation, growth from the melt and grinding two solid co-crystal formers in a ball mill are also suitable methodologies. More often than not, the phase that is obtained is independent of the synthetic methodology. The technique of solvent-drop grinding, addition of a small amount of suitable solvent to the ground mixture to accelerate co-crystallization, appears to be a particularly effective preparation method [66, 67]. That co-crystals can sometimes be prepared in a facile manner does not mean that their synthesis and isolation is routine:

- A detailed understanding of the supramolecular chemistry of the functional groups present in a given molecule is a prerequisite for designing a co-crystal, since it facilitates selection of appropriate co-crystal formers. However, when multiple functional groups are present in a molecule, as is often the case for APIs, the CSD rarely contains enough information to address the hierarchy of the multiple possible supramolecular synthons.
- Mismatched solubility between the components of a co-crystal can preclude their generation from solution.
- The role of solvent in nucleation of crystals and co-crystals remains poorly understood and choice of solvent can be crucial in obtaining a particular co-crystal.

1.3. Why Are Co-crystals of Relevance in the Context of APIs?

The very nature of APIs, molecules that contain exterior hydrogen-bonding moieties, means that they are predisposed to formation of co-crystals. Unfortunately, it also means that APIs are prone to formation of polymorphs, hydrates or solvates. This is of course a matter of concern and interest to the pharmaceutical industry, and the potential role of co-crystals in this context is the primary focus of the remainder of this contribution. However, there are other roles that co-crystals might play in enhancing the processing and delivery of APIs:

- APIs have traditionally been limited to salts, polymorphs, hydrates and solvates. However, there are hundreds or even thousands of potential co-crystal formers that would be complementary for a particular API. Pharmaceutical co-crystals therefore, represent a class of substance that is in principle more numerous than all of these traditional forms combined. Furthermore, the API is not modified from a covalent perspective when it is present in a pharmaceutical co-crystal. It could therefore be anticipated that diversity alone will mean that pharmaceutical co-crystals will afford forms of APIs with improved physical properties such as solubility, stability, hygroscopicity and dissolution rate.
- A co-crystal might be used to isolate or purify an API during manufacturing and the co-crystal former could be discarded prior to formulation.
- Co-crystallization might be used to separate enantiomers.

2. WHAT IS THE ORIGIN OF POLYMORPHISM AND IS IT PREVALENT IN CO-CRYSTALS?

Polymorphism [68] is defined as the phenomenon wherein the same chemical substance exists in different crystal forms and is of fundamental importance to solid-state chemistry. Mitscherlich recognized the phenomenon of polymorphism in 1822 [69]. In 1965, McCrone defined polymorphism as "A solid crystalline phase of a given compound resulting from the possibility of at least two different arrangements of the molecules of that compound in the solid state" [70]. McCrone's definition is particularly appropriate in the context of drugs, since highly functional and conformationally flexible APIs manifest themselves via adopting multiple modes of self-organization or self-assembly with other molecules. Hence, most APIs exist as polymorphic, solvated and/or hydrated forms. Polymorphs exhibit different crystal structures and hence, different physical and chemical properties [71]. The existence of polymorphism implies that kinetic factors are important during crystal nucleation and growth [72] and that free energy differences between different crystalline forms are small (<10 kJ mol^{-1}) [73]. Attaining a desired polymorph is of the utmost importance in drugs, agrochemicals, explosives, dyes, pigments, flavors and confectionery products [74–76].

As revealed by Table 1, polymorphism is more prevalent in single component crystals than in multiple-component crystals. Indeed, there are so few examples of multiple-component crystals known to exhibit polymorphism that it would be appropriate to address examples of such structures individually in order to determine the origin of the polymorphism. There are 22 co-crystals in the CSD that exhibit polymorphism. However, most of these entries do not contain three-dimensional atomic coordinates for one or more of the entries. We present herein an analysis of the seven cases of polymorphism in co-crystals for which atomic coordinates are available.

Table 1. CSD statistics of polymorphic single and multicomponent crystal structures

	No. of entries	% of organic structures
Organic crystal structures	116,686[a]	100[d]
Single component molecular organic structures	83,603	71.6[d]
Single component polymorphic structures	1,525[b]	1.8[e]
Hydrates	8,681	7.4[d]
Co-crystals	1,352[c]	1.1[d]
Polymorphic co-crystals	22[b]	1.6[f]

CSD July'04 update, 322,421 entries.

[a]Structures containing Na, Li, K, As were excluded.

[b]Only one refcode was counted for each polymorphic compound.

[c]Co-crystals sustained by strong hydrogen bonds.

[d]Percentages with respect to organic crystal structures only.

[e,f]Percentages with respect to 83,603 and 1,352 sub-totals respectively.

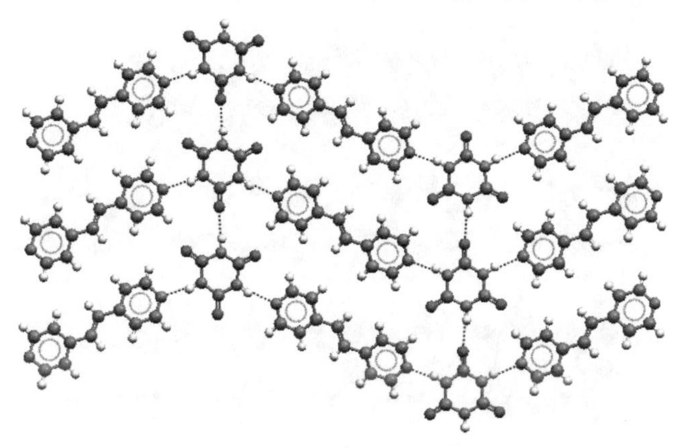

Figure 1. The two-dimensional sheets generated by co-crystal polymorphs of cyanuric acid and 1,2-*bis*(4-pyridyl)ethene. Both forms consist of the same layers but with slightly different crystal packing between the layers.

(a) 1:1 Co-crystals of cyanuric acid/1,2-*bis*(4-pyridyl)ethylene (CSD Refcode: HADKUT – HADKUT01)

The crystal structures of polymorphic forms of 1:1 co-crystals of cyanuric acid and 1,2-*bis*(4-pyridyl)ethene were described by Champness *et al.* in 2003 [77]. Both forms were crystallized in a monoclinic crystal system (space group $C2/c$). Form I comprises one molecule of each component in the asymmetric unit, whereas Form II contains one cyanuric acid in a general position and second on a 2-fold axis. The 1,2-*bis*(4-pyridyl)ethene molecules in Form II lie in general positions and around inversion centers. The crystal packing reveals that both forms exhibit two-dimensional sheets constructed from cyanuric acid infinite one-dimensional N–H$\cdots$O hydrogen-bonded chains that are cross-linked by 1,2-*bis*(4-pyridyl)ethene moieties via N–H$\cdots$N hydrogen bridges (Figure 1). The difference between the asymmetric units of Forms I and II has only a subtle effect on crystal packing. Form I contains a single sheet, whereas Form II contains two crystallographically independent sheets. Thus, A-A-A packing of sheets occurs in Form I, whereas A-A-B-A-A-B stacking is seen in Form II. It is interesting to note that the supramolecular synthons are identical in both forms.

(b) 1:1 Co-crystals of hydroquinone/*p*-benzoquinone (QUIDON – QUIDON02) [78–80]

Quinhydrone is a co-crystal consisting of hydroquinone and *p*-benzoquinone with a 1:1 stoichiometry. Interestingly, quinhydrone is the earliest entry of a co-crystal in the CSD for which coordinates are available. There are monoclinic and triclinic forms of quinhydrone. The monoclinic form ($P2_1/c$ space group, QUIDON02) was observed by Matsuda *et al.* in 1958. The crystal structure reveals that both hydroquinone and *p*-benzoquinone molecules reside on an inversion center. The molecules form linear chains along [120] through O–H$\cdots$O and weak C–H$\cdots$O hydrogen bonds. Such chains are interconnected via C–H$\cdots$O bonds as shown in Figure 2. In the triclinic polymorph ($P\overline{1}$ space group, QUIDON), both molecules sit around an inversion center and form infinite molecular chains through O–H$\cdots$O hydrogen bonds along the [120] direction. However,

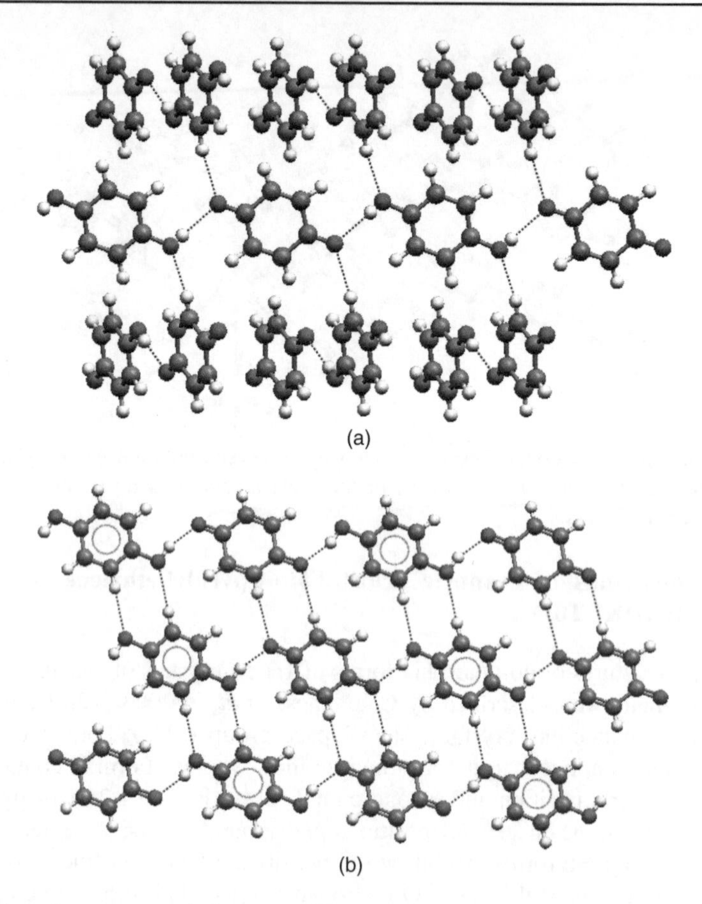

Figure 2. Crystal structures of the monoclinic and triclinic forms of quinhydrone. Figure (a) illustrates the monoclinic form and reveals an infinite chain formed by alternating hydroquinone and benzoquinone molecules that are connected through O–H···O hydrogen bonds. Such chains are orthogonally bonded via weak C–H···O interactions. Figure (b) presents the crystal packing of the triclinic form of quinhydrone. Infinite chains of hydroquinone and benzoquinone molecules generate a two-dimensional layer structure through weak C–H···O interactions.

such chains are packed side by side to form a molecular sheet that is parallel to the (001) plane with C–H···O bonds (Figure 2). Even though there are differences in the crystal packing between QUIDON and QUIDON02 caused by the relative orientation of adjacent chains, the supramolecular synthons are unchanged.

(c) 2:1 Co-crystals of *N*-methylurea/oxalic acid (MUROXA – MUROXA01)

The crystal structures of the orthorhombic and monoclinic polymorphs of MUROXA were reported by Harkema *et al.* in 1979 [81]. Both forms were synthesized by dissolving *N*-methylurea and oxalic acid in 2:1 stoichiometry and crystallized from an aqueous environment. The orthorhombic form (space group *Pnma*) contains eight molecules of *N*-methylurea and four molecules of oxalic acid in the unit cell whereas the monoclinic form ($P2_1/c$) contains four molecules of *N*-methylurea and two molecules of oxalic

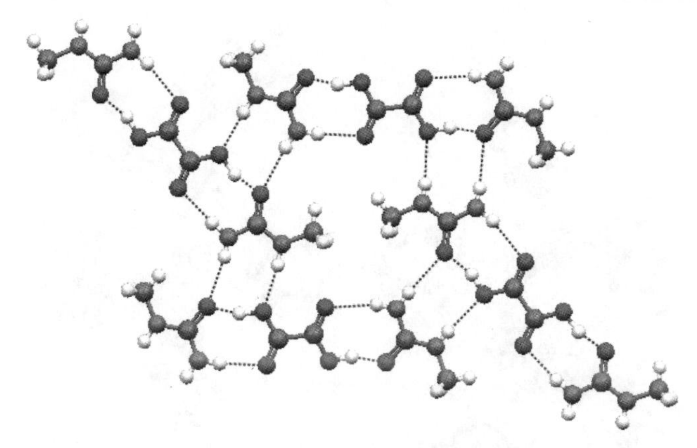

Figure 3. Crystal structure of the orthorhombic form, MUROXA. Note that oxalic acid molecules form carboxylic acid-amide supramolecular heterosynthons with two urea moieties. Similar packing is observed in monoclinic modification.

acid. The crystal structures reveal that oxalic acid molecules form carboxylic acid-amide supramolecular heterosynthons with two molecules of *N*-methylurea. These three component aggregates self-associate via two N–H···O hydrogen bonds into a two-dimensional hydrogen-bonding network that resembles a herringbone pattern. The sheets are parallel to the (010) plane in the orthorhombic form and the (102) in the monoclinic form (Figure 3). Once again, the same supramolecular synthons are observed in both forms.

(d) Concomitant polymorphism in 4:1 co-crystals of triphenylsilanol and 4,4′-bipyridine (MACCID – MACCID01 – MACCID02)

Glidewell *et al.* reported that crystallization of triphenylsilanol and 4,4′-bipyridine in 4:1 stoichiometry in methanol produces three concomitant polymorphs [82]. All three forms of MACCID exist as triclinic crystals (space group $P\overline{1}$) with Z′ values of 0.5, 1 and 4, respectively. The hydrogen-bonded aggregate that exists in MACCID is the same in all forms: a pair of silanol molecules linked to the 4,4′-bipyridyl via O–H···N hydrogen bonds and a further pair of silanol molecules linked to the preceding pair through O–H···O hydrogen bonds. The result of the hydrogen bonding is a five molecule aggregate as presented in Figure 4. Form I contains one such aggregate lying across a center of inversion (Z′ = 0.5), Form II contains two such aggregates, both residing on an inversion center (Z′ = 1) and Form III exhibits six such independent aggregates, four of which reside on inversion centers and two of which occupy general positions (Z′ = 4). The three forms of MACCID differ only in terms of phenyl ring rotation along the Si–C bonds and they therefore, represent examples of conformational polymorphism. Once again, the supramolecular synthons are the same in all three forms of MACCID.

(e) Polymorphism in the 1:1 co-crystal of piperazine-2,5-dione/2,5-dihydroxy-*p*-benzoquinone (AJAJEA – AJAJEA01)

Polymorphism in AJAJEA was reported by Luo and Palmore [83]. AJAJEA contains two half molecules of piperazine-2,5-dione and two half molecules of 2,5-dihydroxy-*p*-benzoquinone in the asymmetric unit, all residing around inversion centers. AJAJEA01,

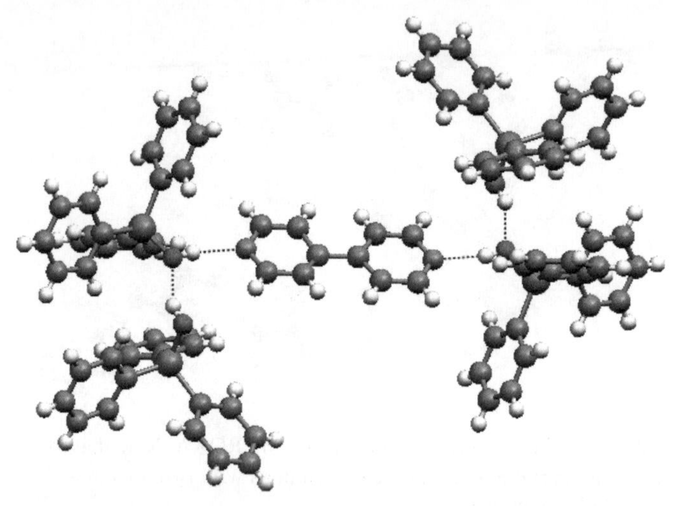

Figure 4. A view of the discrete five molecule aggregates that are present in all three 4:1 polymorphs of triphenylsilanol/4,4′-bipyridine co-crystal, MACCID.

contains one molecule of each co-crystal former and once again all molecules reside around centers of inversion. The crystal packing of both forms indicates that piperazine-2,5-dione molecules form a linear tape via N–H···O carboxamide dimer synthon. Such tapes are interconnected through 2,5-dihydroxy-*p*-benzoquinone moieties, resulting in a molecular sheet. Even though all supramolecular synthons are identical in both structures, there is a subtle difference in the crystal packing caused by the orientation of the 2,5-dihydroxy-*p*-benzoquinone molecules. In AJAJEA, adjacent 2,5-dihydroxy-*p*-benzoquinone molecules alternate in their relative orientation to the piperazine-2,5-dione tapes. By contrast, adjacent 2,5-dihydroxy-*p*-benzoquinone molecules in AJAJEA01 are oriented in the same direction and therefore, related by translation along the tape axis (Figure 5).

(f) Polymorphism in the 1:1 co-crystal of *N,N′-bis*(4-bromophenyl)melamine/5,5′-diethylbarbituric acid (JICTUK01 – JICTUK10)

In 1994, Whitesides and coworkers have reported the existence of polymorphism in 1:1 co-crystals of *N,N′-bis*(4-bromophenyl) melamine and 5,5′-diethylbarbituric acid [49]. One form crystallizes in a monoclinic crystal system ($P2_1/n$, JICTUK01), whereas the other crystallizes in a triclinic system ($P\bar{1}$, JICTUK10). Both polymorphs form linear tapes through 3-point recognition involving N–H···O and N–H···N hydrogen bonds. The supramolecular synthons are the same in both forms and it is the rotation of the phenyl ring along the C–N bond and a different arrangement of linear tapes that results in the observed polymorphism (Figure 6).

(g) Polymorphism in the 1:1 co-crystal of phenyl nitronyl nitroxide radical and phenylboronic acid (ZIGPAG – ZIGPAG01).

Two polymorphs of the co-crystal that forms between the phenyl nitronyl nitroxide radical and phenylboronic acid were reported by Kobayashi and coworkers in 1995 and

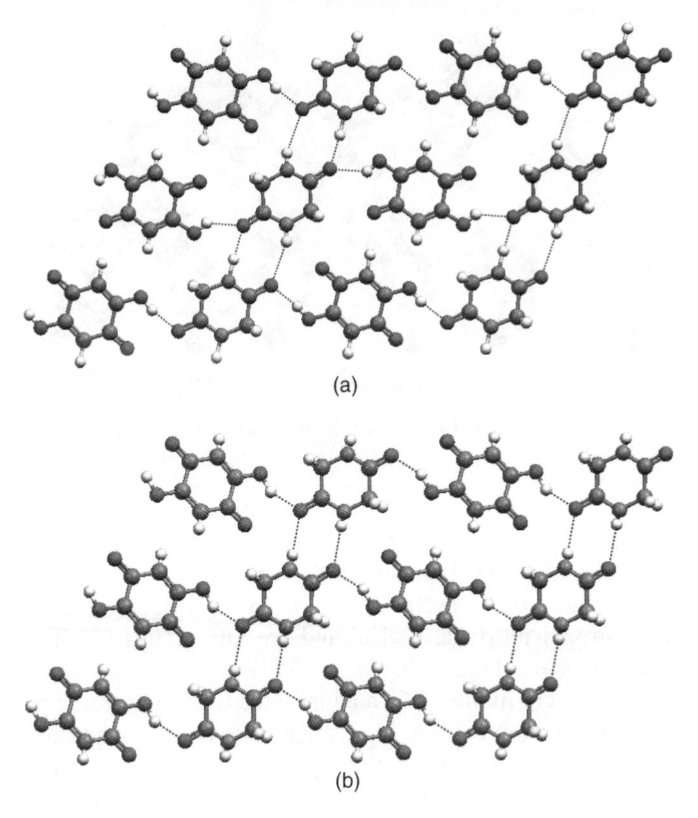

Figure 5. Crystal packing of two reported polymorphs of piperazine-2,5-dione and 2,5-dihydroxy-*p*-benzoquinone co-crystal, AJAJEA (a) and AJAJEA01 (b).

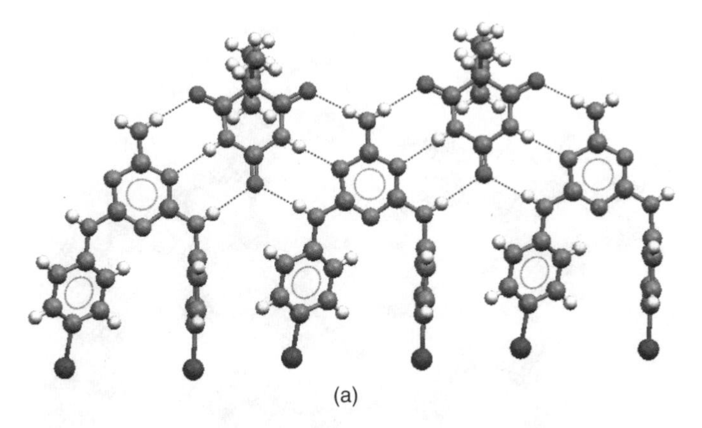

Figure 6. Linear tapes as exhibited by *N*,*N*′-*bis*(4-bromophenyl)melamine and 5,5′-diethylbarbituric acid co-crystals, JICTUK01 (a) and JICTUK10 (b). Note that supramolecular synthons are identical in both forms, whereas the packing of the linear tapes only differs by rotation of the phenyl rings.

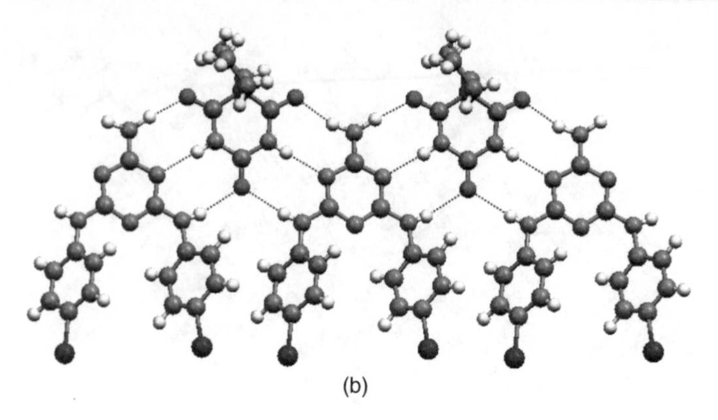

(b)

Figure 6. (*continued*)

1999 [84, 85]. The crystal structures reveal that the monoclinic form (space group $P2_1/n$) contains one molecule of each component, whereas the triclinic modification ($P-1$) contains two molecules of each component in the asymmetric unit. In both modifications, the component molecules are connected by O–H···O hydrogen bonds between the NO groups of phenyl nitronyl nitroxide radical and the OH groups of phenylboronic acid, resulting in an infinite chain (Figure 7).

In summary, there is very limited information in the CSD concerning polymorphism in co-crystals but the information that we have suggests that polymorphism is not linked to the supramolecular synthons that sustain the co-crystal. Rather, it appears to be the result of subtle conformational or packing changes. Indeed, to our knowledge, there have not yet been any examples of "supramolecular isomerism" [5] in hydrogen bonded co-crystals.

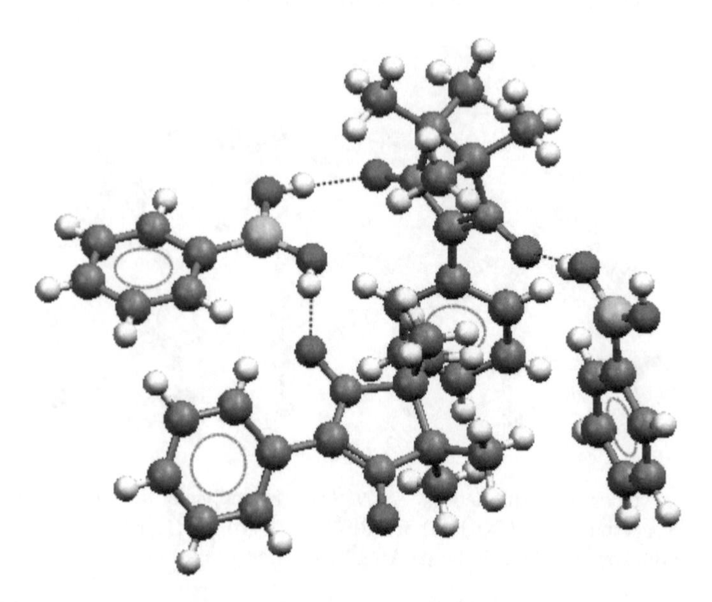

Figure 7. The crystal packing that occurs in both polymorphs of phenyl nitronyl nitroxide radical and phenylboronic acid co-crystal: infinite chains are sustained by O–H···O H-bonds.

If these observations hold true over a broader range of compounds, it would suggest that co-crystals represent a desirable class of compounds from the perspectives of both design and phase stability.

3. WHAT IS A PHARMACEUTICAL CO-CRYSTAL?

Pharmaceutical co-crystals are a subset of the broader group of multicomponent crystals that can be formed by APIs, which includes salts, solvates (pseudo-polymorphs), clathrates, hydrates and inclusion crystals. There is a recent debate on the usage of the term "pseudo-polymorph" for solvates of a compound by Seddon [86] and Desiraju [87, 88]. It is our preference to use the term solvate. In the context of supramolecular chemistry, solvates and pharmaceutical co-crystals are related to one another in that at least two components of the crystal interact by hydrogen bonding or other non-covalent interactions rather than by ionic interactions. Solvates and co-crystal formers can both involve organic acids or bases that remain in their neutral form within the multicomponent crystal. The foremost difference between solvates and co-crystals is the physical state of the isolated pure components. If one component is a liquid (i.e. solvent) at ambient temperature, the crystals are called *solvates* [76], and if both components are solids at room temperature the products are co-crystals. Whereas solvates or hydrates are quite common because they often occur as a serendipitous result during the crystallization process from solution, pharmaceutical co-crystals represent a relatively unexplored class of compounds. However, co-crystals are more prone to rational design than are polymorphs, hydrates or solvates and, as detailed earlier, they appear to be less prone to polymorphism. Furthermore, the number of pharmaceutically acceptable solvents is very small compared to the range of possible co-crystal formers and solvents tend to have higher vapor pressures than co-crystal formers. Modification of the form of APIs using supramolecular synthesis is a relatively low-risk strategy because it employs molecular recognition and self-assembly rather than covalent bond formation or breakage. Pharmaceutical co-crystals therefore have the potential to become broadly useful in the context of pharmaceutical formulations.

How large is the space of pharmaceutical co-crystals? The polymorphic tendency of APIs varies greatly, but the consensus is that most APIs are at some time or other going to display polymorphic behavior. However, the extent of polymorphism of APIs is almost certainly going to be limited to a handful of different crystal forms. Solvates (including hydrates) can be more numerous, and in certain cases, very large numbers of solvates can be observed. Indeed, sulfathiazole is inordinately promiscuous in terms of solvate formation, with over one hundred solvates found [89]. Salt forms can also be numerous, with over 90 acids and 30 bases considered suitable for pharmaceutical salt selection [90]. Examples of compounds possessing a dozen or more crystalline salt forms have been published [91, 92]. However, it is important to remember that salt formation is generally directed at one acidic or basic functional group. In contrast, co-crystals can simultaneously address multiple functional groups in a single API. In addition, the space is not limited to binary combinations (such as acid–base pairs) since tertiary [54] and quaternary co-crystals are realistic possibilities. Co-crystal formers that are suitable for pharmaceutical use remain to be enumerated fully, but there are over a hundred solid materials with "Generally Regarded As Safe" ("GRAS") status (including food additives and other well-accepted substances). One might even consider using subtherapeutic amounts of eminently safe drug substances, such as aspirin and acetaminophen, as legitimate co-crystal formers, thereby expanding the arsenal even further. The space of pharmaceutical co-crystals would therefore appear to be extremely large with thousands of possibilities for any given drug when at least two synthons are present in an API. Such diversity will probably be best addressed with combinatorial methodologies, such as high-throughput crystallization.

3.1. A Case Study: Pharmaceutical Co-crystals of Carbamazepine, 1 (CBZ, 1)

Carbamazepine, [5H-dibenz(b,f) azepine-5-carboxamide] (**1**, CBZ), is known to exist as four well characterized polymorphs [93–100], a dihydrate [101], an acetone solvate [95] and two ammonium salts [102]. CBZ has been in use for over 30 years to treat epilepsy and trigeminal neuralgia even though it poses multiple challenges to oral drug delivery, including a small therapeutic window, autoinduction of metabolism, and dissolution-limited bioavailability [103]. From a supramolecular perspective, CBZ is a simple molecule with only one hydrogen-bonding group, a primary amide. The self-complementary nature of the amide moiety manifests itself in a predictable manner since the CBZ molecules crystallize in all previously known forms of CBZ via the amide-amide supramolecular homosynthon, **I** [104–108] (Figure 8). CBZ would therefore seem to represent an excellent candidate for a detailed case study that addresses how APIs can be converted to pharmaceutical co-crystals and whether or not pharmaceutical co-crystal forms can offer physical, chemical or biochemical advantages over other forms of API.

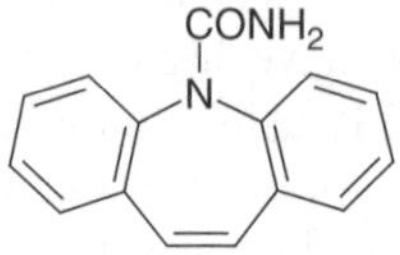

1 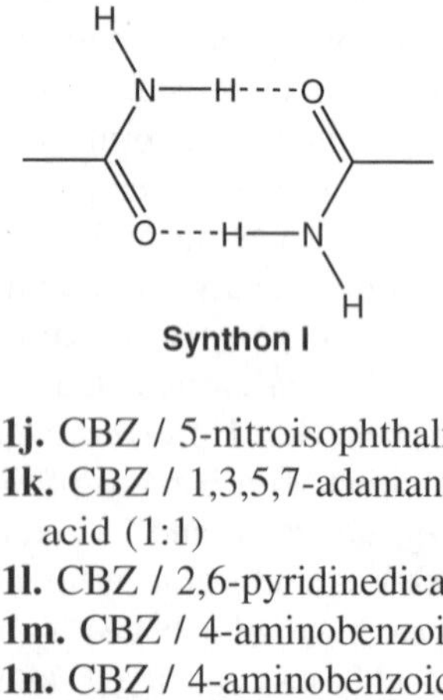 **Synthon I**

1a. CBZ / benzoquinone (2:1)
1b. CBZ / terephthalaldehyde (2:1)
1c. CBZ / 4,4′-bipyridine (2:1)
1d. CBZ / nicotinamide (1:1)
1e. CBZ / saccharin (1:1)
1f. CBZ / acetone (1:1)
1g. CBZ / DMSO (1:1)
1h. CBZ / formamide (1:1)
1i. CBZ / trimesic acid (1:1)
1j. CBZ / 5-nitroisophthalic acid (1:1)
1k. CBZ / 1,3,5,7-adamantanetetracarboxylic acid (1:1)
1l. CBZ / 2,6-pyridinedicarboxylic acid (1:1)
1m. CBZ / 4-aminobenzoic acid (2:1)
1n. CBZ / 4-aminobenzoic acid / H_2O (2:1:1)
1o. CBZ / formic acid (1:1)
1p. CBZ / acetic acid (1:1)
1q. CBZ / butyric acid (1:1)

The primary amide dimer, like the carboxylic acid dimer, is a well-documented supramolecular homosynthon in the CSD. Of the 1231 crystal structures that contain at least one primary amide functional group, the dimer is exhibited in 428 of these structures (35%). In most of these structures, the peripheral NH moiety forms a hydrogen bond to an adjacent amide, thereby generating tapes of sheets, or it hydrogen bonds to a different functional group via a supramolecular heterosynthon. However, this is not the case with

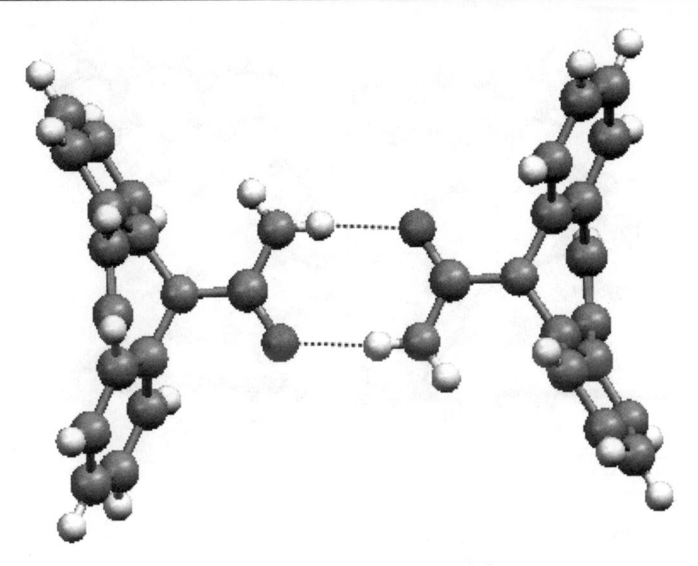

Figure 8. The CBZ dimer structure that exists in all previously reported forms of CBZ. The dimer is sustained by the amide supramolecular homosynthon **I**.

CBZ, in which the peripheral H-bond donors and acceptors are unused, presumably due to steric constraints imposed by the azepine ring of CBZ. That the CBZ dimer does not engage its peripheral hydrogen-bonding capabilities represents one avenue for crystal engineering and CBZ forms a number of co-crystals and solvates that retain the primary amide dimer.

Figure 9 presents the crystal structures of five pharmaceutical co-crystals of CBZ, **1a–e**. In the benzoquinone (**1a**) and terephthalaldehyde (**1b**) co-crystals, the ketone and aldehyde, respectively, hydrogen bond with the anti N–H· · · O hydrogen bond of the CBZ

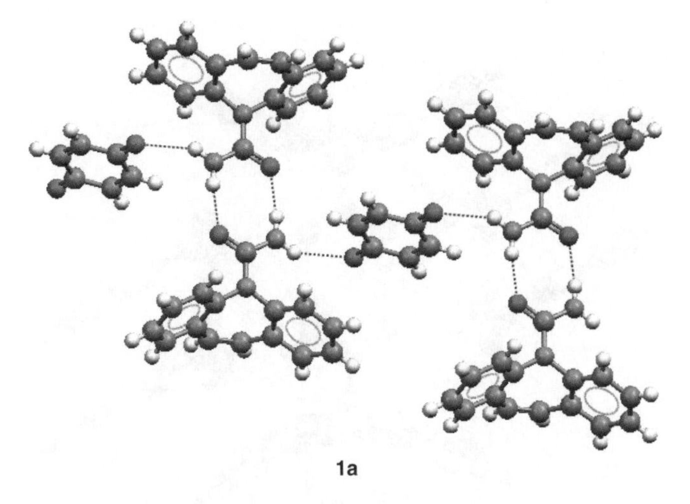

Figure 9. The supramolecular interactions that occur in pharmaceutical co-crystals **1a–e**. **1a–d** can be described a one-dimensional tapes, whereas **1e** exists as a discrete structure.

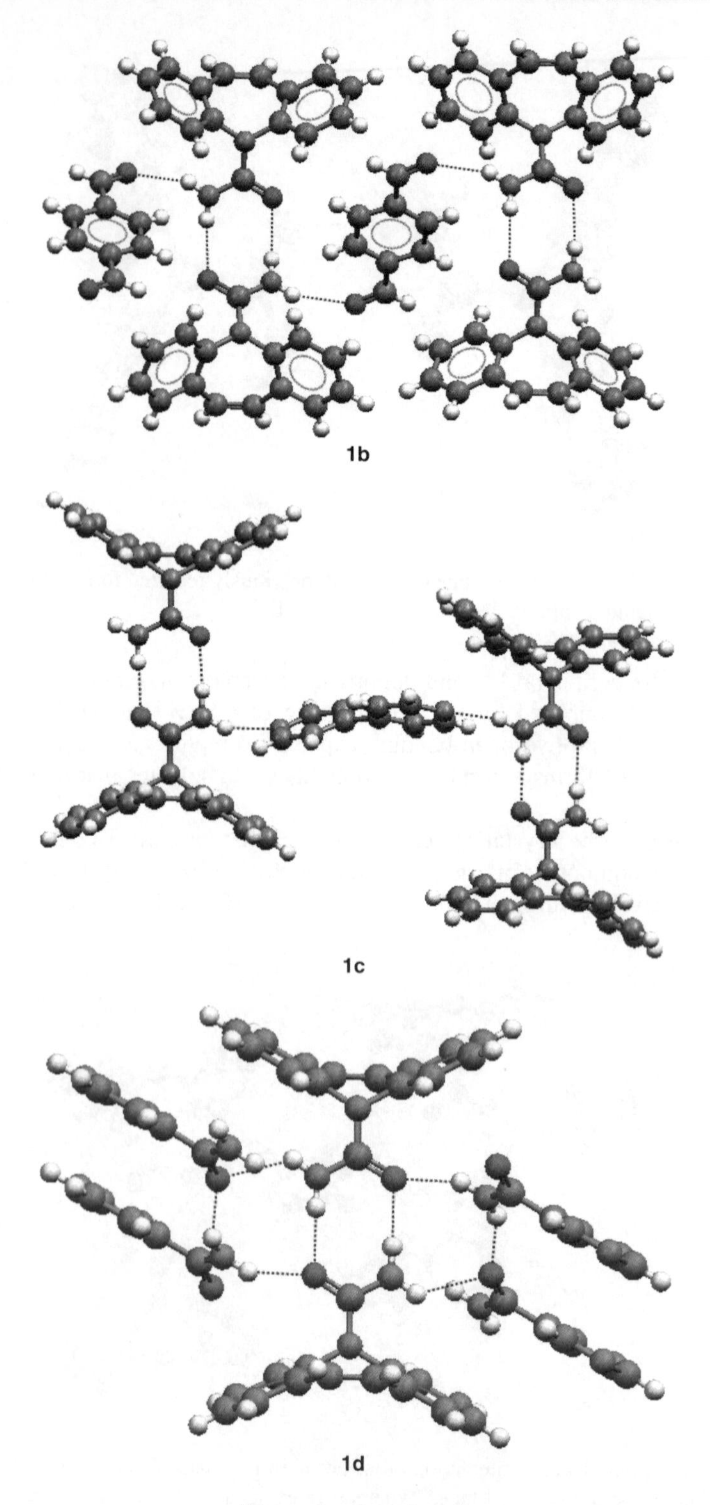

Figure 9. (*continued*)

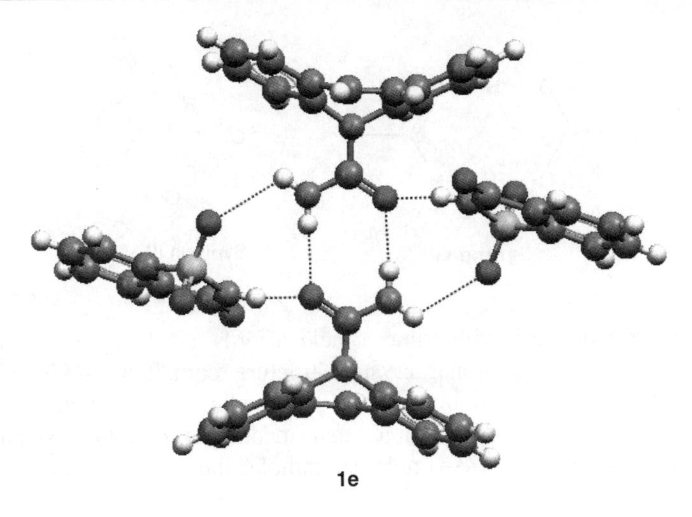

Figure 9. *(continued)*

dimers and sustain one-dimensional hydrogen-bonded chains. A similar result occurs in **1c**, in which the aromatic amines of 4,4′-bipyridine molecules act as hydrogen-bond acceptors to the anti-oriented NH of CBZ. Structure **1d** is a pharmaceutical co-crystal of CBZ and nicotinamide that exists as more complex one-dimensional hydrogen-bonded chain. CBZ dimers H-bond to the *syn* positions of the nicotinamide amide group through an exterior translation-related pattern. The anti-oriented hydrogen-bonding sites of the nicotinamide amide group form a catemer motif with adjacent nicotinamide molecules. The aromatic nitrogen of the nicotinamide is not involved in hydrogen bonding. In **1e** saccharin, molecules serve as hydrogen-bond donors by forming N–H···O hydrogen bonds with CBZ carbonyl groups. They also serve as hydrogen-bond acceptors: the S=O group of the saccharin bonds to the exterior N–H moiety of CBZ. The carbonyl and the second S=O of saccharin molecules are not involved in strong hydrogen bonding.

The CBZ solvates with acetone (**1f**) and dimethylsulfoxide (**1g**) are isostructural and the asymmetric units each contain one molecule of CBZ and one solvent molecule. The CBZ molecules in these structures form homosynthon I, and the void that exists between adjacent CBZ dimers lies around a crystallographic inversion center. The co-crystal formers interact with the anti N–H of the CBZ dimers to form N–H···O=C/S hydrogen bond. The formamide solvate (**1h**) is a rare example of a crystal structure that contains two chemically different amide groups that each form primary amide dimer homosynthons. These dimers interact by peripheral hydrogen bonding, thereby forming an amide–amide’ alternating tape (Figure 10).

A second strategy for co-crystal formation involving CBZ would involve the breakage of the amide–amide supramolecular homosynthon and the formation of a supramolecular heterosynthon. This would be expected to occur with a functional group that is complementary with the amides, that is, a moiety with both a hydrogen-bond donor and a hydrogen-bond acceptor. Carboxylic acids fit this criterion; 71 of the 153 structures in the CSD that contain both a carboxylic acid and a primary amide are sustained by supramolecular heterosynthon II. Co-crystal formation between CBZ and carboxylic acids is indeed facile and Figure 11 presents six examples of pharmaceutical co-crystals, four of which are sustained by supramolecular heterosynthon II.

Synthon II Synthon **III**

Co-crystallization of CBZ with trimesic acid affords a 1:1 supramolecular complex **1i** that exhibits a one-dimensional crystal structure containing both acid–acid and acid–amide supramolecular synthons. Two carboxylic acid groups of the trimesic acid form supramolecular homosynthons that mediate a one-dimensional array and the third acid group forms acid–amide supramolecular heterosynthon II with CBZ (Figure 11). Co-crystallization of CBZ with 5-nitroisophthalic acid from methanol yields **1j**, a co-crystal which exhibits one acid–acid supramolecular homosynthon and one acid–amide supramolecular heterosynthon. The nitro group is not involved in strong hydrogen bonding. The anti NH group forms N–H···O hydrogen bonds to disordered solvent molecules. Compound **1k** was formed via co-crystallization of CBZ with 1,3,5,7-adamantanetetracarboxylic acid and the structure is sustained by two acid–amide supramolecular heterosynthons. The remaining two carboxylic acid groups

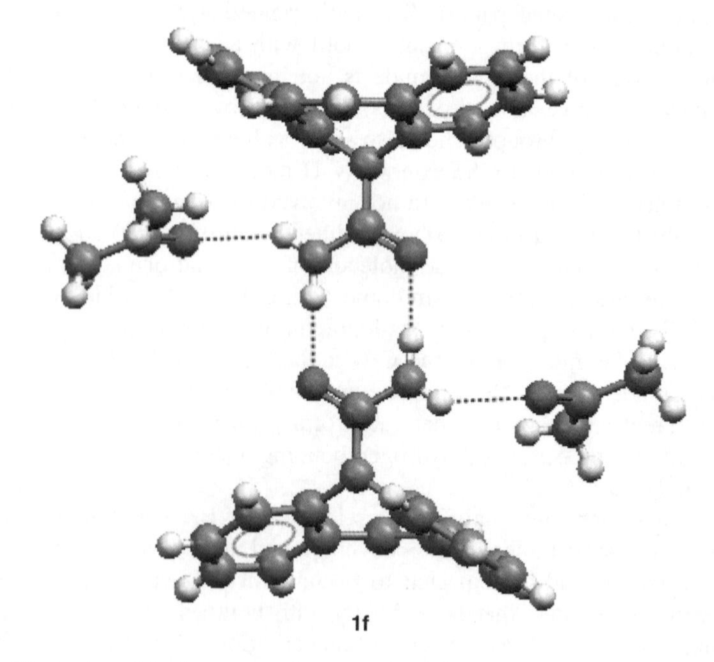

Figure 10. Illustration of the interactions between CBZ dimers and acetone, **1f**, dimethylsulfoxide, **1g**, and formamide, **1h**.

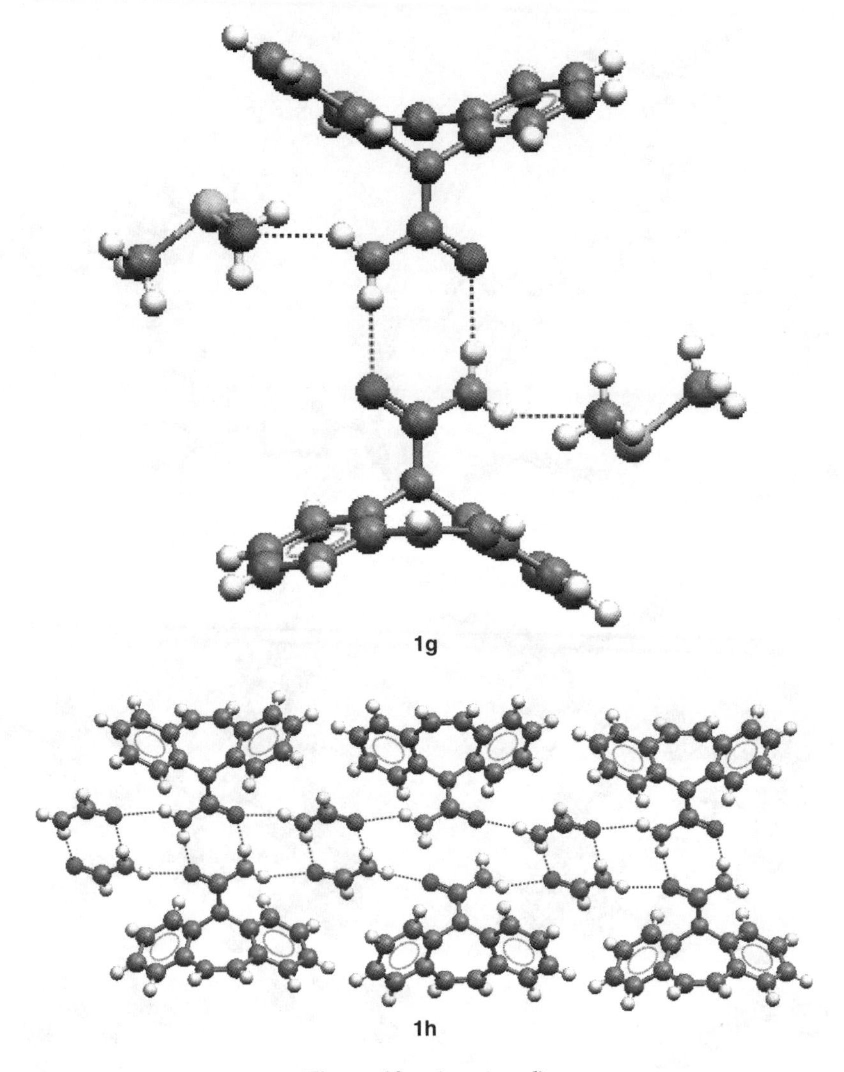

Figure 10. (*continued*)

of adamantanetetracarboxylic acid form hydrogen bonds via N–H···O=C(acid) and O–H(acid)···O=C(amide) interactions. This hydrogen-bond pattern means that all strong H-bond donors/acceptors of both are satisfied. **1l** was formed from the co-crystallization of CBZ with 2,6-pyridinedicarboxylic acid. Interestingly, the expected acid–amide supramolecular heterosynthon II is not seen in this structure. Rather, the co-crystal exhibits an unusual hydrogen-bonding motif III. Only 1-point interactions are present, with each CBZ molecule bonding to three different acid molecules. Carboxylic acid OH donors are also involved in intramolecular O–H···N hydrogen bonds with the pyridine moiety. While it is common to observe a carboxylic acid OH moiety forming an hydrogen bond to a carbonyl group, bifurcation of a carbonyl (amide or simple ketone) to two acid

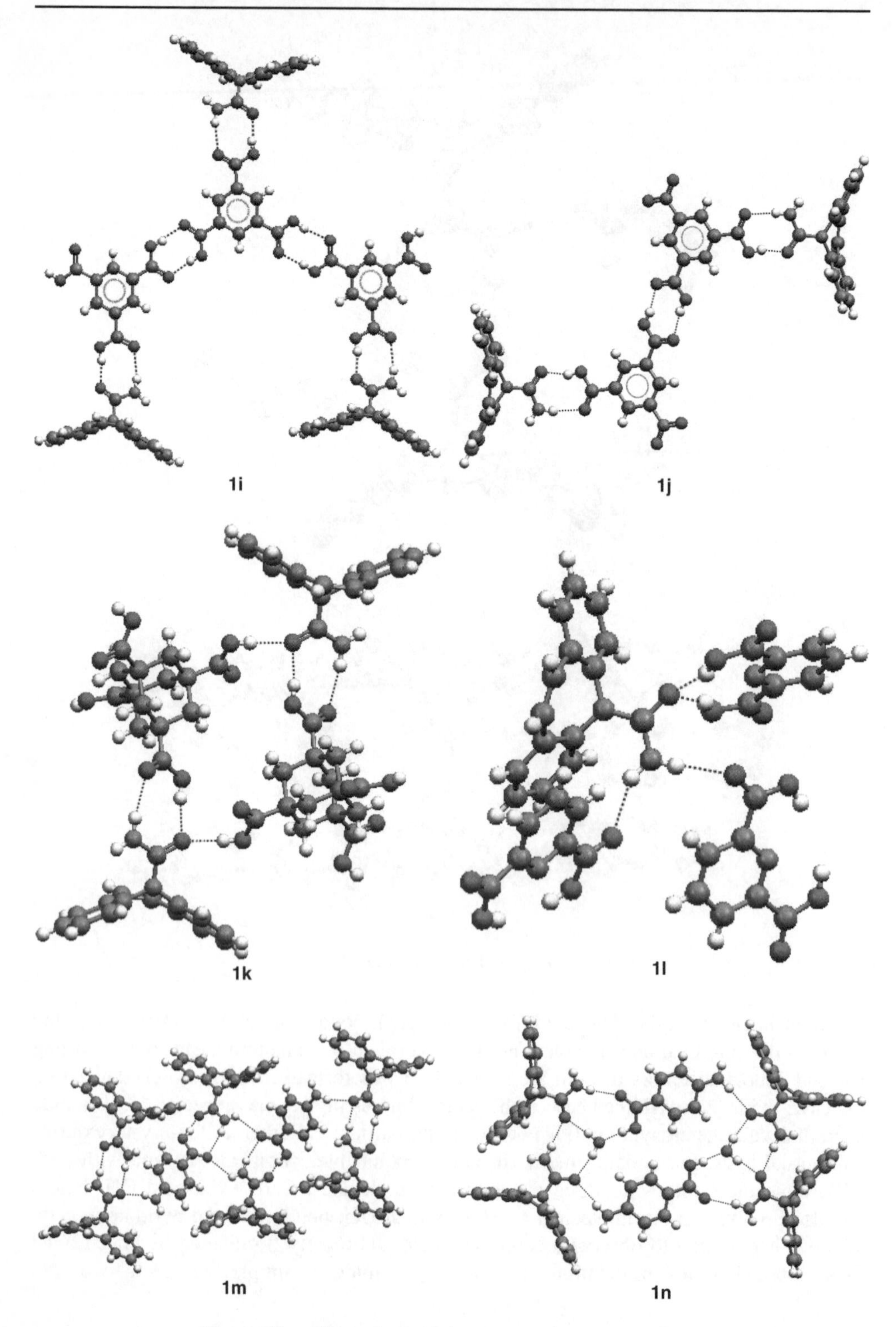

Figure 11. Illustrations of the co-crystal structures **1i–n**

OH groups without a 2-point supramolecular heterosynthon present is seen in only one structure in the CSD [109]. Co-crystal **1m** exhibits a 2:1 stoichiometry that contains both supramolecular heterosynthon II and a primary amide dimer, I. Two acid–amide supramolecular heterosynthons form tetrameric motifs which are in turn hydrogen bonded to adjacent primary amide dimers through (amine) N–H···O (amide) hydrogen bonds. Complex **1n** was obtained from co-crystallization of the same components as **1m** using ethanol rather than methanol. The resulting product, a 2:1:1 co-crystal of CBZ with 4-aminobenzoic acid and adventitious H_2O, exhibits crystal packing that is markedly different from the two-component structure of **1m**. It forms an eight molecule discrete unit through O–H···O and N–H···O hydrogen bonds. This unit consists of four CBZ molecules, two 4-aminobenzoic acid molecules and two water molecules. Insertion of a water molecule into the acid–amide supramolecular heterosynthon to form a different supramolecular heterosynthon, **IV**, is unusual but not unprecedented. Hydration or solvation of carboxylic acids by water or alcohol molecules is a common phenomenon during crystallization [110] and water has been aptly described as a "gluing factor" in organic crystals [111]. A CSD survey reveals that there are 116 structures in which carboxylic acids are hydrogen bonded to another carboxylic acid moiety through one or two water and/or alcohol molecules. Three crystal structures were found in the CSD which exhibit supramolecular heterosynthon **IV** (CSD Refcodes: GASTRN10, IHANIO and OBIZII). The water molecules insert between the primary amide carbonyl and the carboxylic acid OH, thereby forming a 1-point N–H···O acid–amide supramolecular heterosynthon. The anti NH moieties of the CBZ molecules are not involved in hydrogen bonding.

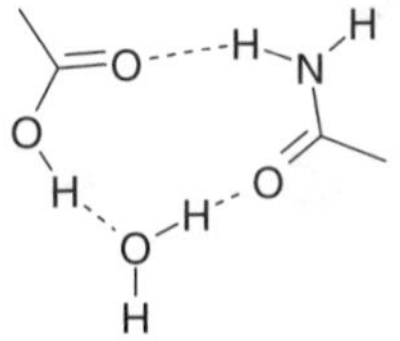

Synthon **IV**

Complexes **1o**, **1p**, and **1q** are solvates of CBZ with formic acid, acetic acid and butyric acid, respectively. In all three structures, carboxylic acid moieties form supramolecular heterosynthon II with CBZ. The anti-NH group of CBZ molecules forms inversion related N–H···O hydrogen bonds with carbonyl moieties of adjacent heterodimers, thereby generating a four-component supramolecular complex (Figure 12). The formic acid and acetic acid solvates are isostructural and the butyric acid motif is structurally similar since the alkyl groups of the butyric acid orient in an appropriate fashion.

That a wide range of pharmaceutical crystals can be crystal engineered is suggested by CBZ and supported by recent studies from other groups that have focused upon other APIs such as Triazole [19], Fluoxetine hydrochloride [21], Trimethoprim [23], Barbital [24], paracetamol [27], and so on. However, there remain several important issues that will only be resolved by further experimentation. For example, can pharmaceutical co-crystals lead to formulations with superior bioavailability when compared to conventional forms of APIs and are they more or less prone to polymorphism? Almarsson and coworkers have investigated the CBZ and saccharin co-crystal (**1e**) in terms of performance attributes, including scale-up, polymorphism, physical stability, *in vitro* dissolution

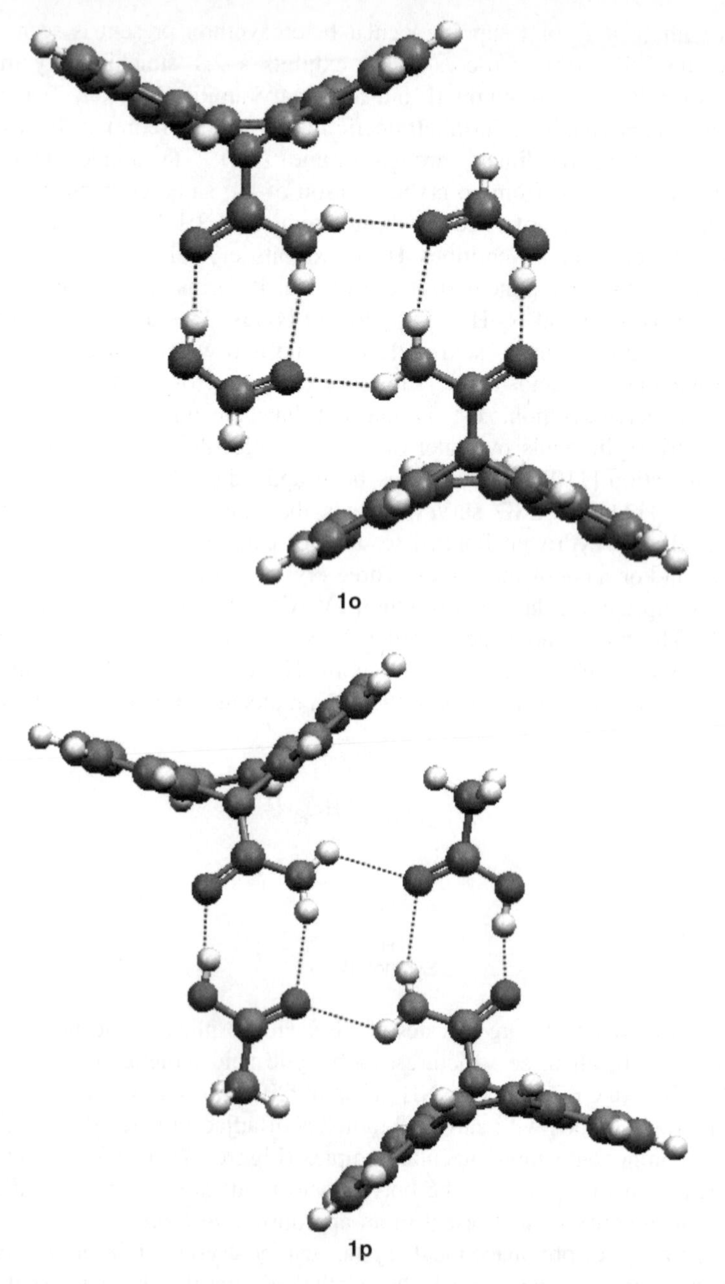

Figure 12. The formic (**1o**), acetic acid (**1p**), and butyric acid (**1q**) solvates of CBZ, all of which exist as four-component supramolecular complexes.

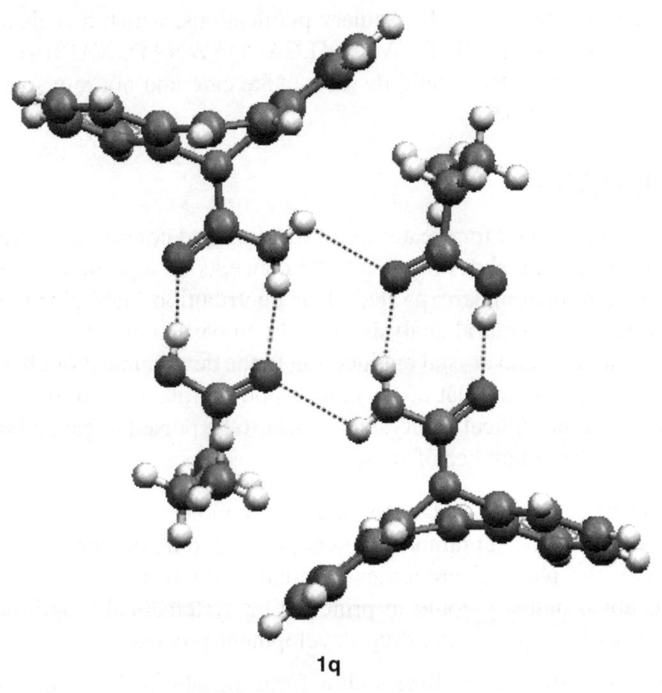

Figure 12. (*continued*)

and oral bioavailability, with the goal of comparing the novel composition with known polymorphs and solvates of the drug (M. B. Hickey, M. L. Peterson, L. Scoppettuolo, A. Vetter, H. Guzmán, J. F. Remenar, J. A. McMahon, M. D. Tawa and Ö. Almarsson. Private Communication). **1e** was synthesized in 30 g scale through cooling a saturated alcohol solution and the physical stability of the co-crystal is superior to that of solvates. A comparison of oral bioavailability of **1e** with Tegretol® tablets in dogs established that **1e** could serve as a practical alternative to anhydrous CBZ in oral formulation. Other observations concerning **1e** include the following: (i) chemical stability appears to be similar to other forms of carbamazepine, (ii) physical stability of **1e** is greater than that of solvates and similar to the marketed forms of CBZ; (iii) **1e** exhibits good PK performance; (iv) Polymorphism was not observed in the co-crystal following over 700 experiments that included high-throughput screening, neat grinding, solvent-drop grinding with several solvents and slurry conversion in different solvents.

In short, if CBZ represents a microcosm of the issues one faces during form selection of APIs then pharmaceutical co-crystals would appear to represent a significant opportunity for improving the diversity and performance of API forms.

3.2. But Beware of "Fake" Pharmaceutical Co-crystals!

Before we conclude it should be emphasized that there can sometimes be ambiguity concerning whether or not a compound is a co-crystal or a salt, especially if x-ray crystallography is the only method of characterization and the difference between the two extremes is the movement of a hydrogen atom by about 1 Å. In the context of APIs, we have noted that CSD searches for co-crystals can retrieve salts. Indeed, in several instances this is the case despite the text in the primary publications, which reveals the ionic nature of the structure. Compounds EBIBEW, PIKLEA, QAWNAD, VAPBAP, VENLUV all come under this category. One should therefore take care and not rely solely on the CSD for identification of co-crystals.

4. CONCLUSIONS

Whereas there is a clear need for greater understanding and control of crystalline phases in the context of pharmaceutical development, the concepts of supramolecular synthesis and crystal engineering remain underexploited. This contribution highlights the need to think "supramolecularly" for structural analysis of APIs. In particular, applying the concepts of supramolecular synthesis and crystal engineering to the development of pharmaceutical co-crystals represents a paradigm that offers many opportunities related to drug development and delivery and pharmaceutical co-crystals appear to be poised to gain a broader foothold in drug formulation for a number of reasons:

- Physical properties are not just dependent upon molecular structure. They are also critically dependent upon supramolecular chemistry and its influence upon crystal structure. Therefore, the physical properties of greatest interest to the formulation side of the pharmaceutical industry could in principle be systematically optimized via crystal engineering at early stages of the drug development process.
- APIs that are difficult to crystallize and/or form unstable solid forms might be recast as pharmaceutical co-crystals as recently illustrated for Itraconazole® [19].
- Pharmaceutical co-crystals are amenable to high-throughput screening technologies. However, unlike polymorphs, which represent the proverbial "needle in haystack", pharmaceutical co-crystals are designable from first principles and it could even be said that they represent many needles in a smaller haystack.
- Pharmaceutical co-crystals do not change the molecular structure of the API. This has important implications for streamlined regulatory approval of new forms.
- The nature of APIs makes them natural candidates for co-crystal design, at least in terms of composition.
- The wide range of possible co-crystal formers means that pharmaceutical co-crystals will inevitably be more ubiquitous than salts, solvates and polymorphs combined.

5. ACKNOWLEDGMENTS

We are grateful for financial support from Transform Pharmaceuticals, 29 Hartwell Avenue, Lexington, MA, 02421.

REFERENCES

1. R. Pepinsky, *Phys. Rev.*, **100**, 971 (1955).
2. G. M. J. Schmidt, *Pure Appl. Chem.*, **27**, 647–678 (1971).
3. G. R. Desiraju, *Crystal Engineering. The Design of Organic Solids*, Elsevier, Amsterdam, 1989.
4. J. M. Lehn, *Supramolecular Chemistry. Concepts and Perspectives*, Wiley-VCH, Weinheim, 1995.

5. B. Moulton and M. J. Zaworotko, *Chem. Rev.*, **101**, 1629–1658 (2001).
6. M. J. Zaworotko, *Chem. Commun.*, 1–9 (2001).
7. M. Eddaoudi, D. B. Moler, L. H. Li, B. L. Chen, T. M. Reineke, M. O'Keeffe and O. M. Yaghi, *Acc. Chem. Res.*, **34**, 319–330 (2001).
8. S. Kitagawa, R. Kitaura and S. Noro, *Angew. Chem., Int. Ed. Engl.*, **43**, 2334–2375 (2004).
9. C. Janiak, *J. Chem. Soc., Dalton Trans.*, **14**, 2781–2804 (2003).
10. S. L. James, *Chem. Soc. Rev.*, 276–288 (2003).
11. M. Oh, G. B. Carpenter and D. A. Sweigart, *Angew. Chem., Int. Ed. Engl.*, **42**, 2026–2028 (2003).
12. D. Braga, F. Grepioni and G. R. Desiraju, *Chem. Rev.*, **98**, 1375–1405 (1998).
13. M. W. Hosseini, *Coord. Chem. Rev.*, **240**, 157–166 (2003).
14. R. E. Melendez and A. D. Hamilton, *Top. Curr. Chem.*, **198**, 97–129 (1998).
15. J. K. Haleblian, *J. Pharm. Sci.*, **64**, 1269–1288 (1975).
16. R. D. B. Walsh, M. W. Bradner, S. G. Fleischman, L. A. Morales, B. Moulton, N. Rodriguez-Hornedo and M. J. Zaworotko, *Chem. Commun.*, 186–187 (2003).
17. S. G. Fleischman, S. S. Kuduva, J. A. McMahon, B. Moulton, R. D. B. Walsh, N. Rodriguez-Hornedo and M. J. Zaworotko, *Cryst. Growth Des.*, **3**, 909–919 (2003).
18. J. A. McMahon, J. A. Bis, P. Vishweshwar, T. R. Shattock, O. L. McLaughlin and M. J. Zaworotko, *Z. Kristallogr.*, **220**, 340–350 (2005).
19. J. F. Remenar, S. L. Morissette, M. L. Peterson, B. Moulton, J. M. MacPhee, H. R. Guzmán and Ö. Almarsson, *J. Am. Chem. Soc.*, **125**, 8456–8457 (2003).
20. S. L. Morissette, Ö. Almarsson, M. L. Peterson, J. F. Remenar, M. J. Read, A. V. Lemmo, S. Ellis, M. J. Cima and C. R. Gardner, *Adv. Drug Deliv. Rev.*, **56**, 275–300 (2004).
21. S. L. Childs, L. J. Chyall, J. T. Dunlap, V. N. Smolenskaya, B. C. Stahly and G. P. Stahly, *J. Am. Chem. Soc.*, **126**, 13335–13342 (2004).
22. Ö. Almarsson and M. J. Zaworotko, *Chem. Commun.*, 1889–1896 (2004).
23. G. Bettinetti, M. R. Caira, A. Callegari, M. Merli, M. Sorrenti and C. Tadini, *J. Pharm. Sci.*, **89**, 478–489 (2000).
24. J. A. Zerkowski, J. C. MacDonald and G. M. Whitesides, *Chem. Mater.*, **9**, 1933–1941 (1997).
25. Ö. Almarsson, M. B. Hickey, M. L. Peterson, S. L. Morissette, C. McNulty, S. Soukasene, M. Tawa, M. MacPhee and J. F. Remenar, *Cryst. Growth Des.*, **3**, 927–933 (2003).
26. S. Datta and D. J. W. Grant, *Nat. Rev. Drug Discov.*, **3**, 42–57 (2004).
27. I. D. H. Oswald, D. R. Allen, P. A. McGregor, W. D. S. Motherwell, S. Parsons and C. R. Pulham, *Acta Crystallogr.*, **B58**, 1057–1066 (2002).
28. J. S. Buck and W. S. Ide, *J. Am. Chem. Soc.*, **53**, 2784–2787 (1931).
29. J. S. Anderson, *Nature*, **140**, 583–584 (1937).
30. F. H. Allen and O. Kennard, *Chem. Des. Automat. News*, **8**, 31–37 (1993).
31. K. Huang, D. Britton, M. C. Etter and S. R. Byrn, *J. Mater. Chem.*, **7**, 713–720 (1997).
32. X. Gao, T. Friščić and L. R. MacGillivray, *Angew. Chem., Int. Ed. Engl.*, **43**, 232–236 (2004).
33. J. Xiao, M. Yang, J. W. Lauher and F. W. Fowler, *Angew. Chem., Int. Ed. Engl.*, **39**, 2132–2135 (2000).
34. B. Q. Ma, Y. Zhang and P. Coppens, *Cryst. Growth Des.*, **2**, 7–13 (2002).
35. B. Q. Ma, Y. Zhang and P. Coppens, *Cryst. Growth Des.*, **1**, 271–275 (2001).
36. L. R. MacGillivray, J. L. Reid and J. A. Ripmeester, *Chem. Commun.*, 1034–1035 (2001).
37. L. R. MacGillivray, P. R. Diamente, J. L. Reid and J. A. Ripmeester, *Chem. Commun.*, 359–360 (2000).
38. L. D. Taylor and J. C. Warner, Process and composition for use in photographic materials containing hydroquinones, US 5338644 A 19940816 Cont. of U.S., 5, 177262, 1994.
39. F. H. Allen, O. Kennard and R. Taylor, *Acc. Chem. Res.*, **16**, 146–153 (1983).
40. G. R. Desiraju, *CrystEngComm*, **5**, 466–467 (2003).
41. J. D. Dunitz, *CrystEngComm*, **4**, 506 (2003).

42. A. S. Cannon and J. C. Warner, *Cryst. Growth Des.*, **2**, 255–257 (2002).
43. B. M. Foxman, D. J. Guarrera, L. D. Taylor, D. Vanengen and J. C. Warner, *Cryst. Eng.*, **1**, 109–118 (1998).
44. K. Hoogsteen, *Acta Crystallogr.*, **16**, 907–916 (1963).
45. E. J. O'Brien, *Acta Crystallogr.*, **23**, 92–106 (1967).
46. M. C. Etter, *J. Phys. Chem.*, **95**, 4601–4610 (1991).
47. V. Videnova-Adrabinska and M. C. Etter, *J. Chem. Crystallogr.*, **25**, 823–829 (1995).
48. C. M. Huang, L. Leiserowitz and G. M. J. Schmidt, *J. Chem. Soc., Perkins Trans. 2*, **5**, 503–508 (1973).
49. J. A. Zerkowski, J. C. MacDonald, C. T. Seto, D. A. Wierda and G. M. Whitesides, *J. Am. Chem. Soc.*, **116**, 2382–2391 (1994).
50. J. P. Mathias, E. E. Simanek, J. A. Zerkowski, C. T. Seto and G. M. Whitesides, *J. Am. Chem. Soc.*, **116**, 4316–4325 (1994).
51. P. Vishweshwar, A. Nangia and V. M. Lynch, *Cryst. Growth Des.*, **3**, 783–790 (2003).
52. B. R. Bhogala, P. Vishweshwar and A. Nangia, *Cryst. Growth Des.*, **2**, 325–328 (2002).
53. C. B. Aakeröy, A. M. Beatty and B. A. Helfrich, *J. Am. Chem. Soc.*, **124**, 14425–14432 (2002).
54. C. B. Aakeröy, A. M. Beatty and B. A. Helfrich, *Angew. Chem., Int. Ed. Engl.*, **40**, 3240–3242 (2001).
55. G. R. Desiraju, *Angew. Chem., Int. Ed. Engl.*, **34**, 2311–2327 (1995).
56. P. Vishweshwar, A. Nangia and V. M. Lynch, *J. Org. Chem.*, **67**, 556–565 (2002).
57. B. R. Bhogala and A. Nangia, *Cryst. Growth Des.*, **3**, 547–554 (2003).
58. A. D. Bond, *Chem. Commun.*, 250–251 (2003).
59. S. Harkema, J. W. Bats, A. M. Weyenberg and D. Feil, *Acta Crystallogr.*, **B28**, 1646–1648 (1972).
60. L. Leiserowitz and F. Nader, *Acta Crystallogr.*, **B33**, 2719–2733 (1977).
61. L. S. Reddy, A. Nangia and V. M. Lynch, *Cryst. Growth Des.*, **4**, 89–94 (2004).
62. C. B. Aakeröy, A. M. Beatty, B. A. Helfrich and M. Nieuwenhuyzen, *Cryst. Growth Des.*, **3**, 159–165 (2003).
63. P. Vishweshwar, A. Nangia and V. M. Lynch, *CrystEngComm*, **5**, 164–168 (2003).
64. G. S. Papaefstathiou and L. R. MacGillivray, *Org. Lett.*, **3**, 3835–3838 (2001).
65. K. Biradha and M. J. Zaworotko, *J. Am. Chem. Soc.*, **120**, 6431–6432 (1998).
66. A. V. Trask, W. D. S. Motherwell and W. Jones, *Chem. Commun.*, 890–891 (2004).
67. N. Shan, F. Toda and W. Jones, *Chem. Commun.*, 2372–2373 (2002).
68. J. Bernstein, *Polymorphism in Molecular Crystals*, Oxford University Press, Oxford, 2002.
69. E. Mitscherlich, *Ann. Chim. Phys.*, **19**, 350–419 (1822).
70. W. C. McCrone, *Polymorphism in Physics and Chemistry of the Solid State*, Vol. 2, (Eds. D. Fox, M. M. Labes and A. Weissberger), Interscience, New York, 725–767, 1965.
71. T. Siegrist, C. Kloc, J. H. Schön, B. Batlogg, R. C. Haddon, S. Berg and G. A. Thomas, *Angew. Chem., Int. Ed. Engl.*, **40**, 1732–1736 (2001).
72. M. R. Caira, *Top. Curr. Chem.*, **198**, 163–208 (1998).
73. A. Gavezzotti and G. Fillippini, *J. Am. Chem. Soc.*, **117**, 12299–12305 (1995).
74. G. Klebe, F. Graser, E. Hädicke and J. Berndt, *Acta Crystallogr.*, **B45**, 69–77 (1989).
75. I. Bar and J. Bernstein, *J. Pharm. Sci.*, **74**, 255–263 (1985).
76. S. R. Byrn, *Solid State Chemistry of Drugs*, Academic press, New York, 79, 1982.
77. S. A. Barnett, A. J. Blake and N. R. Champness, *CrystEngComm*, **5**, 134–136 (2003).
78. H. Matsuda, K. Osaki and I. Nitta, *Bull. Chem. Soc. Jpn.*, **31**, 611–620 (1958).
79. T. Sakurai, *Acta Crystallogr.*, **B24**, 403–412 (1968).
80. T. Sakurai, *Acta Crystallogr.*, **B19**, 320–330 (1965).
81. S. Harkema, J. H. M. Brake and H. J. G. Meutstege, *Acta Crystallogr.*, **B35**, 2087–2093 (1979).

82. K. F. Bowes, G. Ferguson, A. J. Lough and C. Glidewell, *Acta Crystallogr.*, **B59**, 277–286 (2003).
83. T. M. Luo and G. T. R. Palmore, *Cryst. Growth Des.*, **2**, 337–350 (2002).
84. T. Akita, Y. Mazaki and K. Kobayashi, *J. Chem. Soc., Chem. Commun.*, 1861–1862 (1995).
85. Y. Pontillon, T. Akita, A. Grand, K. Kobayashi, E. L. Berna, J. Pécaut, E. Ressouche and J. Schweizer, *J. Am. Chem. Soc.*, **121**, 10126–10133 (1999).
86. K. R. Seddon, *Cryst. Growth Des.*, **4**, 1087 (2004).
87. G. R. Desiraju, *Cryst. Growth Des.*, **4**, 1089–1090 (2004).
88. V. S. S. Kumar, S. S. Kuduva and G. R. Desiraju, *J. Chem. Soc., Perkin Trans. 1*, 1069–1074 (1999).
89. A. L. Bingham, D. S. Hughes, M. B. Hursthouse, R. W. Lancaster, S. Tavener and T. L. Threllfall, *Chem. Commun.*, 603–604 (2001).
90. P. H. Stahl and M. Nakano, in *Handbook of Pharmaceutical Salts: Properties, Selection, and Use*, (Eds. P. H. Stahl and C. G. Wermuth), Wiley-VCH/VCHA, New York, 2002.
91. S. M. Berge, L. D. Bighley and D. C. Monkhouse, *J. Pharm. Sci.*, **66**, 1–19 (1977).
92. J. F. Remenar, J. M. MacPhee, B. K. Larson, V. A. Tyagi, J. H. Ho, D. A. McIlroy, M. B. Hickey, P. B. Shaw and Ö. Almarsson, *Org. Process Res. Dev.*, **7**, 990–996 (2003).
93. V. L. Himes, A. D. Mighell and W. H. De Camp, *Acta Crystallogr.*, **B37**, 2242–2245 (1981).
94. J. P. Reboul, B. Cristau, J. C. Soyfer and J. P. Astier, *Acta Crystallogr.*, **B37**, 844–1848 (1981).
95. M. M. J. Lowes, M. R. Caira, A. P. Lotter and J. G. Van der Watt, *J. Pharm. Sci.*, **76**, 744–752 (1987).
96. J. N. Lisgarten, R. A. Palmer and J. W. Saldanha, *J. Crystallogr. Spectrosc. Res.*, **19**, 641–649 (1989).
97. C. Rustichelli, G. Gamberini, V. Ferioli, M. C. Gamberini, R. Ficarra and S. Tommasini, *J. Pharm. Biomed. Anal.*, **23**, 41–54 (2000).
98. R. Ceolin, S. Toscani, M. F. Gardette, V. N. Dzyabchenko and B. Bachet, *J. Pharm. Sci.*, **86**, 1062–1065 (1997).
99. M. Lang, J. Kampf and A. J. Matzger, *J. Pharm. Sci.*, **91**, 1186–1190 (2002).
100. A. L. Grzesiak, M. Lang, K. Kin and A. J. Matzger, *J. Pharm. Sci.*, **92**, 2260–2271 (2003).
101. G. Reck and G. Dietz, *Cryst. Res. Technol.*, **21**, 1463–1468 (1986).
102. G. Reck and W. Thiel, *Pharmazie*, **46**, 509–512 (1991).
103. Y. Kobayashi, S. Ito, S. Itai and K. Yamamoto, *Int. J. Pharm.*, **193**, 137–146 (2000).
104. L. Leiserowitz and G. M. J. Schmidt, *J. Chem. Soc.*, **A16**, 2372–2382 (1969).
105. J. C. MacDonald and G. T. R. Palmore, in *The Amide Linkage: Selected Structural Significance in Chemistry, Biochemistry, and Materials Science*, (Eds. A. Greenberg, C. M. Breneman and J. F. Liebman), John Wiley & Sons, New York, 291–336, 2000.
106. R. E. Meléndez and A. D. Hamilton, in *Design of Organic Solids*, (Ed. E. Weber), Springer, Berlin, 97–130, 1998.
107. J. Bernstein, M. C. Etter and L. Leiserowitz, in *Structure Correlation*, Vol. 2, (Ed. H.-B. Bürgi and J. D. Dunitz), VCH, Weinheim, 431–507, 1994.
108. J. C. MacDonald and G. M. Whitesides, *Chem. Rev.*, **94**, 2383–2420 (1994).
109. M. Ranjbar, H. Aghabozorg, A. Moghimi and A. Yanovsky, *Z. Kristallogr. – New Cryst. Struct.*, **216**, 626–628 (2001).
110. A. L. Gillon, N. Feeder, R. J. Davey and R. Storey, *Cryst. Growth Des.*, **3**, 663–673 (2003).
111. T. M. Krygowski, S. J. Grabowski and J. Konarski, *Tetrahedron*, **54**, 11311–11316 (1998).

3

Template-controlled Solid-state Synthesis: Toward a General Form of Covalent Capture in Molecular Solids

TOMISLAV FRIŠČIĆ and LEONARD R. MACGILLIVRAY
Department of Chemistry, University of Iowa, Iowa City, IA, USA.

1. INTRODUCTION

1.1. Target-oriented Organic Synthesis

The cornerstone of organic chemistry is the synthesis of molecules for technological, medicinal or purely academic purposes [1]. In an ideal case, a molecule is constructed in a single step, quantitative yield and in large amounts (i.e. grams). However, current limitations of modern organic synthetic chemistry presents the synthesis of a molecule as a multidimensional problem that necessitates a stepwise approach to the construction of covalent bonds and the placement of functional groups [2, 3]. Such synthesis is made further difficult by problems of controlling reaction regiochemistry, which eventually results in by-products and reduced yield. To alleviate such difficulties, Corey has developed the concept of target-oriented synthesis, a systematic approach to molecular synthesis [4]. Target-oriented synthesis introduced the idea of a molecular synthon, a structural feature of a molecule that can be obtained by a chemical reaction, and retrosynthetic analysis, a conceptual tool that enables the recognition of molecular synthons and provides instructions on their construction. The success of retrosynthetic analysis relies on chemical reactions that are generally compatible with and applicable to reactants of differing geometrical (e.g. size and shape) and/or chemical (e.g. functional groups) properties [5]. The stepwise application of synthons, in a sequence dictated by retrosynthetic analysis, should

Frontiers in Crystal Engineering. Edited by Edward R.T. Tiekink and Jagadese J. Vittal

produce the target at the highest yield and in the least number of steps. Consequently, the logic of target-oriented synthesis enables the construction of targets of considerable size and complexity [6], albeit sometimes through multistep and low-yielding procedures (methodologies that address the difficulties of controlling reactivity are continuously being introduced to target-oriented synthesis) [7].

In recent years, advances in the understanding of non-covalent bonds have allowed chemists to transfer the concept of target-oriented synthesis to supramolecular chemistry [8]. Whereas originally intended to provide a rational entry to molecules, the concept has also been applied to supramolecular structures. Such transfer of knowledge has largely been introduced by Desiraju [9] in the context of crystal engineering organic solids [10]. Thus, a retrosynthetic analysis of a solid will allow a chemist to identify supramolecular synthons responsible for its construction. In that way, a supramolecular synthon finds its origins in a molecular synthon and a supramolecular target finds its origins in a molecular target. Crystal engineering defined a supramolecular target as a solid-state network of molecules [11] and supramolecular synthons as structural units that connect network nodes and are based on intermolecular forces [12, 13]. In contrast to a molecular target, a supramolecular target is achieved under conditions of thermodynamic control, where inherent error-correction and convergence provide access to the target with high fidelity and in a single step [14].

1.2. Target-oriented Organic Synthesis and Covalent Capture

That supramolecular chemistry has developed to a level of constructing targets raises a question as to whether principles of target-oriented supramolecular chemistry can be employed to construct molecules [15]. Synthetic benefits of a supramolecular approach to constructing molecules would arise from the ability to assemble and preorganize reactants into positions suitable to form one or more covalent bonds in a single step [16, 17]. Such a design for covalent synthesis would, *de facto*, involve the construction of a supramolecular architecture as a target with subsequent covalent capture of the architecture [18, 19]. If such a design provided synthetic freedoms of molecular synthesis, then the approach would represent a general form of covalent capture that makes available molecular targets with the high level of precision and fidelity realized in supramolecular syntheses.

1.3. Overview

It is with these ideas in mind that we, in the following chapter, describe efforts in our laboratory to construct supramolecular targets, in the form of hydrogen-bonded molecular assemblies, that facilitate covalent capture in the solid state. The design and construction of the targets involves the use of bifunctional molecules, based on resorcinol and derivatives that act as linear templates by assembling and juxtaposing functionalized olefins, via hydrogen bonds, in positions suitable for single and multiple [2+2] photodimerizations. We describe how the method has allowed us to construct molecular targets in the form of a [2.2]-paracyclophane and [n]-ladderanes (where $n = 3$ and 5). We also demonstrate how a concept known as *template switching* allows fine-tuning of solid-state environments so as to increase the yield of the template-controlled solid-state reactions. We will then finish with descriptions of efforts made by us, and others, to develop additional molecules that act as linear templates in the solid state.

2. CONTROLLING REACTIVITY USING LINEAR TEMPLATES

The use of supramolecular chemistry to direct the formation of covalent bonds is realized by chemical species known as templates. Templates are inspired by biological systems where macromolecules such as enzymes or nucleic acids aide and direct the construction of naturally occurring molecules [20]. An expectation that the synthetic efficiency of *Nature* may be expanded to artificial systems and novel chemical transformations [21] has led chemists to design molecules that function as linear templates (Scheme 1). Such bifunctional molecules employ principles of molecular recognition and self-assembly to juxtapose two molecules, via directional non-covalent bonds, to react, in the minimalist case, within a ternary complex to form a covalent bond [22]. In that way, the geometry of the reactants within the ternary complex is covalently captured [23] to give a product

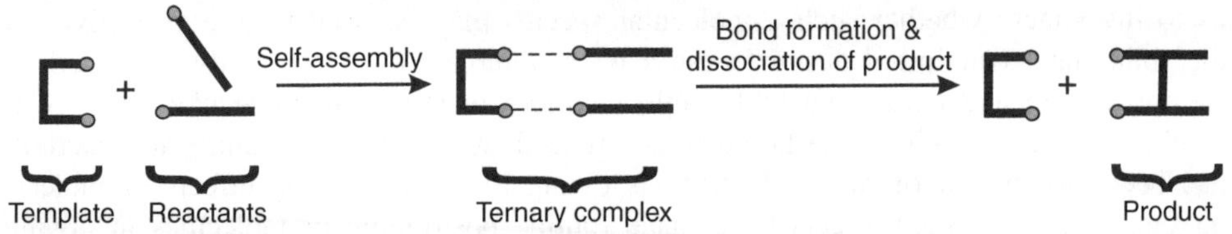

Template + Reactants → Self-assembly → Ternary complex → Bond formation & dissociation of product → + Product

Scheme 1

(a)

(b)

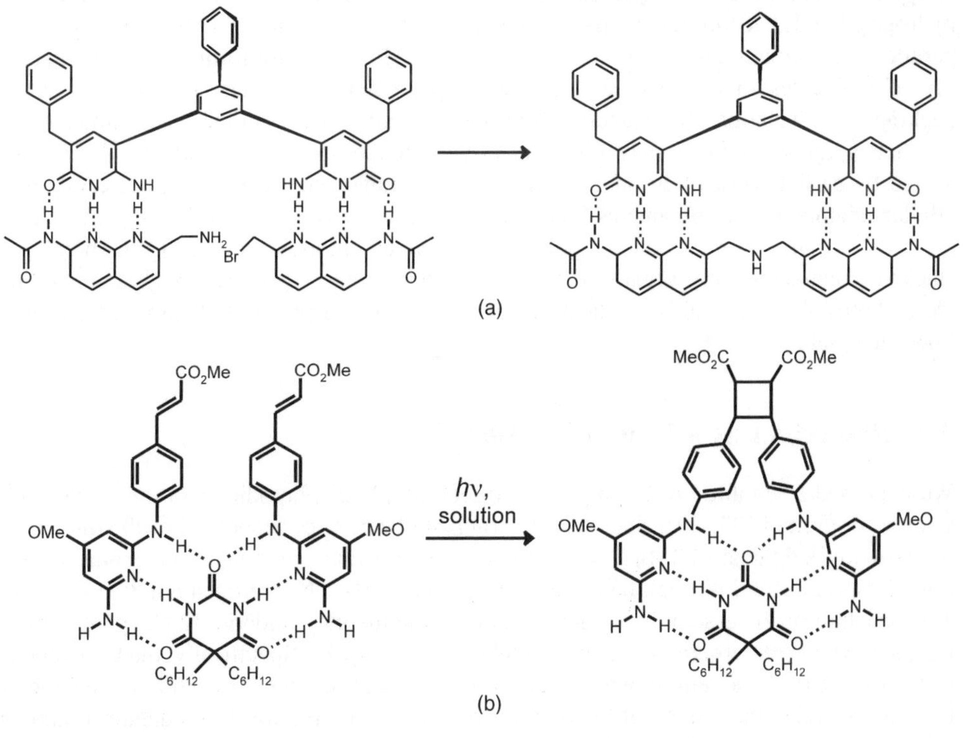

Figure 1. Template-controlled reactivity in the liquid phase. (a) S_N2 reaction and (b) [2+2] photodimerization.

with a predefined regiochemistry. The first linear template was described by Kelly, who devised a template that promotes S_N2 reactions via hydrogen bonds in the liquid phase (Figure 1(a)). The work of Kelly was later followed by Bassani who extended the concept to [2+2] photodimerizations in solution (Figure 1(b)) [24, 25]. Although the templates of Kelly and Bassani exhibited enhanced reaction rates and quantum yields, respectively, the synthetic value of both designs was limited by conformational flexibility and dissociation equilibria of the resulting ternary complexes in solution [26, 27].

3. TEMPLATE-CONTROLLED SOLID-STATE REACTIVITY

3.1. Template-controlled Reactivity in the Solid state

That chemical reactivity can be directed using linear templates in the liquid phase prompted us to investigate whether such a molecular species may be used to direct reactivity in the solid state. Our initial desire to use a linear template to direct solid-state reactivity stemmed from a general goal to be able to direct reactivity in molecular solids with synthetic freedoms akin to the liquid phase. Indeed, such control of solid-state reactivity has been largely out of reach of chemists owing to the frustrating effects of molecular close packing. Such effects have made routine positioning of molecules in organic solids for reaction difficult to achieve. To develop such a template, we focused on the solid-state [2+2] photodimerization of olefins [28] in molecular co-crystals [29–31]. The template would be used to preorganize two carbon–carbon double (C=C) bonds, via hydrogen bonds, according to the geometry criteria of Schmidt for photoreaction (i.e. parallel arrangement and separation <4.2 Å) [10]. Moreover, by forming a finite assembly of molecules that is photoactive, the template would alleviate problems of close packing by "insulating" the reactants from the surrounding crystalline environment.

That a linear template may direct reactivity in the solid state had also been suggested by earlier work of Ito and Scheffer, who rationalized the reactivity of crystalline ethylenediammonium cinnamates based on proximity effects (Figure 2) [32]. Structural analyses of the solids revealed hydrogen-bonded assemblies wherein C=C bonds of stacked cinnamates were organized for [2+2] photodimerization. Although some solids were photostable, the solids displayed features of chemical reactivity being directed by a linear template.

3.2. Resorcinol as a Linear Template

We identified a suitable linear template to direct the [2+2] photodimerization in the solid state as resorcinol [33, 34]. Specifically, co-crystallization of resorcinol with *trans*-1, 2-bis(4-pyridyl)ethylene (4,4′-bpe) was shown to give a four-component molecular assembly, 2(resorcinol)·2(4,4′-bpe), held together by four O–H···N hydrogen bonds (Figure 3). The structure of the assembly suggested that the olefins may undergo [2+2] photodimerization to yield *rctt*-tetrakis(4-pyridyl)cyclobutane (4,4′-tpcb). Specifically, the C=C bonds of the olefins of the assembly were arranged parallel and separated by 3.65 Å, a geometry that conformed to the criteria of Schmidt for photoreaction (Figure 4). As demonstrated by ^{1}H NMR spectroscopy, UV-irradiation of 2(resorcinol)·2(4,4′-bpe) yielded the expected photoproduct, 4,4′-tpcb, in quantitative yield. The identity of the photoproduct was verified

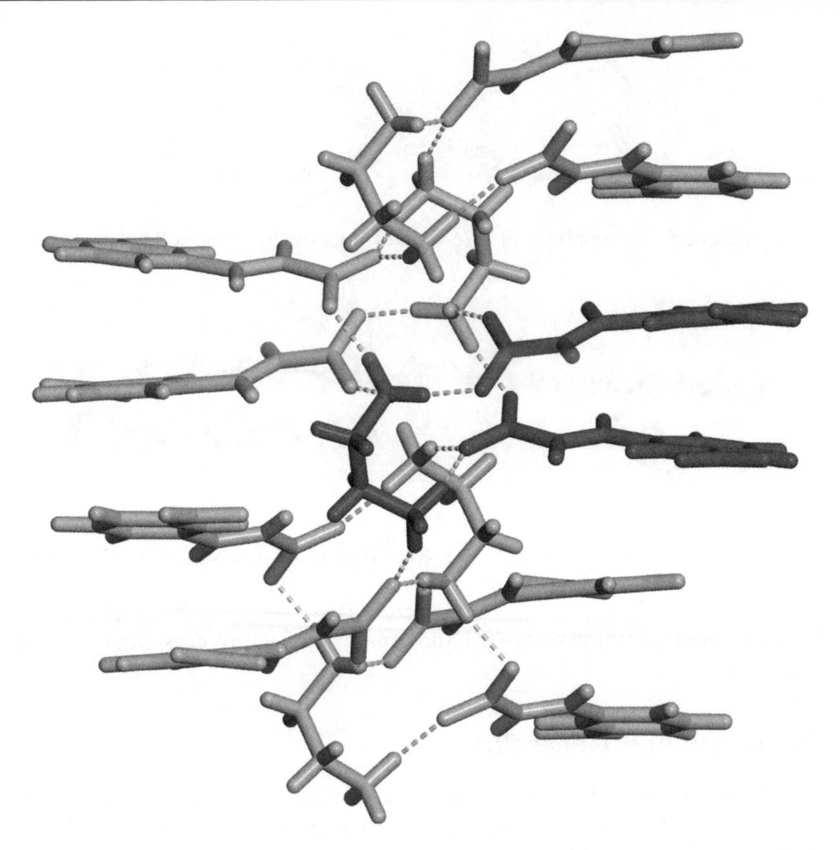

Figure 2. Arrangement of molecules in ethylenediammonium 2-nitrocinnamate. A single assembly is highlighted in bold.

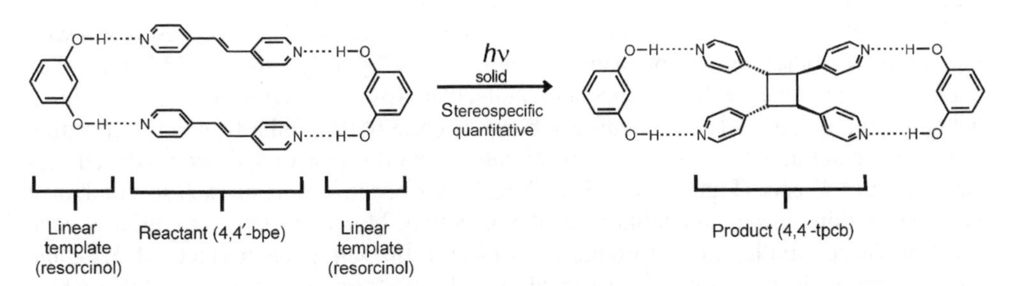

Figure 3. Schematic of the [2+2] photodimerization involving 2(resorcinol)·2(4,4′-bpe).

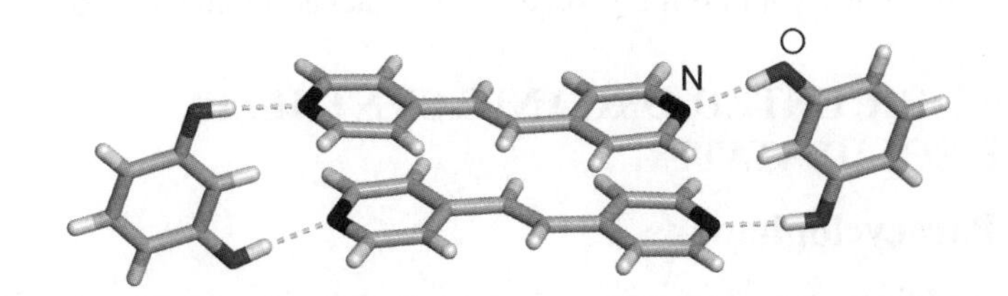

Figure 4. Perspectives of the photoactive assemblies of 2(resorcinol)·2(4,4′-bpe).

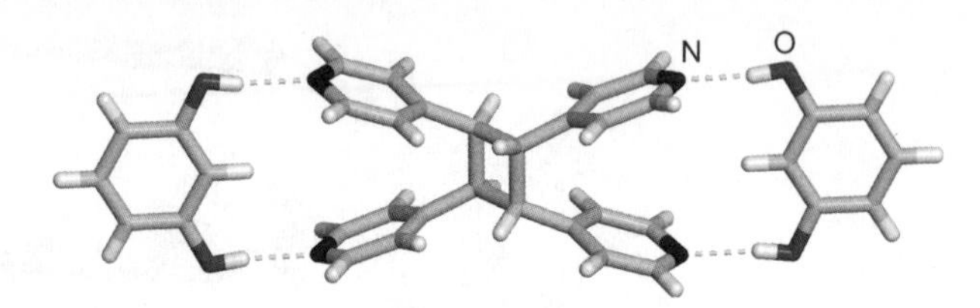

Figure 5. Assemblies of 2(resorcinol)·(4,4′-tpcb) in the solid state.

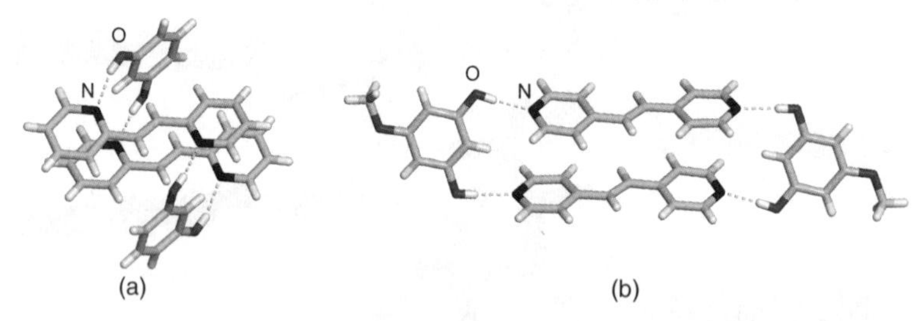

Figure 6. Photoactive assemblies in (a) 2(resorcinol)·2(2,2′-bpe) and (b) 2(5-OMe-res)·2(4,4′-bpe).

by a subsequent single-crystal x-ray structure analysis of the re-crystallized photoreacted solid (Figure 5).

3.3. Modularity and Generality

That resorcinol could be used as a template to direct reactivity in the solid state encouraged us to investigate the ability of resorcinol to serve as a general tool to direct the formation of covalent bonds in organic solids. In particular, we anticipated that the inherent modularity of the template method would allow us to modify the structures of the templates and reactants so as to increase the diversity of the template approach.

The ability to change the reactant was realized by a co-crystallization of resorcinol with the 2-pyridyl isomer of 4,4′-bpe, namely, *trans*-1,2-bis(2-pyridyl)ethylene (2,2′-bpe). The components produced a four-component molecular assembly, 2(resorcinol)·2(2,2′-bpe), held together by four O–H···N hydrogen bonds (Figure 6(a)). Similarly, co-crystallization of the resorcinol derivative 5-methoxyresorcinol (5-OMe-res) with 4,4′-bpe produced 2(5-OMe-res)·2(4,4′-bpe) (Figure 6(b)). The C=C bonds of the olefins of each solid adopted positions within discrete assemblies for photoreaction. Moreover, UV-irradiation of each solid produced, similar to 2(resorcinol)·2(4,4′-bpe), the *rctt* photoproducts of 2,2′- and 4,4′-bpe, respectively, in quantitative yield [34]. These observations suggested that a system of linear templates based on resorcinol may be used, in a more general way, to assemble and covalently capture the geometries of stacked olefins in molecular solids.

4. TARGET-ORIENTED ORGANIC SYNTHESIS IN THE SOLID STATE

4.1. [2.2]-Paracyclophane

That a system of linear templates based on resorcinol could be used with different reactants and templates, prompted us to explore whether the method could provide a synthetic

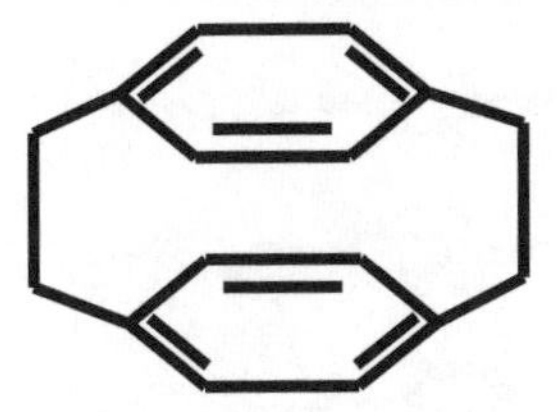

[2.2]-Paracyclophane

Scheme 2

freedom of the liquid phase. Specifically, we anticipated that the ability of resorcinol to assemble along the periphery of the reactants would enable us to construct a lengthened molecule in the form of a [2.2]-paracyclophane (Scheme 2). The cyclophane framework, studied extensively by Cram [35], has relevance to physical organic chemistry and materials science owing to its photo- and electrochemical properties [36, 37]. Despite this, the syntheses of [2.2]-paracylophanes remain a synthetic challenge [38].

A retrosynthetic analysis of the targeted [2.2]-paracyclophane suggested the molecule may be constructed using a linear template from the bifunctional diene 1,4-bis[2-(4-pyridyl)ethenyl]benzene (1,4-bpeb). Co-crystallization of resorcinol, or a derivative, with 1,4-bpeb would yield the four-component molecular assembly, 2(resorcinol)·2(1,4-bpeb), with the two dienes positioned for photoreaction. UV-irradiation of the solid would result in covalent capture of the two dienes to give the cyclophane target, tetrakis(4-pyridyl)-1,2,9,10-diethano[2.2]-paracyclophane (4,4′-tppcp) (Scheme 3). In an ideal case, the target would form in quantitative yield and gram quantities.

As anticipated, co-crystallization of 1,4-bpeb with 5-OMe-res produced the four-component assembly 2(5-OMe-res)·2(1,4-bpeb) (Figure 7). Similar to 2(resorcinol)·2(4,4′-bpe), the components of the assembly were held together by four O–H···N hydrogen bonds with C=C bonds positioned at a separation of 3.70 Å. The assemblies packed with C=C bonds of adjacent assemblies parallel and separated by 3.95 Å (Figure 8). As determined by ^{1}H NMR spectroscopy, UV-irradiation of 2(5-OMe-res)·2(1,4-bpeb) produced 4,4′-tppcp in 60% yield. In addition to the cyclophane, a monocyclized dimer, as well as indefinable products, were shown to form [34].

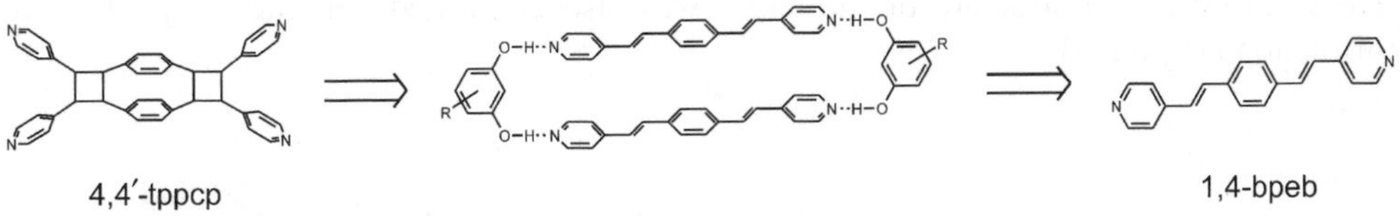

4,4′-tppcp 1,4-bpeb

Scheme 3

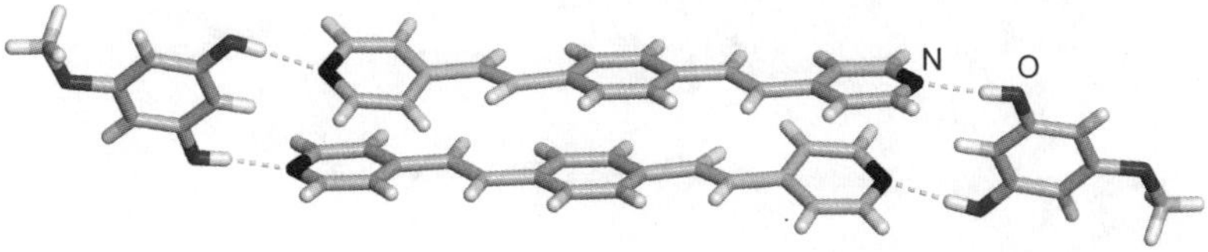

Figure 7. Molecular assemblies of 2(5-OMe-res)·2(1,4-bpeb).

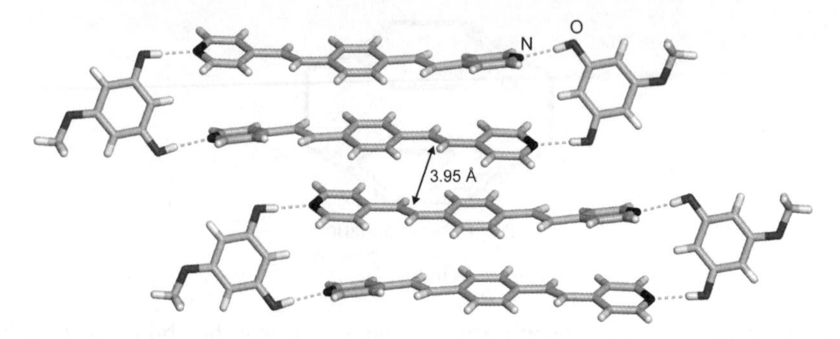

Figure 8. Arrangement of nearest-neighbor molecular assemblies in 2(5-OMe-res)·2(1,4-bpeb).

4.2. Template Switching

In addition to providing access to the targeted cyclophane, we anticipated that the modular nature of the template method could permit the yield of the solid-state reaction to be increased. Specifically, we anticipated that switching the template to a different resorcinol derivative could be used as a means to promote the formation of a finite assembly based on 1,4-bpeb that exhibits a packing environment different from 2(5-OMe-res)·2(1,4-bpeb). Ideally, the C=C bonds of adjacent assemblies in the newly formed solid would not assemble in close enough proximity to react. Although a general ability to predict solid-state packing remains elusive, such "template switching" would exploit unpredictable structural consequences of crystal packing so as to increase the likelihood of achieving a packing environment suitable for the quantitative formation of the targeted photoproduct.

A co-crystallization of 1,4-bpeb with 4-benzylresorcinol (4-bn-res) as the template confirmed our hypothesis. As similar to 2(5-OMe-res)·2(1,4-bpeb), co-crystals of 2(4-bn-res)·2(1,4-bpeb) consisted of four-component assemblies with the two dienes positioned for [2+2] photodimerization. However, in contrast to 2(4-bn-res)·2(1,4-bpeb), the closest separation between C=C bonds of adjacent assemblies was 5.4 Å (Figure 9). As anticipated, UV-irradiation of 2(4-bn-res)·2(1,4-bpeb) produced 4,4′-tppcp in quantitative yield [39]. The structure of the target was also confirmed via single-crystal x-ray diffraction (Figure 10).

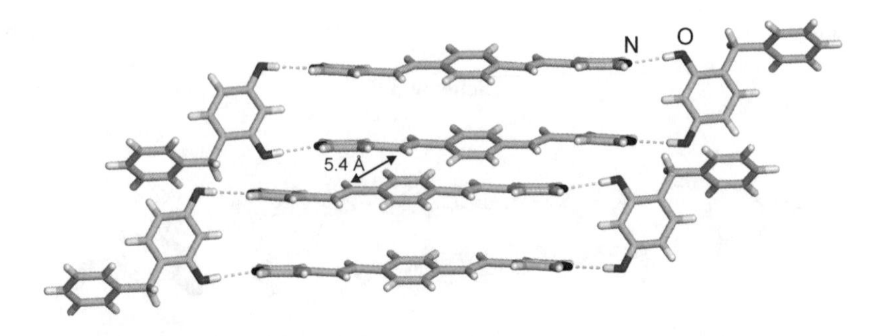

Figure 9. Arrangement of nearest-neighbor molecular assemblies in 2(4-bn-res)·2(1,4-bpeb).

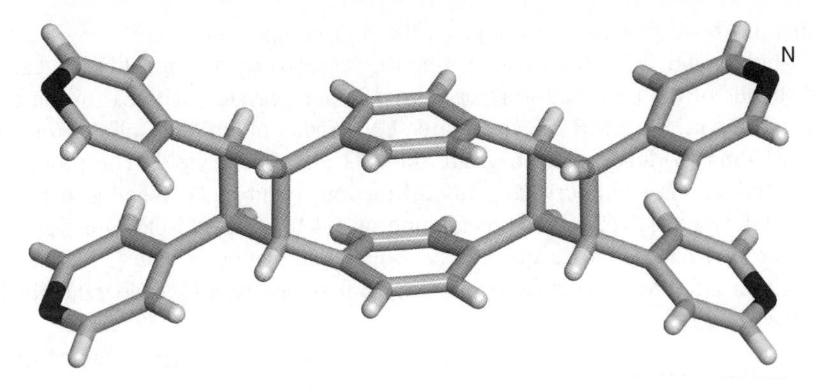

Figure 10. The [2.2]-paracyclophane target as revealed by x-ray crystal structure analysis.

4.3. Ladderanes

In addition to a [2.2]-paracyclophane, we anticipated that a resorcinol could be used to construct [n]-ladderanes (where $n = 3$ or 5) (Scheme 4). Such molecular frameworks had been discovered in intracellular membrane lipids of *anammox* bacteria that participate in the oceanic nitrogen cycle [40]. A retrosynthetic analysis of such a ladder framework [41, 42] suggested a template-controlled [2+2] photodimerization of the conjugated diene *trans,trans*-1,4-bis(4-pyridyl)-1,3-butadiene (1,4-bpbd) would yield the corresponding all-*trans*-tetrakis(4-pyridyl)-[3]-ladderane (4-tp-3-lad). Likewise, a template-controlled [2+2] photodimerization of the triene *trans,trans,trans*-1,6-bis(4-pyridyl)-1,3,5-hexatriene (1,6-bpht) would provide the corresponding all-*trans*-tetrakis(4-pyridyl)-[5]-ladderane (4-tp-5-lad) (Scheme 5).

By way of template switching, we discovered 5-OMe-res to be a suitable template for the construction of the two molecular ladders. Thus, co-crystallization of 5-OMe-res

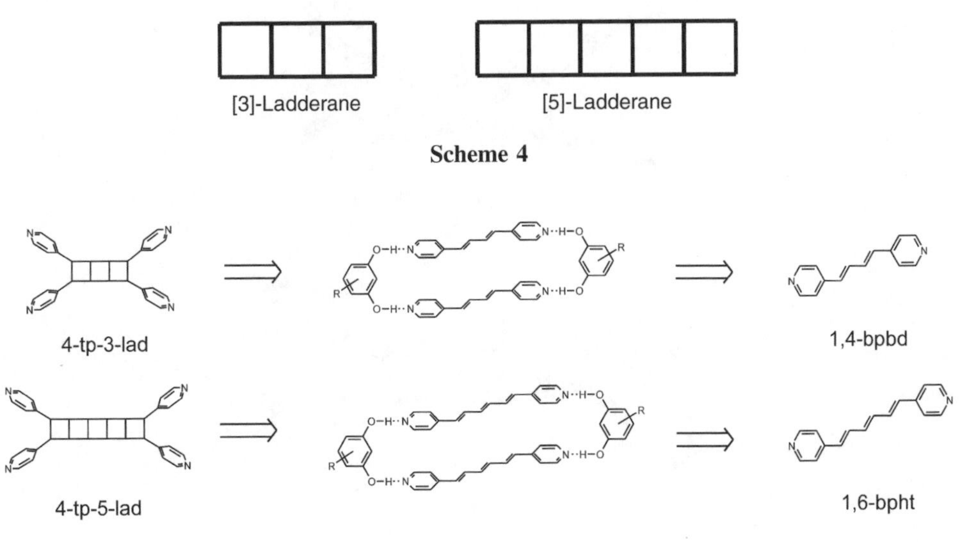

Scheme 4

Scheme 5

with either 1,4-bpbd or 1,6-bpht produced the four-component assemblies of 2(5-OMe-res)·2(1,4-bpbd) and 2(5-OMe-res)·2(1,6-bpht), respectively (Figure 11). In each case, the C=C bonds of the polyenes were organized in appropriate positions for the reaction. As established via ^{1}H NMR spectroscopy, UV-irradiation of the solids produced the targeted [3]- and [5]-ladderanes in gram amounts and 100% yield. The products were also characterized via single-crystal x-ray diffraction (Figure 12). Whereas the template-controlled solid-state [2+2] photodimerization of 1,4-bpbd in 2(5-OMe-res)·2(1,4-bpbd) provided the first example of a quantitative solid-state synthesis of a [3]-ladderane, the reaction within 2(5-OMe-res)·2(1,6-bpht) was the first case of a [5]-ladderane obtained in quantitative yield [43, 44].

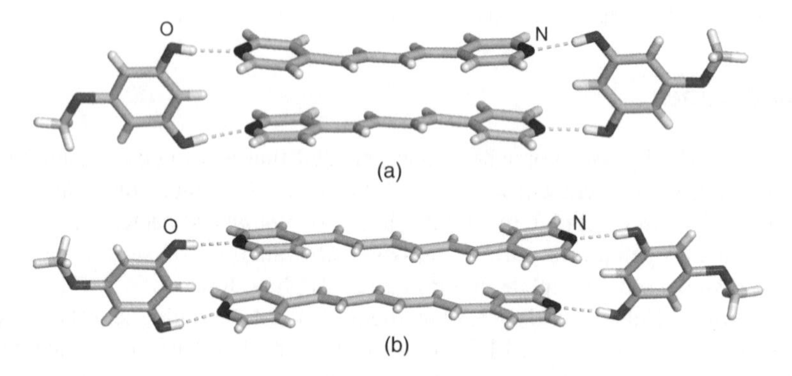

Figure 11. Photoactive molecular assemblies. (a) 2(5-OMe-res)·2(1,4-bpbd) and (b) 2(5-OMe-res)·2(1,6-bpht).

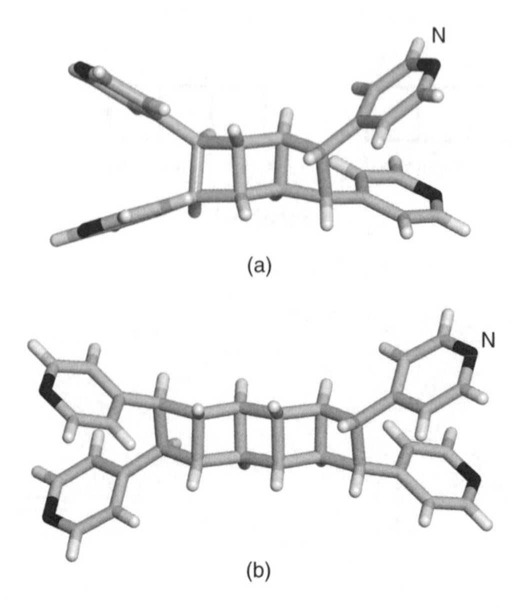

Figure 12. x-ray structures of (a) [3]-ladderane and (b) [5]-ladderane targets.

5. OTHER LINEAR TEMPLATES

That molecular targets, in the form of a paracyclophane and ladderanes, could be constructed using a system of linear templates in the solid state prompted us to further explore the scope of the template method. In particular, we anticipated the development of additional templates to be a worthy goal [45]. Additional templates could increase the synthetic flexibility of the approach by providing a means, for example, to increase a product yield where a system of resorcinols may be less effective. Additional templates could also be used to develop recognition sites, other than pyridyl groups, attached to the reactants, which would lend to increasing the structural diversity of the photoproducts.

5.1. 1,8-Naphthalenedicarboxylic Acid

Our first entry to extend linear templates beyond resorcinol involved 1,8-naphthalenedicarboxylic acid (1,8-nap) (Scheme 6) [46]. Similar to 2(resorcinol)·2(4,4′-bpe), we anticipated that co-crystallization of 1,8-nap with 4,4′-bpe would produce the four-component assembly 2(1,8-nap)·2(4,4′-bpe). Each carboxylic acid group of each diacid would interact with the 4-pyridyl group of 4,4′-bpe via an O–H···N hydrogen bond. As anticipated, the components assembled to form the molecular assembly with C=C bonds of 4,4′-bpe organized parallel at a separation of 3.73 Å (Figure 13). Moreover, UV-irradiation of the solid produced 4,4′-tpcb in quantitative yield. The ability of 1,8-nap to serve as a linear template was also demonstrated using 2,2′-bpe and 3,4′-bpe as reactants [47].

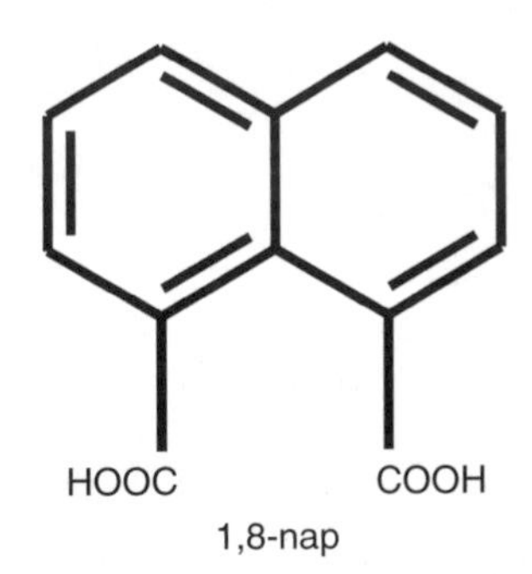

Scheme 6

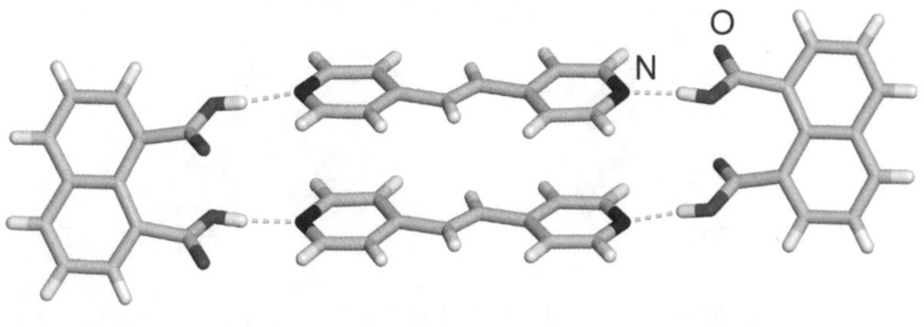

Figure 13. Photoactive molecular assembly in 2(1,8-nap)·2(4,4′-bpe).

5.2. Bis-*p*-phenylene[34]-crown[10]

Garcia-Garibay and Stoddart have demonstrated the ability of the crown ether bis-*p*-phenylene[34]-crown[10] (BPP34C10) to serve as a linear template in the solid state [48]. That BPP34C10 served as a template stems from the ability of the crown ether to encircle a pair of ammonium dications, namely, 4,4′-bis(benzylammonium)stilbene (4,4′-bbs) (Scheme 7). Co-crystallization of 4,4′-bbs hexafluorophosphate with BPP34C10 produced the four-component cationic assembly $[2(4,4'\text{-bbs})\cdot 2(\text{BPP34C10})]^{4+}$. The components of the assembly were held together by eight $N^+–H\cdots O$ hydrogen bonds (Figure 14). The C=C bonds of the two cations were oriented parallel and separated by 4.20 Å. UV-irradiation of the solid produced the corresponding cyclobutane, 1,2,3,4-tetrakis(4-benzyl-ammoniumphenyl)cyclobutane (4,4′-tbac), regioselectively and in approximately 80% yield.

5.3. Carballylic and 1,2,4,5-Benzenetetracarboxylic Acids

Jones has expanded the system of linear templates to tricarballylic (tca) and 1,2,4,5-benzenetetracarboxylic acids (bta) (Scheme 8) [49]. Specifically, co-crystallization of 4,4′-bpe with either tca or bta produced crystalline solids tca·1.5(4,4′-bpe) and bta·2(4,4′-bpe), respectively. In contrast to the examples described above, the components formed infinite, or polymeric, hydrogen-bonded structures. In the case of tca·1.5(4,4′-bpe), the components

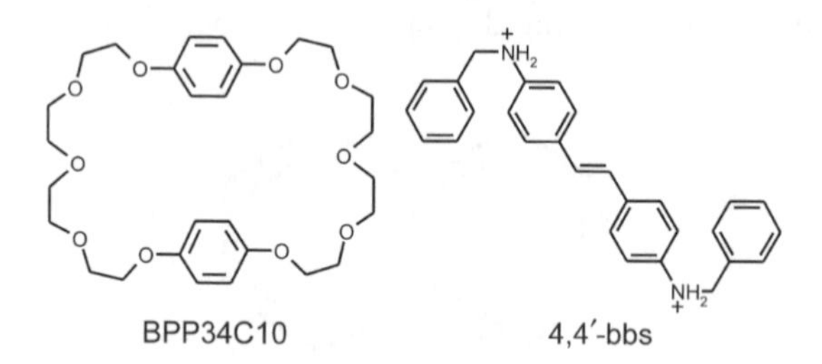

Scheme 7

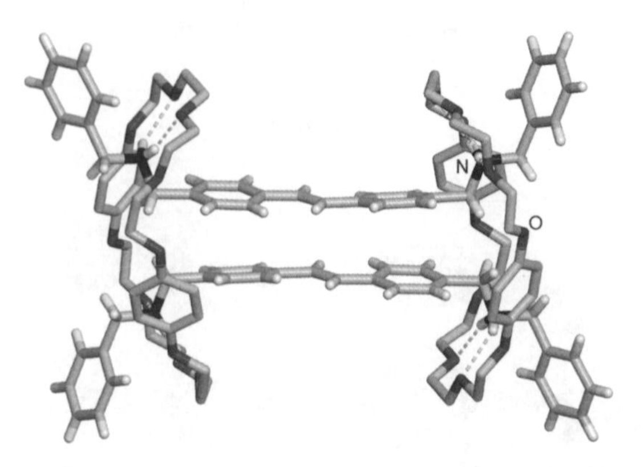

Figure 14. Photoactive cationic assembly of $[2(\text{BPP34C10})\cdot 2(4,4'\text{-bbs})]^{4+}$. Hydrogen atoms of BPP34C10 have been omitted.

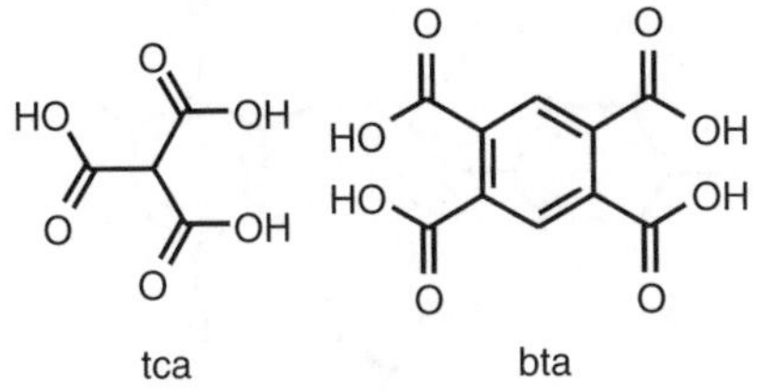

tca bta

Scheme 8

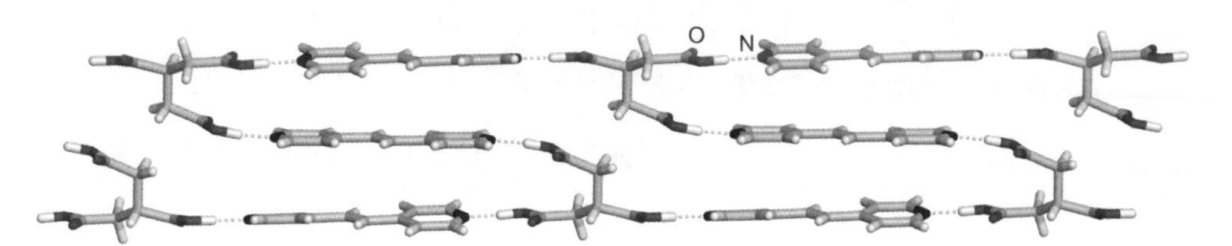

Figure 15. The infinite hydrogen-bonded tape in tca·1.5(4,4′-bpe) demonstrating the triple-stack arrangement of 4,4′-bpe molecules.

produced tapes with reactants organized in triple-decker stacks such that two terminal carboxylic acid groups positioned pairs of C=C bonds in a parallel arrangement and at separations of 3.82 and 3.80 Å (Figure 15). UV-irradiation of tca·1.5(4,4′-bpe) produced 4,4′-tpcb in 90% yield. Similarly, co-crystallization of bta with 4,4′-bpe produced tapes with alternating 4,4′-bpe and bta molecules. The ability of bta to function as linear template was achieved through pairs of ortho-positioned carboxylic acid moieties that aligned the C=C bonds of 4,4′-bpe within each tape in parallel, and at a separation of 3.77 Å (Figure 16). UV-irradiation of bta·2(4,4′-bpe) produced 4,4′-tpcb (Figure 3) in quantitative yield.

5.4. Tetrakis(4-iodoperfluorophenyl)erythritol

Halogen bonding has been used by Metrangolo and Resnati to develop a linear template in the solid state [50]. Specifically, tetratopic tetrakis(4-iodoperfluorophenyl)erythritol (tie) was shown to serve as a template (Scheme 9). That the molecule functioned as a linear template was anticipated from both the halogen-bonding ability of the iodine atoms as well as a tendency of the perfluoroarene groups to participate in face-to-face $\pi \cdots \pi$ stacking. Thus, co-crystallization of tie with 4,4′-bpe produced infinite one-dimensional ribbons with alternating 4,4′-bpe and tie molecules. The molecules were held together by two

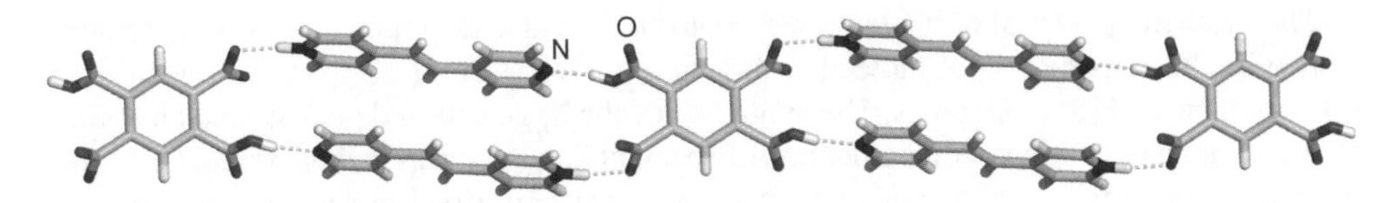

Figure 16. The infinite hydrogen-bonded tape in bta·2(4,4′-bpe) demonstrating the arrangement of 4,4′-bpe molecules into stacked pairs.

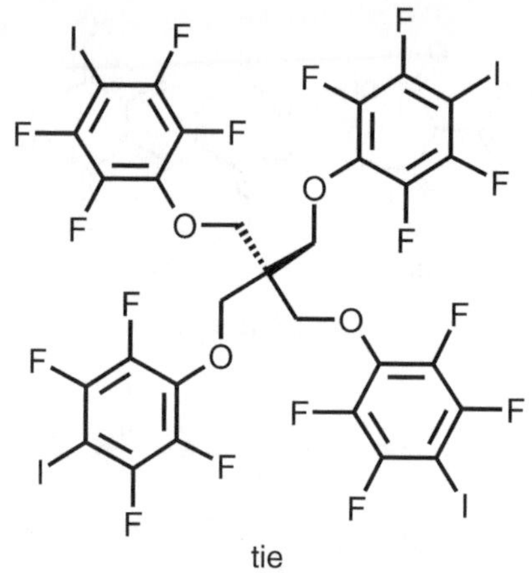

tie

Scheme 9

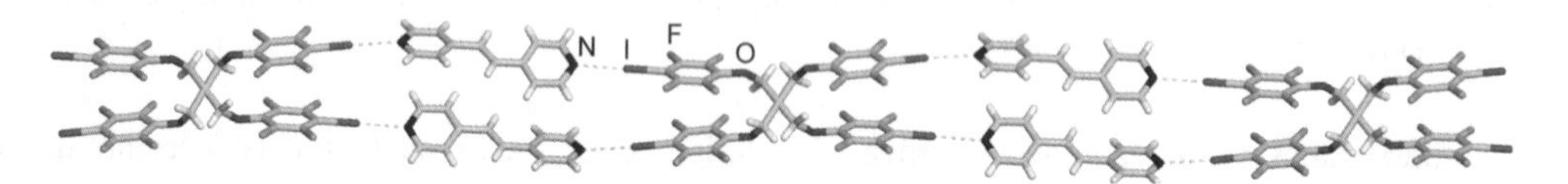

Figure 17. The infinite halogen-bonded tape in tie·2(4,4′-bpe) demonstrating the arrangement of 4,4′-bpe molecules into stacked pairs.

N···I halogen bonds with N···I separations of 2.81 and 2.86 Å (Figure 17). The olefins of each ribbon were aligned parallel and separated by less than 4.51 Å. UV-irradiation of the solid produced 4,4′-tpcb in quantitative yield.

6. SUMMARY AND OUTLOOK

In this chapter, we have demonstrated that molecules that function as linear templates provide a means to control the organization of molecules in the organic solid state, so as to direct chemical reactions and conduct molecular syntheses. The templates have been shown to organize complementary reactants within molecular assemblies for [2+2] photoreaction. The reactions have been demonstrated to proceed regiospecifically and in quantitative yield. In that way, template-controlled [2+2] photodimerizations have resulted in the covalent capture of olefins within supramolecular assemblies to give molecules as products.

The synthetic potential of the template-controlled solid-state approach has been demonstrated by the construction of molecular targets in the form of a [2.2]-paracyclophane and [3]-ladderane and [5]-ladderanes. The syntheses of the targets have demonstrated an ability of the templates to adapt to lengthening of reactants and corresponding products. Moreover, that the templates have been used to synthesize prescribed molecules suggests that the templates may provide a general way to develop targeted supramolecular architectures that may be covalently captured to form targeted molecules.

A key step in the generalization of linear templates as a means to covalently capture molecules in the solid state has involved the development of new templates. The availability of additional templates has enabled the modularity of the method to be exploited to optimize reactivity and may eventually be used to impart additional properties (i.e. magnetic, optical) to such reactive solids [45, 51]. Having achieved the development of diverse sets of reactants and templates, the solid state may eventually serve as a synthetic playground, with a synthetic spirit encountered in the liquid phase, for synthetic supramolecular and molecular chemists.

REFERENCES

1. K. C. Nicolau and S. A. Snyder, *Proc. Natl. Acad. Sci. USA*, **101**, 11929–11936 (2004).
2. E. J. Corey, *Chem. Soc. Rev.*, **17**, 111–133 (1988).
3. E. J. Corey, *Pure Appl. Chem.*, **14**, 19–37 (1967).
4. E. J. Corey, *Angew. Chem. Int. Ed. Engl.*, **30**, 455–465 (1991).
5. T. M. Trnka and R. H. Grubbs, *Acc. Chem. Res.*, **34**, 18–29 (2001).
6. K. C. Nicolaou and P. S. Baran, *Angew. Chem. Int. Ed.*, **41**, 2678–2720 (2002).
7. (a) K. C. Nicolaou, T. Montagnon and S. A. Snyder, *Chem. Commun.*, 551–564 (2003); (b) E. J. Corey and J. O. Link, *Tetrahedron Lett.*, **31**, 601–604 (1990).
8. J.-M. Lehn, *Angew. Chem. Int. Ed. Engl.*, **27**, 89–112 (1988).
9. G. R. Desiraju, *Angew. Chem. Int. Ed. Engl.*, **34**, 2311–2327 (1995).
10. G. M. J. Schmidt, *Pure Appl. Chem.*, **27**, 647–678 (1971).
11. J. Dunitz, *Pure Appl. Chem.*, **63**, 177–185 (1991).
12. (a) G. R. Desiraju, *Crystal Engineering*, Elsevier Science Publishers Inc., Amsterdam, 1989; (b) V. R. Thalladi, B. Satish Goud, V. J. Hoy, F. H. Allen, J. A. K. Howard and G. R. Desiraju, *Chem. Commun.*, 401–402 (1996).
13. (a) L. R. MacGillivray and J. L. Atwood, *Angew. Chem. Int. Ed.*, **38**, 1018–1033 (1999); (b) F. M. Tabellion, S. R. Seidel, A. M. Arif and P. J. Stang, *J. Am. Chem. Soc.*, **123**, 7740–7741 (2001); (c) S. R. Seidel and P. Stang, *J. Acc. Chem. Res.*, **35**, 972–983 (2002); (d) N. C. Seeman, *Acc. Chem. Res.*, **30**, 357–363 (1997).
14. G. M. Whitesides, E. E. Simanek, J. P. Mathias, C. T. Seto, D. N. Chin, M. Mammen and D. M. Gordon, *Acc. Chem. Res.*, **28**, 37–44 (1995).
15. (a) J. S. Lindsey, *New J. Chem.*, **15**, 153–180 (1991); (b) G. M. Whitesides, J. P. Mathias and C. T. Seto, *Science*, **254**, 1312–1319 (1991).
16. D. H. Busch, *Acc. Chem. Res.*, **11**, 392–400 (1978).
17. S. Anderson, H. L. Anderson and J. K. M. Sanders, *Acc. Chem. Res.*, **26**, 469–475 (1993).
18. (a) T. D. Clark, K. Kobayashi and M. R. Ghadiri, *Chem. Eur. J.*, **5**, 782–792 (1999); (b) T. D. Clark and M. R. Ghadiri, *J. Am. Chem. Soc.*, **117**, 12364–12365 (1995).
19. F. Cardullo, M. Crego Callama, B. H. M. Snellink-Ruël, J.-L. Weidmann, A. Biejelewska, R. Fokkens, N. M. M. Nibbering, P. Timmerman and D. N. Reinhoudt, *Chem. Commun.*, 367–368 (2000).
20. S. J. Benkovic and S. Hammes-Schiffer, *Science*, **301**, 1196–1202 (2003).
21. J. T. Goodwin and D. G. J. Lynn, *J. Am. Chem. Soc.*, **114**, 9197–9198 (1992).
22. F. Diederich and P. J. Stang (Eds.), *Templated Organic Synthesis*, Wiley-VCH, Weinheim, Germany, 2000.
23. J. D. Hartgerink, *Curr. Opin. Chem. Biol.*, **8**, 604–609 (2004).
24. T. R. Kelly, C. Zhao and G. J. J. Bridger, *J. Am. Chem. Soc.*, **111**, 3744–3745 (1989).
25. D. M. Bassani, V. Darcos, S. Mahony and J.-P. Desvergne, *J. Am. Chem. Soc.*, **122**, 8795–8796 (2000).

26. (a) T. R. Kelly, G. J. Bridger and C. J. Zhao, *J. Am. Chem. Soc.*, **112**, 8024–8034 (1990); (b) V. Darcos, K. Griffith, X. Sallenave, J.-P. Desvergne, C. Guyard-Duhayon, B. Hasenknopf and D. M. Bassani, *Photochem. Photobiol. Sci.*, **2**, 1152–1161 (2003).
27. D. M. Bassani, X. Sallenave, V. Darcos and J. P. Desvergne, *Chem. Commun.*, 1446–1447 (2001).
28. A. Mustafa, *Chem. Rev.*, **51**, 1–23 (1952).
29. K. Tanaka and F. Toda, *Chem. Rev.*, **100**, 1025–1074 (2000).
30. C. V. K. Sharma, K. Paneerselvam, L. Shimoni, H. Katz, H. L. Carrell and G. R. Desiraju, *Chem. Mater.*, **6**, 1282–1292 (1994).
31. V. Ramamurthy and K. Venkatesan, *Chem. Rev.*, **87**, 433–481 (1987).
32. Y. Ito, B. Borecka, J. Trotter and J. R. Scheffer, *Tetrahedron Lett.*, **36**, 6083–6087 (1995).
33. Y. Aoyama, K. Endo, T. Anzai, Y. Yamaguchi, T. Sawaki, K. Kobayashi, N. Kanehisa, H. Hashimoto, Y. Kai and Y. J. Masuda, *J. Am. Chem. Soc.*, **118**, 5562–5571 (1996).
34. L. R. MacGillivray, J. L. Reid and J. A. J. Ripmeester, *J. Am. Chem. Soc.*, **122**, 7817–7818 (2000).
35. (a) D. J. Cram and H. J. Steinberg, *J. Am. Chem. Soc.*, **73**, 5691–5704 (1951); (b) D. J. Cram and J. M. Cram, *Acc. Chem. Res.*, **4**, 204–213 (1971).
36. (a) F. Diederich, *Cyclophanes*, Royal Society of Chemistry, Cambridge, 1991; (b) G. P. Bartholomew and G. C. Bazan, *Acc. Chem. Res.*, **34**, 30–39 (2001).
37. A. De Meijere and B. Koenig, *Synlett*, **11**, 1221–1232 (1997).
38. J. A. Reiss, In: *Cyclophanes*, Vol. 2 (Eds. P. M. Keehn. and S. M. Rosenfeld., Academic Press, New York, 1983.
39. T. Friščić and L. R. MacGillivray, *Chem. Commun.*, 1306–1307 (2003).
40. (a) M. M. M. Kuypers, A. O. Sliekers, G. Lavik, M. Schmid, B. B. Jorgensen, J. G. Kuenen, J. S. Sinninghe Damste, M. Strous and M. S. M. Jetten, *Nature*, **422**, 608–611 (2003); (b) J. S. Sinninghe Damste, M. Strous, W. I. C. Rijpstra, E. C. Hopmans, J. A. J. Geenevasen, A. C. T. van Duin, L. A. van Niftrik and M. S. M. Jetten, *Nature*, **419**, 708–712 (2002).
41. H. Hopf, *Angew. Chem. Int. Ed.*, **42**, 2822–2825 (2003).
42. H. Hopf, H. Greiving, P. G. Jones and P. Bubenitschek, *Angew. Chem. Int. Ed. Eng.*, **34**, 685–687 (1995).
43. X. Gao, T. Friščić and L. R. MacGillivray, *Angew. Chem. Int. Ed.*, **43**, 232–236 (2004).
44. T. Friščić and L. R. MacGillivray, *Supramol. Chem.*, **17**, 47–51 (2005).
45. T. Friščić, D. Drab and L. R. MacGillivray, *Org. Lett.*, **6**, 4647–4650 (2004).
46. G. S. Papaefstathiou, A. J. Kipp and L. R. MacGillivray, *Chem. Commun.*, 2462–2463 (2001).
47. D. B. Varshney, G. S. Papaefstathiou and L. R. MacGillivray, *Chem. Commun.*, 1964–1965 (2002).
48. D. G. Amirsakis, M. A. Garcia-Garibay, S. J. Rowan, J. F. Stoddart, A. J. P. White and D. J. Williams, *Angew. Chem. Int. Ed.*, **40**, 4256–4261 (2001).
49. (a) N. Shan and W. Jones, *Tetrahedron Lett.*, **44**, 3687–3689 (2003); (b) N. Shan and W. Jones, *Green Chem.*, **5**, 728–730 (2003).
50. T. Caronna, R. Liantonio, T. A. Logothetis, P. Metrangolo, T. Pilati and G. J. Resnati, *J. Am. Chem. Soc.*, **126**, 4500–4501 (2004).
51. G. S. Papaefstathiou, Z. Zhong, L. Geng and L. R. J. MacGillivray, *J. Am. Chem. Soc.*, **126**, 9158–9159 (2004).

4

Interplay of Non-covalent Bonds: Effect of Crystal Structure on Molecular Structure

JONATHAN W. STEED
Department of Chemistry, University of Durham, University Science Laboratories, Durham DH1 3LE, UK.

1. INTRODUCTION

An ordered, periodic crystalline solid represents a *de facto* manifestation of a highly complicated series of energetic compromises between the numerous interactions that bind the crystal together and contribute to its assembly. These interactions are involved in terms of solution association and aggregation, rates of crystal nucleation and growth and in the final overall crystal lattice energy. A simplistic yet powerful picture of crystalline solids is that of an effectively infinite array of rigid molecules or *tectons* [1] linked by a repeating pattern of intermolecular interaction or *supramolecular synthon* [2–4]. Thus, Hosseini *et al.* have promoted a view of molecular tectonics based on systems such as **1**, which allows the controlled spacing of gold ions in anions such as $[Au(CN_2)]^-$ [5], and **2**, which forms solid-state helices based on a simple OH···N(pyridyl) synthon [6]. Similarly, Orpen *et al.* have identified tectons such as 2,2′-bipyridinium (**3**) and $[PtCl_4]^{2-}$ (**4**) that form linear solid-state tapes [7] linked by an $R_1^2(4)$ hydrogen-bonded synthon in graph set terminology [8] (Figure 1). Tectons produced by Wuest have been used to form extremely large, interpenetrating diamondoid networks and porous organic zeolite mimics [1, 9–12].

In many cases, it is appropriate to regard tectons as rigid or well-defined building blocks and indeed the most successful tectons (as judged by the predictability and reproducibility of the supramolecular synthons and structures they form) are molecular species with very few degrees of conformational freedom. As soon as more flexible components are examined, the complexity of the range of possible crystal structures increases

Frontiers in Crystal Engineering. Edited by Edward R.T. Tiekink and Jagadese J. Vittal

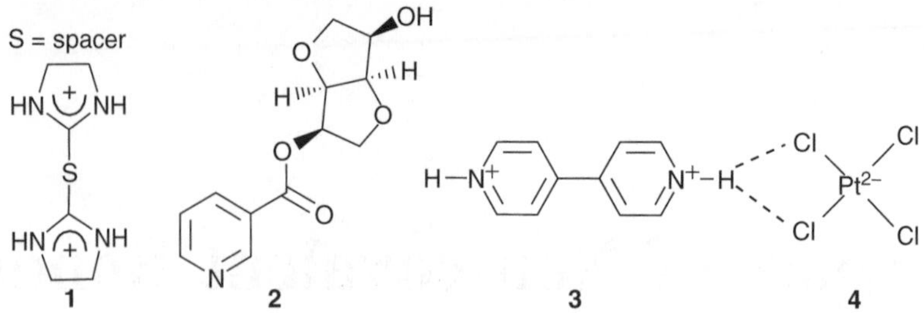

Figure 1. Tecton **1** is capable of controlled spacing of gold atoms depending on the length of spacer "S"; tecton **2** assembles into homochiral, helical chains via an OH···N hydrogen-bonding interaction and, tectons **3** and **4** assemble via an $R_1^2(4)$ hydrogen-bonded synthon to give linear tapes in the solid state.

rapidly, with consequent decrease in the systems' predictability, both in terms of the occurrence of particular supramolecular synthons and the molecular conformation, and hence the topology or geometry of the resultant structure. Thus, molecular conformation is expressed in crystalline structure and is also subject to the constraints placed upon it by the requirements of a viable set of crystal packing interactions. Indeed, sometimes more than one conformation of chemically identical molecules can be observed in the same crystal, or a compound may exhibit conformational polymorphism in which different crystal forms adopt different molecular conformations [13, 14]. For example, in the case of bis(phenylimino)isoindoline (**5**)[13], the triclinic polymorph contains a 3:1 ratio of anti, anti and anti, syn conformers, representing a case of conformational isomorphism. A second polymorph consists entirely of the anti, anti conformer; thus, the compound also exhibits conformational polymorphism. Similarly, coordination compounds can adapt their geometry to optimize crystal packing interactions as exemplified by the remarkable structure of $[Cr(en)_3][Ni(CN)_5]\cdot 1.5H_2O$ (**6**) (en = 1,2-diaminoethane), which contains both trigonal bipyramidal and square pyramidal Ni(II) centers within the same crystal [15].

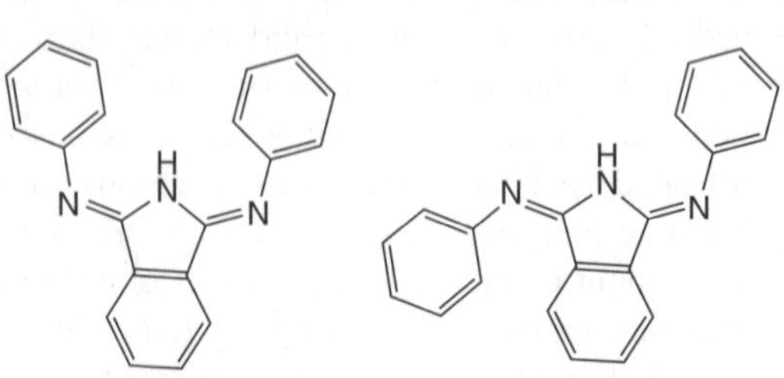

anti, anti-**5** anti, syn-**5**

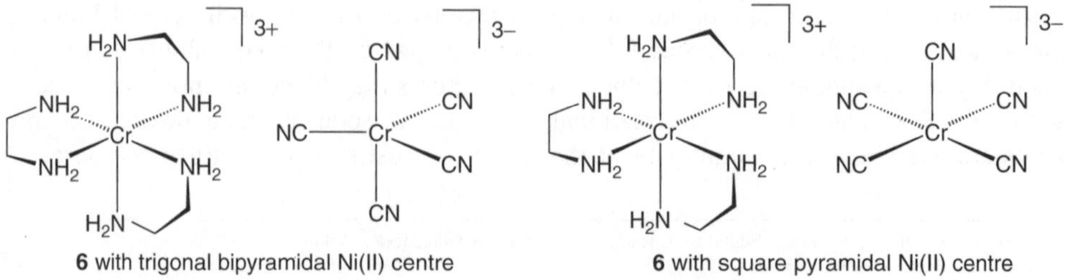

6 with trigonal bipyramidal Ni(II) centre **6** with square pyramidal Ni(II) centre

In addition to molecular conformation (as measured by changes in intramolecular torsion angles), molecular geometry (in terms of bond lengths, angles and even coordination number and nuclearity in inorganic systems) is also subject to the influence of crystal packing effects. In the case of robust organic tectons such as **1** and **2**, this influence can be relatively minimal because the energy of the covalent bonds holding together the atoms within the tecton is very much greater than the energy of the intermolecular interactions holding tectons together within the crystal. Hence, the crystal packing interactions exert very little perturbation on the tecton geometry. However, in other cases, particularly coordination compounds of relatively soft metal or nonmetal centers, intra-tecton bonds, particularly hypervalent bonds, are not dissimilar in strength to the strength of the inter-tecton bonding. This condition is particularly true when there is multiple inter-tecton bonding and hence, a large number of (possibly individually weak) inter-tecton interactions acting in concert.

In a very insightful article, Tiekink has analyzed the role of crystal packing effects on the coordination geometries of a series of Hg(II) and Sn(IV) species [16]. Quantitative comparison between *ab initio* gas-phase structures and observed solid-state geometry allows an estimation of the energetic magnitude of crystal packing effects. Comparison with calculations also demonstrates that crystal packing effects tend to result in a lowering of molecular symmetry and a blurring of the distinction between covalent (coordination) interactions and hypervalent bonding in such systems.

In this chapter, we will examine some representative examples in which there is a subtle and often nontrivial interplay between intra and intermolecular interactions and hence, the systems may be regarded as almost purely supramolecular in the sense that the solid-state structure is, to a lesser or greater extent, a result of the interplay of the forces between some or all of the *atoms* as opposed to between molecules or tectons. Thus, in the extreme, the very definition of what constitutes the tecton is open to debate.

In the final case, a holistic view is required, even to begin to understand the assembly and stability of crystalline systems. This view means a consideration of not just strong intermolecular interactions that perhaps determine the nature of the packing to a first approximation, but also the synergy between stronger and weaker interactions, overall van der Waals attraction and hence shape fit criteria [17], and the ability of the shape and conformation of molecules to deform in order to satisfy the criterion for formation of a regular, tessellating three- dimensional crystal (the *de facto* criterion for observation of a particular structure). It is often remarked just how impressive it is that self-assembly processes can effectively solve these simultaneous equations and lead to a stable crystal nucleus and hence an observable crystal structure in the vast majority of cases.

2. SECOND-SPHERE COORDINATION

Braga, Desiraju and Grepioni have highlighted the thought-process of crystal deconstruction in which the solid-state environment of a reference molecule is analyzed in terms of an ever simplifying series of interactions with its nearest neighbors [18]. In classical terms, in the case of metal complexes, the nearest neighbors constitute the second coordination sphere of a metal atom or ion. While many metal complexes, particularly those of low oxidation states, exhibit well-defined, orbitally controlled coordination geometries, the low strength of some metal–ligand interactions and their relatively weak angular dependence in comparison to covalent carbon–carbon bonds for example, makes the second coordination sphere of a coordination complex a fertile ground for insight into the effects of crystal packing interactions on molecular geometry.

As the strongest and most directional of intermolecular interactions, the hydrogen bond is perhaps the most influential crystal packing force and one that, in conjunction with molecular shape, frequently governs the gross features of crystal packing, particularly when "isolated", that is, when a strong hydrogen-bond donor and acceptor are present in a particular system in which the other possible intermolecular interactions are relatively weak and nondirectional, compare **2**. In such circumstances, Margaret Etter was able to state that in the vast majority of cases the crystal packing will be based on the interaction between the strongest hydrogen-bond donor and acceptor [19].

We have carried out an extensive series of studies of hydrogen-bonded solids of the crown ethers and their role in determining metal ion geometry has been reviewed [20]. While acting as relatively good hydrogen-bond acceptors by virtue of the ether-oxygen atoms, the crown ethers are weak hydrogen-bond donors acting only via CH$\cdots$O interactions [21], with the profusion of mildly hydrogen-bond acidic CH_2 groups promoting crystallinity without imposing significant directionality in most cases. Studies on an isostructural series of complexes of general formula $[M(H_2O)_6][ClO_4]_2 \cdot$18-crown-6 (**7**) (M = Co, Ni and Zn; Figure 2) revealed an infinite linear chain in which the strong hydrogen-bond donor metal(II) hexaaqua ions alternate with the hydrogen-bond acceptor crown ethers, with the crown ether comprising the second coordination sphere of the metal ions. In all three compounds, the same small, but non-random variations were induced in the metal primary coordination sphere as a result of the constraints imposed by the packing of the chain as a whole and the maintenance of optimum multiple hydrogen-bonding interactions. These variations comprised a distortion of the O–M–O vector along the direction of chain propagation and a small compression of one of the M–O bond lengths in comparison to the literature average and to other analogous distances in the same structure [22].

The octahedral geometry in hard, class I [23] metal hexaaqua ions is essentially a result of the optimization of the electrostatic effects associated with the packing of ligands about a charged sphere, and hence, is conceptually no different from crystal packing influences. As a result, it is not surprising that second-sphere hydrogen-bonding interactions should

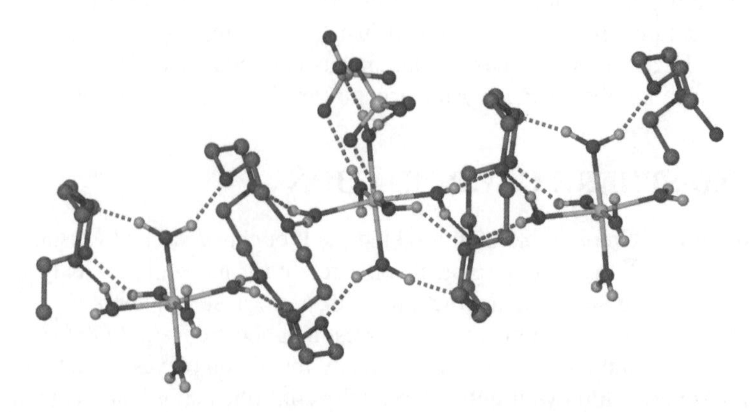

Figure 2. Hydrogen-bonded chain structure in $[Co(H_2O)_6][ClO_4]_2 \cdot$18-crown-6 (**7**) (crown ether hydrogen atoms omitted for clarity) [22].

affect the inner sphere coordination geometry. It is noteworthy, that the same distortions occur across a series of metal ions, irrespective of the metal's electronic nature.

More recently, we have been able to further isolate the effects of hydrogen-bonding interactions on the coordination geometry of M(II) aqua complexes (not involving crown ethers) in a study of a related isostructural series of compounds of formula $[M(L^1)_4(H_2O)_2][SO_4]{\cdot}2H_2O$ (**8**) (M = Co, Ni, Cu, Zn; see Figure 3 for L^1). [24]

Crystals of the Co(II), Ni(II) and Zn(II) homologues of **8** were studied using single-crystal neutron diffraction via the Laué method on the VIVALDI instrument at ILL at a range of temperatures (4, 120 and 293 K) [24]. The Laué diffraction method allows the rapid collection of data even on relatively small crystals of supramolecular systems [25]. At low temperature, compound **8** (Figure 3), is ordered and exhibits strong, near-linear hydrogen bonds from the coordinated water to the enclathrated water. The enclathrated water in turn interacts with the urea carbonyl functionalities, lowering the molecular symmetry from fourfold ($P4/n$ at room temperature) to twofold (P-4 at 4 K). At 120 K, comparison of the x-ray results for all four members of the series reveals that the hydrogen bonding in the water square and its affinity for the ligand carbonyl groups is sufficiently strong to introduce an unusual lengthening of one of the two M–OH_2 distances. The effect

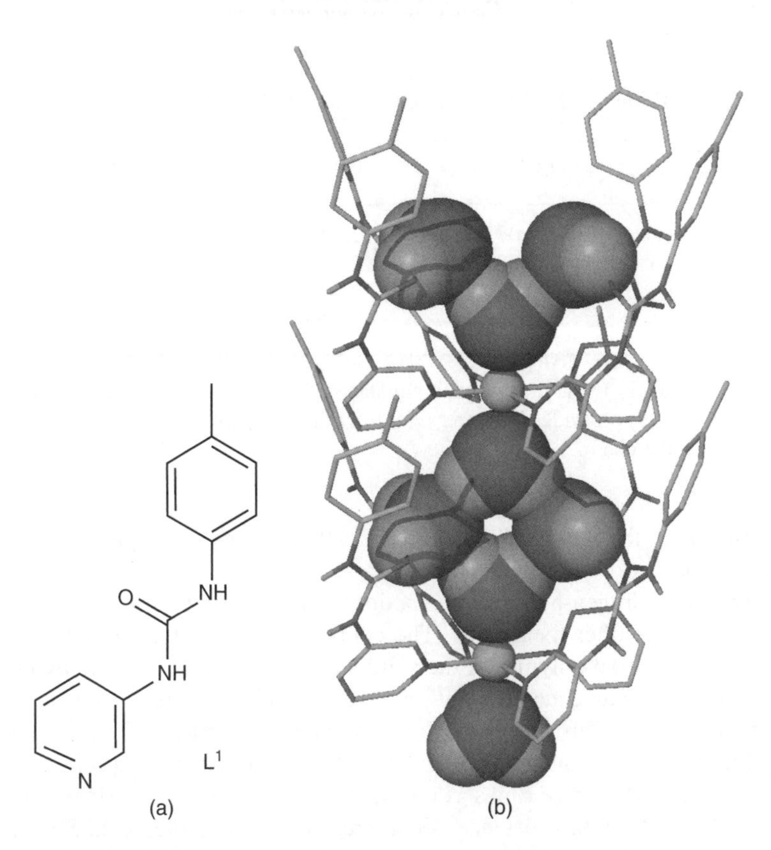

Figure 3. (a) ligand L^1; (b) the linear hydrogen-bonded chain based on a water square in $[M(L^1)_4(H_2O)_2][SO_4]{\cdot}2H_2O$ (**8**); and (c) VIVALDI neutron structure of **8** (M = Co) at 4 K.

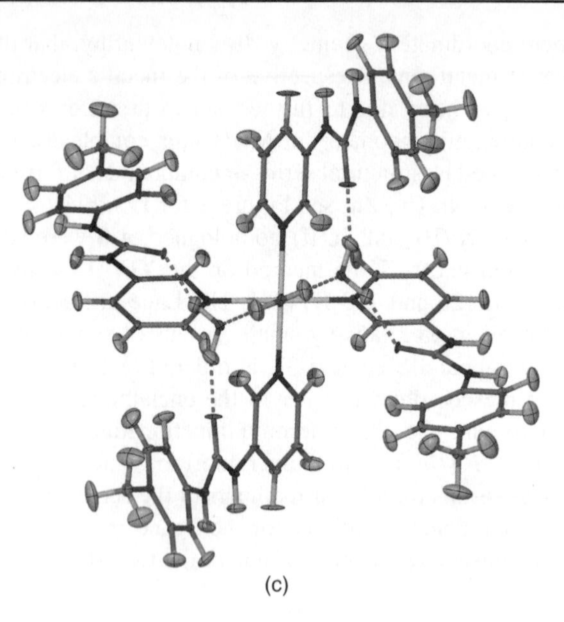

(c)

Figure 3. *(continued)*

Table 1. M–OH_2 bond lengths in complexes **8**

Metal	M–O(1)/Å[a]	M–O(2)/Å[a]	Δ M–O/Å
Co	2.077(3)	2.127(3)	0.050
Ni	2.077(4)	2.121(4)	0.044
Cu	2.281(3)	2.462(4)	0.181
Zn	2.133(4)	2.188(4)	0.055

[a]O(1) is the aquo ligand situated within the cavity, while O(2) protrudes into the cavity from the next molecule along the chain.

is most prominent in the Jahn-Teller distorted Cu(II) complex but is present in all systems studied (Table 1).

Second-sphere hydrogen-bonding interactions can also have an effect on the coordination geometries of softer metal ions. The copper(I) thiourea (tu) complex $[Cu(tu)_3]Cl$ is an infinite coordination polymer in the solid state [26]. Addition of the hydrogen-bond acceptor 18-crown-6 to an aqueous solution of $[Cu(tu)_3]Cl$ however, results in isolation of a discrete binuclear ion $[Cu_2(tu)_4(\mu\text{-tu})_2]^{2+}$ in a crystalline array of formula $[Cu_2(tu)_4(\mu\text{-tu})_2](Cl)_2{\cdot}2tu{\cdot}2H_2O{\cdot}2(18\text{-crown-6})$ (**9**) (Figure 4). The di-copper(I) species $[Cu_2(tu)_4(\mu\text{-tu})_2]^{2+}$ has been reported three times as a simple salt with Cu$\cdots$Cu distances in a narrow range 2.83–2.86 Å. In **9**, this distance is compressed to a remarkable 2.55 Å, despite the fact that the d^{10} Cu(I) ions would not be expected to show a metal–metal bonded interaction. The compression is a result of the strong hydrogen-bonding interactions between

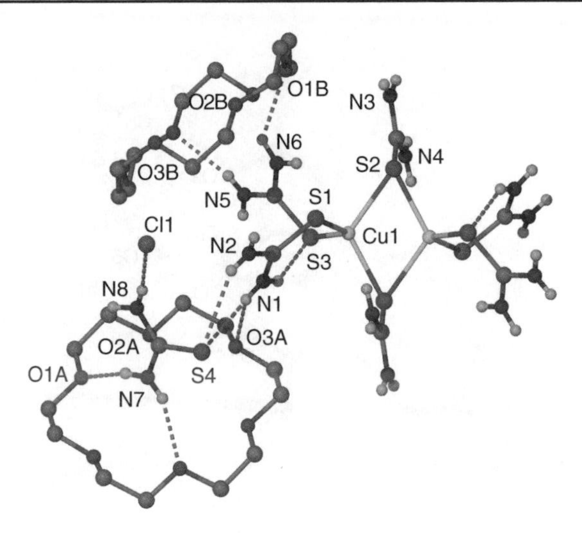

Figure 4. Hydrogen bonding in $[Cu_2(tu)_4(\mu\text{-}tu)_2](Cl)_2{\cdot}2tu{\cdot}2H_2O{\cdot}2(18\text{-crown-}6)$ (**9**) [27].

the thiourea NH groups and the 18-crown-6 units, which requires an approximate size match between cation and crown ether [27].

Such distortions in metal coordination environment might be expected to occur more frequently in series of compounds involving very strong hydrogen bonding by Brönstead acid donor groups. An interesting comparison is found in the two Cu(II) acetate "lantern" (or "paddlewheel") complexes $[Cu_2(\mu\text{-}O_2CCH_3)_4(CH_3CO_2H)_2]$ (**10**), which displays intramolecular hydrogen bonding to coordinated acetic acid, and $[Cu_2(\mu\text{-}O_2CCH_3)_4(H_2O)_2]{\cdot}2CH_3CO_2H$ (**11**), exhibiting intermolecular interactions to an acetic acid molecule in the lattice (Figure 5). Complex **10** exhibits significant distortions in the Cu–Cu–O vector in order to hydrogen bond with a coordinated acetate oxygen atom. This hydrogen-bond acceptor atom exhibits an elongated Cu–O(4) distance compared to the

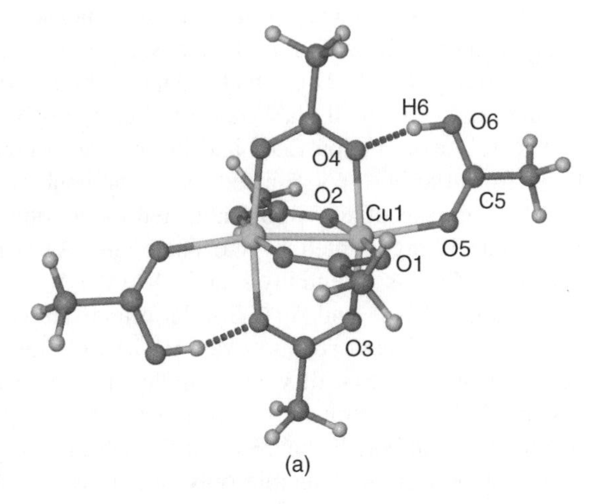

Figure 5. (a) Structure of $[Cu_2(\mu\text{-}O_2CMe)_4(O_2HCMe)_2]$ (**10**) and (b) $[Cu_2(\mu\text{-}O_2CMe)_4(OH_2)_2]$ $\cdot 2MeCO_2H$ (**11**).

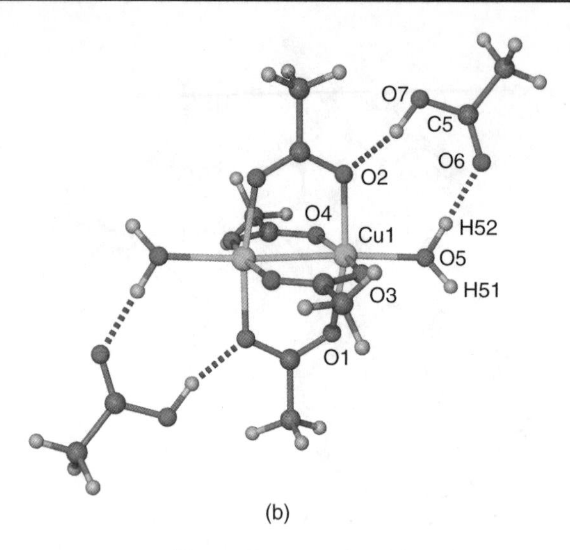

Figure 5. *(continued)*

other analogous bond lengths in the molecule as a result of the attraction to the carboxylic acid proton. A survey of the Cambridge Structural Database (CSD) indicates that this kind of intramolecular distortion in related systems is general, but is much less pronounced for intermolecular interactions as in **11** [28].

In addition to distortion of bond lengths, hydrogen bonds can also introduce distortions to a varying degree in molecular shape in order to optimize hydrogen-bonding interactions. Such distortions can result in a lowering of molecular symmetry and/or a lowering of crystal symmetry. For example, in the case of 18-crown-6, the most common molecular shape in 18-crown-6 complexes is the symmetrical, near-planar D_{3d} conformation (Figure 6(a)). However, the molecule is very flexible and can readily adapt itself in response to the dictates of non-covalent interactions, particularly in order to coordinate a metal cation of a particular ionic radius and in response to hydrogen bonds to the crown ether oxygen atoms. Lowering of ligand symmetry is exemplified in the complex [Na(18-crown-6)(H_2O)(NO_3)] (**12**), which adopts a highly distorted geometry in order to coordinate the relatively small Na^+ cation (18-crown-6 is generally taken to be better suited in size to K^+ coordination [29]) and in order to accommodate an intra-complex hydrogen-bond from the Na^+ coordinated water molecule as well as CH $\cdots$ O interactions (Figure 6(b)). The role of ion-dipole and hydrogen-bonding interactions in determining molecular shape is highlighted by the analogous Ag(I) complex, [Ag(18-crown-6)(H_2O)(NO_3)] [30], which is essentially identical to the Na^+ analog, despite the very different electronic nature of Na^+ and Ag^+. Similar effects are noted in a variety of related compounds, particularly [Na(18-crown-6)($MeOH)_2$](BPh_4) [30].

While being interesting from the point of view of molecular conformation, the supramolecular interactions in **12** do not result in a lowering of crystal symmetry – there is only a single unique molecule (although the D_{3d} conformation is capable of adopting crystallographic threefold symmetry, in principle only one-third need be unique). However, in the case of the hydrogen-bonded complex [$UO_2(H_2O)_3Cl_2$]·15-crown-5 (**13**), the requirements of optimal hydrogen bonding along the infinite donor–acceptor chain are at

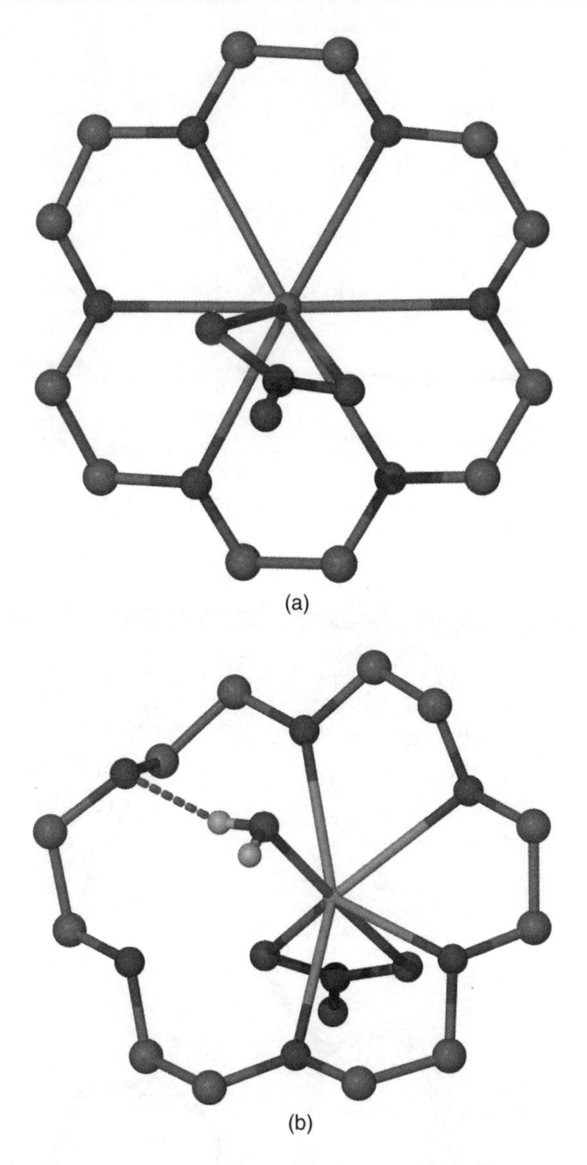

Figure 6. (a) Common D_{3d} conformation of 18-crown-6 as its KNO_3 complex and (b) the distorted $NaNO_3$ analog (**12**).

odds with the demands of crystal close packing, resulting in a kind of frustration, which is resolved by the formation of an extremely large unit cell with no less than 16 crystallographically independent formula units, arranged in a 4×4 grid in space group $P3_2$ (Figure 7) [31, 32].

Hydrogen-bond induced distortions in the very flexible crown ether conformation can also result in interesting changes to the bulk properties of the crystal. In the case of the strong acid hydrogen-bonded complex 18-crown$-6 \cdot 2HNO_2 \cdot 2H_2O$ (**14**), single-crystal

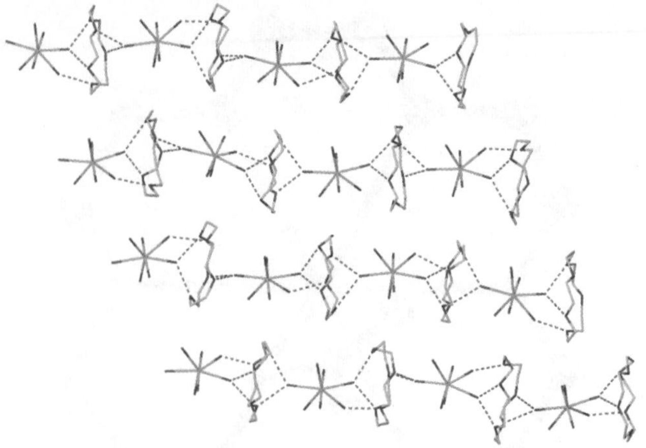

Figure 7. The 16 crystallographically independent uranyl complexes and crown ethers in [$UO_2(H_2O)_3Cl_2$]·15-crown-5 (**13**) [31, 32].

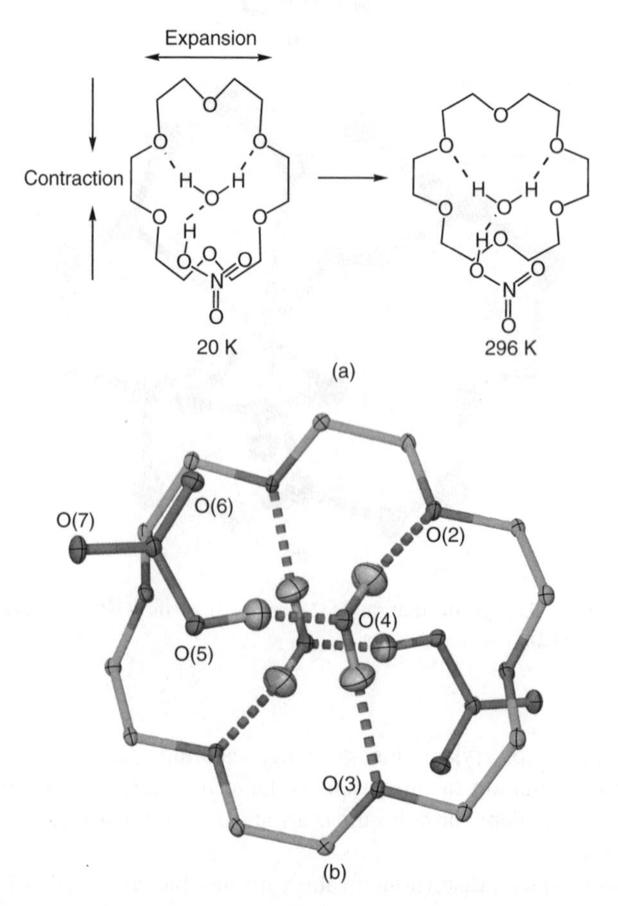

Figure 8. (a) Temperature dependent distortion in 18-crown-6·2HNO_2·2H_2O (**14**) and (b) single-crystal neutron structure of **14** at 20 K.

neutron and x-ray structure determinations at higher temperatures (close to room temperature) indicate a relatively symmetrical conformation close to the ideal D_{3d} [33]. However, as the temperature is lowered down to 20 K, the crown ether conformation becomes increasingly distorted or pinched as the hydrogen bonds from the water molecules to four of the six crown ether-oxygen atoms shorten as water proton thermal motion decreases (O4$\cdots$O2/O3). At 20 K, single-crystal neutron work shows that these interactions are short and linear, and result in a change in the crown ether diameter of some 0.15 Å. The consequence is a marked, smooth, anisotropic change in the crystal shape, with the crystallographic β-angle decreasing by some 2.5° over the temperature range studied (Figure 8) [33].

Even weak intermolecular interactions such as CH$\cdots\pi$ hydrogen bonding can have a marked effect, particularly on molecular conformation. For example, recent work by Junk *et al.* has shown that crystals of N, N'-dibenzyl-4,13-diaza-18-crown-6 grown from protic or acidic solution adopt a conformation that maximizes CH$\cdots\pi$ interactions in a number of complexes with different anions and is quite different from the extended conformation the macrocycle adopts when unprotonated [34]. This example is representative of a vast array of evidence assembled by Nishio, and recently reviewed, showing that CH$\cdots\pi$ interactions are highly pervasive as a force for influencing molecular conformation [35].

3. SOFT COORDINATION ENVIRONMENTS

3.1. Mercury and Tin

The structure of the cation in **9** has already demonstrated that soft metal ion centers with relatively plastic coordination spheres are subject to the influence of crystal packing interactions. The phenomenon is particularly pronounced for formally hypervalent compounds [36] in which the hypervalent interactions are of a strength comparable to hydrogen bonds and other crystal packing influences. The diversity in hypervalent geometries and structures induced by crystal packing forces has been elegantly surveyed by Tiekink [16], who has termed the study of the phenomenon as *syntactic structural chemistry*. Some of Tiekink's extensive work in this area is described in Chapter 6 of this book. Some key results are summarized herein.

A range of structures of triorganotin carboxylates $R_3Sn(O_2CR')$ (**15**) where $O_2CR'H$ = **16** have been determined, which show remarkable crystal and molecular structural diversity, particularly degree of aggregation, arising from ostensibly simple changes in the nature of the R groups and centered around the formally hypervalent interaction to the uncoordinated oxygen atom of the carboxylate. For the series R = Me, Et, *n*-Bu, Ph and c-Hex where O_2CR' = **16**, the Me, Et and *n*-Bu derivatives all exhibit a polymeric structure, whereas the Ph and c-Hex derivatives are both monomeric (Figure 9) [37].

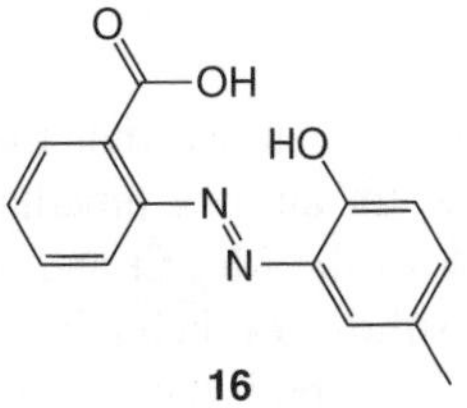

16 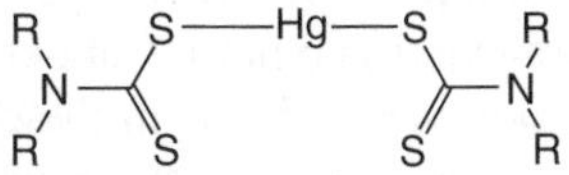 **17**

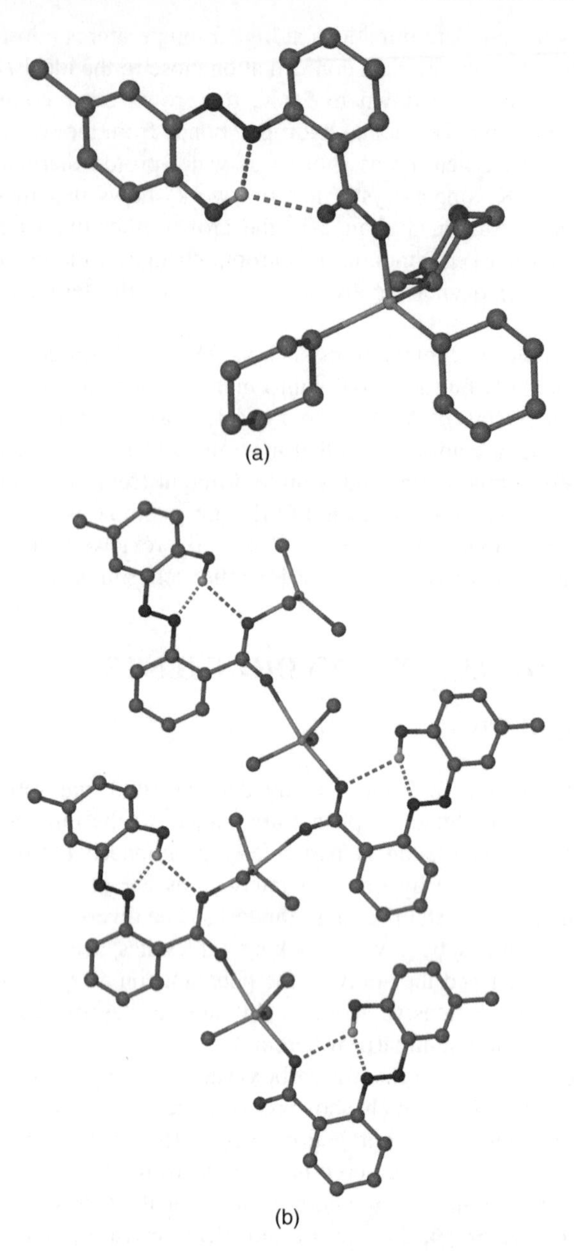

Figure 9. (a) monomer structure of **15** where $O_2CR' = \mathbf{16}$ (R = c-Hex) and (b) polymeric structure of the methyl derivative (CH hydrogen atoms omitted for clarity) [16, 37]

In contrast, in solution, ^{119}Sn NMR spectrometry shows that the complexes are all monomeric with tetrahedral geometries about the tin atoms immediately suggesting that the formation of the polymeric chains in the case of the Me, Et and *n*-Bu derivatives is a solid-state effect. The reason for the surprising solid-state behavior appears to lie in the $Sn\cdots Sn$ distances, which are approximately 5.0 Å in both the monomeric and polymeric structures (in the latter case, they are constrained by the dimensions of the bridging carboxylato ligands). Thus, it has been suggested that the monomers are deposited from solution approximately 5 Å apart, regardless of the nature of the organic substituents, and the tetrahedral complexes subsequently adjust their structure in order to maximize intermolecular interactions and minimize free space in the lattice. In the case of the polymeric species, this adjustment is achieved by formation of hypervalent, intermolecular Sn$\cdots$O interactions that result in a polymeric structure. Such polymerization is prevented by the steric bulk of the cyclohexyl and phenyl substituents. This rationalization is appealing since higher molecular weight oligomeric or polymeric species are unlikely to be very soluble, although there is good evidence in other systems that hypervalent tin is stable in solution [36, 38].

Other soft metal atoms, particularly those capable of exhibiting hypervalent bonding, display a marked structural variation upon apparently minor changes in substituents pattern. Tiekink has also studied a series of mercury(II) bis(dithiocarbamates), $[Hg(S_2CNR_2)_2]$ (**17**), which have been found to exhibit a total of five distinct structural motifs [39, 40] despite the fact that the vast majority of such compounds exhibit either monomeric or dimeric structures in the solid state, in contrast to the often polymeric mercury(II) bis(xanthates) [16]. In the cases where R = *i*-Bu, a four-coordinate distorted tetrahedral geometry is observed (Figure 10(a)). Substituting the butyl groups for methyl results in a coordination geometry

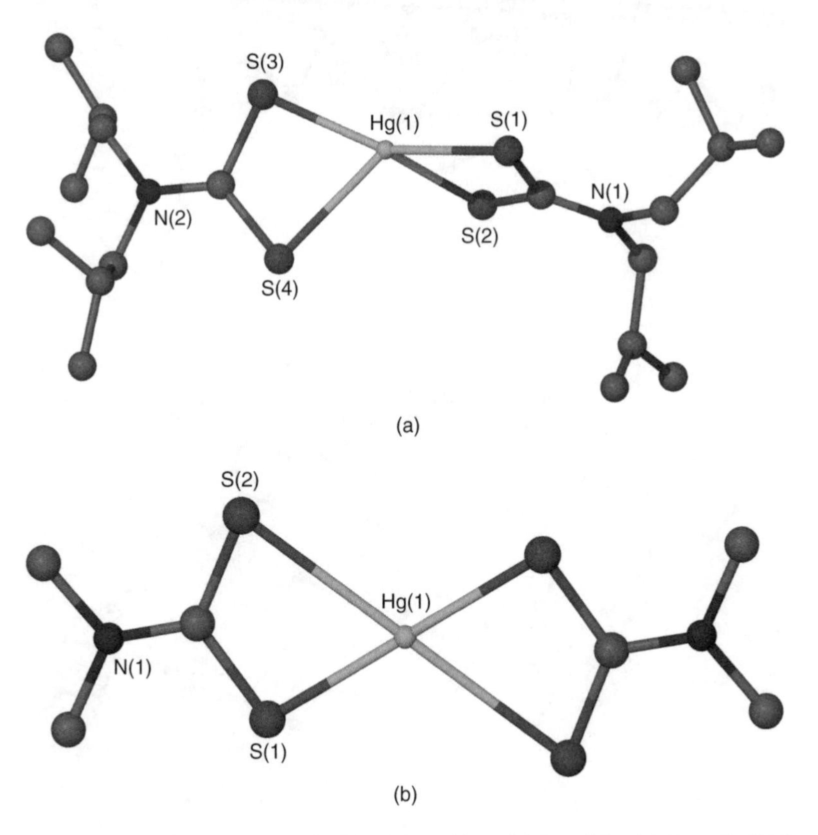

Figure 10. Structures of the compounds **17**. (a) R = *i*-Bu; (b) R = Me; (c) R = Et; (d) R = *n*-Bu; and (e) R = H [16, 39, 40].

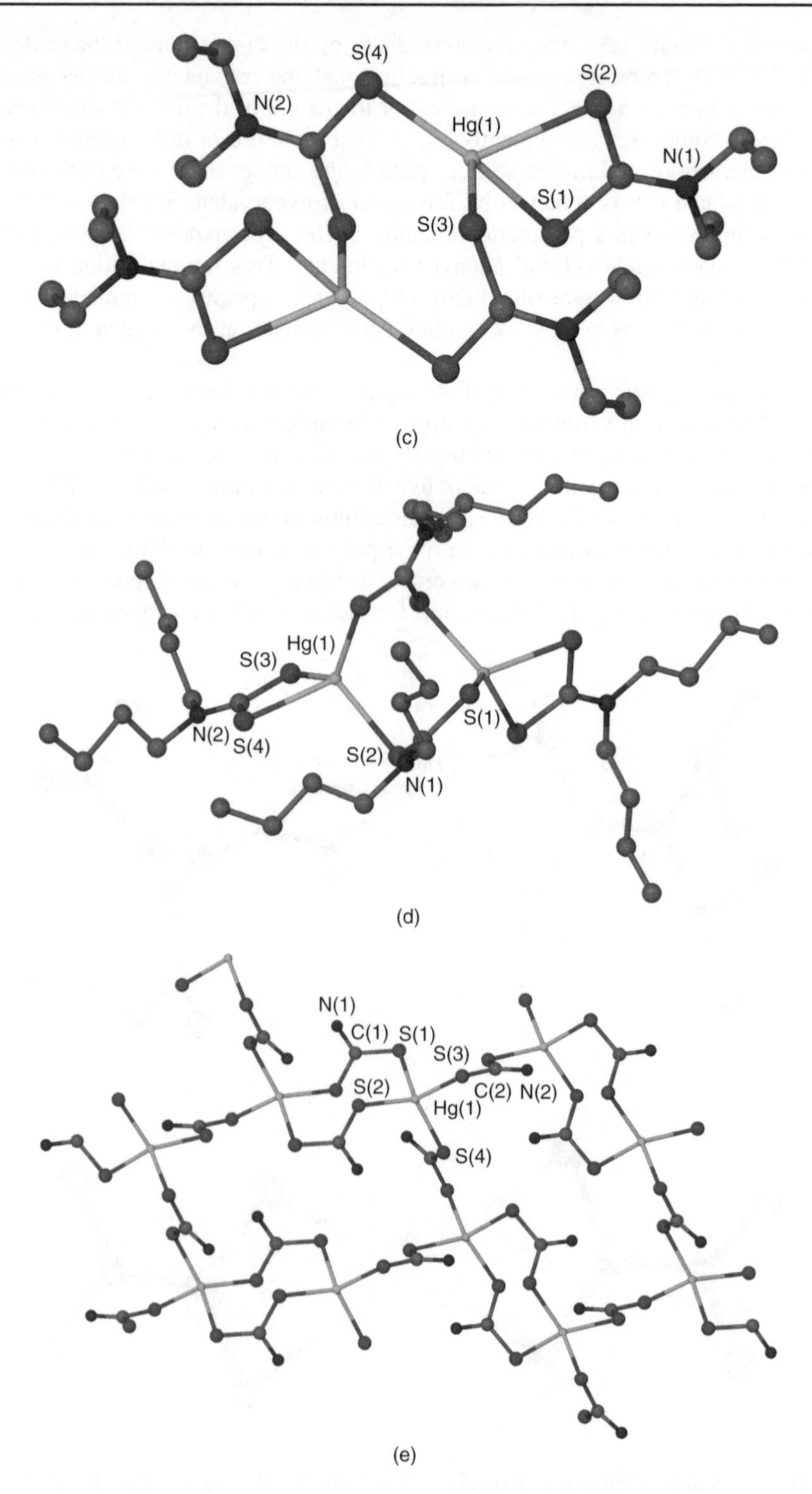

Figure 10. *(continued)*

more resembling a square planar environment, albeit with long Hg···S interactions above and below the plane (Figure 10(b)). For R = Et and *n*-Bu, dimeric structures are observed containing two chelating and two bridging dithiocarbamate ligands and a distorted tetrahedral geometry about the Hg(II) ions. While the ethyl derivative is centrosymmetric, the *n*-Bu compound is situated on a crystallographic twofold axis with the bridging dithiocarbamates on one side of the molecule (Figures 10(c) and 10(d)). The final structural type in this series is represented by the complex with the smallest ligand, S_2CNH_2 (R = H, (Figure 10(e))), which adopts a polymeric layered structure based on 8-membered $–[Hg–S–C–S–]_2$ rings interconnected by large 24-membered $–[Hg–S–C–S–]_6$ motifs. The structure is stabilized by intraring NH···S interactions. The diversity in this series of structures has been qualitatively rationalized. To a first approximation, the valency at Hg(II) is satisfied by just two Hg–S bonds and thus any additional Hg–S interactions represent intra or intermolecular hypervalent bonds and are weaker than a full, covalent Hg–S linkage and subject to distortion, according to the dictates of crystal packing optimization. In forming the crystal lattice, the distinction between the two bonding types is very much blurred and hence, reorganization of the electron density occurs in order to maximize Hg–S interactions, resulting in marked geometric distortions. The origin of the different monomer/dimer motifs is thought to be steric in origin with more bulky alkyl groups giving isolated tetrahedral complexes as in (Figure 10(a)).

As a further example of this phenomenon, the plasticity of the Hg(II) coordination sphere makes the coordination geometry about mercury even susceptible to influence by ostensibly weak intermolecular interactions. Thus, in the series of compounds [Hg(18-crown-6)X_2] (**18**) (X = anionic ligand) the Hg(II) center sits in the model of the crown ether with the two "X" ligands bonding axially above and below the crown ether plane. In the vast majority of cases, the X–Hg–X vector is linear and perpendicular to the crown ether mean plane. In the case where X = NO_3, however, the O–Hg–O vector is inclined at an angle of 68.6° to the crown ether mean plane, apparently as a result of secondary Hg···O interactions to the uncoordinated nitrato oxygen atom and particularly the formation of intramolecular CH···O hydrogen bonds in a way related to Figures 6(b) and 11 [41].

3.2. Comparison with Calculation

It is a criticism that is often leveled at interpretations of solution-structure based on either *ab initio* gas-phase calculations or on crystallographic data that the solid state and *in silico* results may not be relevant to the solution phase and indeed may not even agree with one another. Tiekink has turned this note of caution into a virtue, however, by pointing out that the key difference between gas-phase calculations and solid-state structure must, logically, be the influence of crystal packing forces (assuming the basis set used in the calculations is sufficiently large). It is thus possible to place a lower limit on the energetic influence of crystal packing forces by comparison of crystallographically observed geometry with the (usually more symmetrical) gas-phase structure arising from *ab initio* calculations. For example, the experimental crystal structure of the diphenyl bis(xanthate) complex $Ph_2Sn(S_2COMe)_2$, (**19**) displays a six-coordinate Sn(IV) center with two short Sn–S bonds and two longer, asymmetric, formally hypervalent Sn–S interactions (Figure 12). In contrast, the calculated gas-phase structure is symmetrical and incorporates hypervalent Sn–O interactions instead of Sn–S bonds, consistent with the hard nature

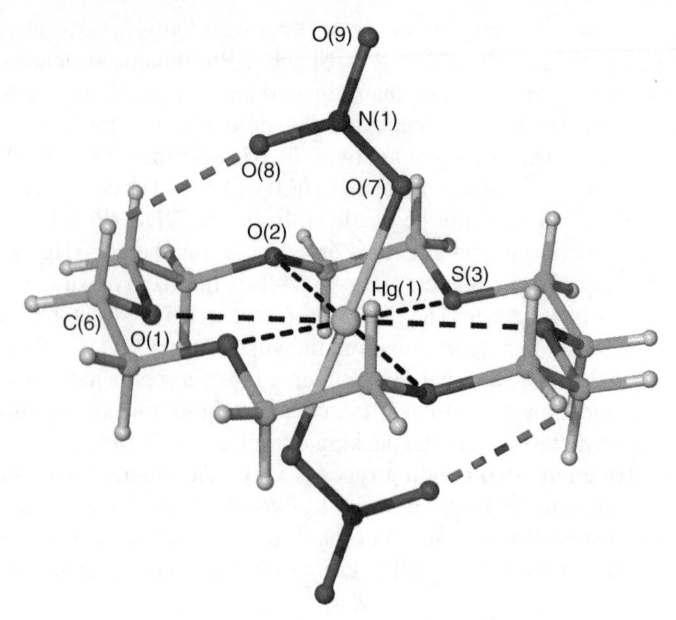

Figure 11. Structure of [Hg(18-crown-6)$(NO_3)_2$] (**18**) showing the secondary CH···O interactions [41].

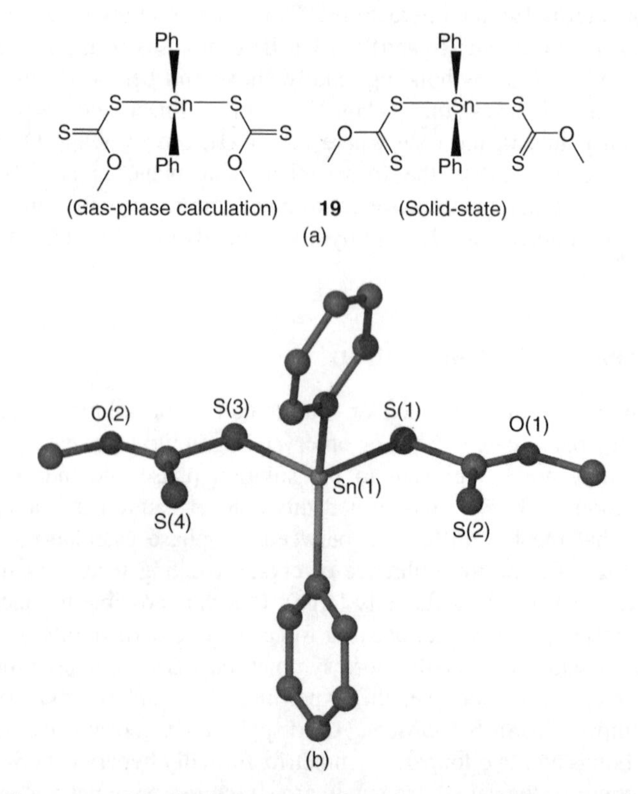

Figure 12. (a) Structures of $Ph_2Sn(S_2COMe)_2$ (**19**) in the gas phase from *ab initio* calculations and in the experimental crystal structure and (b) the experimental crystal structure [16].

of the Sn(IV) center. In the solid state, the oxygen atoms of the xanthate ligands are turned outward and become available to take part in stabilizing intermolecular interactions, which are absent in the gas phase. The energy difference between the gas-phase global minimum structure and that calculated using the x-ray coordinates is about 20 kJ mol^{-1} and hence, at least this much energy must be assumed to arise from crystal packing interactions.

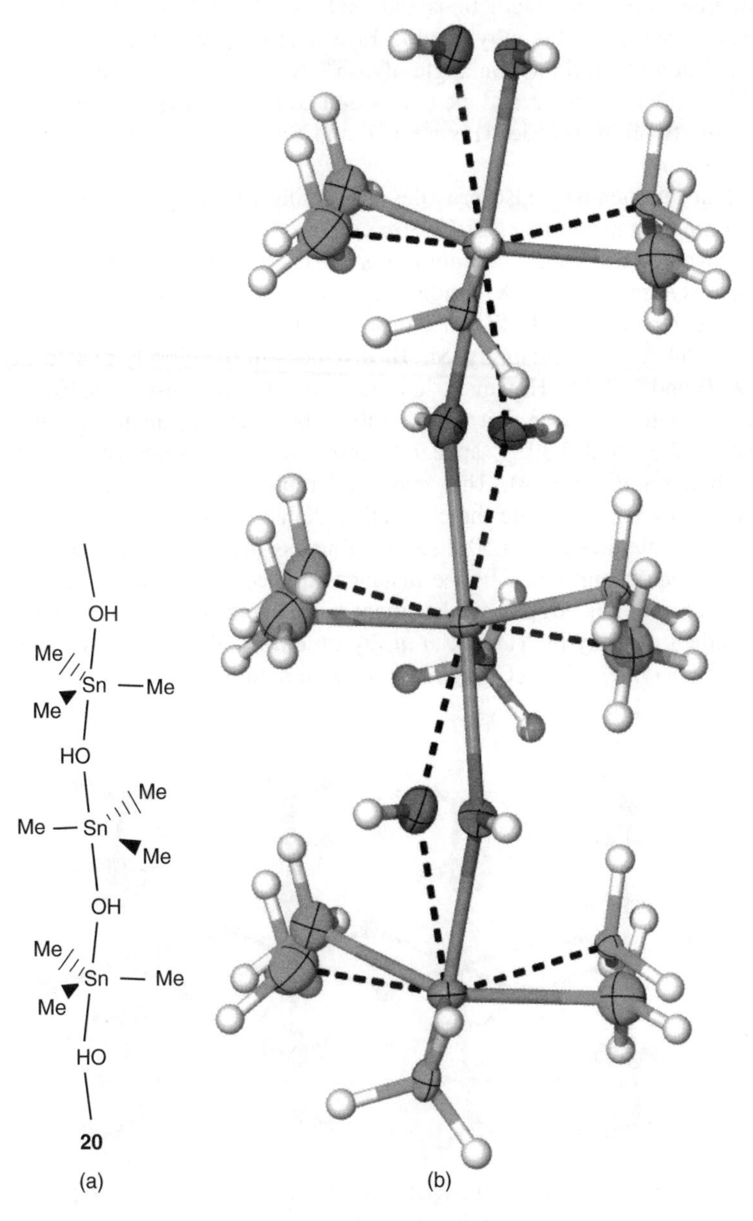

Figure 13. (a) Chemical structure of Me_3SnOH (**20**) and (b) the 120-K x-ray crystal structure exhibiting resolved disorder [43].

3.3. Influence of Disorder

While crystal packing interactions undoubtedly exert a major influence on molecular structure in soft metal coordination complexes, it can occasionally be difficult to disentangle genuine effects from the influence of unresolved disorder. Such a situation might well arise in the polymeric five-coordinate structure of Me_3SnOH (**20**). At room temperature, the compound exhibits a fascinating $Z' = 32$ structure arising from a complicated suite of intra and interchain packing interactions and the Sn–O–Sn vector appears to be essentially linear, in stark contrast to the ethyl analog Et_3SnOH [42], which adopts the same chain arrangement, but with an Sn–O–Sn angle of 145°. At 120 K, the complex crystal packing effects in Me_3SnOH disappear and the compound adopts a much simpler $Z' = 1$ structure with resolvable twofold disorder (Figure 13), and Sn–O–Sn angles of 140 and 141° for the two components [43].

Hypervalent tin chemistry also provides the setting for another apparently mysterious geometric contrast in the simple series R_2SnX_2 ($X_2 = Cl_2$, Br_2 or ClBr; R = *n*-alkyl) that is possibly disorder induced. The complexes all crystallize as infinite halogen bridged chains with unsymmetrical Sn_2X_2 bridges containing two short and two long bonds to each tin(IV) center [44–47]. In the case of *n*-Bu_2SnCl_2, the two unique Sn–Cl distances are 2.37 and 3.54 Å. Unsurprisingly, Sn–Br distances in the closely related Et_2SnBr_2 are longer at 2.50 and 3.78 Å. However, it is the mixed halide *n*-$Pr_2SnClBr$ [46, 47] that is the surprise, with distances that do not fall in between the dichloride and dibromide analogs but, at 2.52 and 3.39 Å, appear to move toward a somewhat more symmetrical bridge situation (Figure 14). This may well represent a genuine adjustment of the crystal structure to accommodate the conflicting demands of the different halide bridges. However, the two halides are 50:50 disordered across the two halide sites in the crystallographic asymmetric unit and thus the distances quoted are to average halide positions, albeit with "normal" anisotropic displacement parameters. Diorganotin dichlorides have also been studied recently by Tiekink *et al.* by *ab initio* calculations, which demonstrate that the long hypervalent Sn···Cl interactions have significant effects on the structure of the "C_2SnCl_2" unit [48].

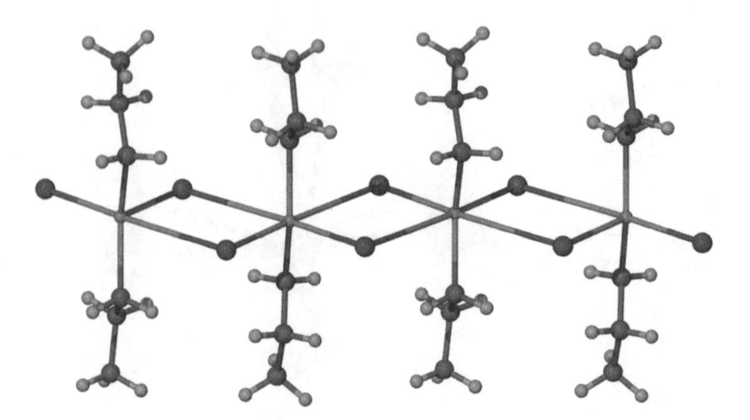

Figure 14. Infinite bridged structure of *n*-$Pr_2SnClBr$ [46, 47].

4. SPECIATION

Somewhat related to the idea that "soft" coordination complexes of metals such as Sn(IV) and Hg(II) are able to radically modify their coordination geometry and nuclearity is the frequent observation that compounds, particularly labile species, characterized by solid-state crystal structures are not necessarily the same species as are present in the solution phase. An extreme example is the isolation of various forms of the hydrated proton (oxonium ion, $H(H_2O)_n{}^+$) in the solid state. The degree of H^+ solvation in solution is in constant flux around a certain average value [49]. However, in the solid state, the oxonium ion observed and its geometry is very much a factor of the crystal environment with a range of x-ray and neutron crystal structures known of $H(H_2O)_n{}^+$ for $n = 1$–6 [50–54]. Indeed, work by Junk and Atwood, and later by our own group, has shown that particular crown ethers of varying ring diameter are capable of selecting out various oxonium ions in accordance with the size of the crown ether cavity. Thus, 18-crown-6 is highly selective for H_3O^+, 21-crown-7 and dibenzo-24-crown-8 binds $H_5O_2{}^+$ while 30-crown-10 derivatives can bind either two equivalents of H_3O^+ or $H_7O_3{}^+$ [50, 52, 55, 56]. However, these relationships are not at all fixed, and smaller crown ethers can also isolate higher oxonium ions according to the lattice spacing dictated by the counter anion [54]. Moreover, as strong hydrogen-bond donors, the geometry of the oxonium ion in the solid state is highly susceptible to the crystal environment. To take one example, in the two single-crystal neutron structures of $H_7O_3{}^+$, the geometry of the oxonium ion is quite different, with two similar O···O distances of 2.438(2) and 2.574(2) Å in $(H_7O_3)[AuCl_4]$·15-crown-5 (**21**) but a single significantly elongated O–H covalent bond associated with the shorter O···O distance (Figure 15) [57]. In contrast, in *o*-sulfobenzoic acid trihydrate, the $H_7O_3{}^+$ unit is better described as $(H_5O_2)^+\cdot H_2O$ with two much more different O···O distances of 2.41 and 2.72 Å [51].

In the case of coordination compounds, work by Stephenson *et al.* in the mid 1980s elucidated an interesting effect involving the reactions of complexes of type $[\{(\eta^6\text{-arene})\text{-}RuCl(\mu\text{-}Cl)\}_2]$ with aqueous base (NaOH or Na_2CO_3) followed by isolation with $NaBPh_4$. In the case where arene = benzene, the reaction produces, as a major product, the binuclear

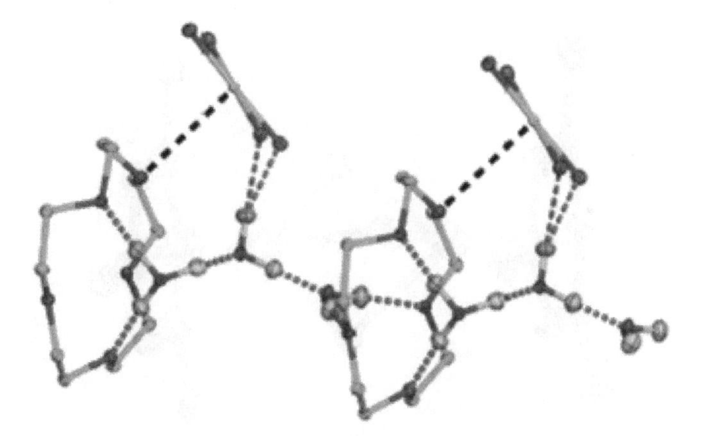

Figure 15. $H_7O_3{}^+$ isolated in the extended structure of $(H_7O_3)[AuCl_4]$·15-crown-5 (**21**) [57].

species $[(\eta^6\text{-}C_6H_6)(OH)Ru(\mu_2\text{-}OH)_2Ru(H_2O)(\eta^6\text{-}C_6H_6)](BPh_4)$ (**22**) along with a smaller amount of the tetranuclear oxo-bridged complex $[Ru_4(\eta^6\text{-}C_6H_6)_4(\mu_2\text{-}OH)_4(\mu_4\text{-}O)](BPh_4)$ (**23**). Recrystallization of **22** from acetone converts it entirely to **23**, while addition of small amounts of water re-forms **22**. In contrast, for all other arenes (e.g. mesitylene, *p*-cymene, etc.), no analog of **23** exists and the reaction yields the triple hydroxo bridged complexes $[Ru_2(\eta^6\text{-arene})_2(\mu_2\text{-}OH)_3]^+$ (**24**). The difference in behavior of the two different classes of compounds was attributed to crystal packing effects arising from the more compact packing of the benzene derivatives selecting the novel μ_4-oxo complex from solution. Further weight is added to the argument, since the behavior of the isomorphous osmium analogs is similar [58]. The structures of **23** and the mesitylene example of **24** are shown in Figure 16.

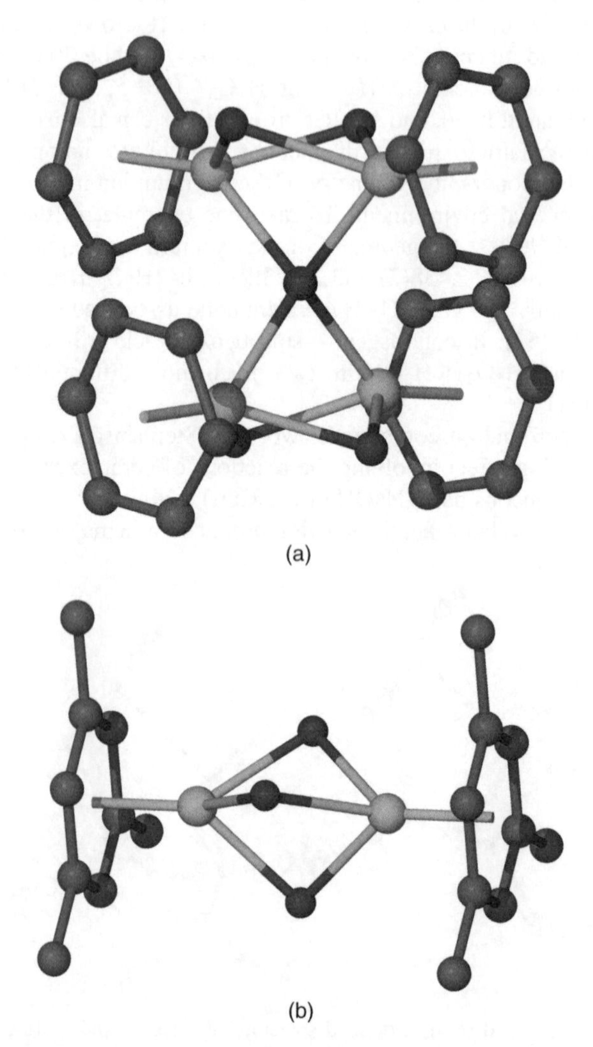

Figure 16. x-ray crystal structures of (a) $[Ru_4(\eta^6\text{-}C_6H_6)_4(\mu_2\text{-}OH)_4(\mu_4\text{-}O)]^+$ (**23**) and (b) $[Ru_2(\eta^6\text{-}1,3,5\text{-}C_6H_3Me_3)_2(\mu_2\text{-}OH)_3]^+$ (**24**) (H atoms omitted for clarity) [58].

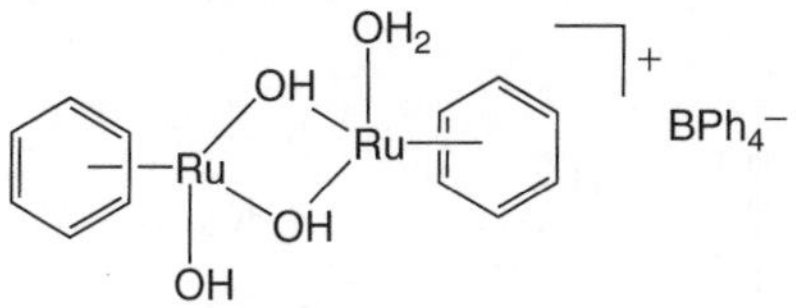

5. MOLECULAR CONFORMATION

In the case of compound **5** (*vide supra*), there is a range of feasible molecular conformations and tautomers that may be potentially present in the liquid phase. The compound exists in three solid-state forms that are obtained as mixtures from hexane with poor reproducibility, namely, the monoclinic and triclinic polymorphs, and a hexagonal hexane-containing *pseudo*-polymorph. On melting solid **5**, the material remains as a viscous liquid even when cooled to ambient temperature, despite a reported melting point of 129 °C [59]. It resolidifies at 0 °C into a mixture of phases with significant amorphous character. This behavior has been interpreted as arising from the presence of a mixture of stereoisomers separated by low stereomutation barriers, which does not crystallize readily. Under such circumstances, it is perhaps unsurprising that influences such as lattice forces will have a significant effect on the observed conformer in the solid state. The more symmetrical $Z' = 1$ monoclinic form, which comprises the anti, anti form of **5** does not form any strong NH···N hydrogen bonds despite the presence of NH groups in the compound. The NH moiety is sterically hindered by the aryl substituents. In contrast, the $Z' = 4$ triclinic polymorph contains a 3:1 mixture of anti, anti and anti, syn isomers, with the latter engaging in a single NH···N interaction. This hydrogen bond is apparently sufficient to either stabilize the anti, syn conformer or at least to effect its crystallization from solution, which has been observed spectroscopically to contain a mixture of the two forms [13].

As with **5**, the simultaneous occurrence of conformational polymorphism and conformational isomorphism has also been observed for 1,4-diethynylcyclohexane-1,4-diol (**25**). The compound crystallizes in two polymorphic forms. Both contain three crystallographically independent molecules arranged in a helical, trimeric hydrogen-bonded motif (a packing mode also shared by the 1:1 hydrate). The two forms that contain either a 2:1 or 1:2 mixture of conformers are **25a** and **25b**. While the form with the two diaxial hydroxyl groups is packed about 3% more efficiently than the other polymorph, this effect is apparently counterbalanced by slightly shorter hydrogen-bonding in the other form [60]. As in the case of **5**, because of the similar stabilities of the two conformers, it is apparently the packing of the crystal lattice that determines the observed form rather than intramolecular considerations.

Slightly more recently, the same workers have studied the remarkable 4,4-diphenyl-2,5-cyclohexanedione system (**26**), which exhibits five polymorphic modifications, three of them concomitant, and nineteen crystallographically independent molecular conformations. Crystals are packed around three supramolecular synthons and individual conformers differ in the torsion angles about the C–C bond between the quinone and the phenyl rings. The fact that so many polymorphs and conformers are observed under very similar conditions is a clear illustration of the flatness of the conformational energy profile and the role of the crystal lattice in selecting conformers that pack efficiently.

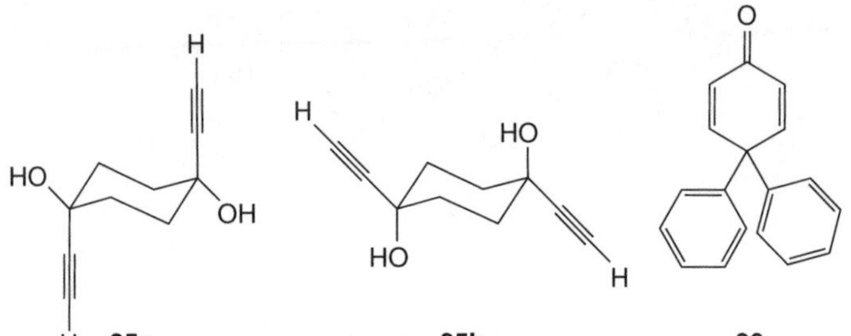

25a **25b** **26**

A more straightforward example, *p*-bromo-*N*-(*p*-dimethylaminobenzylidene) aniline, was observed in 1994 by Ahmet *et al* [61]. The structure consists of two independent molecules with very different conformations. One is more or less planar, whereas the other exhibits an interplanar angle of 146° between the two phenyl rings, while the NMe_2 group is twisted by 10° out of the plane of the phenyl ring to which it is bound. The differences lie in the crystal packing effects and the structure highlights the ability of particular molecular arrangements to stabilize a less favorable molecular conformation.

Finally, there is a wide range of literature on biological systems covering the profound effects of nucleic acid and protein crystal packing and the effect on structure [35, 62, 63]. While being fascinating, such effects are beyond the scope of the present work.

6. CONCLUSIONS

The study of crystal packing effects brings together very diverse fields spanning coordination, supramolecular, organic and biological chemistry. It is satisfying that such a different and exciting range of structural data can be rationalized and at least partially understood with reference to a common pantheon of intermolecular interactions, all acting synergically in order to maximize crystal stability and crystal growth rate. Many examples are relatively esoteric. However, the ability to understand and predict crystal structure and crystal growth is of fundamental importance. Moreover, the degree to which molecular structure is capable of distortion by the influence of neighboring molecules sheds important light on the reactivity and properties of the material in question. Even in the early days of structural systematics, Dunitz was able to use a collated body of this kind of data to firmly establish the elongation of a C–X bond (where X is a leaving group) as a result of the influence of a nearby nucleophile in solid-state geometries reminiscent of steps on the reaction coordinate of a variety of fundamental reaction such as the S_N1 and S_N2 nucleophilic substitution reactions [64–69]. It is to be hoped that further such fundamental insights will result from the field.

REFERENCES

1. D. Su, X. Wang, M. Simard and J. D. Wuest, *Supramol. Chem.*, **6**, 171–178 (1995).
2. V. R. Thalladi, B. S. Goud, V. J. Hoy, F. H. Allen, J. A. K. Howard and G. R. Desiraju, *Chem. Commun.*, 401–402 (1996).
3. G. R. Desiraju, *Angew. Chem., Int. Ed. Engl.*, **34**, 2311–2327 (1995).
4. G. R. Desiraju, 'The Crystal As a Supramolecular Entity', in *Perspectives in Supramolecular Chemistry*, Vol. 2 (Ed. J. M. Lehn), John Wiley & Sons, Chichester, 1996.

5. C. Paraschiv, S. Ferlay, M. W. Hosseini, V. Bulach and J.-M. Planeix, *Chem. Commun.*, 2270–2271 (2004).
6. P. Grosshans, A. Jouaiti, V. Bulach, J. M. Planeix, M. W. Hosseini and J. F. Nicoud, *CrystEngComm*, **5**, 414–416 (2003).
7. B. Dolling, A. L. Gillon, A. G. Orpen, J. Starbuck and X. M. Wang, *Chem. Commun.*, 567–568 (2001).
8. J. Bernstein, R. E. Davis, L. Shimoni and N.-L. Chang, *Angew. Chem., Int. Ed. Engl.*, **34**, 1555–1573 (1995).
9. D. Laliberte, T. Maris and J. D. Wuest, *Can. J. Chem. – Rev. Can. Chim.*, **82**, 386–398 (2004).
10. J. H. Fournier, T. Maris, M. Simard and J. D. Wuest, *Cryst. Growth Des.*, **3**, 535–540 (2003).
11. J. H. Fournier, T. Maris, J. D. Wuest, W. Z. Guo and E. Galoppini, *J. Am. Chem. Soc.*, **125**, 1002–1006 (2003).
12. N. Sauriat-Dorizon, T. Maris, J. D. Wuest and G. D. Enright, *J. Org. Chem.*, **68**, 240–246 (2003).
13. Z. Q. Zhang, S. Uth, D. J. Sandman and B. M. Foxman, *J. Phys. Org. Chem.*, **17**, 769–776 (2004).
14. V. Lozano, O. Moers, P. G. Jones and A. Blaschette, *Z. Naturforsch., B*, **59**, 661–672 (2004).
15. K. N. Raymond, P. W. Corfield and J. A. Ibers, *Inorg. Chem.*, **7**, 1362 (1968).
16. E. R. T. Tiekink, *Rigaku J.*, **19**, 14–24 (2002).
17. A. I. Kitaigorodskii, *Organic Chemical Crystallography*, Iliffe, 1962.
18. D. Braga, F. Grepioni and G. R. Desiraju, *Chem. Rev.*, **98**, 1375–1405 (1998).
19. M. C. Etter, *Acc. Chem. Res.*, **23**, 120–126 (1990).
20. J. W. Steed, *Coord. Chem. Rev.*, **215**, 171–221 (2001).
21. G. R. Desiraju, *Acc. Chem. Res.*, **29**, 441–449 (1996).
22. J. W. Steed, B. J. McCool and P. C. Junk, *J. Chem. Soc., Dalton Trans.*, 3417–3423 (1998).
23. R. G. Pearson, *Chemical Hardness*, John Wiley & Sons, New York, 1997.
24. (a) D. R. Turner, M. B. Hursthouse, M. E. Light and J. W. Steed, *Chem. Commun.*, 1354–1355 (2004); (b) D. R. Turner, M. Henry, C. Wilkinson, G. J. McIntyre, S. A. Mason, A. E. Goeta, and J. W. Steed, *J. Am. Chem. Soc.*, **127**, 11063–11074 (2005).
25. J. M. Cole, G. J. McIntyre, M. S. Lehmann, D. A. A. Myles, C. Wilkinson and J. A. K. Howard, *Acta Crystallogr., Sect. A*, **57**, 429–434 (2001).
26. Y. Okaya and C. Knobler, *Acta Crystallogr.*, **17**, 928 (1964).
27. K. Johnson and J. W. Steed, *J. Chem. Soc., Dalton Trans.*, 2601–2602 (1998).
28. G. Vives, S. A. Mason, P. D. Prince, P. C. Junk and J. W. Steed, *Cryst. Growth Des.*, **3**, 699–704 (2003).
29. J. S. Bradshaw and R. M. Izatt, *Acc. Chem. Res.*, **30**, 338–345 (1997).
30. J. W. Steed, K. Johnson, C. Legido and P. C. Junk, *Polyhedron*, **22**, 769–774 (2003).
31. H. Hassaballa, J. W. Steed, P. C. Junk and M. R. J. Elsegood, *Inorg. Chem.*, **37**, 4666–4671 (1998).
32. H. Hassaballa, J. W. Steed and P. C. Junk, *Chem. Commun.*, 577–578 (1998).
33. M. Calleja, S. A. Mason, P. D. Prince, J. W. Steed and C. Wilkinson, *New J. Chem.*, **27**, 28–31 (2003).
34. D. J. Evans, P. C. Junk and M. K. Smith, *New J. Chem.*, **26**, 1043–1048 (2002).
35. M. Nishio, *CrystEngComm*, **6**, 130–158 (2004).
36. M. Lattman, 'Hypervalent Compounds', *Encyclopedia of Inorganic Chemistry*, Wiley, New York, 1994.
37. R. Willem, I. Verbruggen, M. Gielen, M. Biesemans, B. Mahieu, T. S. B. Baul and E. R. T. Tiekink, *Organometallics*, **17**, 5758–5766 (1998).
38. M. J. Bearpark, G. S. McGrady, P. D. Prince and J. W. Steed, *J. Am. Chem. Soc.*, **123**, 7736–7737 (2001).
39. M. J. Cox and E. R. T. Tiekink, *Z. Kristallogr.*, **214**, 571–579 (1999).
40. C. Chieh and S. K. Cheung, *Can. J. Chem. – Rev. Can. Chim.*, **59**, 2746–2749 (1981).

41. M. Calleja and J. W. Steed, *J. Chem. Crystallogr.*, **33**, 609–612 (2003).
42. G. B. Deacon, E. Lawrenz, K. T. Nelson and E. R. T. Tiekink, *Main Group Met. Chem.*, **16**, 265 (1993).
43. J. W. Steed, *CrystEngComm*, 169–179 (2003).
44. J. F. Sawyer, *Acta Crystallogr., Sect. C*, **44**, 633–636 (1988).
45. N. W. Alcock and J. F. Sawyer, *J. Chem. Soc., Dalton Trans.*, 1090–1095 (1977).
46. D. A. Armitage and J. W. Steed, King's College London, unpublished work.
47. D. A. Armitage and A. Tarassoli, *Inorg. Chem.*, **14**, 1210–1211 (1975).
48. M. A. Buntine, F. J. Kosovel and E. R. T. Tiekink, *CrystEngComm*, **5**, 331–336 (2003).
49. C. I. Ratcliffe and D. E. Irish, in *Water Science Reviews 2*, (Ed. F. Franks), Cambridge University Press, 1986.
50. M. Calleja, K. Johnson, W. J. Belcher and J. W. Steed, *Inorg. Chem.*, **40**, 4978–4985 (2001).
51. R. Attig and J. M. Williams, *Inorg. Chem.*, **15**, 3057–3061 (1976).
52. J. L. Atwood, S. G. Bott, K. D. Robinson, E. J. Bishop and M. T. May, *J. Crystallogr. Spectrosc. Res.*, **21**, 459–462 (1991).
53. D. Steinborn, O. Gravenhorst, H. Hartung and U. Baumeister, *Inorg. Chem.*, **36**, 2195–2199 (1997).
54. J. L. Atwood and P. C. Junk, *Polyhedron*, **19**, 85–91 (2000).
55. P. C. Junk and J. L. Atwood, *J. Chem. Soc., Dalton Trans.*, 4393–4399 (1997).
56. J.-P. Behr, P. Dumas and D. Moras, *J. Am. Chem. Soc.*, **104**, 4540–4543 (1982).
57. M. Calleja, S. A. Mason, P. D. Prince, J. W. Steed and C. Wilkinson, *New J. Chem.*, **25**, 1475–1478 (2001).
58. R. O. Gould, C. L. Jones, T. A. Stephenson and D. A. Tocher, *J. Organomet. Chem.*, **264**, 365–378 (1984).
59. P. F. Clark, J. A. Elvidge and R. P. Linstead, *J. Chem. Soc.*, 3593–3601 (1953).
60. C. Bilton, J. A. K. Howard, N. N. L. Madhavi, A. Nangia, G. R. Desiraju, F. H. Allen and C. C. Wilson, *Chem. Commun.*, 1675–1676 (1999).
61. M. T. Ahmet, J. Silver and A. Houlton, *Acta Crystallogr., Sect. C*, **50**, 1814–1818 (1994).
62. S. A. Shah and A. T. Brunger, *J. Mol. Biol.*, **285**, 1577–1588 (1999).
63. V. Tereshko and J. A. Subirana, *Acta Crystallogr., Sect. D-Biol. Crystallogr.*, **55**, 810–819 (1999).
64. J. D. Dunitz, *Philos. Trans. R. Soc. London, Ser. B-Biol. Sci.*, **272**, 99–108 (1975).
65. H. B. Burgi, J. D. Dunitz and E. Shefter, *Acta Crystallogr., Sect. A*, **B 30**, 1517–1527 (1974).
66. P. Murray-Rust, H. B. Burgi and J. D. Dunitz, *J. Am. Chem. Soc.*, **97**, 921–922 (1975).
67. H. B. Burgi, E. Shefter and J. D. Dunitz, *Tetrahedron*, **31**, 3089–3092 (1975).
68. D. Britton and J. D. Dunitz, *J. Am. Chem. Soc.*, **103**, 2971–2979 (1981).
69. E. Bye, W. B. Schweizer and J. D. Dunitz, *J. Am. Chem. Soc.*, **104**, 5893–5898 (1982).

5

Crystal Engineering of Halogenated Heteroaromatic Clathrate Systems

ROGER BISHOP
School of Chemistry, The University of New South Wales, Sydney, NSW, Australia

1. INTRODUCTION

1.1. Clathrates

One of the earliest texts on inclusion compounds defined these fascinating chemical materials in the following terms: *Clathrates are complex compounds composed of two or more components. They differ from other complex compounds in that the molecules of their components are associated without ordinary chemical bonding. In each case there is complete enclosure of the molecules of one component in a suitable structure which has been formed by the molecules of the other* [1].

It therefore follows that clathrate compounds are ideal systems for exploration of the non-covalent relationships present between the different chemical species, host and guest. This is the subject of the research work described in this chapter.

Major historical problems in the clathrate chemistry area have been the unpredictability of these inclusion properties and the difficulty in knowing in advance which new molecules will function as hosts. In a crystalline solid, factors such as molecular size, shape, symmetry, functionality, forces of repulsion and attraction, and chirality, are all involved in complex interrelationships that determine the resultant solid-state structure [2, 3]. In the early days of inclusion chemistry, these factors were overwhelming, and new examples were only found by means of pleasant serendipitous discoveries.

Frontiers in Crystal Engineering. Edited by Edward R.T. Tiekink and Jagadese J. Vittal

1.2. New Clathrand Inclusion Hosts

As our knowledge has grown over the years, more systematic approaches to the preparation of clathrand hosts have been developed. One method that has achieved particular success has been to synthesize molecules with awkward shapes that consequently fit together poorly by themselves [4]. Such molecules tend to have relatively low solid-state packing volumes and to form amorphous or microporous solids. Some of these compounds will include guest molecules as crystalline clathrates that have greater packing efficiency [5]. The heteroaromatic hosts that are the subject of this chapter fall into this category. These compounds provide families of new clathrates that are ideal for the systematic study of weak host–guest interactions.

1.3. Halogenated Heteroaromatic Hosts

The general features of the initial host design are illustrated in Figure 1. Each molecule comprises three structural parts, all of which perform a clearly defined functional role. In the first synthetic step, two planar aromatic wings are conjoined with an alicyclic linker group to give a racemic product with actual or pseudo C_2 symmetry. This central linker additionally provides a mechanism by which twisting and/or folding of the molecule can occur in response to differing guest inclusion requirements. The wings offer the possibility of host–host aromatic interactions [6], plus a variety of potential host–guest contacts. Two benzylic *exo*-bromine substituents are added in the second reaction. These attenuate the Offset Face–Face (OFF) interactions, cause packing difficulties for the apo-host (guest free compound), and provide hot spots for host–guest attractions such as halogen$\cdots$halogen [7, 8].

In a development of this synthetic approach, the halogen atoms are present as substituents on the aromatic wings instead of the alicyclic substructure. These may be introduced by condensing appropriate halogen-containing building blocks with the central linker or, alternatively, by use of a subsequent electrophilic aromatic substitution reaction. The two approaches can also be combined as sequential synthetic steps. Scheme 1

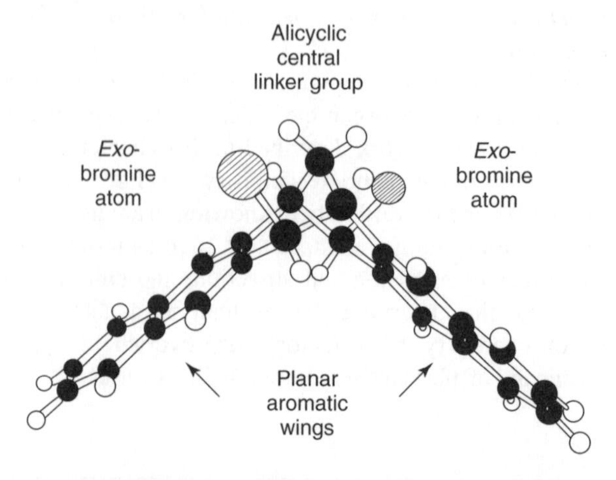

Figure 1. The general design for preparing heteroaromatic host molecules, such as **4** and **6**, with an alicyclic linker group substituted with *exo*-bromine atoms and aromatic wings.

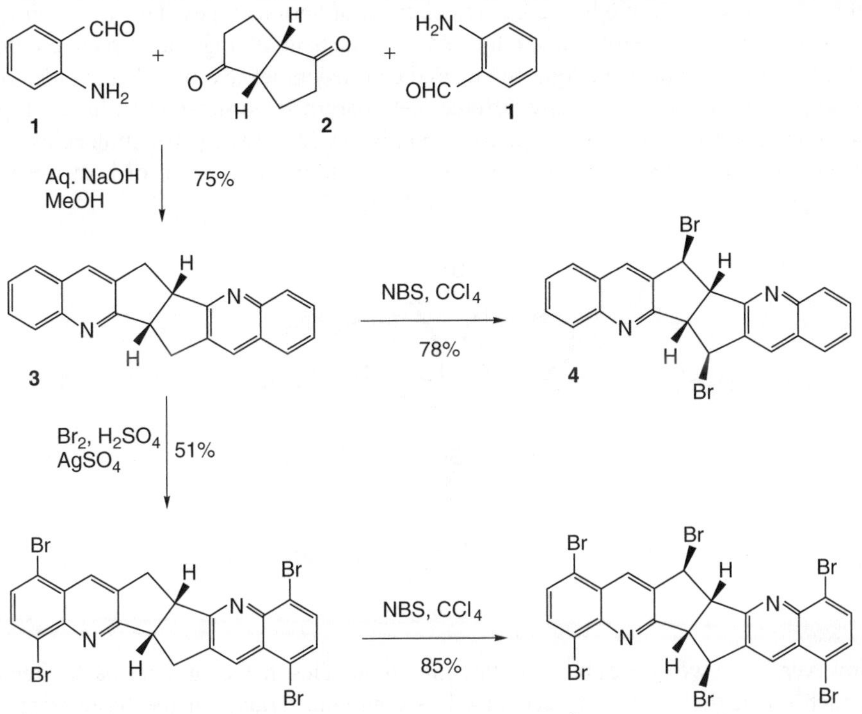

Scheme 1. Illustration of the synthetic methodology used for preparation of three typical brominated diquinoline inclusion hosts **4**–**6**.

illustrates these procedures using compounds **1**–**3** for the preparation of the dibromide **4**, tetrabromide **5**, and hexabromide **6**, which are representative examples of these new heteroaromatic hosts.

It will be appreciated that these syntheses are both modular in nature and simple to carry out. Thus, use of different precursors for the aromatic wings or alicyclic linker can easily provide a range of potential host molecules. In addition, strong hydrogen-bonding interactions are unlikely to be present as supramolecular synthons [9] in the resulting clathrate structures. Hence, these new hosts provide excellent templates for both the study of weaker intermolecular interactions, and the discovery of previously unrecognized types of packing motifs.

2. AROMATIC EDGE–EDGE C–H···N DIMERS

It is well known that planar aromatic hydrocarbons commonly associate in the solid state by means of two packing arrangements, the OFF and Edge–Face (EF) interactions [10–13]. There is no common mechanism for such molecules to associate in an edge–edge manner.

Unsurprisingly, both OFF and EF interactions are observed in the crystal structures of the hosts described here. However, the introduction of heteroatoms into an aromatic structure can now result in edge–edge associations of various types. This is certainly true here, where weak hydrogen-bonded edge–edge contacts involving nitrogen are commonly observed. The archetype is an eight-membered centrosymmetric dimer that is encountered in the solid-state structures of many different heteroaromatic systems [14]. In our diquinoline compounds, this robust synthon may simply occur between two molecules, or the dimer may be subtended by both wings, thereby resulting in a chain of host molecules, as illustrated for (**7**)•(tetrahydrofuran) [15] in Figure 2.

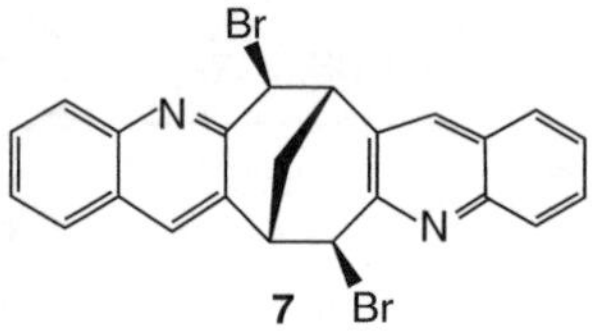

However, since we are deliberately targeting molecules that should not pack effectively with each other, they can be forced into less common variants of the basic edge–edge dimer. For example, Ar–H···N associations may become replaced by BrC–H···N interactions as shown by the diquinoxaline example $(\mathbf{8})_2$•(1,1,2,2-tetrachloroethane) [16] in Figure 3. The formation of both motifs is understandable in terms of electron density, since both Ar–H and BrC–H hydrogen atoms are electron poor relative to the electron rich nitrogen atom. However, note that this simple change in hydrogen association causes

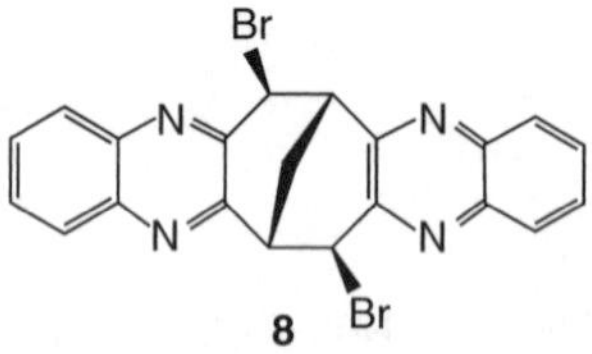

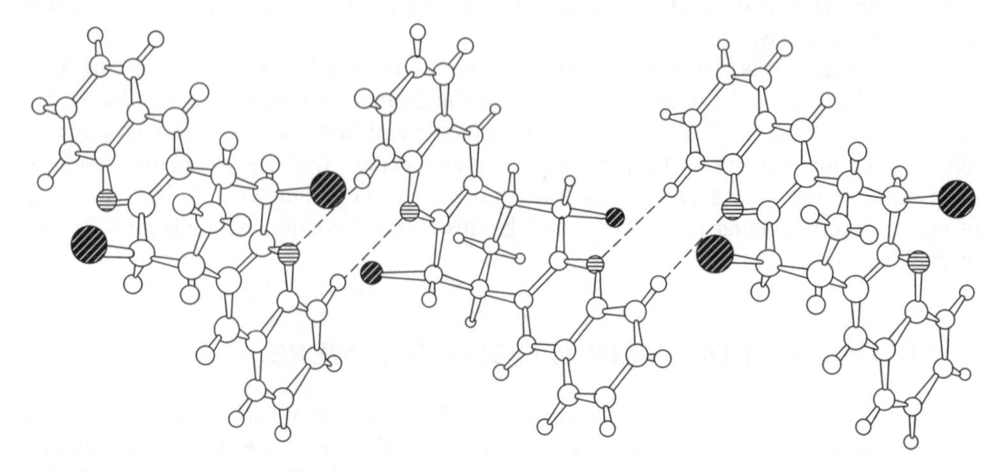

Figure 2. Part of a chain of host molecules in (**7**)•(tetrahydrofuran) linked by centrosymmetric edge–edge Ar–H···N dimers (indicated by dashed lines). In this, and subsequent Figures, nitrogen atoms are indicated by horizontal hatching and the bromine atoms by black spheres with diagonal hatching.

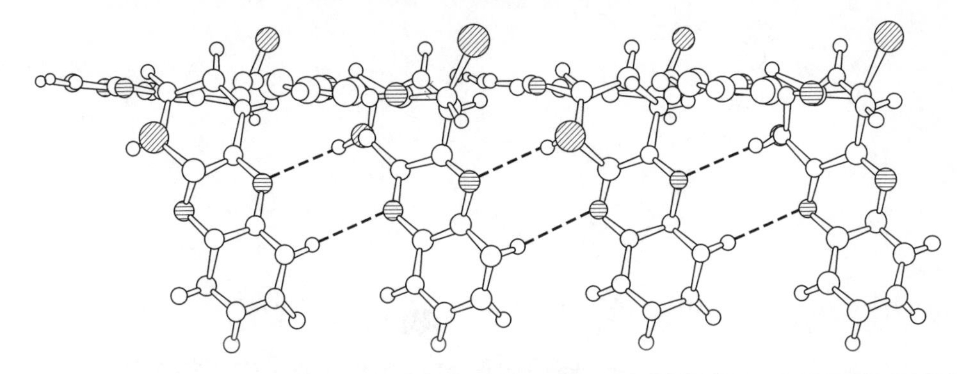

Figure 3. Part of a chain of host molecules in $(\mathbf{8})_2$•(1,1,2,2-tetrachloroethane). The host molecules are now linked edge–edge in an asymmetric manner by means of both Ar–H···N and BrC–H···N attractions.

fundamental changes to the way in which the host molecules pack, and also that more complex variations of edge–edge Ar–H···N dimeric associations can be observed [17].

3. HETEROATOM-1,3-*PERI* INTERACTIONS

3.1. The Ether–1,3-*Peri* Aromatic Hydrogen Interaction

In light of our successful molecular design (Section 1.3), it was surprising that the oxa-bridged dibromide **9** showed no host properties. However, its crystal structure revealed the likely cause of this anomalous behavior [18]. A very efficient interaction between the ether oxygen of one molecule and two 1,3-*peri* hydrogen atoms of its neighbor, C–*H*···*O* distances 2.57 and 2.69 Å, was present (Figure 4). A similar motif (2.57 and 2.66 Å) was also observed for the x-ray structure of its non-bromo precursor. A search of the Cambridge Structural Database (CSD) [19] revealed that this motif is not especially common, but that these two particular ether–1,3-*peri* interactions were easily the shortest on record. It is therefore assumed that this favorable attraction dominates the crystal packing of **9** to the exclusion of alternative inclusion arrangements.

Br N O N Br

9

R N S N R

10 R = H
11 R = Br

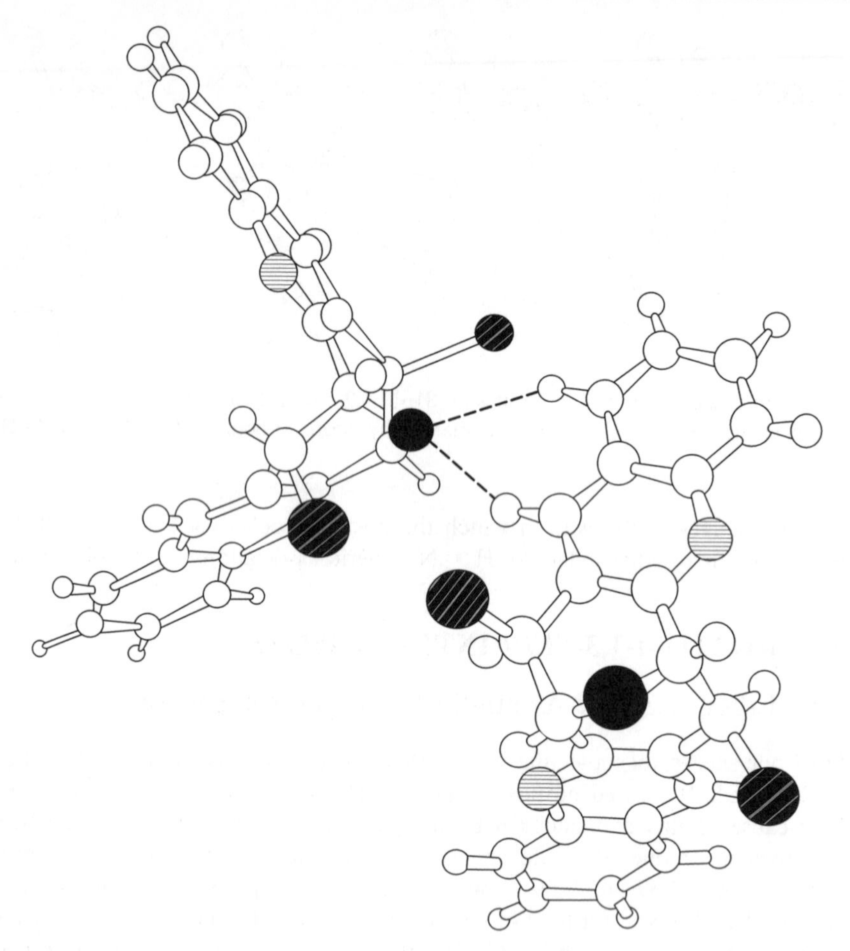

Figure 4. The efficient ether–1,3-*peri* interaction present in the crystal structure of dibromide **9**. Dashed lines indicate the C–*H*···*O* interactions (2.57 and 2.69 Å). The oxygen atoms are shown as black spheres.

3.2. The Thioether–and Aza–1,3-*Peri* Aromatic Hydrogen Interactions

A further CSD search revealed a small number of examples of the thioether–1,3-*peri* aromatic hydrogen interaction. However, when the thia-bridged compound **10** and its dibromide **11** were synthesized, they were both found to form inclusion structures that did not contain this motif [20].

In contrast, the compound $(\mathbf{10})_6$•(methanol) uses a centrosymmetric aza–1,3-*peri* aromatic hydrogen interaction as part of its crystal packing (Figure 5).

Aza–1,3-*peri* aromatic hydrogen interactions were also located in the CSD, but the dimensions of the example described above are particularly favorable [20]. It turns out that this bifurcated C–H···N synthon is highly important in construction of the molecular pen structures described in the Section 4.

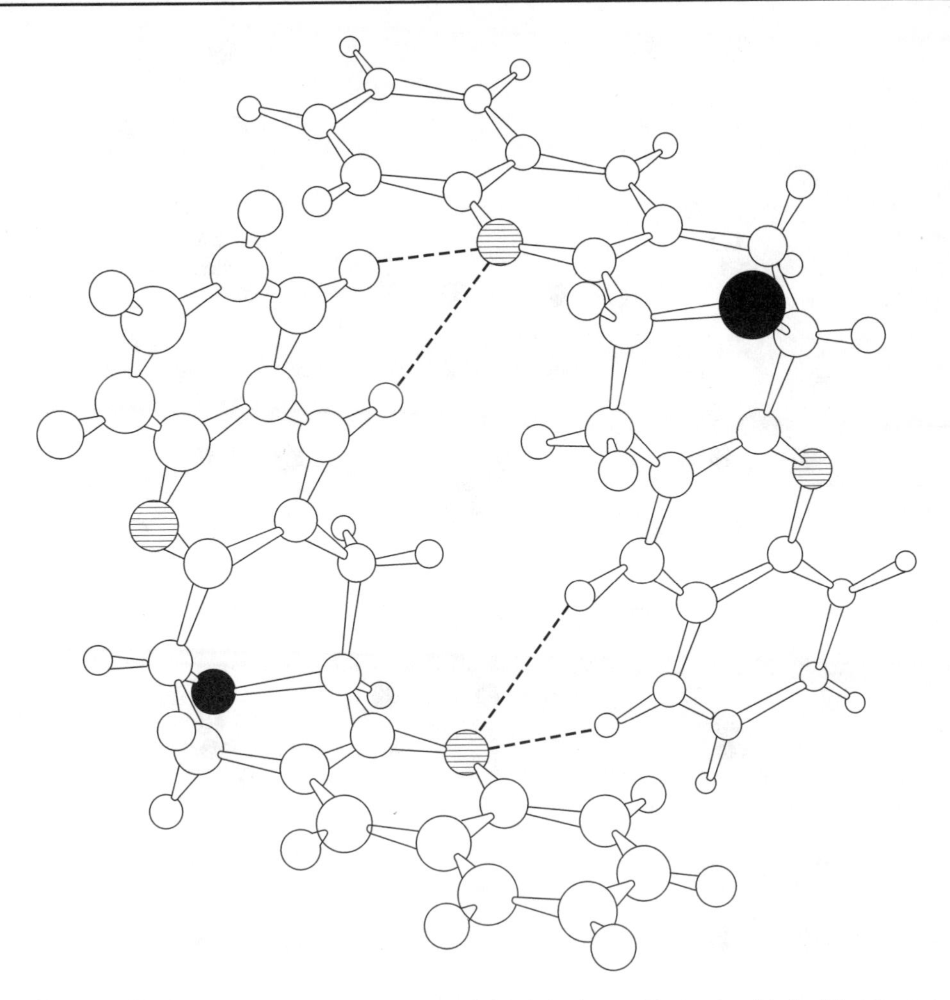

Figure 5. The centrosymmetric aza−1,3-*peri* aromatic hydrogen interaction (dashed lines) present in solid $(\mathbf{10})_6$•(methanol). The sulphur atoms are indicated by the solid black spheres.

4. MOLECULAR PEN STRUCTURES

Several dibromo diquinolines crystallize in an efficient manner, whereby two host molecules wrap around a single guest to produce a quadrilateral repeat unit. These units are not covalently bonded at two of the corners and therefore form a penannular structure [21]. A useful means of visualizing this arrangement is to imagine the four aromatic wings as providing fences that enclose the guest like an animal in a stockyard. These molecular pens then pack in layers, assisted by aromatic OFF interactions.

A number of different variants of molecular pen structures are possible. The one described here is that of $(\mathbf{4})_2$•(1,1,2,2-tetrachloroethane), which crystallizes in space group $P2_1/c$ [22]. Two independent host molecules (A,B) are present in this crystal structure, and each layer contains two centrosymmetric pens A+A* and B+B* with different geometries (where A* and B* are the enantiomers of A and B). In this particular example,

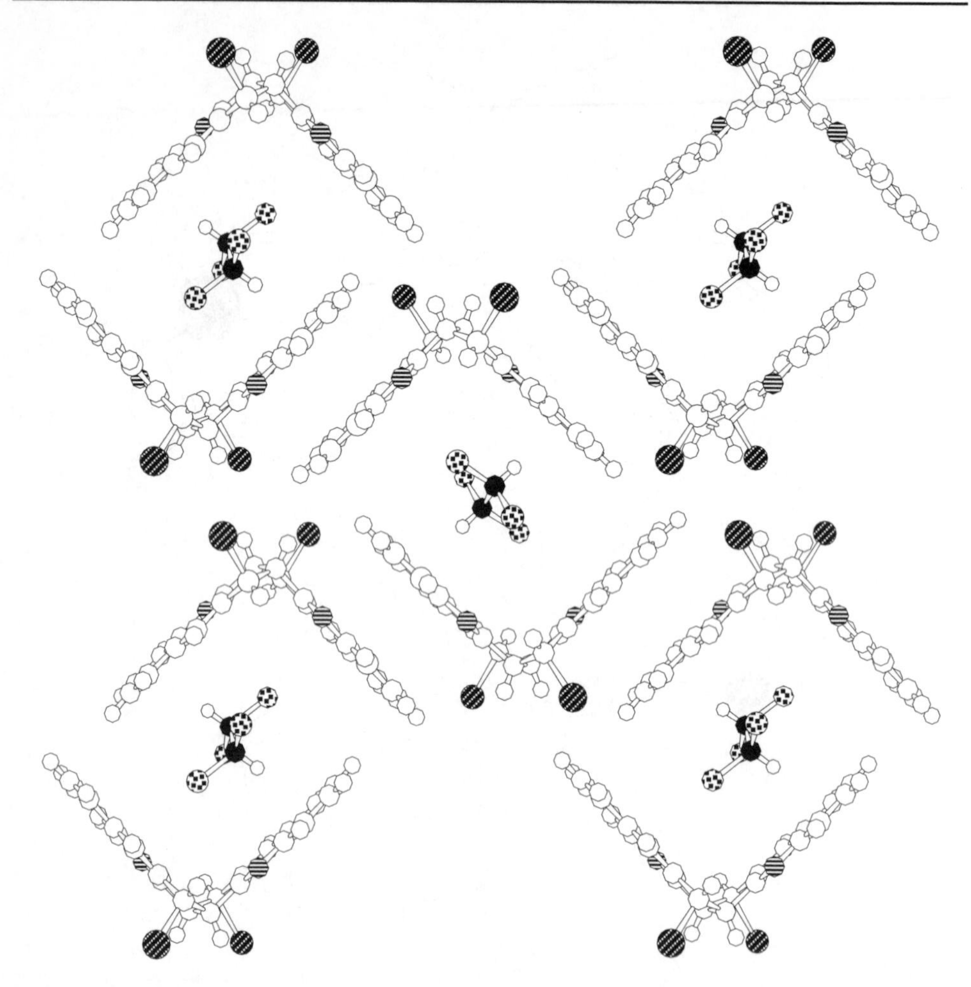

Figure 6. Part of a layer structure in solid $(\mathbf{4})_2\bullet$(1,1,2,2-tetrachloroethane) showing the assembly of four A+A* and one B+B* molecular pens and their guest molecules. Chlorine atoms are stippled.

the layers pack directly over each other to produce channels that contain the guest molecules (see Figure 6).

Although the aromatic OFF attractions within layers are immediately obvious, the interactions between the layers are arguably of greater significance. These comprise a combination of aromatic edge–edge C–H···N dimers (Section 2) and aza–1,3-*peri* hydrogen interactions (Section 3.2). The example of solid $(\mathbf{4})_2\bullet$(1,1,2,2-tetrachloroethane) is shown in Figure 7. In this particular case, one C–H···N dimer, and two different aza–1,3-*peri* hydrogen bifurcated interactions, link adjacent layers. As usual, the C–H···N dimer involves opposite enantiomers, but this time only between A and A* (and not B and B*) molecules. Unusually, this example of the motif is not symmetric. By contrast, the two bifurcated dimers link A and B molecules of opposite chirality. These and related results indicate that double C–H···N motifs are quite commonplace. They are much more robust and useful construction motifs in crystal engineering than the corresponding single C–H···N interaction [23].

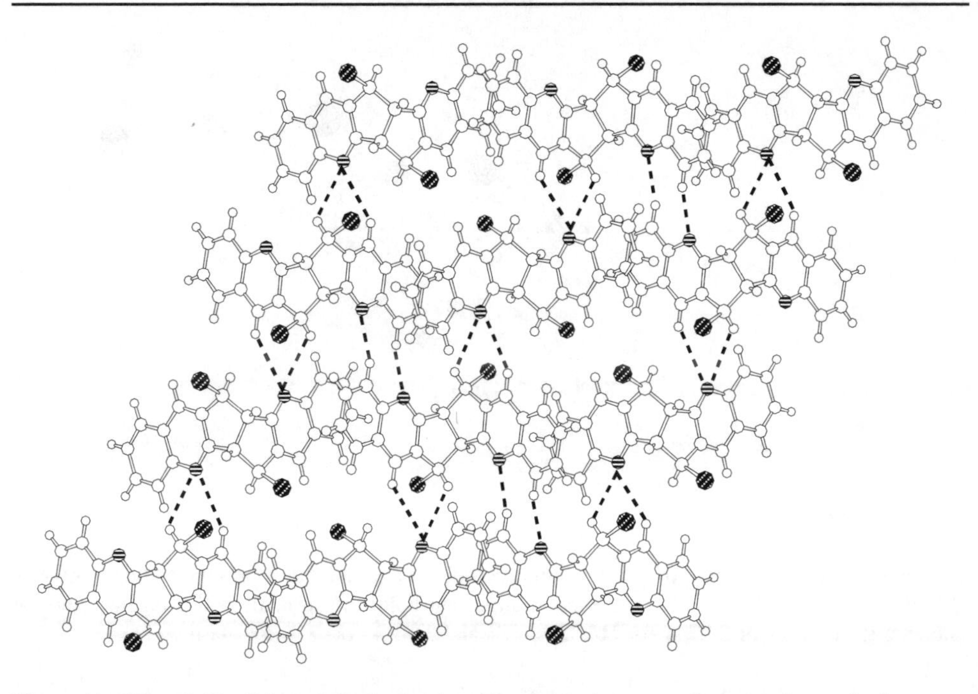

Figure 7. The C–H···N weak hydrogen-bonding network present between layers in the structure of $(\mathbf{4})_2$•(1,1,2,2-tetrachloroethane). One edge–edge aromatic edge–edge C–H···N dimer and two bifurcated aza–1,3-*peri* hydrogen weak hydrogen bonds are repeated in this structure. These interactions are indicated by the dashed lines.

5. HALOGENATED EDGE–EDGE INTERACTIONS

In Sections 2–4, the important structural role of aromatic edge–edge C–H···N interactions was discussed in detail. Very recent work has revealed additional edge–edge associations, involving bromine atoms, which complement these earlier ones [24].

The host lattice of the thia-bridged tetrabromide in compound $(\mathbf{12})_2$•(chloroform) crystallizes as chains of **12** molecules that are joined by two alternating centrosymmetric

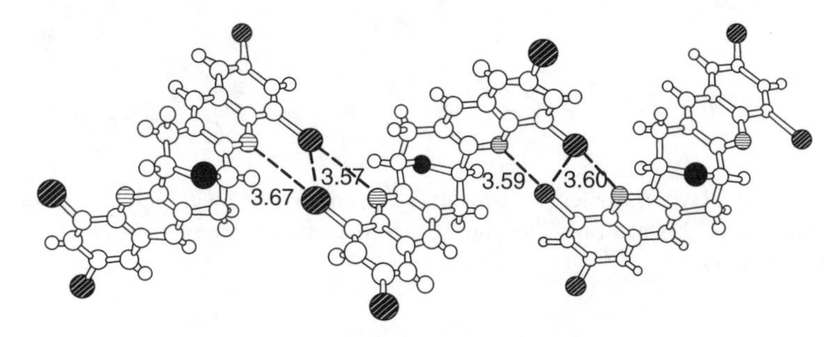

Figure 8. Part of a chain of host molecules in the structure $(\mathbf{12})_2$•(chloroform). This diagram shows how they are joined edge–edge into an infinite chain by two alternating centrosymmetric C–*Br*···*N* dimers (3.67 and 3.59 Å). The two bromine atoms within each motif are also in close contact (3.57 and 3.60 Å, respectively). These atomic contacts are indicated by the dashed lines.

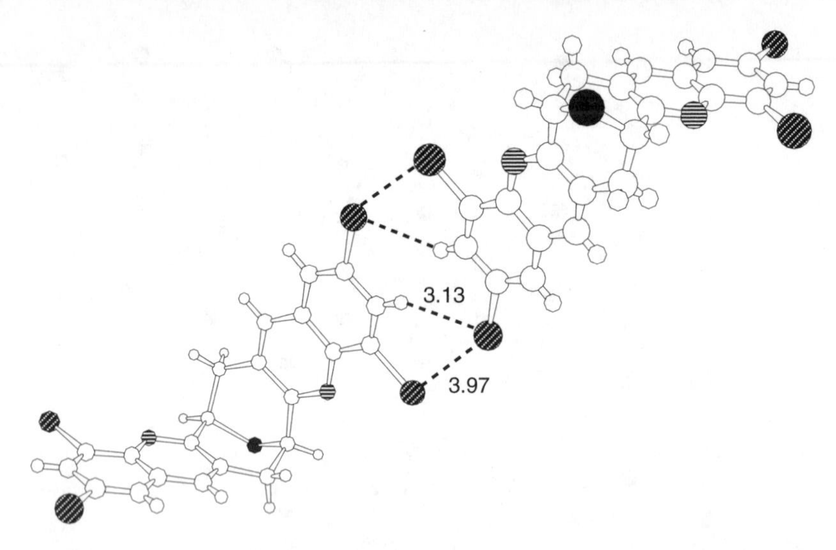

Figure 9. The second type of centrosymmetric edge–edge attraction present in the compound (**12**)$_2$•(chloroform). This double bifurcated arrangement holds the aromatic wings together and has interatomic distances of 3.97 Å for H··· *Br*··· *Br* and 3.13 Å for *H*··· *Br*··· Br.

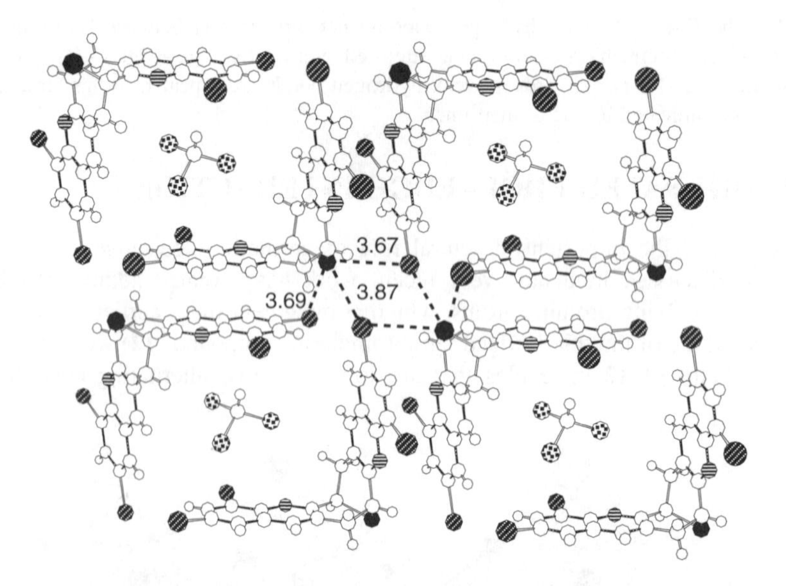

Figure 10. Part of a layer of molecular pens in (**12**)$_2$•(chloroform) with only one disorder component of the guest drawn. Dashed lines indicate the centrosymmetric interpen motif (Br··· S distances 3.67, 3.69 and 3.87 Å).

C–Br··· N dimers with different dimensions (Figure 8). These motifs are direct analogs of the C–H··· N dimers described in Section 2. However, since the halogen atoms are in close contact, these are best regarded as comprising two fused five-membered rings, rather than as being just a simple eight-membered structure.

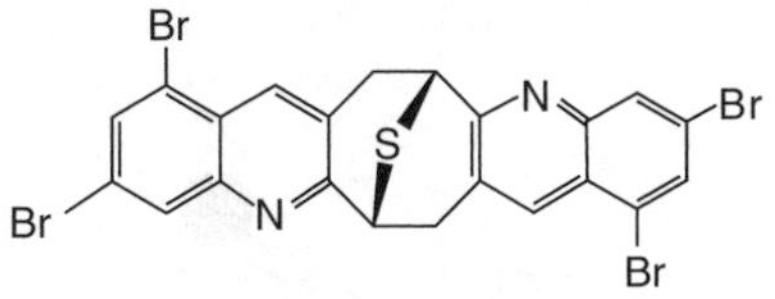

12

These molecular chains are then cross-linked by a second type of edge–edge attraction, the double bifurcated H··· Br··· Br arrangement illustrated in Figure 9.

These edge–edge packing motifs provide a particularly neat method of self-assembly for **12**. This tetrabromide is a versatile inclusion host that encloses its various guests within molecular pens that are arranged in layers. A novel motif using multiple S··· Br interactions is involved in holding these pens together as shown in Figure 10.

6. PI–HALOGEN DIMER (PHD) INTERACTIONS

6.1. A New Aromatic Building Block

As noted earlier, our heteroaromatic molecules have a slightly twisted Vshape. Frequently, interhost OFF interactions are present in their clathrate compounds and, less so, interhost EF arrangements. These attractions are unsurprising and, indeed, expected. However, we have found that a new type of aromatic interaction can often occur. This is the Pi–Halogen Dimer (PHD) motif [25, 26], which is a combination of one OFF plus four aromatic pi–halogen interactions (Figure 11). The latter is a relatively common packing motif that involves a halogen atom, positioning itself above the pi–cloud of an aromatic ring [27, 28].

To date, we have determined 18 crystal structures that contain a total of 26 PHD motifs. In all cases, the PHD unit is present in an inclusion, rather than an apohost, crystal structure and the building block always forms between opposite enantiomers. Twenty-three of these motifs are centrosymmetric [29].

The PHD interaction can be considered as arising when the aromatic *endo*-faces of two molecules stack face–face, and then a small mutual rotation occurs. One halogen atom from each molecule then becomes positioned within the V-shaped cleft of its partner and over the two electron deficient pyridine rings (Figure 12). The effect is to generate a compact unit that self-assembles with other PHD units as a one-directional staircase, or with other non-PHD forming molecules to create a two-dimensional layer, inclusion structure.

6.2. Staircase Inclusion Compounds

In 12 of the 18 crystal structures, the PHD units stack by means of OFF interactions to form molecular staircases. These unidirectional assemblies pack parallel to each other in the crystal, with the guest molecules occupying interstitial sites.

Different stacking modes can result in different space groups and stoichiometries for these compounds, but the most common of these is the host arrangement present in $(\mathbf{13})_4$•(toluene) [30]. This crystal structure in space group $P\overline{1}$ contains two crystallographically independent host molecules (A,B), plus their enantiomers (A*,B*), and hence both A–A* and B–B* PHD units are present and these alternate along the staircases

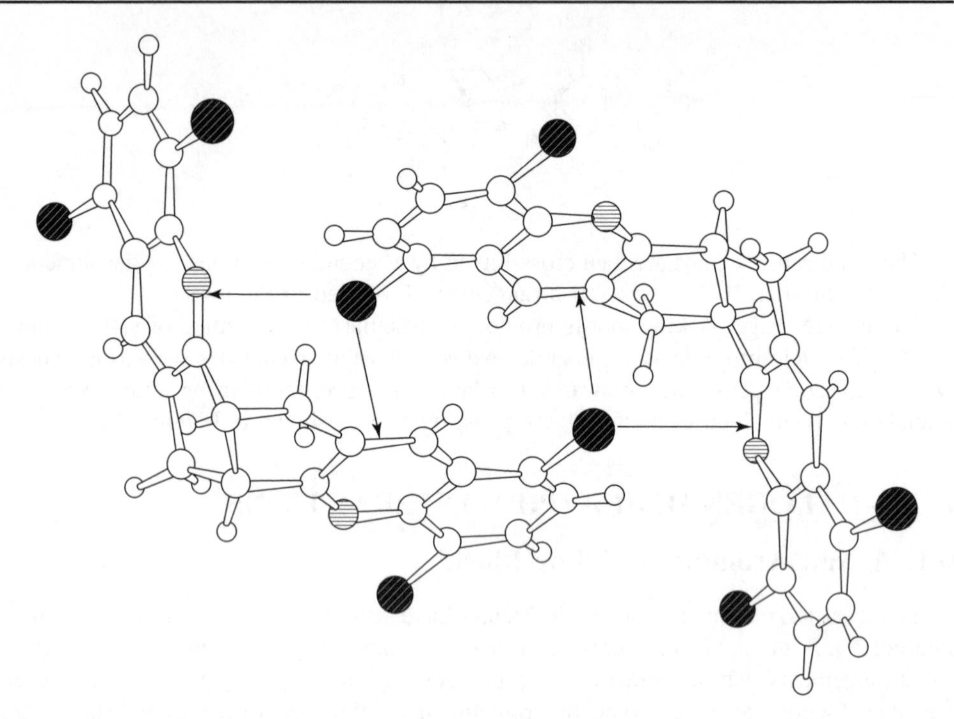

Figure 11. A typical centrosymmetric PHD motif, illustrating how two molecules of opposite chirality self-assemble into a compact building block. The OFF interaction is in the center and the four pi–halogen interactions are represented by arrows.

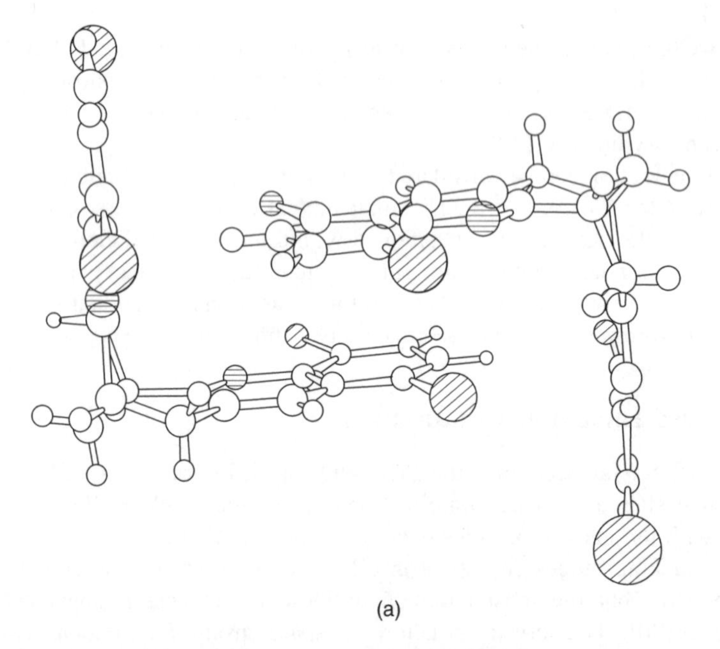

Figure 12. Two views of the PHD unit present in $(\mathbf{13})_4$•(toluene) showing: (a) the OFF, and (b) the π-halogen overlap. Iodine atoms are indicated by diagonal hatching.

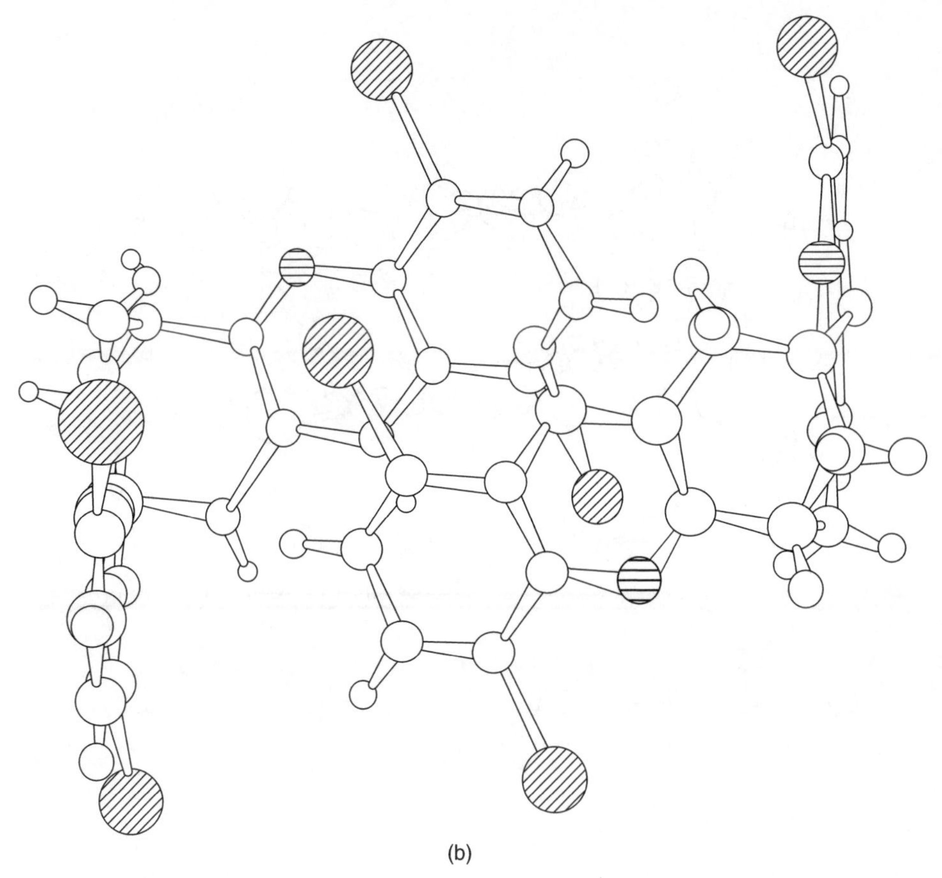

(b)

Figure 12. *(continued)*

I I N N I I

13

Br Br N N Br Br

14

(Figure 13). There is no symmetry relationship between adjacent PHD units in this structure. A projection view of the staircases and their associated guests is shown in Figure 14

A second, simpler example is shown in Figure 15. This compound, $(\mathbf{14})_2$•(dichloromethane), crystallizes in space group $C2/c$ [30, 31]. There is only one crystallographically independent host in this structure and the adjacent PHD units are related by twofold axes (Figure 15).

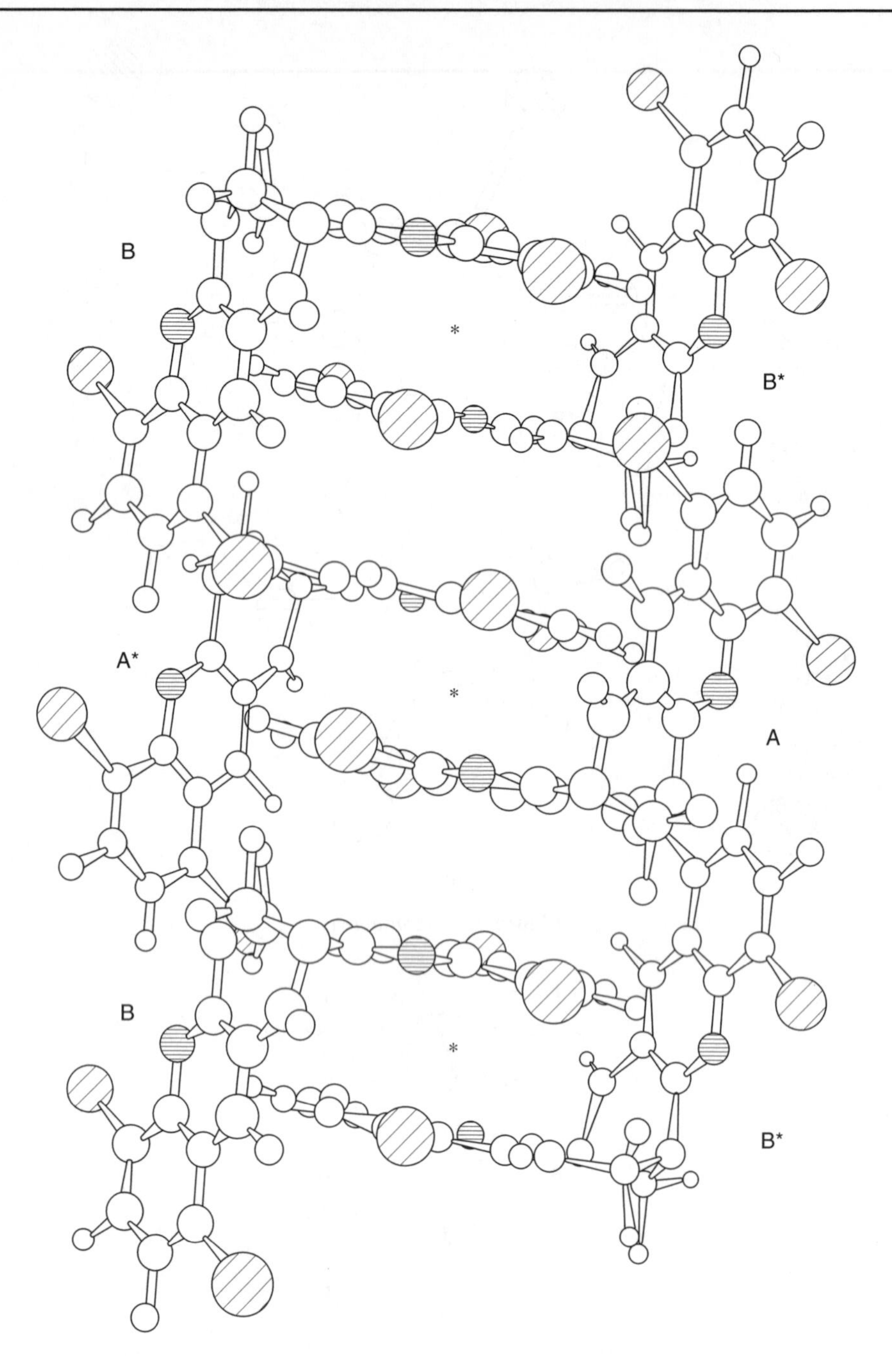

Figure 13. Side view of part of a staircase structure in $(\mathbf{13})_4$•(toluene). The asterisks in the center of the diagram designate the centers of symmetry present within PHD units.

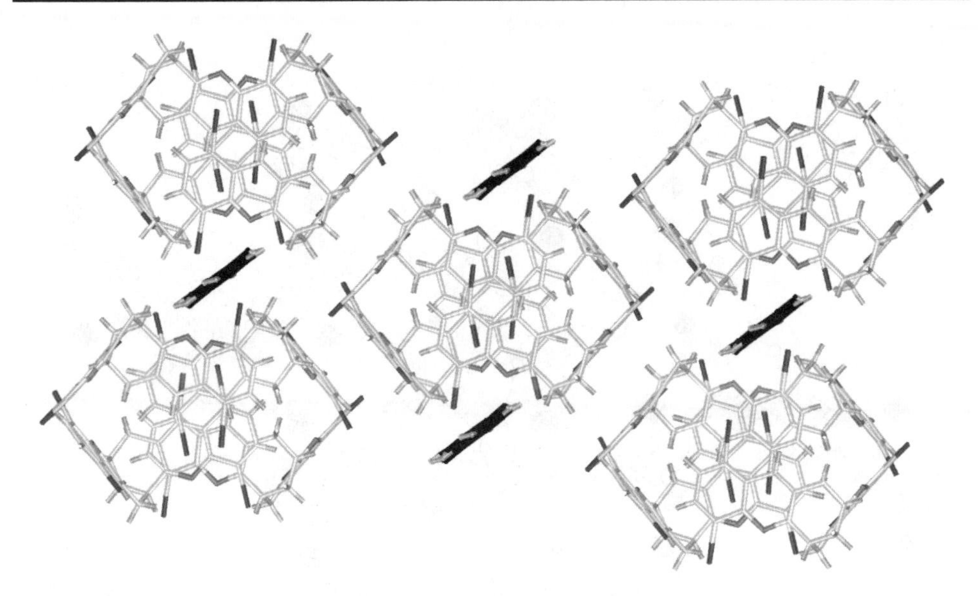

Figure 14. A projection view of $(\mathbf{13})_4$•(toluene) showing five molecular staircases with the toluene guest molecules occupying interstitial sites.

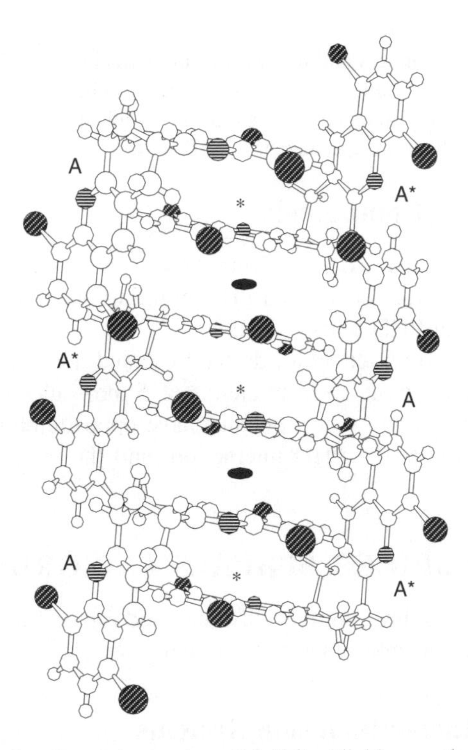

Figure 15. Part of a molecular staircase in solid $(\mathbf{14})_2$•(dichloromethane). In this case, all the PHD units are identical. Adjacent dimers are related by a twofold axis (designated here by a solid ellipse).

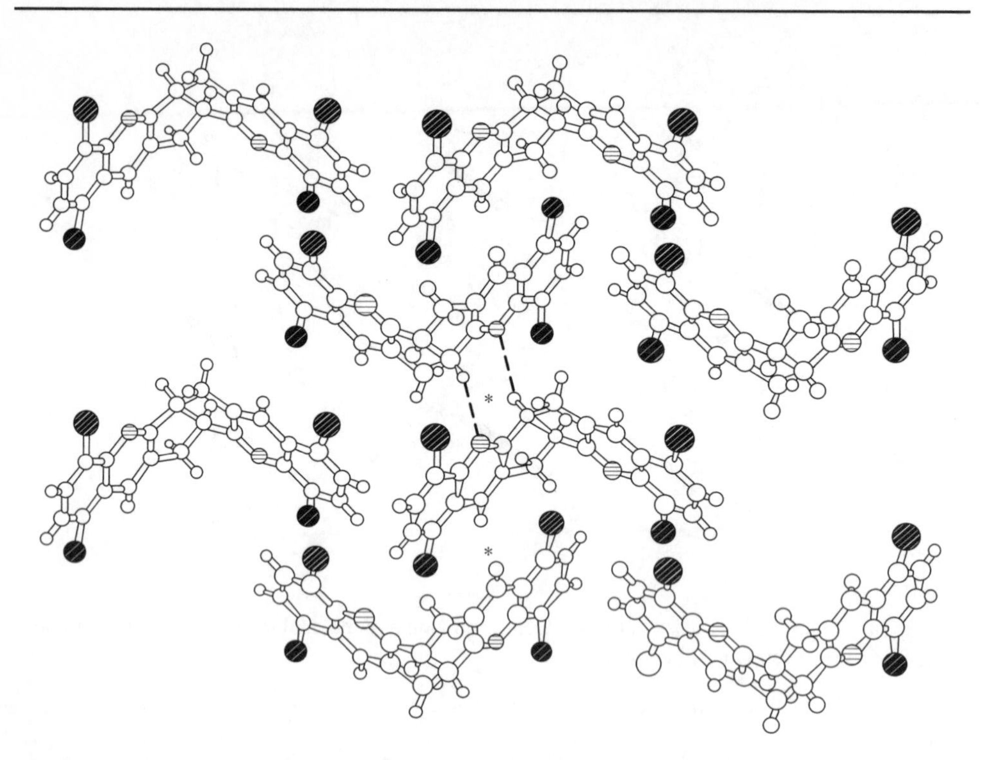

Figure 16. Part of a layer in crystalline (**5**)•(carbon tetrachloride). This is constructed from homochiral chains of **5** of alternating handedness. Asterisks indicate inversion centers where adjacent chains of opposite chirality interact. The dashed lines indicate an edge–edge C–H···N dimer, while the lower asterisk indicates the center of a PHD unit.

6.3. Layer Inclusion Compounds

The remaining 6 PHD structures, out of the 18 crystal structures determined, involve layer structures. These are formed when a PHD building block associates with other host molecules not involved in PHD units.

The example of (**5**)•(carbon tetrachloride) is presented here [25, 32]. In this solid, the host molecules form layers built up from chains of **5** molecules of the same handedness and joined by EF associations. Adjacent chains have opposite chirality and are linked by two centrosymmetric motifs: the PHD interaction, and the edge–edge C–H···N dimer (Figure 16).

7. MOLECULAR BRICKS, SPHERES AND GRIDS

This final section discusses three entirely different, and rather unusual, types of molecular assembly formed by our heteroaromatic compounds.

7.1. Bricks and Mortar Inclusion Systems

The concept of molecular self-assembly involves the spontaneous aggregation of molecular building blocks into a larger and more complex structure. Each building block carries functional groups whose ability to carry out non-covalent bonding drives the assembly process. These attractive forces act as the supramolecular mortar, creating and stabilizing the aggregate. The diphenyl host molecule **15** is a rather interesting example of these general principles.

When racemic **15** is crystallized, a centrosymmetric dimeric unit is produced. Two molecules of opposite handedness interact by means of three different aromatic EF interactions to produce the bricklike $(EF)_6$ unit illustrated in Figure 17. This repeat unit is present in solid-state structures of the apohost and all known lattice inclusion compounds of **15** [33].

The brick dimensions are almost constant across the complete series of structures. However, differences in translation or tilting of the bricks, in conjunction with changes in host–guest association, result in variations of host packing and the incorporation of guest molecules of various types. The overall outcome is that pairs of **15** molecules (the bricks) pack into rows by means of aromatic OFF interactions, and guest molecules occupy sites between the rows and/or at the corners of the bricks. Host–guest interactions provide stabilization (the mortar) between neighboring rows, and host–guest layers (walls) are formed as a result.

In (**15**)•(carbon disulphide), for example, the host molecules form different parallel rows along the *a* and *c* directions. The relatively small guests only associate with the

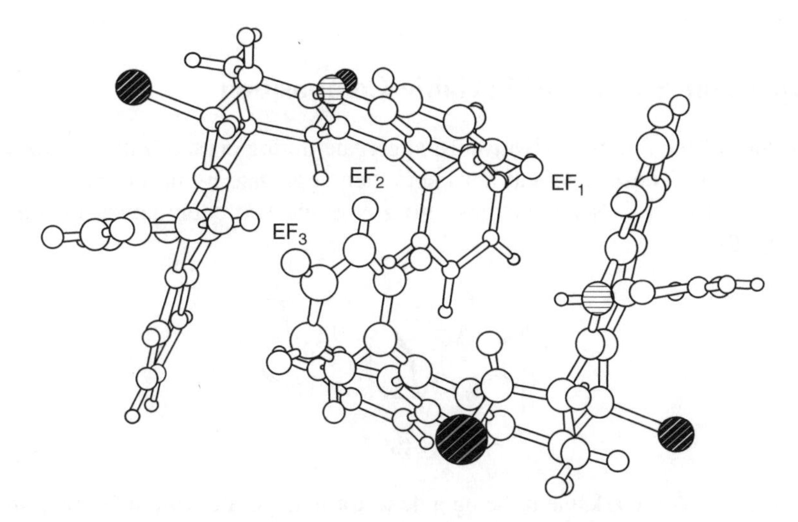

Figure 17. A typical centrosymmetric $(EF)_6$ dimer produced when the diphenyl diquinoline **15** is crystallized. The three different types of aromatic edge–face interactions are marked EF_1–EF_3.

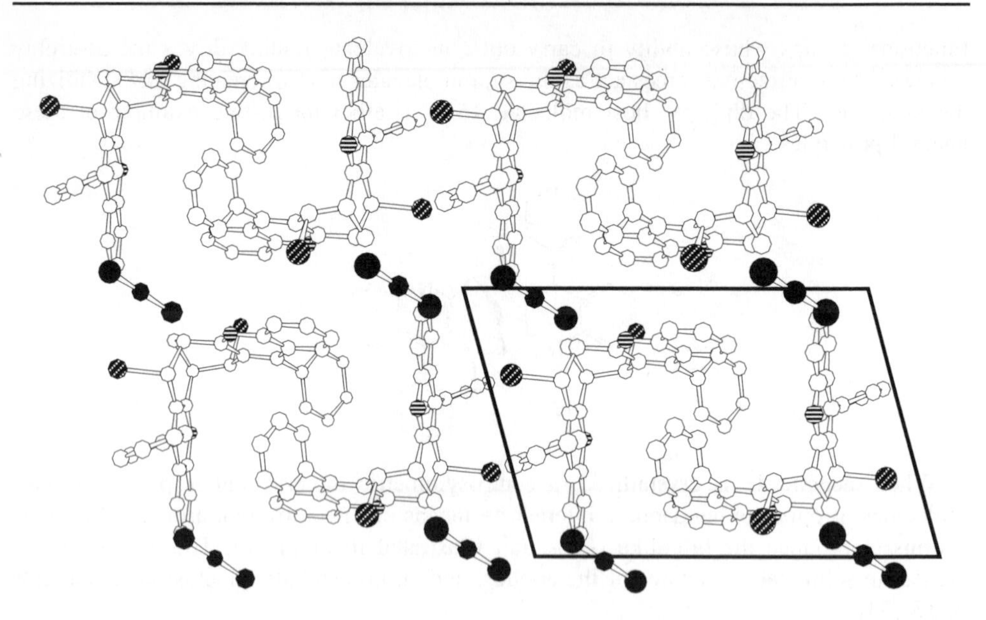

Figure 18. The layer structure of (**15**)•(carbon disulphide) projected onto the *ac* plane. This shows the different rows of **15** dimers (bricks) along *a* and along *c*, that form a wall, and the guests interacting (mortar) at the corners of the dimeric host building blocks. The atoms of the carbon disulphide molecules are shown in black.

bricks where the two types of rows abut, namely, between the corners of the molecular bricks (Figure 18). A combination of host–guest Br$\cdots$Cl, N$\cdots$Cl, Br$\cdots$H–CCl_3, plus guest–guest Cl$\cdots$Cl interactions provide the intermolecular mortar.

To extend this analogy further, crystals of the apohost **15** assemble into rows without the benefit of included guest (mortar-free structure). This type of assembly parallels the construction of a dry-stone wall.

7.2. Molecular Spheres of Variable Composition

Six molecules of the dinitro derivative **16** aggregate during crystallization to form a ball-like hexamer [34]. These spherical hexamers then aggregate by means of a combination of OFF, EF and C–H$\cdots$N interactions into a close-packed cubic lattice in space group $R\overline{3}$ (Figure 19).

NO_2 N N NO_2

16

This substance is remarkable in being able to form a hydrate with the same host molecular structure, ranging all the way from the apohost (host:guest ratio = 6:0) (Figure 20(a)), through to full occupancy (host:guest ratio = 6:1) or (**16**)•$(water)_{0.17}$ (Figure 20(b)). The

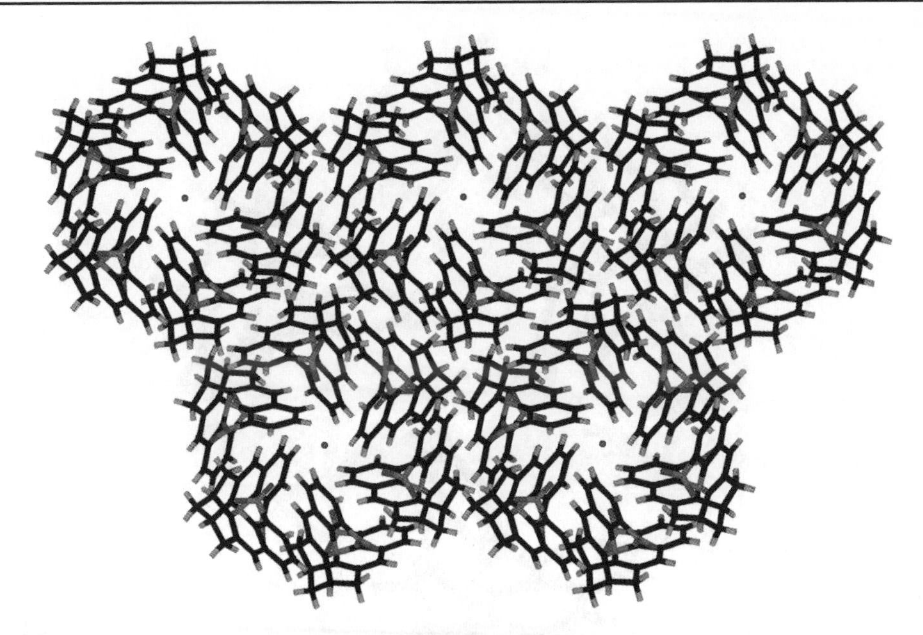

Figure 19. Part of the crystal structure of solid **16** showing the packing arrangement of the spherical hexamers. The sites that can be occupied by water molecules are indicated by the black dots.

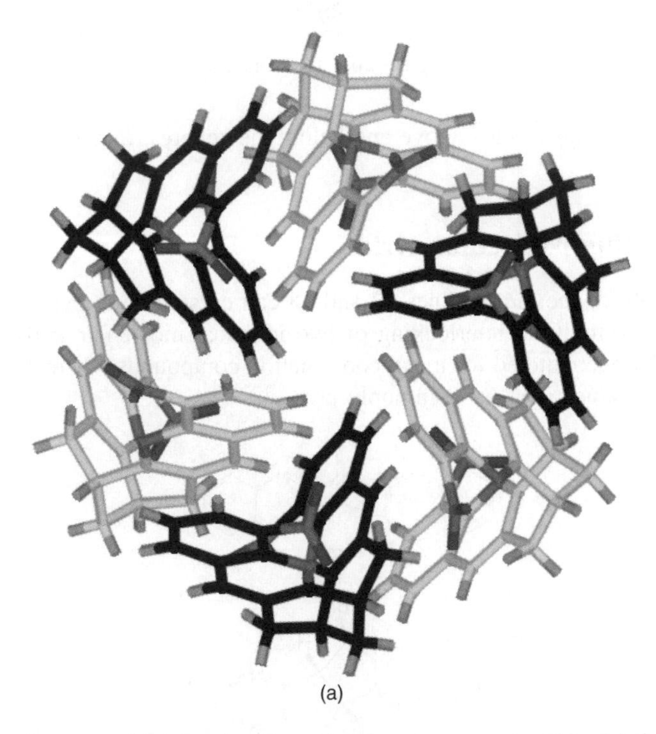

Figure 20. One hexamer of **16** showing the molecular arrangement within: (a) the empty apohost structure, and (b) the fully occupied (**16**)•(water)$_{0.17}$ structure. Alternate molecules of **16** in the hexamer are colored light and dark for clarity. The water molecule is indicated by a black dot.

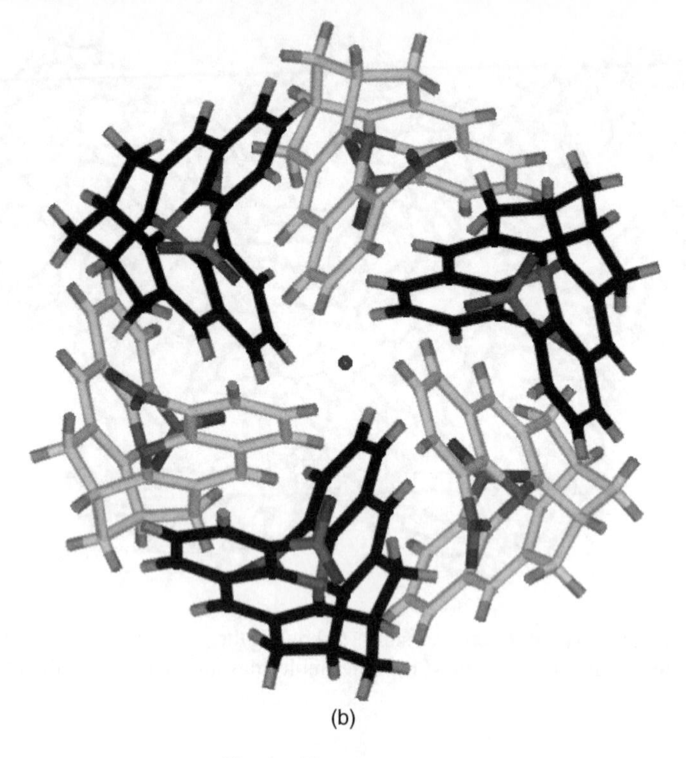

(b)

Figure 20. *(continued)*

water molecules occupy $\overline{3}$ sites above and below the spheres, rather than being enclosed within them.

7.3. Interlocking Molecular Grids

Remarkably, both of the diquinolines **17** and **18** can crystallize in the form of structures resulting from the multiple interlocking of two identical molecular grids. This behavior is more typically encountered for metal coordination compounds where strong and highly directional grid connections are commonly present.

17

18

Both interlocking grid structures are rather similar, but rather complex and not easy to describe fully. Detailed analyses (including the role of the enantiomers) are presented in color in our original paper [35]. Briefly, however, half of the molecules are present as a set of identical but offset parallel layers. A second identical set of layers intersects the first at an incline of 86° for **17**, or 84° in the case of **18**. However, the angles within the actual grids are quite different (acute angles of 83 and 70°, respectively). The Figures 21 and 22 illustrate the overall structures of these two respective cases.

These crystal lattices do not utilize edge–edge eight-membered C–H···N dimers, and only poorly efficient OFF interactions are involved. Other types of C–H···N weak hydrogen bonds are present, as are H···π (for **17**) and Cl···π (for **18**) interactions.

In each crystal structure, two molecules of opposite handedness in one layer associate by means of their endo-faces, leaving a small void. Two further molecules belonging to the intersecting layer are inserted into this space to produce a centrosymmetric tetramer. Each resulting inversion center is defined as a point belonging to one of the molecular grids. Infinite repetition of this arrangement in three dimensions creates the interlocked molecular grid structure. Figure 23 shows a pair of molecules of **17** from one layer, and their association with an identical pair from an intersecting layer. The comparable structures for compound **18** are shown in Figure 24.

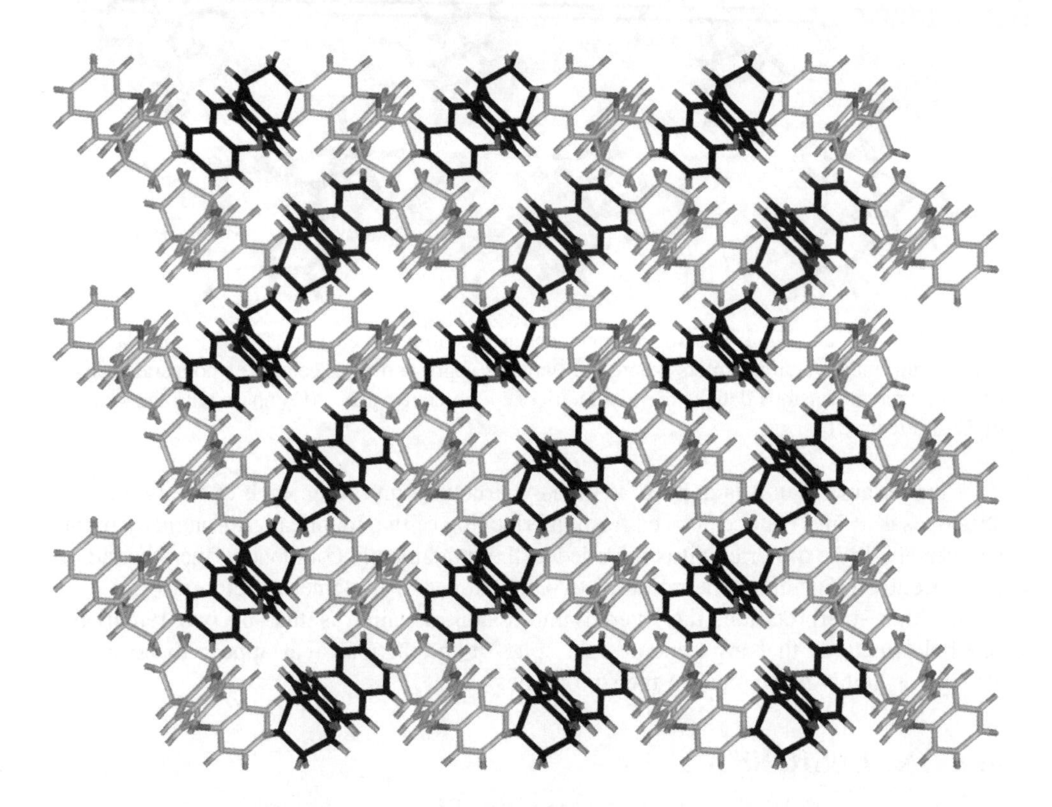

Figure 21. The packing arrangement in solid **17** projected onto the *bc* plane, showing how the two sets of parallel planes intermesh. Each place where the nonparallel layers cross is a centrosymmetric connection motif. Connection of these inversion centers in three dimensions defines the two interlocked molecular grids.

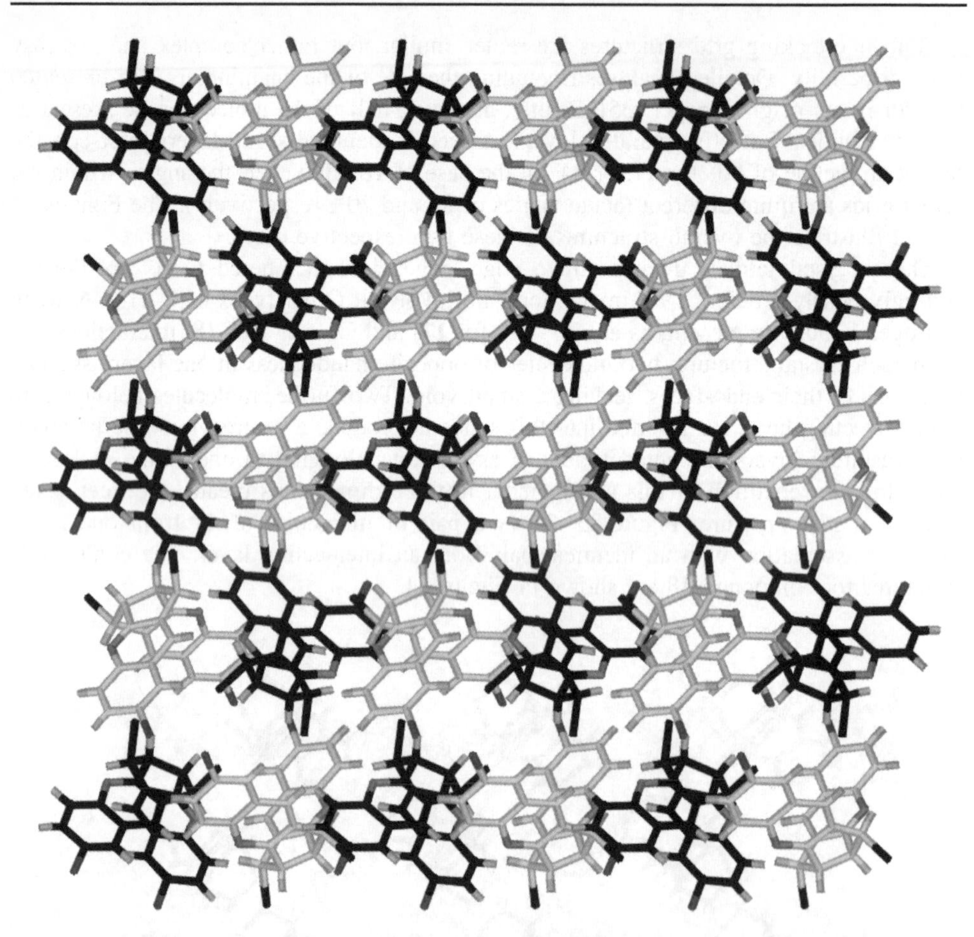

Figure 22. The analogous packing arrangement in crystalline **18** also projected onto the *bc* plane. Once again, the inversion centers where the two sets (light and dark colors) of nonparallel layers cross in three dimensions defines the two interlocked molecular grids. (Compare with the structure of **17** in Figure 21).

If the tetrameric unit is defined as the key structural unit, then each of the two crystal structures is a single lattice derived from interlocking of two identical grids at many points. Conversely, if the dimeric unit is regarded as the fundamental component, then the lattices are the outcome of interpenetration of two sublattices of (4,4) net structure [36].

In either interpretation, the overall impression remains as that of two remarkably ordered structures that are generated in spite of interactions that appear to be neither strong nor highly directional in their properties.

8. CONCLUSIONS

A major aim when commencing this study of halogenated diquinoline compounds was to demonstrate that new lattice inclusion compounds could be designed in a rational manner. The wide selection of examples discussed here demonstrates that this target has been amply met.

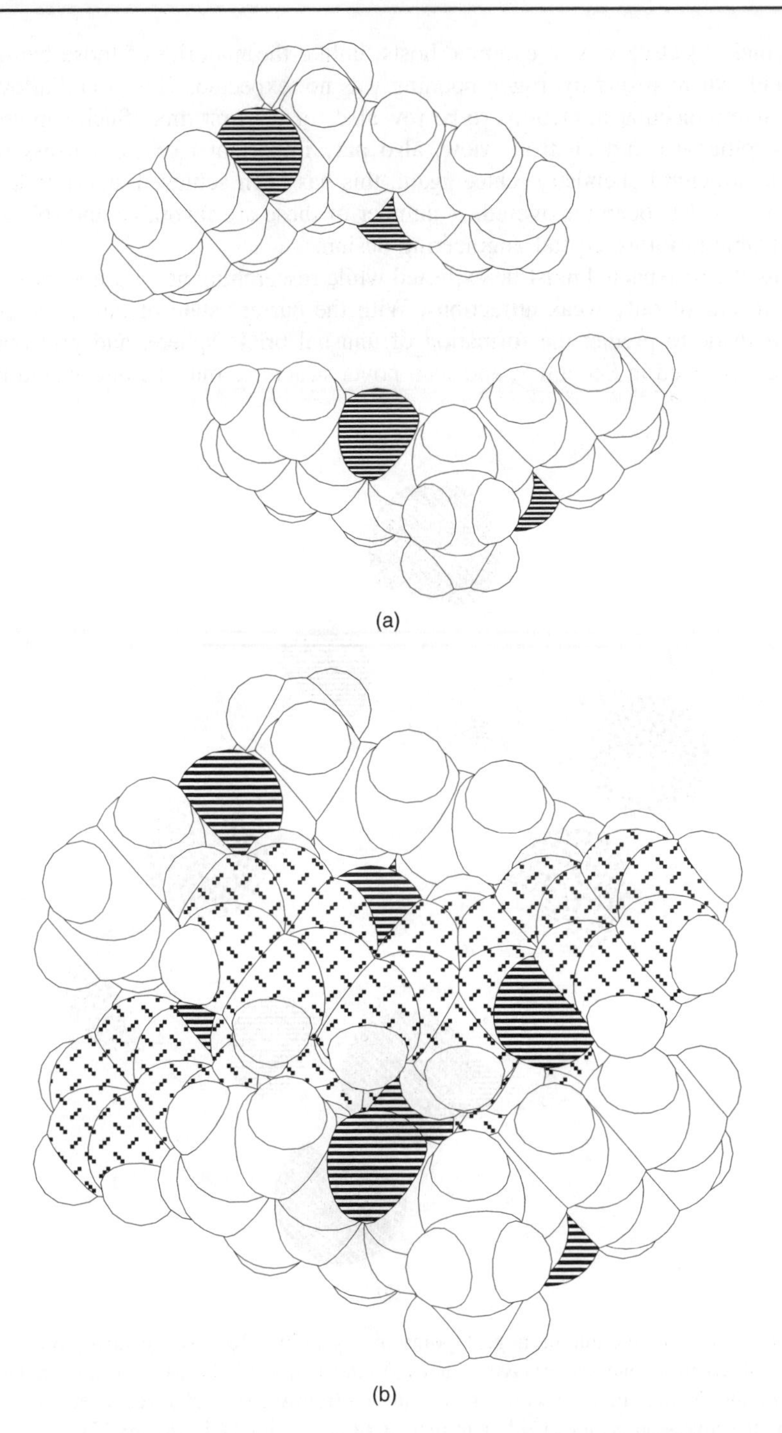

Figure 23. The molecular construction of a molecular grid site in solid **17**. (a) centrosymmetric arrangement of two molecules of **17** from one set of layers, and (b) the same arrangement (white C atoms), with addition of an identical pair (hatched C atoms) from the second intersecting set of layers. Each grid point location is the inversion center of such a tetramer.

A second objective was to examine hosts, unlike the majority of those being studied in the field, where strong hydrogen bonding was not expected. This would allow weaker types of intermolecular attractions to be revealed for the first time. Such supramolecular synthons, otherwise hidden from view, also play important roles in self-assembly and solid-state structural chemistry. Once again, this has been achieved and a wide range of new attractions has been uncovered. A number of these are favorable and robust enough to be valuable in future crystal engineering design.

Finally, the unexpected must be expected while researching new compounds that associate by means of only weak attractions. With the current state of our knowledge, it is not yet realistic to predict the formation of unusual brick, sphere, and grid compounds like those described in Section 7, and such novel structures must be discovered by experiment.

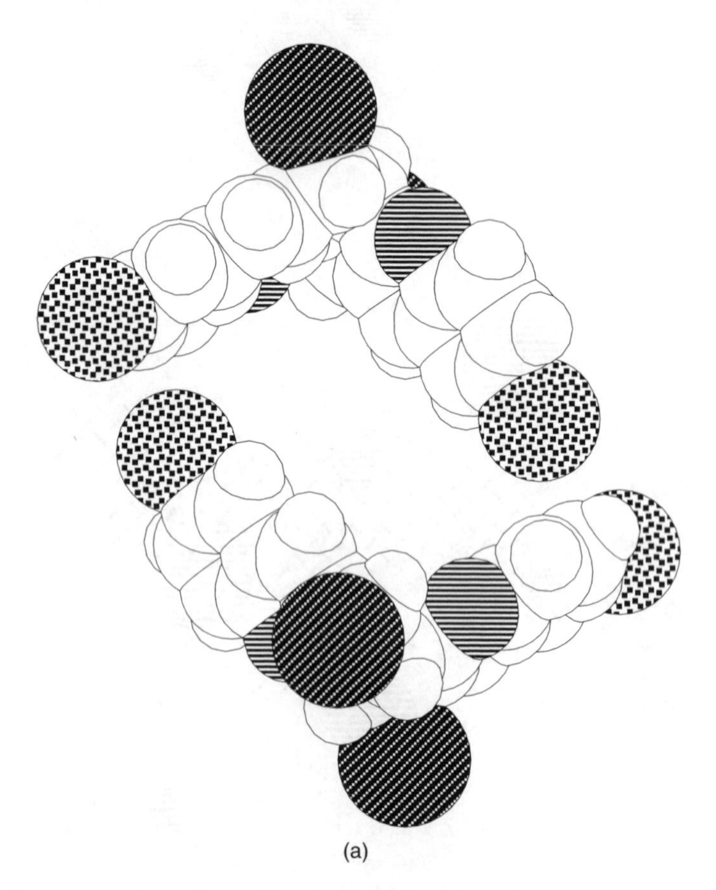

Figure 24. The construction of a grid point in crystalline **18**. (a) a centrosymmetric pair of molecules of **18** from one set of layers, and (b) intersection of the same arrangement (white C atoms) with an identical pair (cross-hatched C atoms) from the second set of layers. The grid point location is the inversion center of this tetramer. (Compare with **17** in Figure 23).

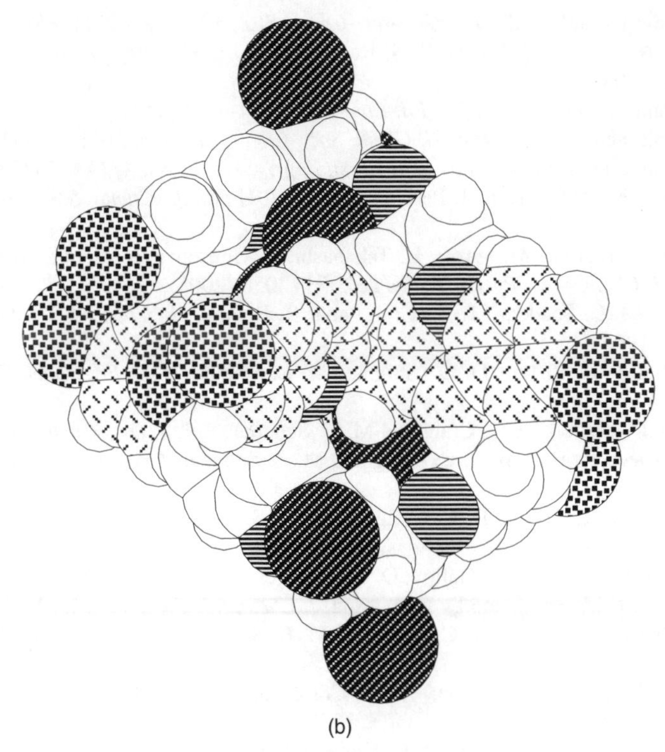

(b)

Figure 24. *(continued)*

9. ACKNOWLEDGMENTS

I wish to thank Dr Marcia Scudder for preparing the crystallographic figures used to illustrate this article and the University of New South Wales for financial support of this crystal engineering research.

REFERENCES

1. M. Hagan, *Clathrate Inclusion Compounds*, Reinhold Publishing Corporation, New York, Preface, page vii, 1962.
2. D. D. MacNicol and G. A. Downing, in *Comprehensive Supramolecular Chemistry, Vol. 6: Solid-State Supramolecular Chemistry: Crystal Engineering*, (Eds. D. D. MacNicol, F. Toda and R. Bishop), Pergamon Press, Oxford, Ch. 14, 421–464, 1996.
3. E. Weber, in *Comprehensive Supramolecular Chemistry, Vol. 6: Solid-State Supramolecular Chemistry: Crystal Engineering*, (Eds. D. D. MacNicol, F. Toda and R. Bishop), Pergamon Press, Oxford, Ch. 17, 535–592, 1996.
4. R. Bishop, *Chem. Soc. Rev.*, **25**, 311–319 (1996).
5. A. I. Kitaigorodskii, *Molecular Crystals and Molecules*, Academic Press, New York, 1973.
6. G. R. Desiraju, *Crystal Engineering: The Design of Molecular Solids*, Elsevier, Amsterdam, 1989.

7. J. A. R. P. Sarma and G. R. Desiraju, *Acc. Chem. Res.*, **19**, 222–228 (1986).
8. S. L. Price, A. J. Stone, J. Lucas, R. S. Rowland and A. Thornley, *J. Am. Chem. Soc.*, **116**, 4910–4918 (1994).
9. G. R. Desiraju, *Angew. Chem., Int. Ed. Engl.*, **34**, 2311–2327 (1995).
10. G. R. Desiraju and A. Gavezzotti, *J. Chem. Soc., Chem. Commun.*, 621–623 (1989).
11. G. R. Desiraju and A. Gavezzotti, *Acta Crystallogr., Sect. B*, **45**, 473–482 (1989).
12. C. A. Hunter, K. R. Lawson, J. Perkins and C. J. Urch, *J. Chem. Soc., Perkin Trans. 2*, 651–669 (2001).
13. H. Suezawa, T. Yoshida, M. Hirota, H. Takahashi, Y. Umezawa, K. Honda, S. Tsuboyama and M. Nishio, *J. Chem. Soc., Perkin Trans. 2*, 2053–2058 (2001).
14. C. E. Marjo, M. L. Scudder, D. C. Craig and R. Bishop, *J. Chem. Soc., Perkin Trans. 2*, 2099–2104 (1997).
15. S. F. Alshahateet, R. Bishop, D. C. Craig and M. L. Scudder, *CrystEngComm*, **3**(48), 225–229 (2001).
16. C. E. Marjo, R. Bishop, D. C. Craig and M. L. Scudder, *Eur. J. Org. Chem.*, 863–873 (2001).
17. S. F. Alshahateet, R. Bishop, D. C. Craig and M. L. Scudder, *Cryst. Growth Des.*, **4**, 837–844 (2004).
18. S. F. Alshahateet, R. Bishop, D. C. Craig and M. L. Scudder, *CrystEngComm*, **3**(25), 107–110 (2001).
19. F. H. Allen, J. E. Davies, J. J. Galloy, O. Johnson, O. Kennard, C. F. Macrae, E. M. Mitchell, G. F. Mitchell, J. M. Smith and D. G. Watson, *J. Chem. Inf. Comput. Sci.*, **31**, 187–204 (1991).
20. S. F. Alshahateet, R. Bishop, D. C. Craig and M. L. Scudder, *CrystEngComm*, **3**(55), 264–269 (2001).
21. A. N. M. M. Rahman, R. Bishop, D. C. Craig and M. L. Scudder, *Chem. Commun.*, 2389–2390 (1999).
22. A. N. M. M. Rahman, R. Bishop, D. C. Craig and M. L. Scudder, *Eur. J. Org. Chem.*, 72–81 (2003).
23. G. R. Desiraju and T. Steiner, *The Weak Hydrogen Bond in Structural Chemistry and Biology*, Oxford University Press, 1999.
24. S. F. Alshahateet, R. Bishop, M. L. Scudder, C. Y. Hu, E. H. E. Lau, F. Kooli, Z. M. A. Judeh, P. S. Chow and R. B. H. Tan, *CrystEngComm*, **7**(21), 139–142 (2005).
25. A. N. M. M. Rahman, R. Bishop, D. C. Craig and M. Scudder, *CrystEngComm*, **4**(84), 510–513 (2002).
26. A. N. M. M. Rahman, R. Bishop, D. C. Craig and M. L. Scudder, *CrystEngComm*, **5**(75), 422–428 (2003).
27. M. D. Prasanna and T. N. Guru Row, *Cryst. Eng.*, **3**, 135–154 (2000).
28. R. K. R. Jetti, A. Nangia, F. Xue and T. C. W. Mak, *Chem. Commun.*, 919–920 (2001).
29. R. Bishop, M. L. Scudder, A. N. M. M. Rahman, S. F. Alshahateet and D. C. Craig, *Mol. Cryst. Liq. Cryst.*, **440**, 173–186 (2005).
30. A. N. M. M. Rahman, R. Bishop, D. C. Craig, C. E. Marjo and M. L. Scudder, *Cryst. Growth Des.*, **2**, 421–426 (2002).
31. C. E. Marjo, A. N. M. M. Rahman, R. Bishop, M. L. Scudder and D. C. Craig, *Tetrahedron*, **57**, 6289–6293 (2001).
32. A. N. M. M. Rahman, R. Bishop, D. C. Craig, and M. L. Scudder, *Org. Biomol. Chem.*, **2**, 175–182 (2004).
33. J. Ashmore, R. Bishop, D. C. Craig and M. L. Scudder, *CrystEngComm*, **6**(100), 618–622 (2004).
34. J. Ashmore, R. Bishop, D. C. Craig and M. L. Scudder, unpublished results.
35. S. F. Alshahateet, A. N. M. M. Rahman, R. Bishop, D. C. Craig and M. L. Scudder, *Cryst-EngComm*, **4**(97), 585–590 (2002).
36. S. R. Batten and R. Robson, *Angew. Chem., Int. Ed. Engl.*, **37**, 1460–1494 (1998).

6

Steric Control over Supramolecular Aggregation: A Design Element in Crystal Engineering?

EDWARD R. T. TIEKINK
Department of Chemistry, The University of Texas at San Antonio, San Antonio, Texas, USA.

1. INTRODUCTION

The pivotal aim of crystal engineering is to rationally design crystal structure starting from molecules (tectons). Tectons may be linked via a variety of intermolecular interactions ranging from covalent bonds to contacts mediated by hydrogen bonding. Indeed, perhaps the most exploited supramolecular "glue" between tectons is the hydrogen bond owing to its strength and directionality. The obvious requirement for hydrogen-bonding is the presence of hydrogen-bond donors and acceptors. Thus, hydrogen-bonding connections between molecules may be designed to occur between purely organic molecules as well as between metal complexes containing ligands with suitable hydrogen-bonding functionality. In the realm of crystal structure design involving coordination complexes, tectons may also be connected via coordinate or dative bonds to give rise to the desired arrangement. A subset of these types of interactions is secondary interactions (bonds) [1]. These typically occur between a Lewis acid (metal center) and a Lewis base (e.g. nitrogen, oxygen, halide, etc.) and are in addition to the bonding interactions defining the immediate coordination environment of the metal center. Such secondary interactions are frequently found in main group element chemistry and inevitably contribute to the organization and stabilization of the resultant crystal structures. In this chapter, the control of such interactions, leading to specific supramolecular architectures is investigated. It will be demonstrated that it is possible, at least in favorable circumstances, to control the formation of secondary interactions

Frontiers in Crystal Engineering. Edited by Edward R.T. Tiekink and Jagadese J. Vittal

and therefore, the nature of supramolecular aggregation. In essence, and put simply, by judicious choice of metal-bound and/or ligand-bound substituents, it is possible to tune secondary interactions and therefore, supramolecular architecture. The obvious analogy to this approach is found in coordination chemistry, where the presence of small ligands allows for high coordination numbers and, conversely, bulky ligands can be exploited to give low-coordinate complexes. In the following, a series of structures, incorporating tin, zinc, mercury and bismuth centers, with carboxylate and 1,1-dithiolate ligands, will be described in which the above "steric control" principle will be examined.

2. DIORGANOTIN CARBOXYLATES

The principle of steric control of supramolecular association is no better illustrated than for the diorganotin dicarboxylates, $R_2Sn(O_2CR')_2$, where the carboxylate ligand is derived from 2-picolinic acid or 2-quinaldic acid [2]; see Figure 1. The ligands are closely related to each other in that 2-quinaldic acid comprises two fused six-membered rings but have a common disposition of the ligand donor set. The molecular structure of Me_2Sn(2-quinaldate)$_2$ is illustrated in Figure 2(a). The tin atom is coordinated by two chelating 2-quinaldate ligands, each coordinating via the ring-nitrogen atom and one of the carboxylate oxygen atoms, so that an approximately planar SnN_2O_2 arrangement results. As the Sn–O bonds (approximately 2.1 Å) are significantly shorter than the Sn–N bonds (approximately 2.6 Å), the arrangement of the N_2O_2 donors is trapezoidal. The tin-bound methyl groups are disposed over the weaker Sn–N bonds and C–Sn–C is approximately 147°, so that the overall coordination geometry is best described as trapezoidal bipyramidal. The structure is monomeric as any possible approach of a potential donor atom from a neighboring molecule is precluded by the presence of the aromatic groups of the 2-quinaldate ligands that occupy the potentially "accessible" region surrounding the nitrogen donor atoms. Reducing the steric bulk of the carboxylate ligand to 2-picolinate, with a single six-membered ring, gives rise to a different structural motif as illustrated in Figure 2(b). In the structure of polymeric $[Me_2Sn(2\text{-picolinate})_2]_n$ [2, 3], both 2-picolinate ligands chelate the tin atom via the nitrogen atom and one of the carboxylate oxygen atoms as in the 2-quinaldate analog. The difference arises as one of the ligands also forms an additional Sn–O interaction but, with a symmetry related tin center resulting in the formation of a polymer, generated by 2_1 symmetry. The tin-bound methyl groups are disposed above and below the SnN_2O_3 plane and the overall coordination geometry is best described as being based on a pentagonal bipyramid. Consistent with the increased coordination number of the tin atom in the 2-picolinate structure, the Sn–O bond distances are significantly longer than those in the 2-quinaldate structure [2]. Thus, qualitatively, the valency requirements of the tin atoms in the two structures are equally satisfied, but by different coordination geometries that are dictated by the accessibility

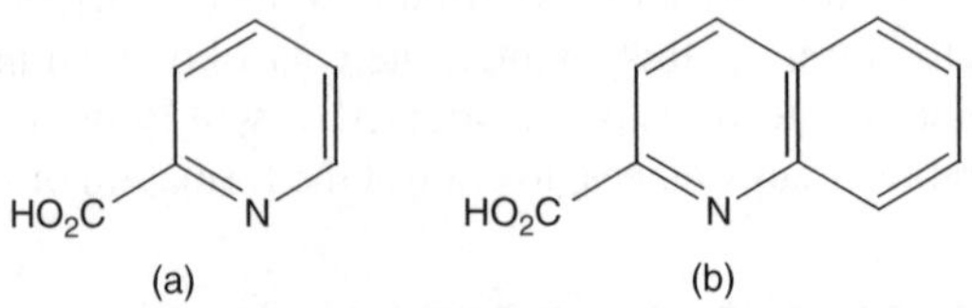

Figure 1. Chemical structures of (a) 2-picolinic acid and (b) 2-quinaldic acid.

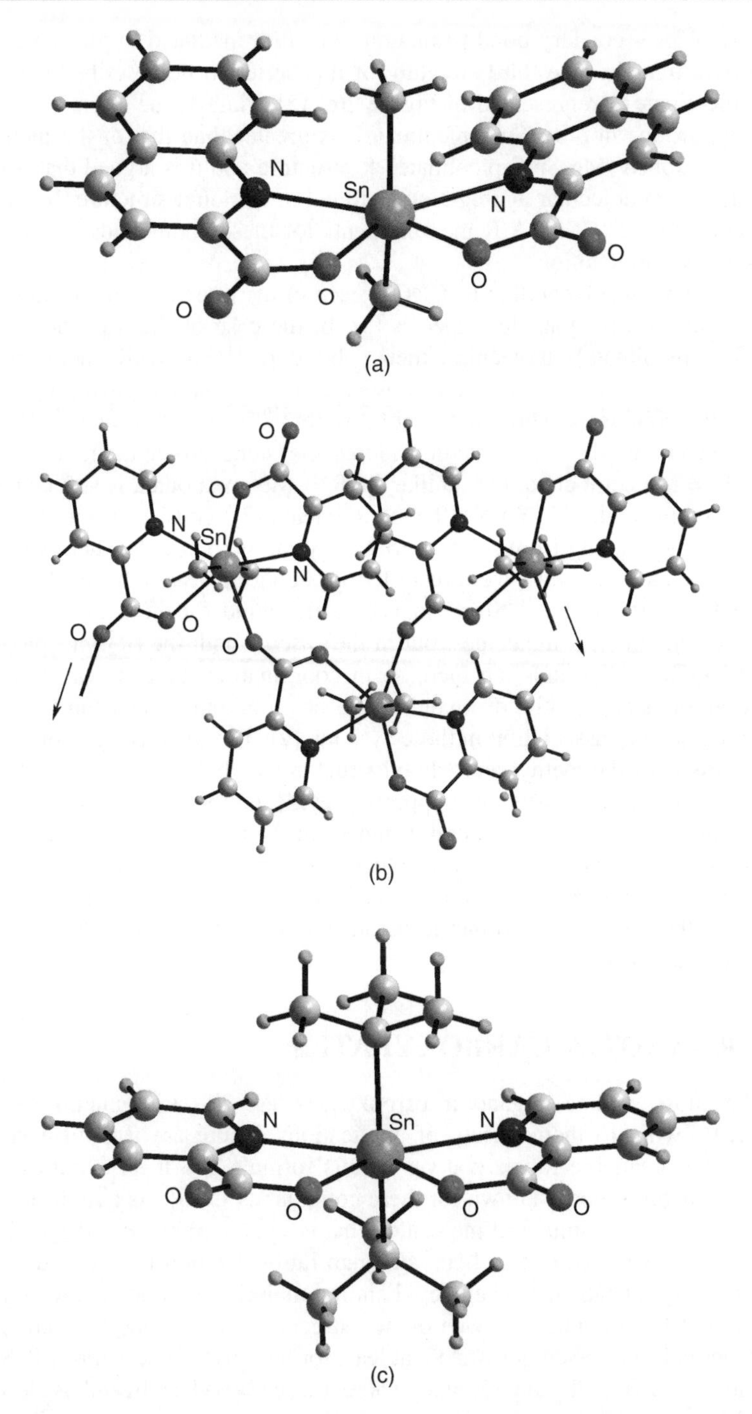

Figure 2. Molecular structures of (a) $Me_2Sn(2\text{-quin})_2$; (b) polymeric $[Me_2Sn(2\text{-pic})_2]_n$ and (c) $t\text{-}Bu_2Sn(2\text{-pic})_2$.

of the tin atom for secondary bond formation, a feature mediated by the steric profile of the carboxylate ligands. The third structure of this series to be described here is that of t-Bu_2Sn(2-picolinate)$_2$, represented in Figure 2(c) [2]. Plainly, the steric bulk of the tin-bound t-butyl groups in t-Bu_2Sn(2-picolinate)$_2$ is greater than that of the methyl groups in the aforementioned [Me_2Sn(2-picolinate)$_2$]$_n$ structure and it is argued that it is this fact that precludes supramolecular aggregation as found in the latter structure. It is apposite to consider the results of ^{119}Sn NMR measurements for these compounds, recorded in both the solid-state and in solution.

In noncoordinating solvent, that is, $CDCl_3$, each of the three compounds have chemical shifts consistent with mononuclear species [2]. In the case of the Me_2Sn(2-quinaldate)$_2$ and t-Bu_2Sn(2-picolinate)$_2$ molecules, their solid state ^{119}Sn NMR chemical shifts are consistent with the mononuclear species observed in solution and proven by X-ray crystallography. By contrast, a shift in the solid-state ^{119}Sn NMR resonance is found for [n-Bu_2Sn(2-picolinate)$_2$]$_n$ (this structure adopts the same structural motif in the solid state as the R = Me compound, but, unlike the R = Me compound, is sufficiently soluble in $CDCl_3$ solution to allow ^{119}Sn NMR measurements [2]) consistent with an increase in coordination number going from solution to the crystalline phase. Further, when the solution ^{119}Sn NMR measurements were recorded in a coordinating solvent, that is, DMSO, no significant changes in the ^{119}Sn resonances were found for the Me_2Sn(2-quinaldate)$_2$ and t-Bu_2Sn(2-picolinate)$_2$ molecules, but in the case of both the [R_2Sn(2-picolinate)$_2$]$_n$, R = Me and n-Bu compounds, an increase in coordination number was evident. Thus, the increase in the steric profile of ligand-bound or tin-bound groups can be exploited to restrict supramolecular association in these systems [2]. The underlying assumption of the above comparison of the methyl and t-butyl structures is that the Lewis acidity of the tin centers is not significantly different. Supporting evidence that there is *no* significant difference is found by comparing the bond distances in the two t-butyl structures containing 2-picolinate and 2-quinaldinate ligands, which are experimentally indistinguishable [2]. Also, theoretical studies, based on *ab initio* molecular orbital calculations, showed that in the gas phase, the geometries/geometric parameters of R_2SnCl_2, R = Me and t-Bu, are essentially identical [4].

3. TRIORGANOTIN CARBOXYLATES

Perhaps the most studied organotin carboxylates are the triorganotin carboxylates, $R_3Sn(O_2CR')$, owing to their variety of applications. There are well in excess of 100 crystal structures available for derivatives of this formula and it turns out that there are four distinct structural motifs known for these compounds [5, 6]. As two of the motifs are adopted by a very small number of molecules, that is, cyclotetrameric and cyclohexameric, these will not be discussed further here, although rationalization for their adoption in the solid-state remains a challenge. The overwhelming majority of structures with the general formula $R_3Sn(O_2CR')$ conform to one of two structural motifs, one monomeric and the other, polymeric. It will be demonstrated, at least for the series of structures, $R_3Sn(O_2CR')$ for R = Me, Et, n-Bu, Ph and Cy and where the carboxylate ligand is derived from 2-[(E)-2-(2-hydroxy-5-methylphenyl)-1-diazenyl]-benzoic acid, shown in Figure 3, that steric control over supramolecular aggregation exists [6].

The polymeric structure of $Me_3Sn(O_2CR')$ is shown in Figure 4(a); the R = Et and n-Bu structures also adopt the same structural motif. The carboxylate ligand is bidentate

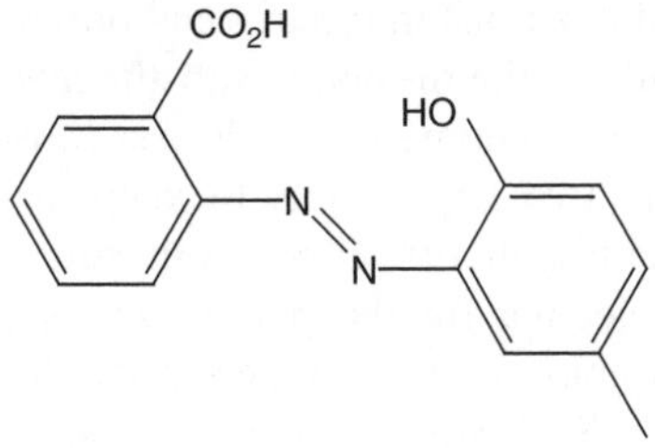

Figure 3. Chemical structure of 2-[(*E*)-2-(2-hydroxy-5-methylphenyl)-1-diazenyl]-benzoic acid.

(a)

(b)

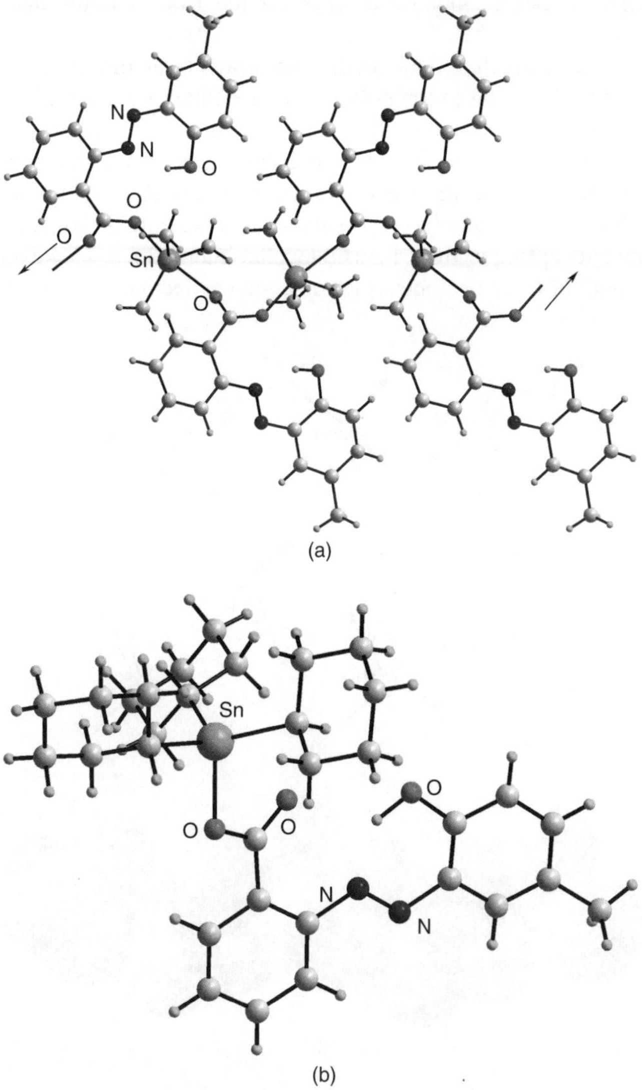

Figure 4. Molecular structures of (a) polymeric $[Me_3Sn(O_2CR)]_n$ and (b) $Cy_3Sn(O_2CR)$ where HO_2CR is 2-[(*E*)-2-(2-hydroxy-5-methylphenyl)-1-diazenyl]benzoic acid.

bridging and defines the axial positions in the trigonal bipyramidal geometry about the tin atom. Increasing the steric bulk of the tin-bound substituents to phenyl [7] and cyclohexyl [6] results in the formation of monomeric species as illustrated in Figure 4(b) for the R = Cy derivative. There are two keys points to keep in mind when rationalizing the appearance of the two structural motifs. First, both structural types feature additional Sn···O interactions, intermolecular for the polymeric, as for the R = Me species, and intramolecular, as for the monomeric R = Cy derivative. The second salient observation relates to the solution state ^{119}Sn NMR of all five species that indicates that these exist as monomers in solution [6]. With this information, it is possible to conceptualize what happens in the crystallization process for these compounds. In the case of the R = Me compound, molecules can be thought of crystallizing from solution and adjusting their positions so as to maximize their intermolecular interactions, the basic premise of crystal packing. In this case, with the small methyl substituents at tin, intermolecular Sn···O interactions are formed leading to the polymer, represented in a simplified form in the top view of Figure 5(a). The backbone of the polymer comprises the [Sn–O–C–O] repeat unit and, not surprisingly, it has been demonstrated that the repeat distance in the polymeric $R_3Sn(O_2CR')$ structures lies in the narrow range 5.19 ± 0.21 Å [8]. When the tin-bound substituent is bulky, such as cyclohexyl, the molecules precipitate from solution and in the same way attempt to optimize their intermolecular interactions. However, owing to the combined steric bulk of three cyclohexyl groups, intermolecular Sn···O interactions, akin

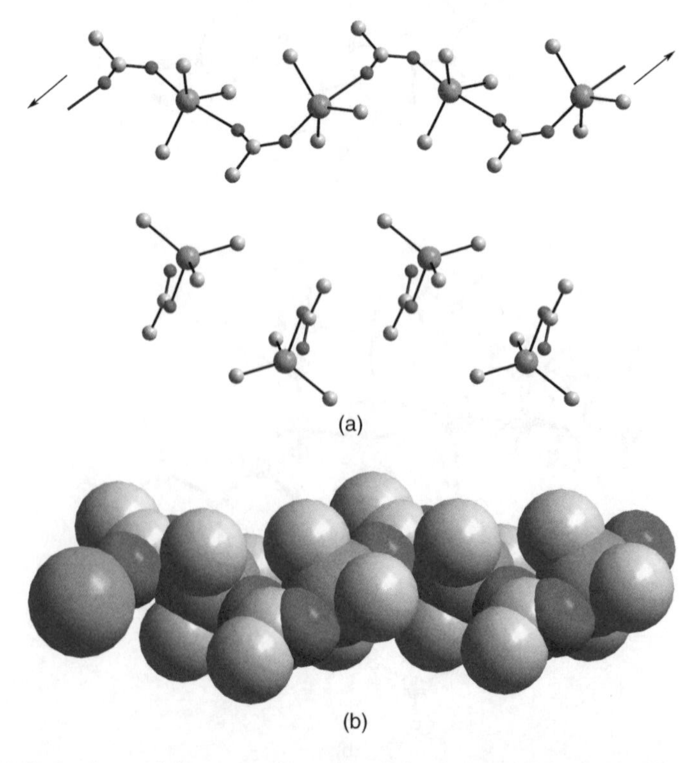

Figure 5. (a) Projections of the crystal structures of polymeric $[Me_3Sn(O_2CR)]_n$ (upper view) and $Cy_3Sn(O_2CR)$, and (b) space-filling projection for the structure of polymeric $[Me_3Sn(O_2CR)]_n$ where HO_2CR is 2-[(*E*)-2-(2-hydroxy-5-methylphenyl)-1-diazenyl]-benzoic acid.

to those in the R = Me derivative, cannot form. Thus, intramolecular Sn···O interactions are formed instead and a monomeric species results. The lower view of Figure 5(a) shows a sequence of the C_3SnO_2CC portions of the $Cy_3Sn(O_2CR')$ molecules and, clearly, the resemblance between this arrangement and that found in the polymeric species is self-evident. The difference between the two arrangements is that the molecules in the monomeric structure are somewhat offset with respect to each other owing to the absence of the intermolecular Sn···O interactions. Indeed, the "repeat" unit, calculated as the distance between successive tin atoms, assuming no displacement, is 5.15 Å, that is, similar to that seen for the polymeric species. Another way of considering this lack of polymer formation for the R = Ph and Cy structures is to consider the space-filling diagram shown in Figure 5(b) for the R = Me structure, but, showing only the C_3SnOCC atoms. While it may be possible to envisage a situation where the R′ groups of the carboxylate ligand are accommodated in such an arrangement, as indeed they are for the R = Me, Et and *n*-Bu structures, this, plus the incorporation of bulky tin-bound substituents, would lead to impossible steric interactions, and so, monomeric structures are found for those species with R = Ph [7] and Cy [6]. The idea of steric control of molecular aggregation is by no means restricted to organotin compounds as may be seen from a consideration of the structures observed for some zinc and mercury 1,1-dithiolate compounds as discussed below.

4. BINARY ZINC XANTHATES

The remaining structures to be discussed in this overview contain 1,1-dithiolate ligands. The chemical structures of the ligands to be described herein are given in Figure 6, that is, xanthate, ^-S_2COR, dithiophosphate, $^-S_2P(OR)_2$, and dithiocarbamate, $^-S_2CNR_2$. The structural chemistry of the binary zinc-triad 1,1-dithiolates is rich and diverse and has been reviewed recently [9]. A careful examination of the crystal structures of closely related species, when available, has allowed for the qualitative rationalization for the appearance of different structural motifs in the solid state. The first series of structures to be described are the binary xanthates of zinc(II), of general formula $Zn(S_2COR)_2$. Here, three very different structural motifs are found that are dependent on the nature of the R group of the xanthate ligand [10]. A portion of the two-dimensional structure of the $[Zn(S_2COEt)_2]_n$ structure is shown in Figure 7(a) [11]. The striking feature of the structure, and one that is found in the higher congenors, is the presence of a 16-membered ring. Thus, a square of zinc atoms has each edge bridged by a bidentate xanthate ligand

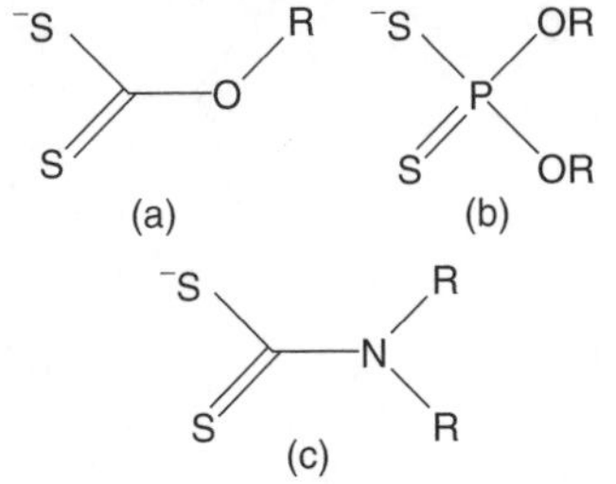

Figure 6. Chemical structures of the (a) xanthate; (b) dithiophosphate and (c) dithiocarbamate anions.

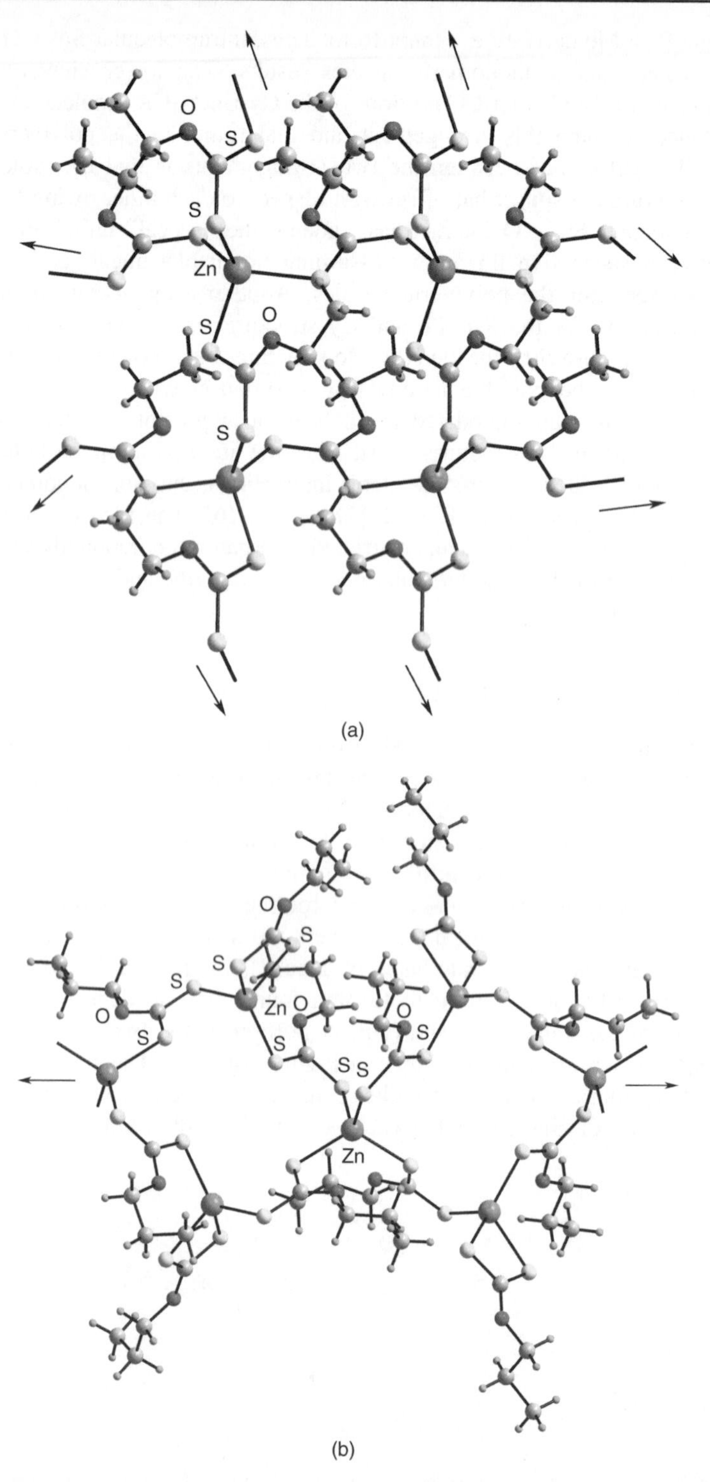

Figure 7. Portions of the crystal structures of (a) polymeric $[Zn(S_2COEt)]_n$; (b) polymeric $[Zn(S_2CO\text{-}n\text{-}Pr)]_n$ and (c) tetrameric $[Zn(S_2CO\text{-}i\text{-}Pr)]_4$.

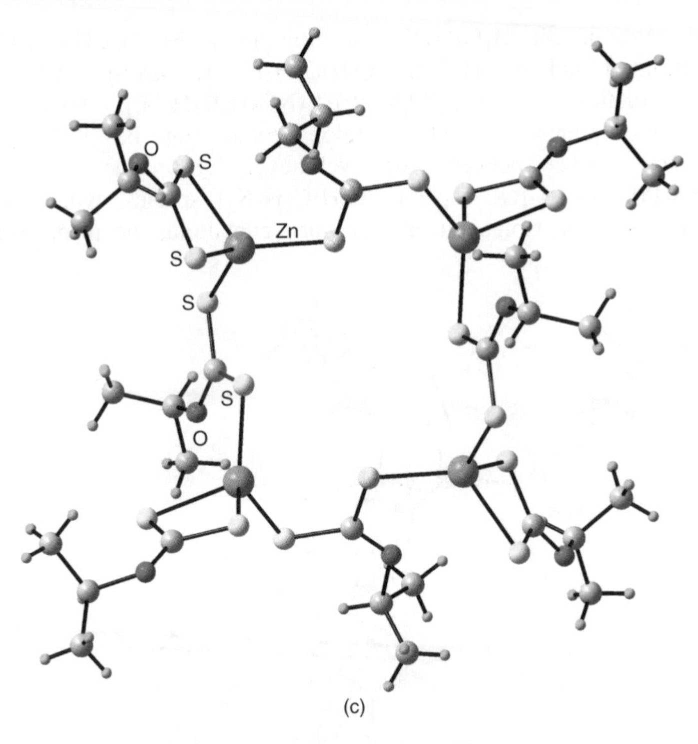

Figure 7. (*continued*)

leading to $[ZnSCS]_4$ rings. As all xanthate ligands are bridging in this structure and the bridges are coplanar, a two-dimensional layer structure results; alternatively, this structure could be described as a (4,4) net. Changing the R substituent of the xanthate ligand from ethyl to *n*-propyl results in a reduction of supramolecular aggregation in the structure. In the structure of $[Zn(S_2CO\text{-}n\text{-}Pr)_2]_n$, represented in Figure 7(b), two-thirds of the xanthate ligands are bridging and one-third are chelating. The 16-membered rings noted for the $[Zn(S_2COEt)_2]_n$ structure persist, but these are now linked only in one dimension, resulting in the formation of a ribbon. Introducing branching in the xanthate-bound R group results in a new motif where the ratio of bridging to chelating xanthate ligands is now 1:1 [12]. The resultant structure, illustrated in Figure 7(c), is cyclotetrameric, being constructed about an isolated 16-membered $[ZnSCS]_4$ ring. Systematic crystallographic studies on the binary nickel(II) [13] and tellurium(II) [14] xanthates as well as theoretical calculations on the uncoordinated xanthate ligands [15] indicate that there are no differences in the coordinating abilities of the xanthate ligands as R is varied, and so, the increase in the steric profile of the xanthate ligand is correlated with the reduction in supramolecular association in these structures. Next, attention is directed to adducts of zinc dithiophosphates.

5. BIPYRIDINE ADDUCTS OF ZINC DITHIOPHOSPHATES

A recent study of the principles directing the formation of polymers and the topology of the resultant chains has shown a definite influence, that is, steric, of remote R substituents in compounds of the general formula, $\{Zn[S_2P(OR)_2]_2(N{\cap}N)\}_n$, where $R = i\text{-Pr}$

and Cy and N∩N is $NC_5H_4C_5H_4N$ (4,4′-bipyridine), $NC_5H_4CH_2CH_2C_5H_4N$ (1,2-bis[4-pyridyl]ethane) and $NC_5H_4C(H){=}C(H)C_5H_4N$ (1,2-bis[4-pyridyl]ethylene) [16]. Whereas the structure of $\{Zn[S_2P(O\text{-}i\text{-}Pr)_2]_2(NC_5H_4C_5H_4N)\}_n$ exists as a zigzag polymer [17], the putative R = Cy analog could not be isolated owing to impossible steric clashes between the cyclohexyl substituents. In fact, only the dimeric $\{Zn[S_2P(OCy)_2]_2\}_2[NC_5H_4C(H){=}C(H)C_5H_4N)]$ species, where steric clashes are minimized, could be isolated from solutions containing the respective precursors

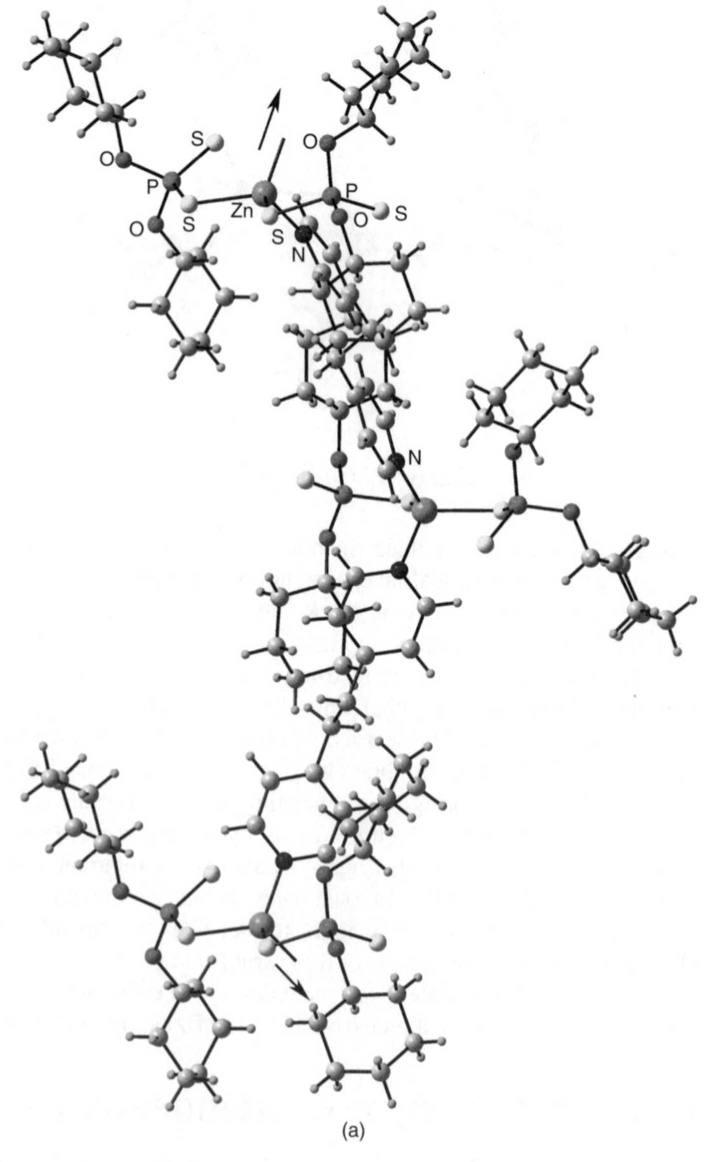

Figure 8. Portions of the polymeric structures of (a) $\{Zn[S_2P(OCy)_2]_2(NC_5H_4CH_2CH_2C_5H_4N)\}_n$ and (b) $\{Zn[S_2P(OCy)_2]_2(NC_5H_4C(H){=}C(H)C_5H_4N)\}_n$.

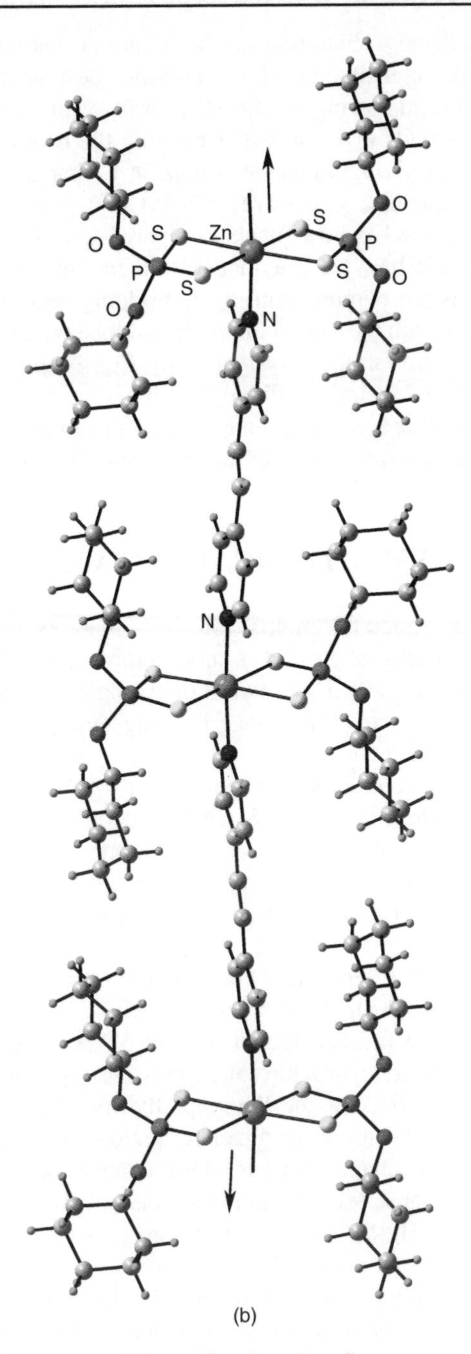

Figure 8. (*continued*)

in different ratios [17]. When the distance between the nitrogen donor atoms of the bipyridine ligands was sufficient to allow for polymer formation, polymeric chains for the R = Cy species could be formed. As illustrated in Figure 8(a) for $\{Zn[S_2P(OCy)_2]_2(NC_5H_4CH_2CH_2C_5H_4N)\}_n$, a zigzag polymeric chain is formed for this structure. What is immediately obvious from this figure is that significant steric clashes already exist between the constituents of the polymer despite the fact that there is a kink in each bridging ligand, owing to the ethyl bridge, that reduces the interactions between successive $Zn[S_2P(OCy)_2]_2$ entities. Changing the nature of the bridging ligand in this system to 1,2-bis[4-pyridyl]ethylene results in a change of the topology of the polymeric chain. In this structure, successive $Zn[S_2P(OCy)_2]_2$ molecules are forced closer together as the bridging ligand is now planar compared with the ethyl analog. So as to avoid unfavorable steric clashes, the polymer straightens up into the chain represented in Figure 8(b). Thus, it is the combination of the bridging requirements of the bridging bipyridine ligands coupled with the steric bulk of the cyclohexyl groups that determines the topology of polymer formation in these systems [16]. Consistent with the above findings, in the analogous adducts of cadmium dithiophosphates, with longer Cd–S and Cd–N bonds, steric considerations are no longer an issue in dictating polymer formation [18]. The next structures to be described are the binary mercury dithiocarbamates.

6. BINARY MERCURY DITHIOCARBAMATES

There are basically four structural motifs for the structures of the general formula $Hg(S_2CNR_2)_2$ [9]. The majority of structures adopt a dimeric motif shown in Figure 9(a) for the R = Et compound [19, 20]. Here, two dithiocarbamate ligands are chelating (2.5 and 2.6 Å) and the other two are tridentate, chelating one mercury atom (2.4 and 3.1 (dashed bond) Å) and at the same time bridging to a symmetry related mercury atom (2.7 Å). The resultant dimer is constructed about a centrosymmetric Hg_2S_2 ring. The other major structural form adopted for the $Hg(S_2CNR_2)_2$ compounds is, ostensibly, isolated with highly distorted tetrahedral geometries about the mercury atom. However, a close inspection of the crystal lattices for these compounds reveals a salient point. That is, while there are no conventional intermolecular Hg$\cdots$S interactions of note, the molecules are orientated so as to *potentially* form such Hg$\cdots$S interactions, leading to dimer formation, as seen for $Hg(S_2CNEt_2)_2$ in Figure 9(a). For example, in the structure of $Hg(S_2CN\text{-}i\text{-}Bu_2)_2$ [22], shown in Figure 9(b), there are indeed Hg$\cdots$S interactions between centrosymmetric pairs, but these are considered long at 3.7 Å, especially when compared to the 2.7 Å found in the dimeric structure of $Hg(S_2CNEt_2)_2$ shown in Figure 9(a). It is proposed that it is the steric bulk of the R groups that preclude closer approach of the $Hg(S_2CN\text{-}i\text{-}Bu_2)_2$ molecules to allow for dimer formation. While a consistent theme has been developed so far in this chapter, that is, steric control over molecular aggregation, the structure illustrated in Figure 9(c) shows that this is not always the only guiding principle in these systems. The $Hg(S_2CNEt_2)_2$ compound crystallizes into two polymorphs, one already discussed and illustrated in Figure 9(a), and the other that is illustrated in Figure 9(c). The coordination geometry for the mercury atom in the second polymorph of $Hg(S_2CNEt_2)_2$ is based on a square planar arrangement of four sulfur atoms, and above and below this plane are supramolecular Hg$\cdots$S interactions leading to a loosely associated chain [19, 21]. A consistent feature in the structures of the two polymorphs of $Hg(S_2CNEt_2)_2$ is the presence of Hg$\cdots$S interactions, but the resultant supramolecular structures are distinct, being based on (4 + 1) and (4 + 2) mercury to sulphur contacts. The addition of hydrogen-bonding functionality, not surprisingly, results in the adoption of a different structure for compounds of the general formula $Hg(S_2CNR_2)_2$ [9].

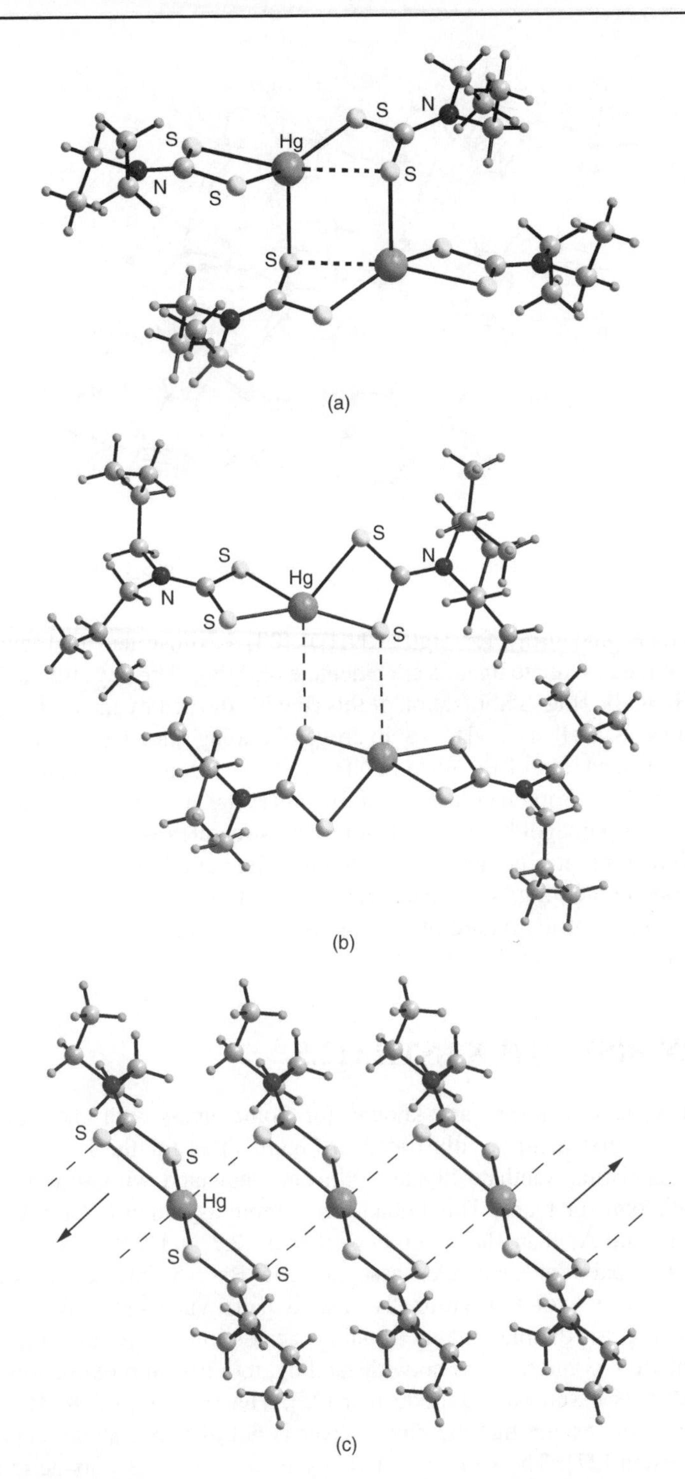

Figure 9. Molecular structures of (a) $[Hg(S_2CNEt_2)_2]_2$; (b) $Hg(S_2CN\text{-}i\text{-}Bu_2)_2$ and (c) a polymorphic form of $Hg(S_2CNEt_2)_2$.

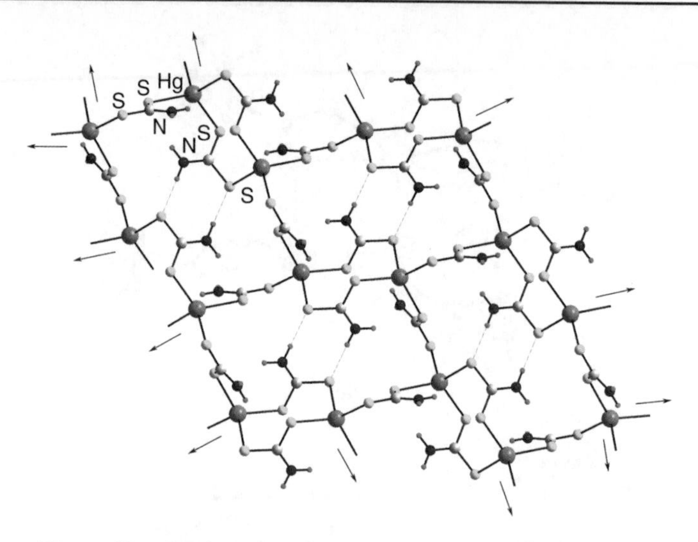

Figure 10. Portion of the layer structure of $[Hg(S_2CNH_2)_2]_n$.

The supramolecular array for $Hg(S_2CNH_2)_2$ [23] is illustrated in Figure 10. In this structure, all dithiocarbamate ligands are bidentate bridging. The key structural unit is a 24-membered $[HgSCS]_6$ ring. Stabilization of this ring is afforded by internal eight-membered $[HgSCS]_2$ rings as well as N–H···S hydrogen-bonding interactions, again leading to eight-membered rings, but of the form $[CNH\cdots S]_2$. A recent contribution describing the importance of N–H···S interactions in stabilizing crystal structures of related organomercury [24] and phosphinegold(I) [25] 1,1-dithiolate structures is available.

From the foregoing, it is perhaps clear that a principle guiding the nature of supramolecular association is emerging. The final series of structures reinforces this hypothesis but, importantly, shows that the nature of supramolecular association can be crucial in determining chemical and physical properties.

7. BINARY BISMUTH XANTHATES

Two distinct structural forms are known for compounds with the general formula $Bi(S_2COR)_3$. The first of these, illustrated in Figure 11(a) for the R = Me species [26], features three chelating xanthate ligands with one, indicated with an α in Figure 11(a), defining a pseudomirror plane. This ligand forms more asymmetric Bi–S bond distances, that is, 2.6 and 3.0 Å, than the other two, that is, 2.7 and 2.9 Å and 2.8 and 3.0 Å. Centrosymmetric pairs of molecules associate via Bi···S interactions of 3.4 Å, and these pairs associate with translationally related pairs via weaker Bi···S interactions of 3.6 Å leading to double chains running across the page in Figure 11(a). The intermolecular Bi···S interactions notwithstanding, to a first approximation, the structure of $Bi(S_2COMe)_3$ is based on a dimeric pair [26]. The structure of the R = Et analogue is virtually the same except that the dimeric pair is not disposed about a crystallographic center of inversion [27]. The structure of R = *i*-Pr derivative is polymeric as illustrated in Figure 11(b). The polymeric chain lies on a mirror plane and the repeat unit features two chelating xanthate ligands (2.7 and 2.9 Å) and a tridentate xanthate ligand forming

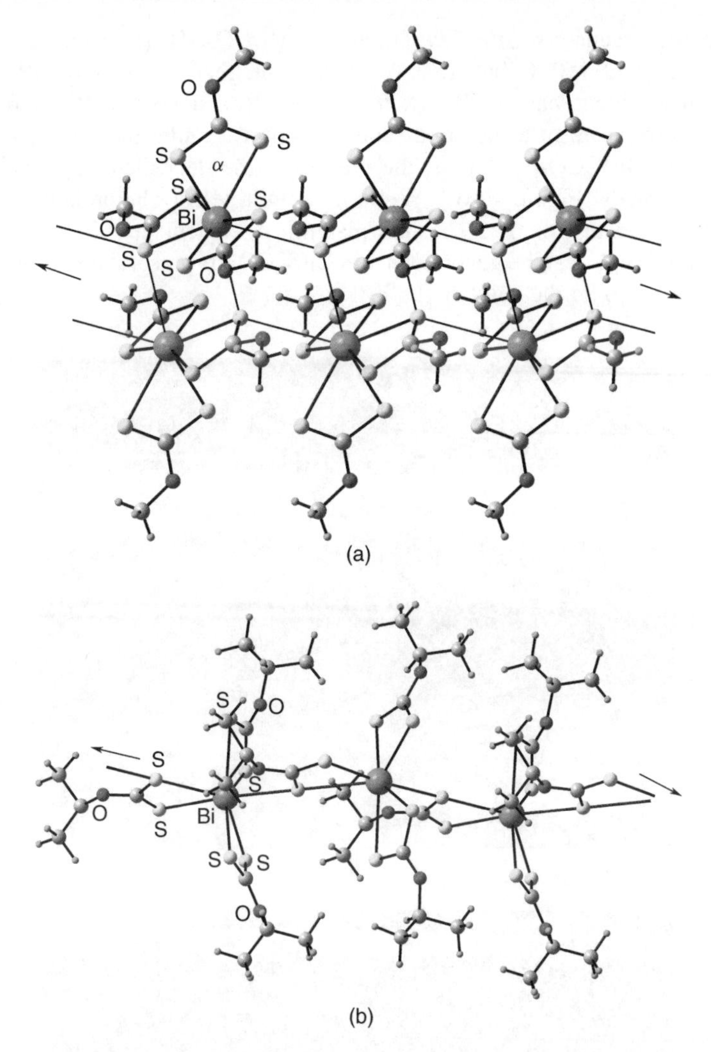

Figure 11. (a) Supramolecular association in the structure of $Bi(S_2COMe)_3$ and (b) polymeric structure of $[Bi(S_2CO\text{-}i\text{-}Pr)_3]_n$.

a chelate (2.8 and 3.2 Å) and a bridge (2.8 Å) to a neighboring repeat unit [28]. The differences between the two structural motifs also arises owing to steric effects in that the presence of the bulky *i*-propyl groups precludes dimerization as found for the smaller congenors, but owing to the propensity of the bismuth atom to increase its coordination number, a polymeric array is found instead. Over and above being simply another example of steric control influencing molecular aggregation/crystal structure, the adoption of different structural motifs for the $Bi(S_2COR)_3$ compounds appears to influence their utility as precursors for nanoparticle generation [29].

(Nano)crystals of Bi_2S_3 can be generated from the controlled decomposition of the $Bi(S_2COR)_3$, via both solvothermal methods (e.g., refluxing in ethylene glycol solution) and chemical vapor deposition (on a silicon substrate) [29]. Thermal gravimetric analyses indicate that the precursors $Bi(S_2COEt)_3$ and $[Bi(S_2CO\text{-}i\text{-}Pr)_3]_n$ each decompose in a single step at around 150 °C but give very different, that is, in morphology and size, decomposition products, namely Bi_2S_3. In Figure 12(a), deposited Bi_2S_3 from dimeric $Bi(S_2COEt)_3$ is in the form of nanorods and by contrast, under the same conditions, but using polymeric $[Bi(S_2CO\text{-}i\text{-}Pr)_3]_n$ as the precursor, the deposited Bi_2S_3 is in the form of microsized crystals (Figure 12(b)). While no definitive correlation between the nature of the precursor and that of the deposited material is evident, it is interesting that such different forms of Bi_2S_3 can be generated from $Bi(S_2COR)_3$ precursors that adopt different supramolecular arrays in the solid state [29].

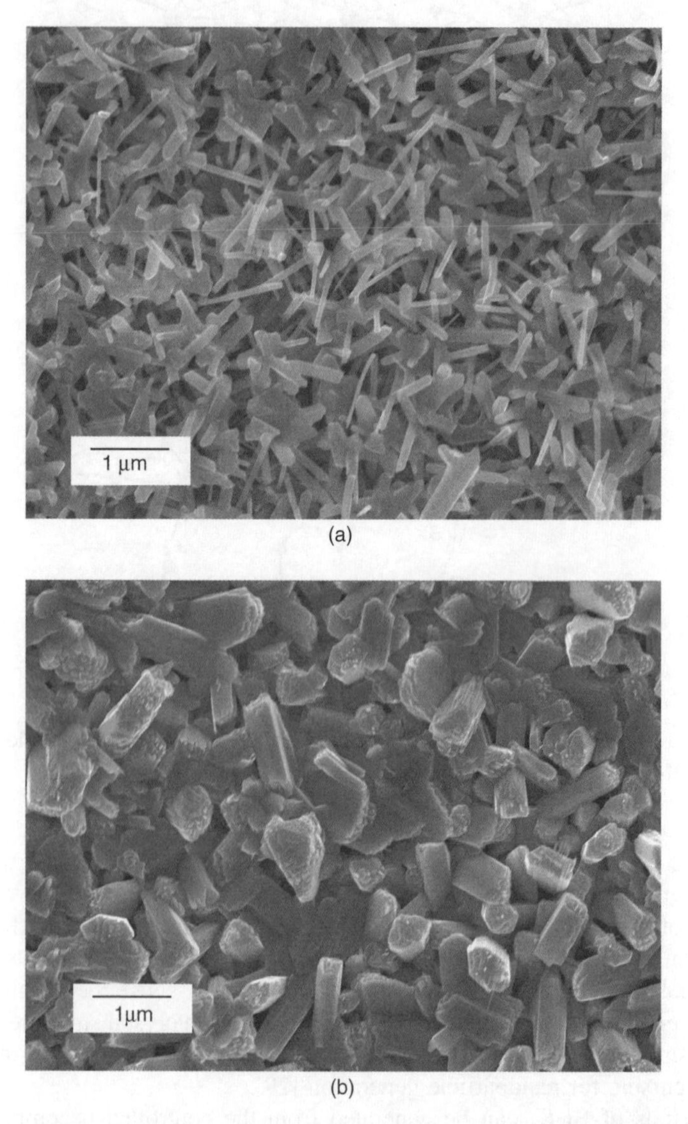

Figure 12. Morphology of Bi_2S_3 particles deposited by chemical vapor deposition on silicon substrates from precursors (a) $Bi(S_2COEt)_3$ and (b) $Bi(S_2CO\text{-}i\text{-}Pr)_3$.

8. CONCLUSIONS AND OUTLOOK

From the foregoing description of coordination compounds involving organotin, zinc, mercury and bismuth centers with carboxylate and 1,1-dithiolate ligands, a design principle for crystal engineering in this systems is emerging. It appears, at least for the systems described, that it is possible to control the degree of supramolecular association by judicious choice of the metal-bound (in the case of tin) and ligand-bound organic substituents. Thus, the presence of bulky groups can preclude supramolecular association in the solid state and conversely, smaller groups can allow aggregation between individual molecules to form supramolecular aggregates of one form or another via secondary interactions. As demonstrated for the $Bi(S_2COR)_3$ compounds that were evaluated for their potential as precursors for Bi_2S_3 nanoparticle generation, different forms of Bi_2S_3 were deposited, leading to the question whether the nature of the precursor structure influences the nature of the deposited Bi_2S_3. The challenge of (molecular) crystal engineering, over and beyond the task of generating specific arrays of molecules, is to provide specialized technology in terms of new materials, and so on and it remains to be seen whether results obtained for the $Bi(S_2COR)_3$ system [29] will find general applicability.

In conclusion and from the preceding discussion, it appears likely that additional examples of supramolecular control of aggregation, exerted by "steric factors", will be recognized and/or generated, thereby adding another tool to the armory of the crystal engineer that can be exploited, especially, in circumstances where conventional intermolecular forces, such as hydrogen bonding, are absent.

9. ACKNOWLEDGMENTS

It is with pleasure that the author thanks students and colleagues who have contributed to this work – their names can be found in the reference listing.

REFERENCES

1. (a) N. W. Alcock, *Adv. Inorg. Chem. Radiochem.*, **15**, 1–58 (1972); (b) N. W. Alcock, *Bonding and Structure: Structural Principles in Inorganic and Organic Chemistry*, Ellis Horwood, New York, 1990; (c) I. Haiduc and F. T. Edelmann, *Supramolecular Organometallic Chemistry*, Wiley-VCH, Weinheim, 1999; (d) I. Haiduc, 'Secondary Bonding' in (Eds. J. L. Atwood and J. Steed), *Encyclopedia of Supramolecular Chemistry*, Marcel Dekker Inc., New York, 1215–1224, 2004.
2. D. Dakternieks, A. Duthie, D. R. Smyth, C. P. D. Stapleton and E. R. T. Tiekink, *Organometallics*, **22**, 4599–4603 (2003).
3. T. P. Lockhart and F. Davidson, *Organometallics*, **6**, 2471–2478 (1987).
4. M. A. Buntine, F. J. Kosovel and E. R. T. Tiekink, *CrystEngComm*, **5**, 331–336 (2003).
5. (a) E. R. T. Tiekink, *Appl. Organomet. Chem.*, **5**, 1–23 (1991); (b) E. R. T. Tiekink, *Trends Organomet. Chem.*, **1**, 71–116 (1994).
6. R. Willem, I. Verbruggen, M. Gielen, M. Biesemans, B. Mahieu, T. S. Basu Baul and E. R. T. Tiekink, *Organometallics*, **17**, 5758–5766 (1998).
7. (a) P. G. Harrison, K. Lambert, T. J. King and B. Majee, *J. Chem. Soc., Dalton Trans.*, 363–369 (1983); (b) T. S. Basu Baul and E. R. T. Tiekink, *Z. Kristallogr.*, **211**, 489–490 (1996).
8. S. W. Ng, C. Wei and V. G. Kumar Das, *J. Organomet. Chem.*, **345**, 59–64 (1988).
9. E. R. T. Tiekink, *CrystEngComm*, **5**, 101–113 (2003).

10. C. S. Lai, Y. X. Lim, T. C. Yap and E. R. T. Tiekink, *CrystEngComm*, **4**, 596–660 (2002).
11. T. Ito, *Acta Crystallogr.*, **B28**, 1697–1704 (1972).
12. T. Ikeda and H. Hagihara, *Acta Crystallogr.*, **21**, 919–927 (1972).
13. M. J. Cox and E. R. T. Tiekink, *Z. Kristallogr.*, **214**, 242–250 (1999).
14. M. J. Cox and E. R. T. Tiekink, *Z. Kristallogr.*, **214**, 584–590 (1999).
15. M. A. Buntine, M. J. Cox, Y. X. Lim, T. C. Yap and E. R. T. Tiekink, *Z. Kristallogr.*, **218**, 56–61 (2003).
16. C. S. Lai, S. Liu and E. R. T. Tiekink, *CrystEngComm*, **6**, 221–226 (2004).
17. L. A. Glinskaya, V. G. Shchukin, R. F. Klevtsova, A. N. Mazhara and S. V. Larionov, *Zh. Struckt. Khim.*, **41**, 772–780 (2000).
18. C. S. Lai and E. R. T. Tiekink, *CrystEngComm*, **6**, 593–605 (2004).
19. H. Iwasaki, *Acta Crystallogr.*, **B29**, 2115–2124 (1973).
20. C. S. Lai and E. R. T. Tiekink, *Z. Kristallogr. NCS*, **217**, 593–594 (2002).
21. P. C. Healy and A. H. White, *J. Chem. Soc., Dalton Trans.*, 284–287 (1973).
22. M. J. Cox and E. R. T. Tiekink, *Z. Kristallogr.*, **214**, 571–579 (1999).
23. C. Chieh and S. K. Cheung, *Can. J. Chem.*, **59**, 2746–2749 (1981).
24. C. S. Lai and E. R. T. Tiekink, *CrystEngComm*, **5**, 253–261 (2003).
25. S. Y. Ho and E. R. T. Tiekink, *Z. Kristallogr.*, **219**, 513–518 (2004).
26. M. R. Snow and E. R. T. Tiekink, *Aust. J. Chem.*, **40**, 743–750 (1987).
27. E. R. T. Tiekink, *Main Group Metal Chem.*, **17**, 727–736 (1994).
28. (a) B. F. Hoskins, E. R. T. Tiekink and G. Winter, *Inorg. Chim. Acta.*, **81**, L33–L34 (1984); (b) B. F. Hoskins, E. R. T. Tiekink and G. Winter, *Inorg. Chim. Acta.*, **99**, 177–182 (1985).
29. Y. W. Koh, C. S. Lai, A. Y. Du, E. R. T. Tiekink and K. P. Loh, *Chem. Mat.*, **15**, 4544–4554 (2003).

7

Incorporating Molecular Hosts into Network Structures

MICHAELE J. HARDIE
School of Chemistry, University of Leeds, Leeds, LS2 9JT, UK.

1. INTRODUCTION

Molecular hosts are single molecules that are capable of complexing other molecules to form a noncovalently bound host–guest complex. Typically, molecular hosts are cyclic molecules with an internal cavity where guest molecules are bound, often through electrostatic interactions. The internal cavity may be the center of a near-planar macrocycle, such as for the ubiquitous crown ethers; an enclosed space where the guest molecule is surrounded by the host in 3-D, such as in cryptands, sepulcrates and other cryptophanes; or the cavity of a cone- or bowl-shaped host molecule where the guest is contained within the molecular cavity but not completely surrounded by it, such as with cone-conformation calixarenes and other cavitands [1]. Host–guest complex formation with cavitand hosts is often promoted by the hydrophobic effect as the internal cavities of cavitands tend to be hydrophobic.

One of the central tenets of crystal engineering is that the ordering of molecules in the crystalline state can be directed or influenced by the chemist, and that this will produce materials with tuneable properties [2]. Molecular components can be manipulated into infinite framework or network structures by utilizing interactions controlled by dynamic equilibria such as hydrogen-bonding and other electrostatic interactions, and/or metal–ligand coordination. Many of the properties of crystal-engineered materials are zeolitic properties that rely on these materials having an ordered 2-D or 3-D framework structure with cavities or channels that can contain other molecules. The inclusion chemistry shown by such materials is frequently termed lattice-type inclusion and is often not site specific, and rather relies on the natural tendency for space to be filled. Such complexes are often termed *clathrates*. Applications of such materials include molecular separations [3], anion or guest exchange [4], and gas storage [5]. We are interested in

Frontiers in Crystal Engineering. Edited by Edward R.T. Tiekink and Jagadese J. Vittal

using molecular hosts as building components, or tectons, for infinite framework materials. This will impart the material with molecular recognition sites embedded into the framework structure and thus, open up the possibility of creating a single material displaying multiple inclusions modes.

Although examples are not plentiful, some of the archetypal molecular hosts have been organized into crystalline network structures. Hydrogen-bonding network structures of crown ethers, calixarenes, calixresorcinarenes, cyclodextrins or other molecular hosts have recently been reviewed [6], and can involve 1-D, 2-D or 3-D network structures. Notably, the lattice inclusion chemistry of calixresorcinarene hydrogen-bonded networks have been used for time-resolved crystallographic studies of the photochemical properties of guest molecules [6, 7]. Coordination polymers have also been reported [8–16]. Crown ethers can be tectons in simple 1-D chain structures [8], and more complicated network structures can be formed from calixarenes [9–15], and cucurbituril [16] molecular hosts. Most calixarene-based examples involve calix[4]arenes with attached functional groups that may bind to a metal ion such as sulfonate [9], phosphonate [12], phosphate [13], cyano [14], or quinone groups [15]. The larger sulfonated calix[5]-[10] and calix[6]arenes [11] can also form complicated coordination polymers with lanthanide metals. Host–guest interactions may also be employed to build up network structures involving calixarenes [17]. The molecular host that we are particularly interested in is the cyclic trimer, CycloTriVeratrylene (CTV) (**1**). As will be discussed in this chapter, CTV can be incorporated into hydrogen-bonded network structures or coordination polymers, and offers considerable scope for extension of the basic CTV building block into novel multifunctional ligands.

Unlike the flexible calixarenes, CTV has a relatively rigid bowl shape [1]. Despite having a bowl shape reminiscent of the smaller calixarenes, CTV itself is a poor host for small organic molecules and, until recently, examples of CTV inclusion chemistry were restricted to crystalline clathrate materials [18, 19]. There are two main phases of clathrates, both of which feature lattice-type guest molecules contained in channels created by the packing of roughly linear stacks of CTV molecules. There is also one example of a γ-phase known, where acetone is complexed between two CTV molecules which form a head-to-head capsule-like dimer around the acetone guest [20], and intracavity complexation of chlorinated guests has recently been reported [21]. In the past decade, the inclusion chemistry of CTV has been expanded to show intracavity complexation of large guest molecules such as fullerenes [22], *o*-carborane (**2**) [23], $[FeCp(arene)]^+$ species [24] and $[Na[2.2.2]cryptate]^+$ [25]. CTV can be a chelating ligand through its dimethoxy moieties, complexing Ag(I) [26] or group 1 metal ions such as Cs^+ [27]. It can also act as a π-type ligand forming organometallic complexes with transition metals [28]. These positively charged complexes are able to include small neutral and anionic guest

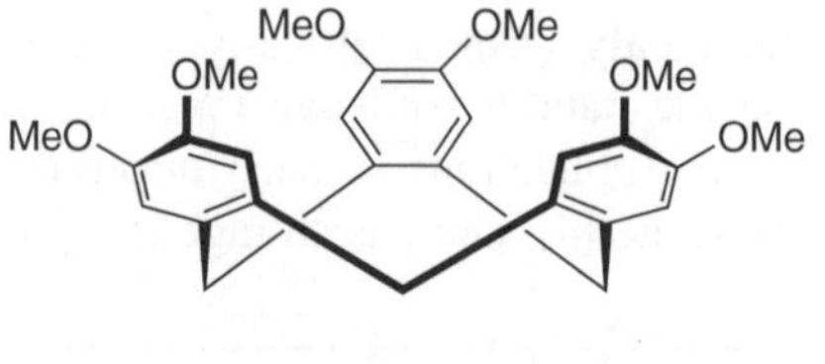

CTV, **1**

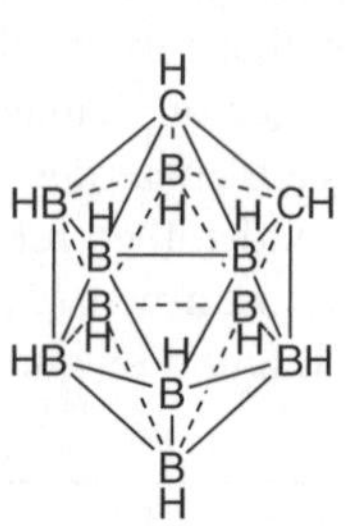

2

molecules. The inclusion chemistry of CTV is altered somewhat by the extension of the basic CTV core into an extended-arm cavitand or dimeric cryptophane [29]. Extended-arm CTV-based cavitands have been mooted for applications in liquid crystals [30], biological systems [31], ion-pair recognition [32] and optical resolution [33].

A little exploited attribute of CTV is the trigonal arrangement of dimethoxy moieties. These may act as hydrogen-bond acceptors, as has been seen in some CTV clathrates, and as potential ligation sites for hard metal ions. In terms of network structures, CTV may be a three-connecting center, although it is not usually a trigonal connector because of its rigid bowl shape. For example, in the tris-Pt(II) complex of a CTV-catechol analog, the angles between dimethoxy groups range from 87 to 89° [34]. This leads to up–down isomerism possibilities that do not occur with flat three-connecting centers, and the use of such a ligand may generate network structures with unusual topologies.

2. HYDROGEN-BONDED STRUCTURES WITH CTV

We have characterized a number of hydrogen-bonded structures with CTV as the hydrogen-bond acceptor. Hydrogen-bond donors are the acidic C–H group of carboranes, hydroxide or water O–H groups, and N–H donors. The types of structures formed have ranged from simple helical chains to very complicated low-symmetry 3-D networks. In general, the known host–guest chemistry of CTV is altered by its incorporation into a hydrogen-bonding network and intracavity complexation of small molecular guests is observed. Hydrogen-bonding modes for CTV include single X–H··· OMe interactions as well as bifurcated donor X–H··· $(OMe)_2$ modes.

2.1. Carborane C–H Donors

o-Carborane or 1,2-dicarbadodecaborane (**2**) is an icosahedral cage cluster compound that is known to be both a good hydrogen-bond donor [35] and a spherical guest molecule for molecular [36] and supramolecular [37] hosts including CTV [23]. In the complex (*o*-carborane)$_2$(CTV) (**3**), carborane molecules act in both capacities [38]. There are two crystallographically distinct *o*-carborane molecules, one of which is included as an intra-cavity guest molecule to form a ball-and-socket assembly. The other carborane molecule is disordered with three vertices on one face having mixed carbon and boron character. Each of these C/B–H sites forms a bifurcated hydrogen bond to the methoxy groups of a CTV molecule, to give a puckered hexagonal 2-D hydrogen-bonded network (Figure 1). The hydrogen-bond interaction is out-of-plane, allowing the CTV to be a strict trigonal connecting center in this network. The networks pack together in a space-filling manner with π–π stacking interactions between CTV molecules of adjacent layers.

The complex (C_{70})(*o*-carborane)(CTV)(1,2-dichlorobenzene) (**4**) also features a hydrogen-bonding network between CTV and *o*-carborane. However, in this material, the carborane is crystallographically ordered and hydrogen bonds to only two symmetry equivalent CTV molecules [38]. Each CTV interacts with two carboranes forming an infinite helical chain, as represented in Figure 2. The crystals are chiral, despite being composed of achiral molecular components. Each CTV receptor site within the helices binds a molecule of C_{70} with π–π stacking interactions evident between them. The dichlorobenzene molecules within the crystal lattice fit snugly into the grooves of each

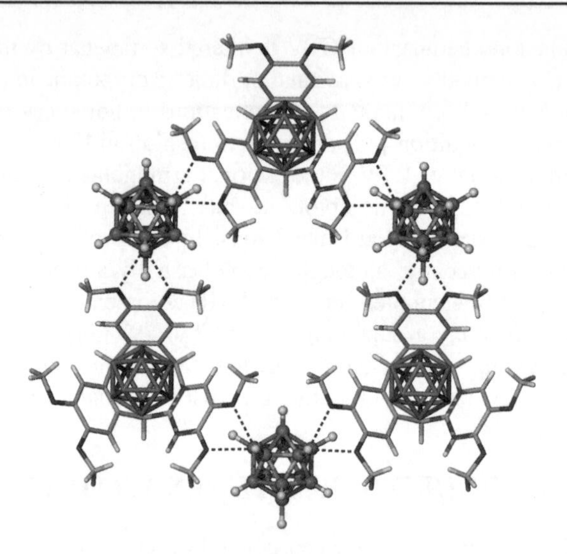

Figure 1. Section of the 2-D hexagonal hydrogen-bonded network structure of (*o*-carborane)$_2$ (CTV) **3** illustrating different *o*-carborane molecules acting as either H-bond donors or spherical guests.

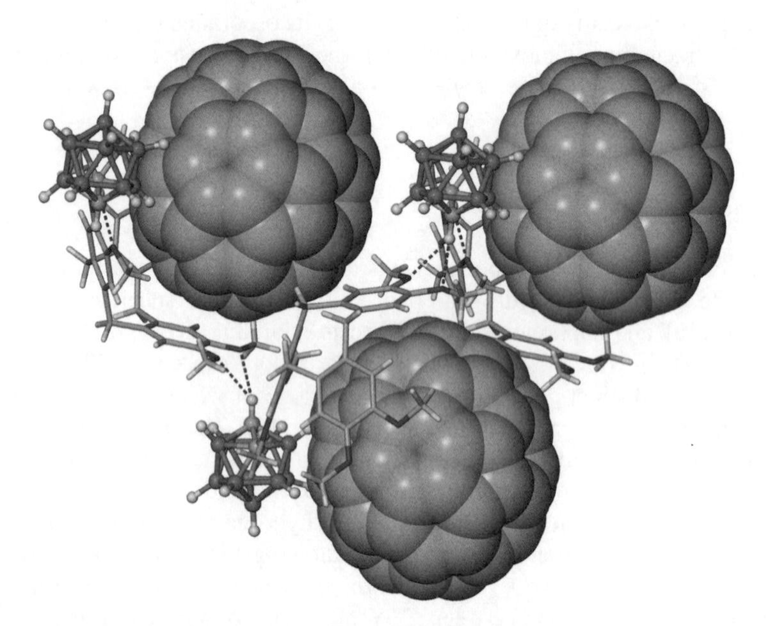

Figure 2. Helical hydrogen-bonded *o*-carborane-CTV chain of complex (C_{70})(*o*-carborane)(CTV) (1,2-dichlorobenzene) **4** with C_{70} guest molecules.

helix, with the plane of the solvent molecule lying normal to the direction of the helices. An isostructural toluene solvate is also known. A notable feature of this material is that crystalline complexes do not result from simple mixtures of CTV and fullerene C_{70}, unlike the case with fullerene C_{60} [22]. Unit cell measurements indicate that the higher fullerene C_{76} is likewise complexed within this helical structure [39].

2.2. N–H and O–H Donor Groups

More traditional hydrogen-bond donors, such as N–H and O–H groups, may also interact with CTV. In the complex $[(NH_2)_3C]\{(CH_3CN)(CTV)\}[Co(C_2B_9H_{11})_2]$ (**5**), a hydrogen-bonded network with N–H···OMe hydrogen bonding is found [6, 40]. The guanidinium cation forms a hydrogen bond to a methoxy group of a CTV through each of its three $-NH_2$ groups. CTV accepts hydrogen bonds from two guanidinium cations, forming a 1-D ladder structure (Figure 3). Each CTV cavity hosts a molecule of CH_3CN. The ladders pack together in the *ab* plane through $\pi-\pi$ stacking between the two of the arene rings on each CTV. Layers of the hydrogen-bonded ladders are separated in the *c* direction by layers of $[Co(C_2B_9H_{12})_2]^-$ anions, which form weak C–H···π hydrogen bonds to CTV molecules.

The complex $[(DMF)(CTV)]_2(H_2O)_4$(*o*-carborane) (**6**), where DMF = N, N'-dimethylformamide, features water as a hydrogen-bond donor within a 2-D network (Figure 4(a)) [41]. A square tetramer of hydrogen-bonded water molecules hydrogen bonds to six CTV molecules through O–H···OMe interactions. Each CTV hydrogen bonds to three water tetramers to create an unusual 2-D network of 3,4-connectivity (Figure 4(b)). These networks pack within the crystal lattice via π stacking between aromatic rings of the CTV molecules with *o*-carborane as a lattice-type guest molecule, contained in hydrophobic channels. The host–guest behavior of the CTV shows DMF as an intracavity guest, which is quite distinct to the host–guest behavior of native CTV with similar small organic guests, where clathrate materials are formed.

A series of complexes with very similar geometric features to **6** are known that feature a combination of coordination interactions to the dimethoxy groups of the CTV and hydrogen-bonding interactions with water or hydroxide as the hydrogen-bond donor [41, 42]. These complexes may be formed with either *o*-carborane or the sterically similar icosahedral anion $(CB_{11}H_{12})^-$ as templating molecules, and are discussed in more detail on the coordination polymers section.

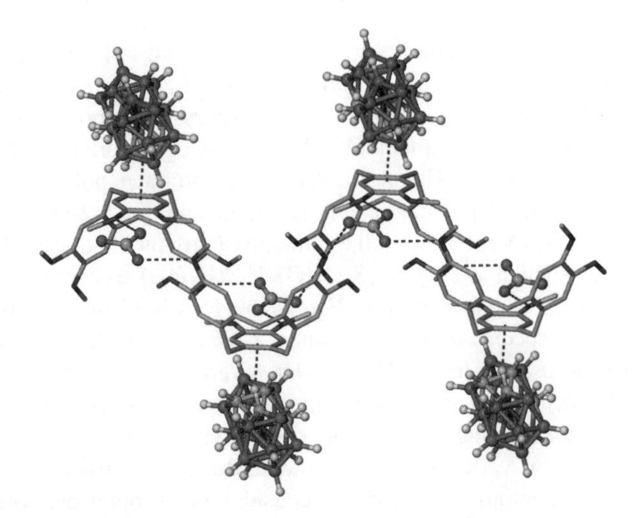

Figure 3. 1-D hydrogen-bonded ladder structure formed by $[(NH_2)_3C]^+$ and CTV in the complex $[(NH_2)_3C]\{(CH_3CN)(CTV)\}[Co(C_2B_9H_{11})_2]$ (**5**). $[Co(C_2B_9H_{11})_2]^-$ anions form C–H···π hydrogen bonds to an arene face of each CTV.

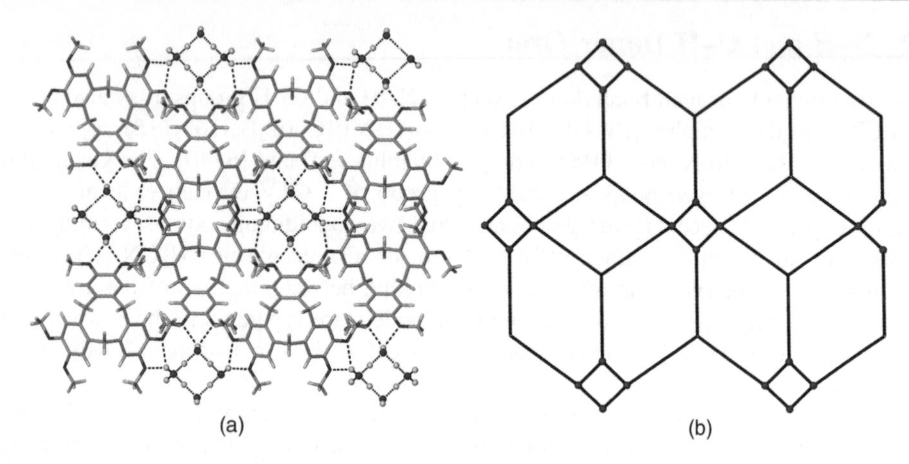

Figure 4. Structure of $[(DMF)(CTV)]_2(H_2O)_4$(*o*-carborane) (**6**). (a) 2-D hydrogen-bonded network formed by water and CTV; (b) connectivity or topological diagram showing three-connecting CTV centers and three- and four-connecting waters (shown as spheres).

Coordinated water tends to be a better hydrogen-bond donor than free water as the metal–ligand interaction effectively increases the polarization of the O–H bond. Furthermore, metal aquo complexes allow for water hydrogen-bond donors to be imparted with the steric constraints inherent in the geometry of the complex. We have used six-, eight- and nine-coordinate metal aquo complexes as hydrogen-bond donors combined with CTV acceptors, which has lead to the formation of complicated 3-D network structures.

In the complex $[Sr(H_2O)_8][(CH_3CN)(CTV)]_4(H_2O)_4[Co(C_2B_9H_{11})_2]_2$ (**7**), the primary hydrogen-bond donor is the complex ion $[Sr(H_2O)_8]^{2+}$ with a triangular dodecahedral geometry [43]. The $[Sr(H_2O)_8]^{2+}$ ion hydrogen bonds to four water molecules to give a $\{[Sr(H_2O)_8]^{2+}/(H_2O)_4\}$ assembly, which acts as an expanded hydrogen-bond donor assembly. This hydrogen bonds to 12 CTV molecules. Each CTV molecule hydrogen bonds to three $\{[Sr(H_2O)_8]^{2+}/(H_2O)_4\}$ assemblies to form a highly unusual 3,12-connected network shown schematically in Figure 5(a). A further feature of the structure is the back-to-back stacking of CTV molecules into tetrameric clusters with approximately tetrahedral geometries. This packing arrangement of CTV is notably similar to the back-to-back packing of 12 *p*-sulfonatocalix[4]arene molecules to produce polyhedral clusters [44], and back-to-back stacking of phosphonated calixarenes within a 2-D coordination polymer [12]. Within each $[CTV]_4$ tetramer, the dimethoxy groups of adjacent CTV molecules are roughly aligned, so that a single $\{[Sr(H_2O)_8]^{2+}/(H_2O)_4\}$ assembly hydrogen bonds to two CTV molecules at these points. For each $[CTV]_4$ cluster, there are six such $\{[Sr(H_2O)_8]^{2+}/(H_2O)_4\}$ species arranged in a near perfect octahedron; see Figure 5(b) for a projection down the pseudo threefold axis. The overall arrangement of the $Sr_6[CTV]_4$ assembly is a deformed adamantoid cage, and the overall network topology shown in Figure 5(a) is of vertex-sharing adamantoid cages, where six cages radiate from each vertex. Other examples of vertex-sharing adamantoid network structures involve sharing between only two adamantoid cages [45]. A considerably simpler network structure can be conceptualized by considering the tetrameric cluster of back-to-back CTV molecules as a single six-connecting center. As each $\{[Sr(H_2O)_8]^{2+}/(H_2O)_4\}$ entity hydrogen bonds to six $[CTV]_4$ clusters, they are also six-connecting centers. Hence, the simplified view

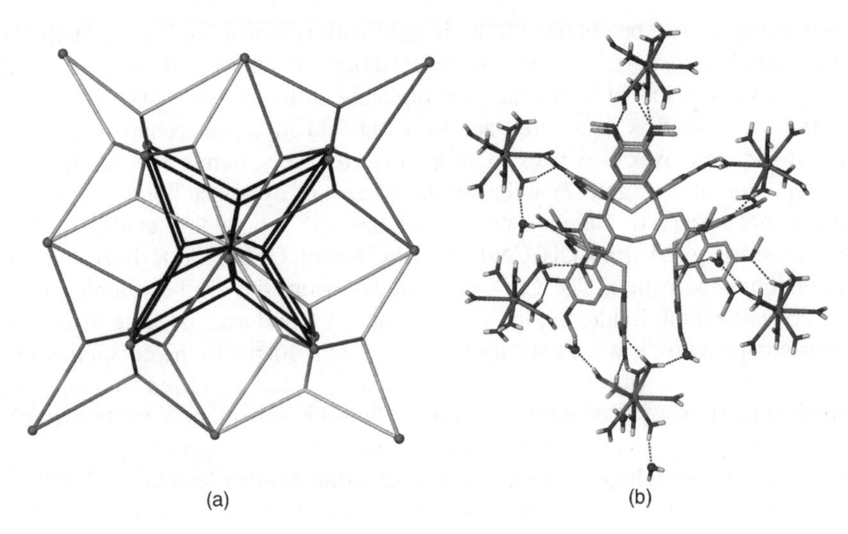

Figure 5. Structure of $[Sr(H_2O)_8][(CH_3CN)(CTV)]_4(H_2O)_4[Co(C_2B_9H_{11})_2]_2$ (**7**). (a) topology diagram showing hydrogen-bond connections between $\{[Sr(H_2O)_8{}^{2+}/(H_2O)_4\}$ clusters (Sr positions shown as spheres) and CTV (shown as three-connecting center) giving a 3,12-connected 3-D network. Six adamantane-shaped prisms that share a single Sr vertex are shown, and (b) detail of one adamantane-shaped prism composed of a tetrahedral cluster of four CTV molecules packed back-to-back, surrounded by six $\{[Sr(H_2O)_8{}^{2+}/H_2O)_4\}$ clusters forming an octahedron.

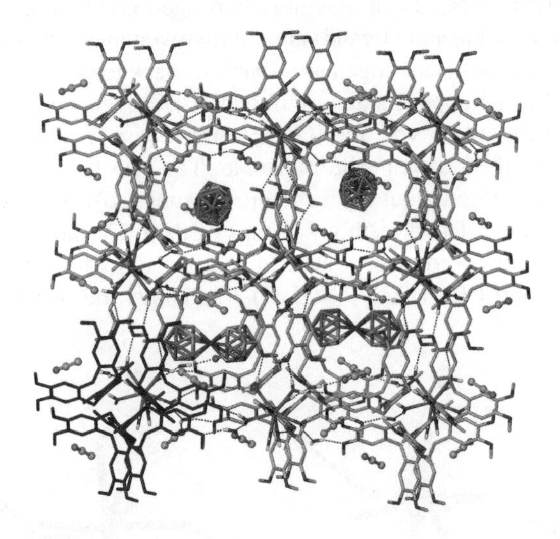

Figure 6. Structure of $[Sr(H_2O)_8][(CH_3CN)(CTV)]_4(H_2O)_4[Co(C_2B_9H_{11})_2]_2$ (**7**). Full packing diagram with one $[CTV]_4$ tetramer highlighted in black.

of the network is of an α-Po 4^8 topology with alternating $[CTV]_4$ and Sr centers. The rectangular channels are filled by $[Co(C_2B_9H_{11})_2]^-$ anions (Figure 6). Each CTV also hosts a molecule of acetonitrile.

Complexes that are isostructural with **7** may be obtained with other 2+ metal ions including the other group 2 metal ions Ca^{2+} and Ba^{2+}, and the transition metal Fe^{2+} [46].

The Fe(II) complex can be obtained in high yield with composition $[Fe(H_2O)_6][(CH_3CN)(CTV)]_4(H_2O)_4[Co(C_2B_9H_{11})_2]_2$ (**8**). The $[Fe(H_2O)_6]^{2+}$ complex in **8** has the same charge as the $[Sr(H_2O)_8]^{2+}$ ion in **7** but has a completely different geometry. The octahedral $[Fe(H_2O)_6]^{2+}$ ion in **8** is disordered across a fourfold inversion center to give a complex ion disordered over two sites in order to effectively mimic the geometry of the $[Sr(H_2O)_8]^{2+}$ ion of **7** (Figure 7) and a similar pattern of hydrogen bonds and host–guest interactions are observed. Notably, crystals of the Ca^{2+}-containing analog of **7** can be obtained in small yield from CH_3CN/H_2O solutions of CTV, $[Co(C_2B_9H_{11})_2]^-$ and an organic or inorganic cation. The Ca^{2+} is abstracted from the CaO-containing glassware used for crystallization. Evidently, the structure of **7** is a preferred packing mode for these molecular components. It would appear that the system adapts to the crystallization mixture, and precipitates whatever component that will give this preferred structure, whether it is the ideal $[M(H_2O)_8]^{2+}$ cation from solution or environment, or a crystallographically disordered $[M(H_2O)_6]^{2+}$ cation that mimics the ideal metal ion geometry.

Metal aquo ions with higher charge and coordination number lead to more complicated hydrogen-bonding network structures as can be seen with the complex $[Eu(H_2O)_9]_{1.5}(CTV)_6(CH_3CN)_{5.5}(H_2O)_{7.5}[Co(C_2B_9H_{11})_2]_{4.5}$ (**9**) [47]. The structure of **9** is remarkably complicated, given that is composed of only small molecular building blocks, and has a unit cell volume >58,000 Å^3. There are two types of $[Eu(H_2O)_9]^{3+}$ cation within the structure occurring in 2:1 proportions, each with distorted capped triangular dodecahedral geometry. As for **7**, each metal aquo ion hydrogen bonds to four water molecules creating two distinct $\{[Eu(H_2O)_9]^{3+}/(H_2O)_4\}$ assemblies with differing spatial arrangements of the water molecules (Figure 8(a)). Each assembly hydrogen bonds to eight CTV molecules, and, unlike for **7**, some of the aquo ligands are not involved in any hydrogen-bonding interactions. The CTV molecules stack together as back-to-back tetramers with two structurally distinct types. One has an approximately tetrahedral arrangement of the CTV molecules as was seen in **7**. In the other type, the $[CTV]_4$ unit is less regular, having been pushed apart by the insertion of an acetonitrile molecule. The two types of $[CTV]_4$ assemblies exist in 2:1 proportions throughout the structure (Figure 8(b)). Each CTV molecule also hosts an acetonitrile guest. The dimethoxy groups of the CTV molecules accept hydrogen bonds from either free or coordinated water. In addition to hydrogen-bonding interactions between the $[CTV]_4$ and $\{[Eu(H_2O)_9]^{3+}/(H_2O)_4\}$ subunits, the two types of $[CTV]_4$ also

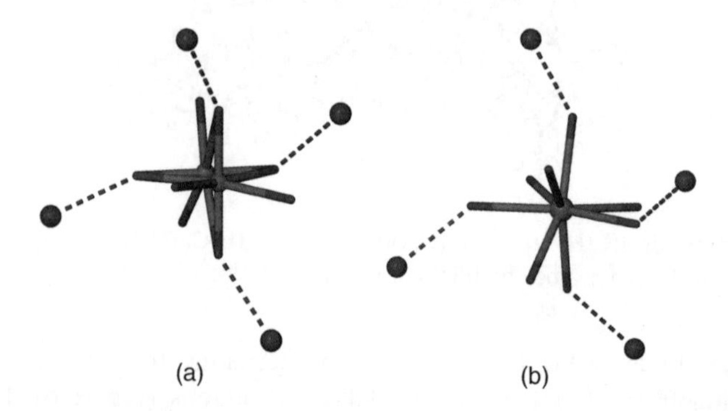

Figure 7. (a) Disordered $[Fe(H_2O)_6]^{2+}$ and four hydrogen-bonded water molecules from the structure of $[Fe(H_2O)_6][(CH_3CN)(CTV)]_4(H_2O)_4[Co(C_2B_9H_{11})_2]_2$ (**8**), which mimics the $\{[Sr(H_2O)_8(H_2O)_4\}^{2+}$ clusters of **7** shown in (b).

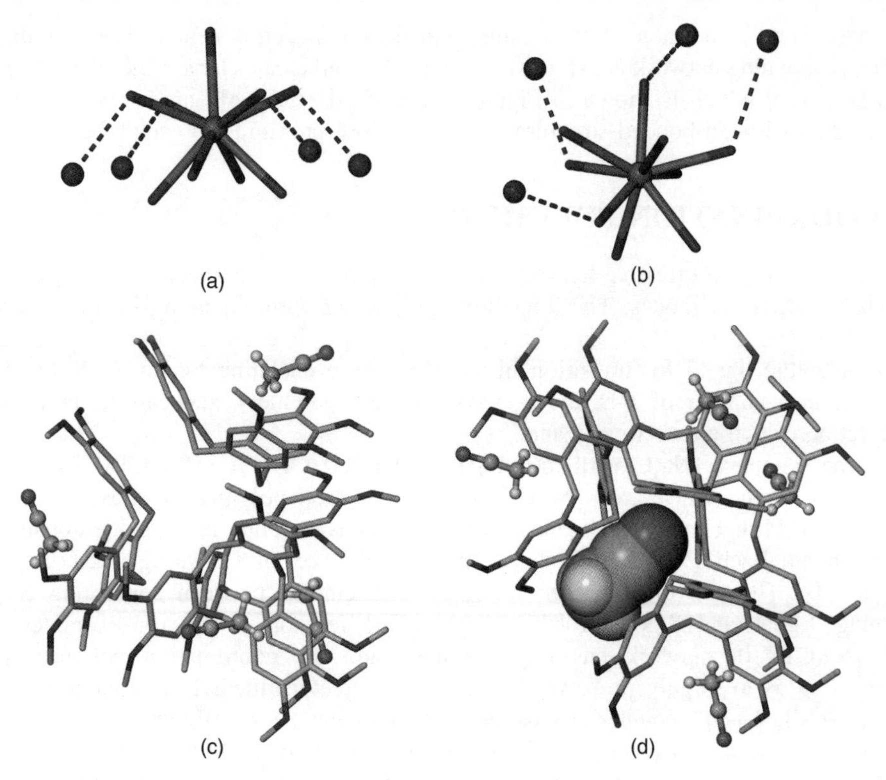

Figure 8. Components of the structure of $[Eu(H_2O)_9]_{1.5}(CTV)_6(CH_3CN)_{5.5}(H_2O)_{7.5}$ $[Co(C_2B_9H_{11})_2]_{4.5}$ (**9**). (a) and (b) The two distinct types of $\{[Eu(H_2O)_9]^{3+}/(H_2O)_4\}^{3+}$ clusters; (c) back-to-back tetramer of CTV molecules with CH_3CN guest molecules and (d) distorted $[CTV]_4$ tetramer with an additional CH_3CN (shown in space filling) inserted into the tetramer.

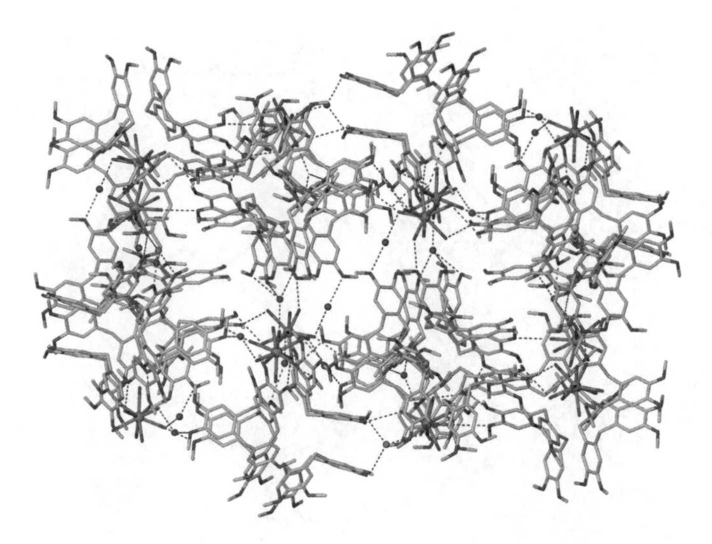

Figure 9. The hydrogen-bonded cage of **9** with anions and guest CH_3CN omitted. These cages tessellate in 3-D to form the overall network.

hydrogen bond to a water cluster acting as a three-connecting center. The overall 3-D hydrogen-bonding network is extremely complicated and creates large cages that tessellate in 3-D, one of which is shown in Figure 9. The $[Co(C_2B_9H_{12})]^-$ anions occupy spaces within the hydrogen-bonded structure and are in close proximity to each other.

3. COORDINATION POLYMERS

As well as being an effective hydrogen-bond acceptor, CTV can behave as a ligand in a number of different modes. The dimethoxy groups can form chelating interactions with Ag(I) [26] and group 1 metals [27, 41, 42, 48–50], while the arene faces can be η^5 organometallic ligand for transition metals [28]. As a chelating ligand, CTV has been a tecton in a number of 1-D and 2-D coordination polymers, and can act as a three-connecting or two-connecting ligand.

In the complex $[Na(CTV)][Co(C_2B_9H_{11})_2](CF_3CH_2OH)_{0.25}$ (**10**), CTV is a three-connecting ligand and a 2-D $[Na(CTV)]_\infty$ coordination polymer of unusual topology is generated [48]. Three CTV ligands chelate to each Na^+ center in a distorted octahedral environment. Each CTV bridges between three Na^+ centers, creating a 2-D network (Figure 10). The network has 4.8^2 topology with circuits between connecting centers forming 4-gons and 8-gons, for which there are few other reported examples [51]. The $[Na(CTV)]^+$ network has large cavities, and the coordination polymers pack together so as to largely preserve these cavities, creating methyl-lined cavities where the $[Co(C_2B_9H_{11})_2]^-$ counterions reside as lattice-type guests (Figure 11). Each CTV molecule acts as a host for a partially occupied molecule of guest CF_3CH_2OH.

A series of complexes with identical topology and very similar geometric features to the hydrogen-bonded network of **6** have been reported that feature a combination of

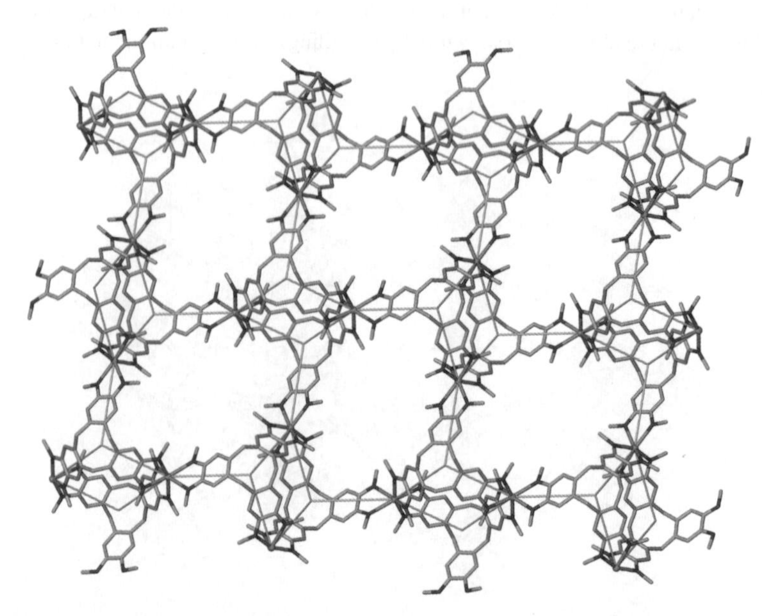

Figure 10. 2-D $[Na(CTV)]^+$ coordination polymer of $[Na(CTV)][Co(C_2B_9H_{11})_2](CF_3CH_2OH)_{0.25}$ (**10**). Connectivity diagram showing the 4.8^2 topology is superimposed as thin lines.

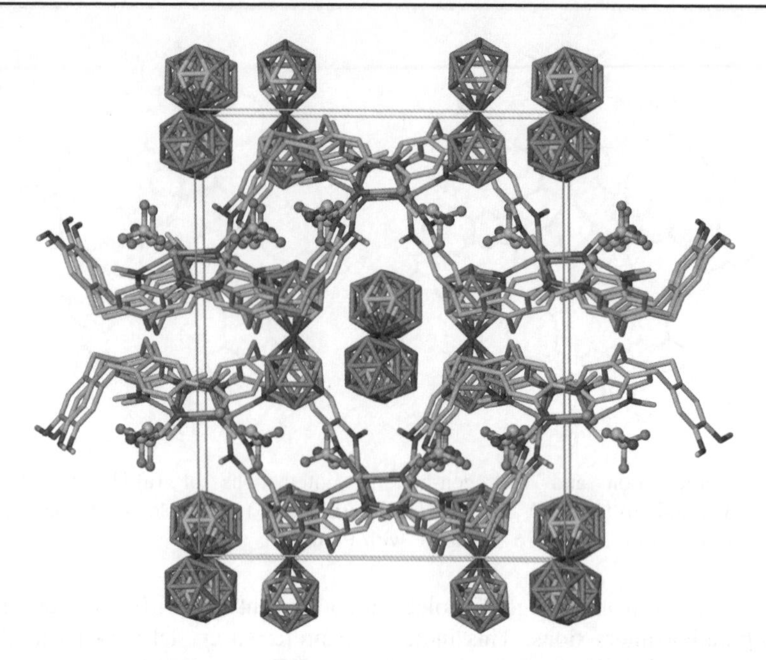

Figure 11. Unit cell diagram of complex **10** viewed down the *c*-axis, showing how the 2-D coordination polymers pack around $[Co(C_2B_9H_{11})_2]^-$ anions, and the guest CF_3CH_2OH molecules.

coordination interactions to the dimethoxy groups of the CTV and hydrogen-bonding interactions with water or hydroxide as the hydrogen-bond donor [41, 42]. These complexes may be formed with either *o*-carborane **2**, or the similar icosahedral anion $[CB_{11}H_{12}]^-$ as steric templating molecules. In the latter case, charge balance is achieved with some of the hydroxide ligands being replaced by water. An example is the complex $[K(OH)(CTV)(DMF)]_2(C_2B_{10}H_{12})$ (**11**) [41], for which the network structure is shown in Figure 12(a). In **11**, K^+ ions are six-coordinate with two chelating CTV ligands and two cis hydroxide ligands. All ligands are two-connecting and a $-K-\mu-(OH)_2-K-\mu-(CTV)_2-K-\mu-(OH)_2-K-$ coordinate chain is observed. Adjacent coordinate chains are hydrogen bonded together via the hydroxide protons forming a bifurcated hydrogen bond to the dimethoxy unit of CTV not involved in coordinate bonding. The overall topology and crystal packing is the same as that for **6**, with the $-K-\mu-(OH)_2-K-$core giving the same connectivity as the water tetramer of **6**. The Na^+ analogue achieves essentially the same result in a slightly different manner. In $[Na(CTV)_2(H_2O)_3(DMF)_2](CB_{11}H_{12})$ (**12**), a coordination chain is not formed [42]. Instead, each Na^+ is six coordinate with two chelating CTV ligands and two cis aquo ligands. The aquo ligands form hydrogen bonds to a water molecule (Figure 12(b)), and both the aquo ligands and this water form hydrogen bonds to adjacent CTV molecules to form the same network topology as for **6** and **11**. Similar complexes with Cs^+ and Rb^+ are also known [42]. In all these cases, DMF molecules act as intracavity guest molecules for the CTV hosts, and pseudopolymorphic complexes with trifluoroethanol or acetonitrile guest molecules in the CTV cavity have also been isolated.

As was seen with the hydrogen-bonded complexes **7** and **8**, and related materials that have the same structure despite containing different types of metal ion, essentially the same

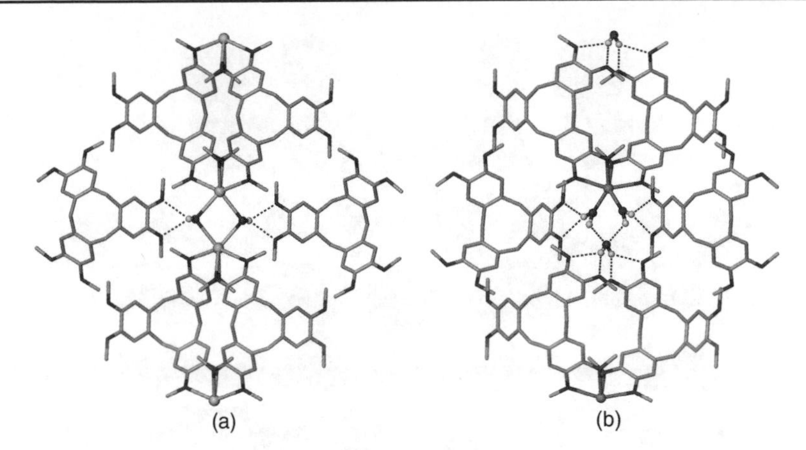

Figure 12. Coordination and hydrogen-bonding interactions of (a) $[K(OH)(CTV)(DMF)]_2(C_2B_{10}H_{12})$ (**11**) and (b) $[Na(CTV)_2(H_2O)_3(DMF)_2](CB_{11}H_{12})$ (**12**), which give an identical network structure to that of complex **6** (compare with Figure 4).

structure is obtained from a range of molecular components and different coordination and hydrogen-bonding interactions. This indicates a preferred crystal packing mode for the major components within these systems, namely, CTV and *o*-carborane or isostructural $[CB_{11}H_{12}]^-$.

The examples above indicate that the anion has a very strong influence over the overall crystalline assembly. The types of coordination polymers that can be generated can be varied by changing the counterion. The carborane anions $(CB_{11}H_{12})^-$ and $(CB_9H_{10})^-$ can be selectively halogenated at the B vertices meta and para to the C–H apex to give anions such as $(CB_{11}X_6H_6)^-$, X = Cl, Br [52], and $(CB_9Br_5H_5)^-$ [53], and these were employed with group 1 metals and CTV to generate coordination polymers.

Mixtures of CTV, $(CB_{11}H_6X_6)^-$ and group 1 metals in CH_3CN/CF_3CH_2OH yield crystalline materials with one of two types of closely related structures [49]. Use of the small Na^+ cation gives a chiral coordination chain in the complex $[Na(CTV)(H_2O)(CB_{11}H_6Cl_6)](CF_3CH_2OH)$ (**13**). The Na^+ has a distorted octahedral geometry with two chelating CTV ligands, a terminal aquo ligand and the $(CB_{11}H_6Cl_6)^-$ anion that coordinates through one chloro atom. This anion is directed into the molecular cavity of the CTV with the host–guest interaction augmented by a C–H$\cdots\pi$ hydrogen bond between the carborane and CTV. Each CTV is a two-connecting bridging ligand and a chiral coordinate chain is propagated (Figure 13). The terminal aquo ligand is a hydrogen-bond donor to a solvent CF_3CH_2OH molecule and an OMe group of an adjacent coordination chain, and overall, a 2-D network structure is generated. The network is 3-connected with a hexagonal 6^3 topology. An isostructural complex is found with $(CB_{11}H_6Br_6)^-$ as the counterion. While carborane anions are renowned to be weakly coordinating [52], there have been a handful of previously reported examples of halogenated carborane anions binding to group 1 metals [54] or transition metals [55] via their halo sites.

Using the larger Rb^+ or Cs^+ cations give essentially isostructural structures that are related to the structure of **13**. In $[Cs(CTV)(CB_{11}H_6Cl_6)(CH_3CN)]$ (**14**) [49], for instance, a 2-D coordination polymer is found with hexagonal 6^3 topology. In this case, each CTV is a three-connecting ligand binding to three Cs^+ centers, and each Cs^+ center is coordinated by three CTV ligands (Figure 14(a)). The network is topologically the same as

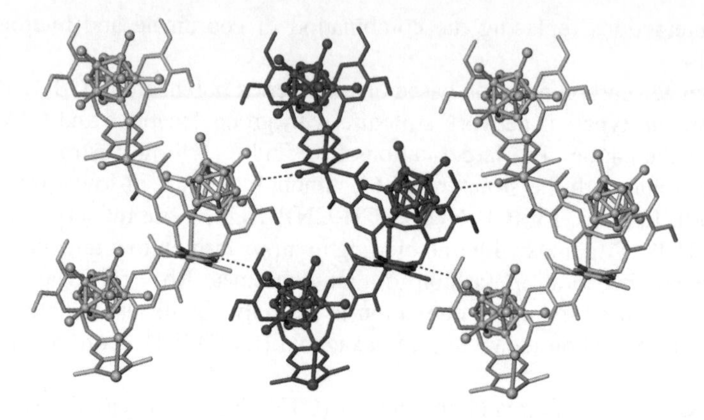

Figure 13. Structure of $[Na(CTV)(H_2O)(CB_{11}H_6Cl_6)](CF_3CH_2OH)$ (**13**). Coordination interactions generate a chiral chain with coordinated $(CB_{11}H_6Cl_6)^-$ anions as guest molecules for CTV. Three chains are shown (highlighted by different shading) and hydrogen bond together through $Na–OH_2 \cdots OMe$ interactions to form a 2-D network with distorted hexagonal topology.

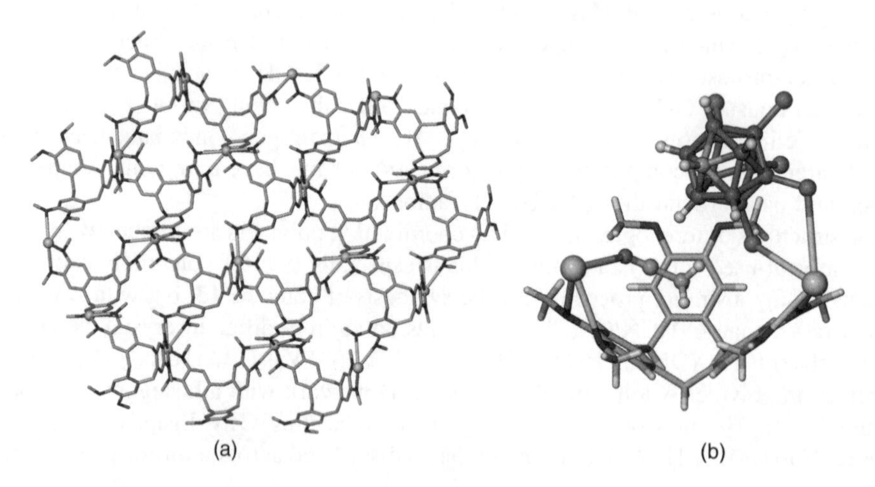

Figure 14. Structure of $[Cs(CTV)(CB_{11}H_6Cl_6)(CH_3CN)]$ (**14**). (a) $[Cs(CTV)]^+$ coordination polymer with distorted hexagonal topology similar to complex **13** (Figure 13), and (b) detail showing host–guest associations with coordinated CH_3CN as the primary guest and coordinated $(CB_{11}H_6Cl_6)^-$ anion perched above it.

that for **13**. However, there are differences in the host–guest behavior with the coordinated acetonitrile solvent becoming the principal guest molecule, and the carborane anion, which coordinates through two chloro atoms to the nine-coordinate metal center, is perched above it (Figure 14(b)). Analogous structures are found for the $(CB_{11}H_6Br_6)^-$ anion and in the complex $[Rb(CTV)(CB_{11}H_6Br_6)(H_2O)]$ [49]. Presumably, the high coordination numbers and longer interatomic distances to the metal center associated with Cs^+ and Rb^+ exclude the formation of structures isostructural with that of the Na^+ complex. A metal cation of intermediate size, K^+, gives a poorly defined and highly disordered intermediate structure, in the complex $[K(CTV)(CB_{11}H_6Cl_6)(CF_3CH_2OH)_{0.5}]$ (**15**). The structure of **15** is essentially isostructural with that of the Na^+ complex **13**, but with long

coordinate interactions replacing the combination of coordinate and hydrogen-bonding interactions [49].

The brominated carborane anion based on a 10-vertex polyhedron, $(CB_9H_5Br_5)^-$, forms distinctly different types of network structure with group 1 cations and CTV depending on the size of the cation. The larger cations Rb^+ and Cs^+ form isostructural complexes that are isomorphic with the structure of **14**, although they are of lower crystallographic symmetry [50]. In $[Rb(CTV)(CB_9H_5Br_5)(CH_3CN)]$ **16** there are two crystallographically distinct $(CB_9H_5Br_5)^-$ anions with one binding in an η^2 fashion to a nine-coordinate Rb^+, and the other binding in an η^3 fashion to a ten-coordinate Rb^+ [50]. The identity of the anions and the higher binding of one anion is the only significant deviation of **16** from the structure of **14**, and an identical 2-D hexagonal $[Cs(CTV)]^+$ coordination polymer is formed.

By contrast, a 1:2 metal:ligand complex $[K(CTV)_2](CB_9H_5Br_5)(CF_3CH_2OH)_2$ (**17**) is formed under similar conditions to **16** but in the presence of an excess of K^+ [50]. Complex **17** has an entirely different type of structure with a $[K(CTV)_2]^+$ coordination chain and no cation–ligand interactions between the K^+ and anion. The K^+ center is eight coordinate with four chelating CTV ligands. Each CTV ligand bridges between two K^+ centers, propagating a chain (Figure 15). There are two crystallographically distinct CTV ligands, one of which has a single methyl group bent out-of-plane. Each type of CTV shows different host–guest interactions with the CF_3CH_2OH intracavity guest, with one showing an unusual O–H$\cdots\pi$ hydrogen bond, rather than being forming more expected hydrophobic interactions. Notably, similar O–H$\cdots\pi$ hydrogen bonds have been reported for calixarene–water complexes [56]. Coordination chains have a propeller-shaped cross section and pack around the $(CB_9H_5Br_5)^-$ anions.

The structures or topologies adopted by coordination polymers are frequently influenced by the solvent used for crystallization. This is especially true for potentially coordinating solvents. Using a similar procedure for the synthesis of complex **13**, but with DMF as the solvent rather than CH_3CN/CF_3CH_2OH, yields an entirely different crystalline complex, namely, that of $[Na_2(DMF)_4(H_2O)_2(CTV)]\{(DMF)_{0.5}(CTV)\}(CB_{11}H_6Br_6)_2$ (**18**) [49]. The coordination polymer within complex **18** is a 2-D network with a hexagonal 6^3 topology (Figure 16(a)). The network is formed by three-connecting CTV ligands that bind to a dimeric $[Na_2(DMF)_4(H_2O)_2]^{2+}$ fragment that is disordered across a mirror plane such that

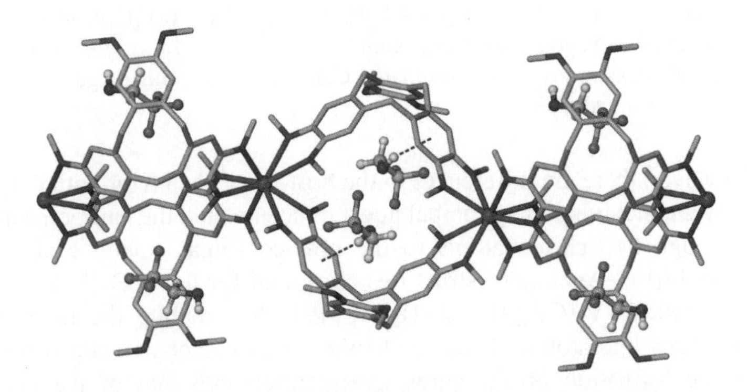

Figure 15. Section of the structure of $[K(CTV)_2](CB_9H_5Br_5)(CF_3CH_2OH)$ (**17**) highlighting the $[K(CTV)_2]^+$ coordination chain and host–guest behavior. Dotted lines show the unusual CF_3CH_2O–H$\cdots\pi$ hydrogen-bonding interaction.

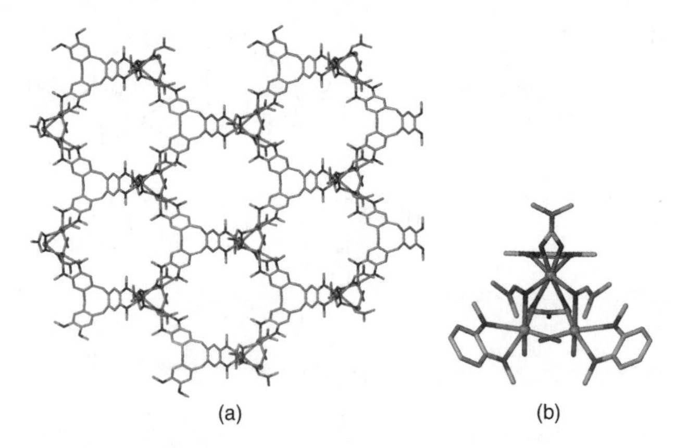

Figure 16. (a) Hexagonal $[Na_2(DMF)_4(H_2O)(CTV)]^{2+}$ 2-D coordination polymer from the structure of $[Na_2(DMF)_4(H_2O)_2(CTV)]\{(DMF)_{0.5}(CTV)\}(CB_{11}H_6Br_6)_2$ (**18**) and (b) detail showing the dimeric $[Na_2(DMF)_4(H_2O)]^{2+}$ fragment, which is disordered across a mirror plane so that the two lower Na positions have 50% occupancy (dynamically disordered aquo ligand shown in one position only).

it appears to be a trimer (Figure 16(b)). Each CTV host within a coordination polymer acts as a host for a secondary molecule of CTV that, in turn, is a host for a DMF guest. Self-hosting stacking motifs are common for clathrates of CTV [18, 19]. The $(CB_{11}H_6Br_6)^-$ counteranions forms hydrogen-bonded chains that run through cavities created by packing of the coordination polymers.

4. EXTENDED-ARM CTV DERIVATIVES AND THEIR COORDINATION POLYMERS

While we have been able to synthesize a range of interesting network structures using CTV as a hydrogen-bond donor or chelating ligand for group 1 metal ions, such chemistry does not engender a high degree of predictability. Furthermore, the preponderance of preferred packing modes indicates that we will not be able to change the nature of the network structure formed by subtle changes to the molecular components. Hence, we have synthesized a range of CTV-based N-donor ligands in order to create ligands which will bind to metal ions in a more predictable fashion, and to allow for the possibility of using metal ions that may have interesting redox or magnetic behavior [57, 58].

There have been a small number of previously reported examples of N-donor ligand functionality being attached to a CTV core. These include a tris-pyridyl derivative that can be assembled into chiral cryptophanes with Pd(II) [59], extended-arm cavitands with bipyridine [60, 61] and phenyl-pyridine functionality [62]. Iron and copper complexes have been characterized for the bipyridine-functionalized cavitands including a Cu(II/I) conformational switch [61]. Other metal complexes of CTV-based ligands include the trinuclear Pt(II) complex of the simple catechol analogue [34], Fe(II/III) complexes of extended-arm catechols [60], a Ni(II) complex of a salicylaldiminato derivative [63] and Fe–S clusters of thiol derived cavitands [64].

The CTV derivatives that we have synthesized are shown in Figure 17 and include hexakis- and tris-substituted molecular hosts, all with nitrogen donor atoms.

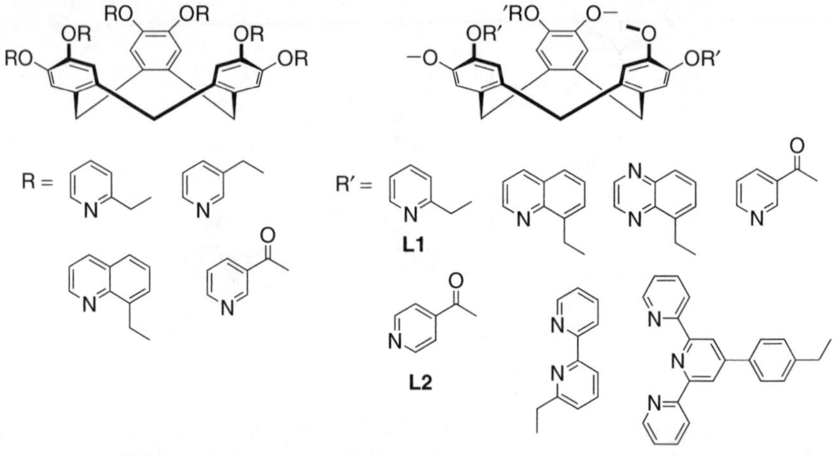

Figure 17. CTV-derived cavitands with N-type donor sites.

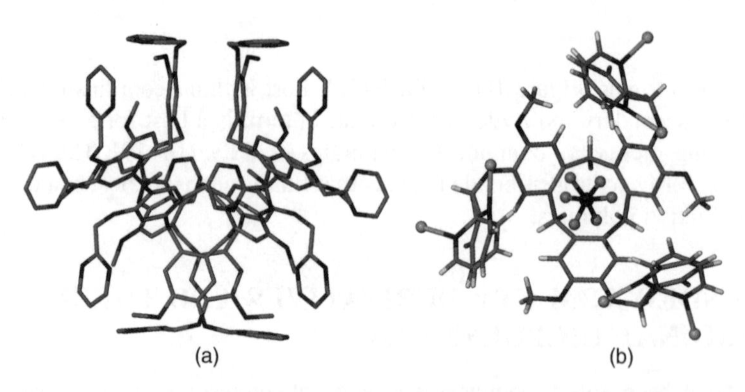

Figure 18. Details from the disordered structure of $[Ag_3(\mathbf{L1})_2](PF_6)_3$ (**19**). (a) Tetrahedral back-to-back tetramer of **L1** ligands and (b) coordination behavior of **L1** with one PF_6^- anion shown as an intracavity guest. Each Ag center (shown as large sphere) and each pyridyl position has 1/2 occupancy. An averaged pyridyl position is shown in part (a).

Initial experiments have given Ag(I) coordination polymers with tris(2-pyridylmethyl)-substituted **L1** and tris(isonicotinyl)-substituted **L2**.

The complex $[Ag_3(\mathbf{L1})_2](PF_6)_3$ (**19**) is a cubic 3-D coordination polymer, and its SbF_6^- analog is also known [65]. Crystals of **19** did not diffract to high angles even with intense x-ray sources, and the structure is extensively disordered. The disorder is manifest in partial occupancies of Ag(I), anion sites and split pyridyl positions. Such disorder is not unusual for high symmetry structures, and the main structural features of complex **19** can be elucidated. Within the structure, molecules of **L1** pack into back-to-back tetramers of tetrahedral symmetry (Figure 18(a)), similar to the $[CTV]_4$ tetramers found in hydrogen-bonding networks of complexes **7–9**. Each pyridyl group of **L1** is disordered over two positions and binds to two symmetry-related Ag(I) sites through Ag–N coordination interactions. There is also a long contact to a methoxy O for each Ag(I) site. Each ligand binds to six Ag(I) sites, each at 1/2 occupancy (Figure 18(b)). Each Ag(I) site

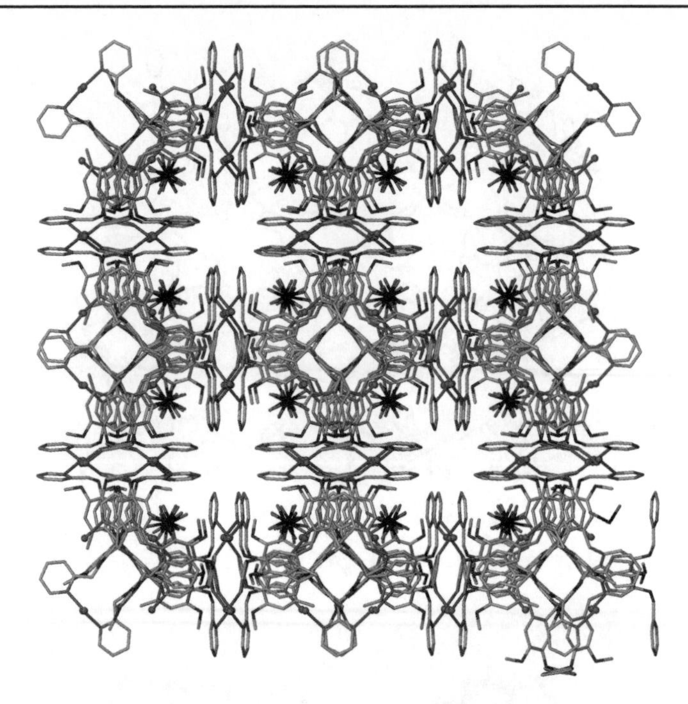

Figure 19. Packing diagram of $[Ag_3(\mathbf{L1})_2](PF_6)_3$ (**19**) showing a cubic 3-D coordination polymer, and some PF_6^- positions. An averaged pyridyl position is shown.

interacts with two ligands. The resultant 3-D coordination polymer is shown in Figure 19. The topology of the network is complicated, and can be regarded as having each ligand acting as an averaged 6-connector and each metal as a topologically trivial 2-connector. The ligands are more properly considered as disordered 3-connectors. Provided some assumptions are made, the disorder can be pulled apart to give a fourfold interpenetrating (10,3)-a motif. As for complex **7**, the network can be simplified by considering the $[\mathbf{L2}]_4$ tetramer as a single connecting center, in which case the network can be described as a simple 6-connected α-Po related net of 4^8 topology. Notably, some of the PF_6^- counterions are intracavity guest molecules, with other anion sites throughout the lattice.

The complex $[Ag(\mathbf{L2})_2][Co(C_2B_9H_{11})_2].9(CH_3CN)$ (**20**) is isolated as yellow-orange crystalline plates from acetonitrile solution [58]. Complex **20** has a 1-D chain structure where adjacent chains interweave through a combination of host–guest and π–π stacking interactions to create an overall 2-D assembly (Figure 20). The coordination sphere features roughly tetrahedral Ag(I) centers coordinated by four **L2** ligands. Each ligand bridges two Ag(I) centers through two of its three isonicotinyl groups. The coordination polymer thus formed is straight and doubly bridged, with two parallel but slipped strands of ligands. The molecular cavities of these ligands alternate in orientation both along the chain and across the strands. Within the crystal structure, adjacent coordination chains interlock through host–guest associations, whereby two molecules of guest acetonitrile are contained within a dimeric capsule-like motif formed by the head-to-head arrangement of two **L2** fragments from different chains (Figure 20(a)). There are two crystallographically distinct types of **L2** ligand within **20**, and both are involved in such dimeric interactions (Figure 20(b)), leading to the interweaving of 1-D polymers. This

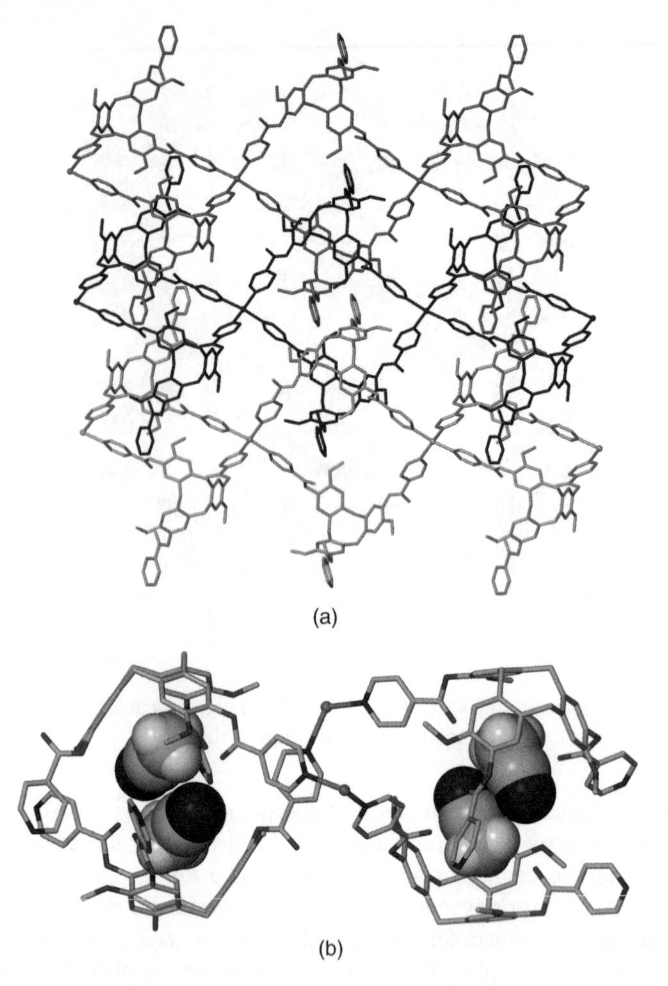

Figure 20. Structure of $[Ag(\mathbf{L1})_2][Co(C_2B_9H_{11})_2]$ (**20**). (a) Three 1-D $[Ag(\mathbf{L2})_2]^+$ coordination chains (each shown with different shading) interweave to form a 2-D network through host–guest associations and π stacking interactions and (b) detail showing the interlocking host–guest associations where head-to-head **L2** capsule-like dimers form around CH_3CN guest molecules.

interweaving is enhanced by $\pi-\pi$ stacking, where chains that do not interact through the host–guest associations form coplanar $\pi-\pi$ stacking interactions between the pyridyl rings that do not coordinate to Ag(I) (Figure 20(a)).

5. CONCLUSIONS

A number of infinite network structures involving the host molecule CTV can be synthesized with the CTV acting as a two-connecting or three-connecting center to give chain, 2-D or 3-D networks. The CTV can act as a hydrogen-bond acceptor or a chelating ligand for group 1 metals, and acts in both capacities in some materials. In general, the host–guest characteristics of the CTV molecule are changed by incorporation into crystalline network structures and intracavity binding of small guest molecules, a rarity for native CTV, commonly occurs. More typical host–guest behavior of native CTV – the formation of perched ball-and-socket assemblies with large guests – is also observed, but often for guest molecules such as C_{70} and $(CB_{11}H_6Cl_6)^-$ that do not form complexes with CTV by themselves. While hexagonal 2-D network structures are common, the rigid bowl character of CTV and its derivatives also lead to unusual network topologies that have not been previously seen, such as the 3,12-connected 3-D network of complex **7**.

The preponderance of preferred packing modes indicates that the engineering of network structures involving CTV by attempting to control the network topology with small changes to the other molecular components, may not be possible. Instead, the structure obtained is largely influenced by the anion or a similar steric template. The synthesis of a range of CTV-based ligands capable of interaction with transition metal, offers considerable scope for creating new, infinite framework materials.

6. ACKNOWLEDGMENTS

Prof. Colin Raston, Ruksanna Ahmad, Dr Christopher Sumby, Bradley Wells, Antonio Salinas and Prof. John Kennedy are thanked. The Australian Research Council, EPSRC and the University of Leeds are thanked for providing equipment and funding.

REFERENCES

1. J. W. Steed and J. L. Atwood, *Supramolecular Chemistry*, Wiley, Chichester, 2000.
2. D. Braga, F. Grepioni and A. G. Orpen (Eds.), *Crystal Engineering: From Molecules and Crystals to Materials*, Kluwer Academic Publishers, Dordrecht, 1999.
3. M. Fujita, Y. J. Kwon, S. Washizu and K. Ogura, *J. Am. Chem. Soc.*, **116**, 1151–1152 (1994).
4. (a) G. J. Halder, C. J. Kepert, B. Moubaraki, K. S. Murray and J. D. Cushion, *Science*, **298**, 1762–1765 (2002); (b) R. Kitaura, K. Fujimoto, S. Noro, M. Kondo and S. Kitagawa, *Angew. Chem., Int. Ed. Engl.*, **41**, 133–135 (2002).
5. M. Eddaoudi, J. Kim, N. Rosi, D. Vodak, J. Wachter, M. O'Keeffe and O. M. Yaghi, *Science*, **295**, 469–472 (2002).
6. M. J. Hardie, *Struct. Bonding*, **111**, 139–174 (2004) and references therein.
7. B.-Q. Ma, L. F. V. Ferreira and P. Coppens, *Org. Lett.*, **6**, 1087–1090 (2004).
8. (a) J. Muehle and W. S. Sheldrick, *Z. Anorg. Allg. Chem.*, **629**, 2097–2102 (2003); (b) N. Malic, P. J. Nichols and C. L. Raston, *Chem. Commun.*, 16–17 (2002); (c) D.-L. Long, Y. Cui, J.-T. Chen, W.-D. Cheng and J.-S. Huang, *Polyhedron*, **17**, 3969–3975 (1998).
9. (a) S. J. Dalgarno and C. L. Raston, *Chem. Commun.*, 2216–2217 (2002); (b) J. L. Atwood, L. J. Barbour, S. Dalgarno, C. L. Raston and H. R. Webb, *J. Chem. Soc., Dalton Trans.*, 4351–4356 (2002); (c) H. R. Webb, M. J. Hardie and C. L. Raston, *Chem. Eur. J.*, **7**, 3616–3620 (2001); (d) S. J. Dalgarno, M. J. Hardie and C. L. Raston, *Cryst. Growth Des.*, **4**, 227–234 (2004).
10. S. J. Dalgarno, M. J. Hardie, J. E. Warren and C. L. Raston, *Dalton Trans.*, 2413–2416 (2004).
11. S. J. Dalgarno, M. J. Hardie, J. L. Atwood and C. L. Raston, *Inorg. Chem.*, **43**, 6351–6356 (2004).
12. J. Plutnar, J. Rohovec, J. Kotek, Z. Zak and I. Lukes, *Inorg. Chim. Acta*, **335**, 27–35 (2002).
13. C. B. Dieleman, D. Matt and A. Harriman, *Eur. J. Inorg. Chem.*, 831–834 (2000).
14. E. Elisabeth, L. J. Barbour, G. W. Orr, K. T. Holman and J. L. Atwood, *Supramol. Chem.*, **12**, 317–320 (2000).
15. P. D. Beer, M. G. B. Drew, P. A. Gale, M. I. Ogden and H. R. Powell, *CrystEngComm*, **2**, 164–168 (2000).

16. D. G. Samsonenko, A. A. Sharonova, M. N. Sokolov, A. V. Virovets and V. P. Fedin, *Russ. J. Coord. Chem.*, **27**, 12–17 (2001).
17. (a) C. Kleina, E. Graf, M. W. Hosseini, A. De Cian and J. Fischer, *Chem. Commun.*, 239–240 (2000); (b) W. Jaunky, M. W. Hosseini, J. M. Planeix, A. De Cian, N. Kyritsakas and J. Fischer, *Chem. Commun.* 2313–2314 (1999); (c) A. Bottino, F. Cunsolo, M. Piattelli, E. Gavuzzo and P. Neri, *Tetrahedron Lett.*, **41**, 10065–10069 (2000).
18. J. W. Steed, H. Zhang and J. L. Atwood, *Supramol. Chem.*, **7**, 37–45 (1996).
19. J. A. Hyatt, E. N. Duesler, D. Y. Curtin and I. C. Paul, *J. Org. Chem.*, **45**, 5074–5079 (1980).
20. B. T. Ibragimov, K. K. Makhkamov and K. M. Beketov, *J. Inclusion Phenom. Macro. Chem.*, **35**, 583–593 (1999).
21. M. R. Caira, A. Jacobs and L. R. Nassimbeni, *Supramol. Chem.*, **16**, 337–342 (2004).
22. (a) J. W. Steed, P. C. Junk, J. L. Atwood, M. J. Barnes and C. L. Raston, *J. Am. Chem. Soc.*, **116**, 10346–10347 (1994); (b) J. L. Atwood, M. J. Barnes, M. G. Gardiner and C. L. Raston, *J. Chem. Soc., Chem. Commun.*, 1449–1450 (1996); (c) D. V. Konarev, S. S. Khasanov, G. Saito, A. Otsuka, Y. Yoshida and R. N. Lyubovskaya, *J. Am. Chem. Soc.*, **125**, 10074–10083 (2003); (d) D. V. Konarev, I. S. Neretin, G. Saito, Y. L. Slovokhotov, A. Otsuka and R. N. Lyubovskaya, *Dalton Trans.*, 3886–3891 (2003).
23. R. J. Blanch, M. Williams, G. D. Fallon, M. G. Gardiner, R. Kaddour and C. L. Raston, *Angew. Chem., Int. Ed. Engl.*, **36**, 504–506 (1997).
24. K. T. Holman, J. W. Steed and J. L. Atwood, *Angew. Chem., Int. Ed. Engl.*, **36**, 1736–1738 (1997).
25. M. J. Hardie and C. L. Raston, *Chem. Commun.*, 905–906 (2001).
26. R. Ahmad and M. J. Hardie, *Cryst. Growth Des.*, **3**, 493–499 (2003).
27. D. V. Konarev, S. S. Khasanov, I. I. Vorontsov, G. Saito, M. Y. Antipin, A. Otsuka and R. N. Lyubovskaya, *Chem. Commun.*, 2548–2549 (2003).
28. (a) K. T. Holman, M. M. Halihan, S. S. Jurisson, J. L. Atwood, R. S. Burkhalter, A. R. Mitchell and J. W. Steed, *J. Am. Chem. Soc.*, **118**, 9567–9576 (1996); (b) K. S. B. Hancock and J. W. Steed, *Chem. Commun.*, 1409–1410 (1998); (c) J. A. Gawenis, K. T. Holman, J. L. Atwood and S. S. Jurisson, *Inorg. Chem.*, **41**, 6028–6031 (2002).
29. A. Collet, *Tetrahedron*, **43**, 5725–5759 (1987)
30. Y. Rio and J.-F. Nierengarten, *Tetrahedron Lett.*, **43**, 4321–4324 (2002).
31. D. Felder, B. Heinrich, D. Guillon, J.-F. Nicoud and J.-F. Nierengarten, *Chem. Eur. J.*, **6**, 3501–3507 (2000).
32. A. Arduini, F. Calzavacca, D. Demuru, A. Pochini and A. Secchi, *J. Org. Chem.*, **69**, 1386–1388 (2004).
33. T. Brotin, R. Barbe, M. Darzac and J.-P. Dutasta, *Chem. Eur. J.*, **9**, 5784–5792 (2003).
34. D. S. Bohle and D. J. Stasko, *Chem. Commun.*, 567–568 (1998).
35. M. J. Hardie and C. L. Raston, *CrystEngComm*, **3**, 162–164 (2001) and references therein.
36. for review see M. J. Hardie and C. L. Raston, *Chem. Commun.*, 1153–1163 (1999).
37. T. Kusukawa and M. Fujita, *Angew. Chem., Int. Ed. Engl.*, **37**, 3142–3144 (1998).
38. M. J. Hardie, P. D. Godfrey and C. L. Raston, *Chem. Eur. J.*, **5**, 1828–1833 (1999).
39. M. J. Hardie and C. L. Raston, (unpublished results).
40. R. Ahmad and M. J. Hardie, (unpublished results).
41. M. J. Hardie, C. L. Raston and B. Wells, *Chem. Eur. J.*, **6**, 3293–3298 (2000).
42. M. J. Hardie and C. L. Raston, *Cryst. Growth Des.*, **1**, 53–58 (2001).
43. M. J. Hardie, C. L. Raston and A. Salinas, *Chem. Commun.*, 1850–1851 (2001).
44. (a) G. W. Orr, L. J. Barbour and J. L. Atwood, *Science*, **285**, 1049–1052 (1999); (b) J. L. Atwood, L. J. Barbour, S. J. Dalgarno, M. J. Hardie and C. L. Raston, *J. Am. Chem. Soc.*, **126**, 13170–13171 (2004).
45. B. F. Abrahams, S. R. Batten, H. Hamit, B. F. Hoskins and R. Robson, *Angew. Chem., Int. Ed. Engl.*, **35**, 1690–1692 (1996).
46. R. Ahmad and M. J. Hardie, *CrystEngComm*, **4**, 227–231 (2002).

47. R. Ahmad, I. Dix and M. J. Hardie, *Inorg. Chem.*, **42**, 2182–2184 (2003).
48. M. J. Hardie and C. L. Raston, *Angew. Chem., Int. Ed. Engl.*, **39**, 3835–3839 (2000).
49. R. Ahmad, A. Franken, J. D. Kennedy and M. J. Hardie, *Chem. Eur. J.*, **10**, 2190–2198 (2004).
50. R. Ahmad and M. J. Hardie, *New J. Chem.*, **28**, 1315–1319 (2004).
51. (a) S. A. Barnett, A. J. Blake, N. R. Champness, J. E. B. Nicolson and C. Wilson, *J. Chem. Soc., Dalton Trans.*, 567–573 (2001); (b) D.-L. Long, A. J. Blake, N. R. Champness and M. Schröder, *Chem. Commun.*, 1369–1370 (2000); (c) S. G. Ang, B. W. Sun and S. Gao, *Inorg. Chem. Commun.*, **7**, 795–798 (2004); (d) S.-Y. Wan, J. Fan, T. Okamura, H.-F. Zhu, X.-M. Ouyang, W.-Y. Sun and N. Ueyama, *Chem. Commun.*, 2520–2521 (2002).
52. Z. Xie, T. Jelínek, R. Bau and C. A. Reed, *J. Am. Chem. Soc.*, **116**, 1907–1913 (1994).
53. Z. Xie, D. J. Liston, T. Jelínek, V. Mitro, R. Bau and C. A. Reed, *J. Chem. Soc., Chem. Commun.*, 384–386 (1994).
54. (a) C.-W. Tsang, Q. Yang, E. T.-P. Sze, T. C. W. Mak, D. T. W. Chan and Z. Xie, *Inorg. Chem.*, **39**, 5851–5858 (2000); (b) C.-W. Tsang, Q. Yang, E. T.-P. Sze, T. C. W. Mak, D. T. W. Chan and Z. Xie, *Inorg. Chem.*, **39**, 3582–3589 (2000).
55. (a) N. J. Patmore, C. Hague, J. H. Cotgreave, M. F. Mahon, C. G. Frost and A. S. Weller, *Chem. Eur. J.*, **8**, 2088–2098 (2002); (b) D. R. Evans and C. A. Reed, *J. Am. Chem. Soc.*, **122**, 4660–4667 (2000); (c) N. J. Patmore, M. J. Ingleson, M. F. Mahon and A. S. Weller, *Dalton Trans.*, 2894–2904 (2003).
56. (a) J. L. Atwood, F. Hamada, K. D. Robinson, G. W. Orr and R. L. Vincent, *Nature*, **349**, 683–684 (1991); (b) A. Drljaca, M. J. Hardie and C. L. Raston, *J. Chem. Soc., Dalton Trans.*, 3639–3642 (1999).
57. M. J. Hardie, R. M. Mills and C. J. Sumby, *Org. Biomol. Chem.*, **2**, 2958–2964 (2004).
58. M. J. Hardie and C. J. Sumby, *Inorg. Chem.*, **43**, 6872–6874 (2004).
59. A. Zhong, A. Ikeda, S. Shinkai, S. Sakamoto and K. Yamaguchi, *Org. Lett.*, **3**, 1085–1087 (2001).
60. G. Vériot, J.-P. Dutasta, G. Matouzenko and A. Collet, *Tetrahedron*, **51**, 389–400 (1995).
61. J. A. Wytko, C. Boudon, J. Weiss and M. Gross, *Inorg. Chem.*, **35**, 4469–4477 (1996).
62. J. A. Wytko and J. Weiss, *Tetrahedron Lett.*, **49**, 7261–7264 (1991).
63. D. S. Bohle and D. J. Stasko, *Inorg. Chem.*, **39**, 5768–5770 (2000).
64. G. P. F. van Strijdonck, J. A. E. H. van Haare, J. G. M. van der Linden, J. J. Steggerda and R. J. M. Nolte, *Inorg. Chem.*, **33**, 999–1000 (1994).
65. C. J. Sumby and M. J. Hardie, *Cryst. Growth & Des.*, **5**, 1321–1324 (2005).

8
Interpenetrating Networks

STUART R. BATTEN
School of Chemistry, Monash University, Victoria 3800, Australia.

1. INTRODUCTION

A net approach to both the design and analysis of new crystalline materials, particularly those containing coordination polymers or hydrogen-bonded networks, is a particularly useful method and is widely used [1, 2]. Complicated structures can be broken down into simple networks of nodes and links, facilitating easier and quicker understanding of connectivity between the components of the crystal structure. The design of new structures is also encouraged by this approach.

One particular design target is new open, porous networks, for applications such as gas and liquid storage, molecular sieving, ion exchange and heterogeneous catalysis. However, once a single network occupies less than 50% of the crystal volume, the possibility arises of two or more interpenetrating networks forming simultaneously, reducing the openness of the structure [3–12].

The formation of interpenetrating networks arises because of the axiom "Nature abhors a vacuum". Wherever possible, crystal structures with large volumes of truly empty space will not form, although they may be produced after crystal formation by, for example, evacuation of guest molecules. Note that these guest molecules are not necessarily ordered – at the extreme, it can be essentially liquid solvent. But, nonetheless, the cavities are not empty, at least not at first.

There are three main methods in which a network crystal can maximize its packing efficiency – interdigitation (for 1-D and 2-D nets), interpenetration and intercalation. These three phenomena, and their unpredictability, are illustrated by three related structures that all contain 2-D (4,4) sheets of Cu(tcm)L, tcm = tricyanomethanide, $C(CN)_3^-$, L = 2-connecting N-donor ligand [13]. The structure of Cu(tcm)(hmt), hmt = hexamethylenetetramine, contains undulating (4,4) sheets that interdigitate by directing the uncoordinated C–CN "arms" of the tcm ligands into the windows of adjacent sheets, and

Frontiers in Crystal Engineering. Edited by Edward R.T. Tiekink and Jagadese J. Vittal

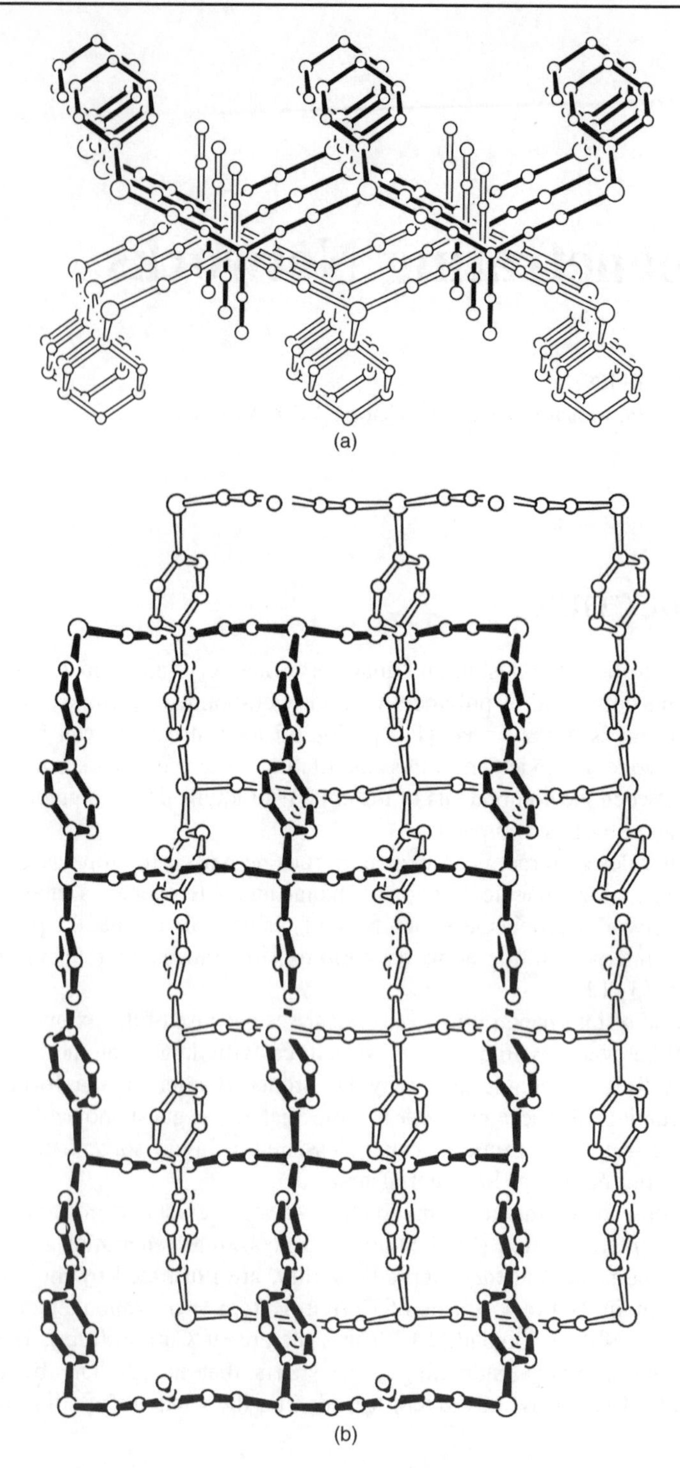

Figure 1. Cu(tcm)L (4,4) sheets showing (a) interdigitation (L = hmt); (b) interpenetration (L = bipy) and (c) intercalation of guest molecules (L = bpee) [13].

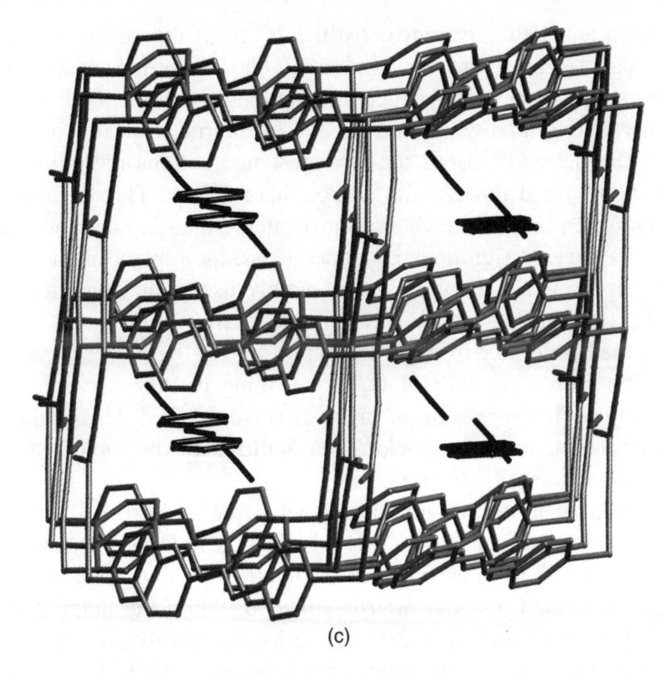

(c)

Figure 1. *(continued)*

vice versa (Figure 1(a)). The structure of Cu(tcm)(bipy), bipy = 4,4′-bipyridine, contains 2-D layers of doubly interpenetrating (4,4) sheets; these layers then interdigitate with each other (Figure 1(b)). Finally, the structure of Cu(tcm)(bpee)·1/4bpee·1/2MeCN, bpee = 1,4-bis(4-pyridyl)ethene contains (4,4) sheets that stack to give 1-D channels that contain intercalated bpee and MeCN molecules (Figure 1(c)). Thus, while each structure contains similar 2-D sheets, a different method of increasing the packing efficiency is manifested each time. Interpenetration is just one method, but a particularly interesting one.

2. NOTATION

The networks examined here are described in terms of their network topologies using a variety of naming methods, including the notation of Wells (e.g. (4,4) or (6,3) sheets), the names of simple structures with the same topology (e.g. diamond, rutile, α-Po), or their Schläfli symbols (e.g. 4.8^2). For further background to this nomenclature, the reader is referred elsewhere [2, 12, 14–16]. However, it is worth emphasizing that these are just different methods of giving a widely recognized name to the underlying net, and for the purposes of this article, they can be treated only as names.

It is also worth noting, in the context of nets, that a net is a topological description and not a geometric one. Nets describe only the connectivities between the nodes and not the geometries of those nodes (or links). Thus a hexagonal 2-D sheet (with trigonal nodes) and a "brick-wall" 2-D sheet (with T-shaped nodes) are topologically identical (both are called (6,3), with three-connecting nodes).

One important challenge in the examination of interpenetrating systems is how to define the networks. This may initially sound trivial, but many crystals show a range of interactions between their covalent constituents, from the strong (coordinate bonds) through to the weak (weak hydrogen bonds, $\pi \cdots \pi$ interactions, van der Waals interactions). The question arises as to which interactions are used to define the networks (intranetwork interactions) and which are classed as internetwork interactions. Generally, we have defined the networks using the strongest interactions (coordination bonds and strong hydrogen bonds) and ignored the weaker interactions. This is somewhat arbitrary (especially for hydrogen-bonded nets), and most interpenetrating networks show internetwork interactions of varying significance. Nonetheless, the separation of crystal structures into interpenetrating networks remains an extremely useful and productive tool for the analysis of crystal structures, which is the ultimate goal.

Consequently, the occurrence of interpenetration means that, for a "crystal engineer" to completely describe such a crystal structure, describing just the network topology is not enough. The *topology of interpenetration* must also be described. This is akin to describing the intermolecular interactions and packing, in addition to the molecular geometry, for structures containing discrete molecules.

Networks can interpenetrate in a number of ways, particularly 1-D and 2-D networks. One general distinction that can be made is whether the 1-D nets that interpenetrate are all parallel, or whether they interpenetrate each other in two or more inclined directions. The former is termed *parallel interpenetration*, while the latter is called *inclined interpenetration*. The same applies to 2-D networks. For parallel interpenetration, all the interpenetrating networks have their mean planes parallel, while for inclined interpenetration the mean planes of the networks are in two or more inclined directions.

One consequence of inclined interpenetration is that the dimensionality of the entanglement is greater than that of the constituent networks. Thus, inclined interpenetration of 1-D nets generates entanglements that are overall 2-D or 3-D in nature. Inclined interpenetration of 2-D nets will always give 3-D entanglements. For parallel interpenetration, the overall entanglement will be of the same or of higher dimension as the nets. Thus, 1-D nets can form 1-D entanglements in addition to those of the 2-D or 3-D variety. The formation of 2-D or 3-D entanglements requires the mean directions of propagation to be parallel but not coincident. There is no such requirement for overall 1-D entanglements, although in all the examples reported to date, the mean propagation axes of the interpenetrating1-D nets are coincident. For parallel interpenetration of 2-D nets, the most common result is discrete 2-D layers of interpenetrating nets. The formation of 3-D entanglements from interpenetrating parallel 2-D nets is unusual, but a significant number have been reported recently. Again, to obtain a 3-D entanglement, the mean planes of the 2-D nets must be parallel, but not coincident. For 2-D entanglements, the mean planes are usually (but are not necessarily required to be) coincident.

These different types of entanglements can also be represented using the convenient shorthand notation $m\mathrm{D} \rightarrow n\mathrm{D}$, where $m\mathrm{D}$ represents the dimensionality of the constituent nets, and $n\mathrm{D}$ represents the dimensionality of the overall entanglement [5]. This notation will be used throughout, with examples to follow. Obviously, for 2-D inclined interpenetration, the only possibility is a 3-D entanglement, and thus the 2-D $\rightarrow$ 3-D notation is redundant.

This notation is a useful and convenient tool for describing important aspects of interpenetration, but it is often not sufficient. For example, structures showing 2-D $\rightarrow$ 2-D parallel interpenetration of (4,4) sheets may still show topologically different modes of

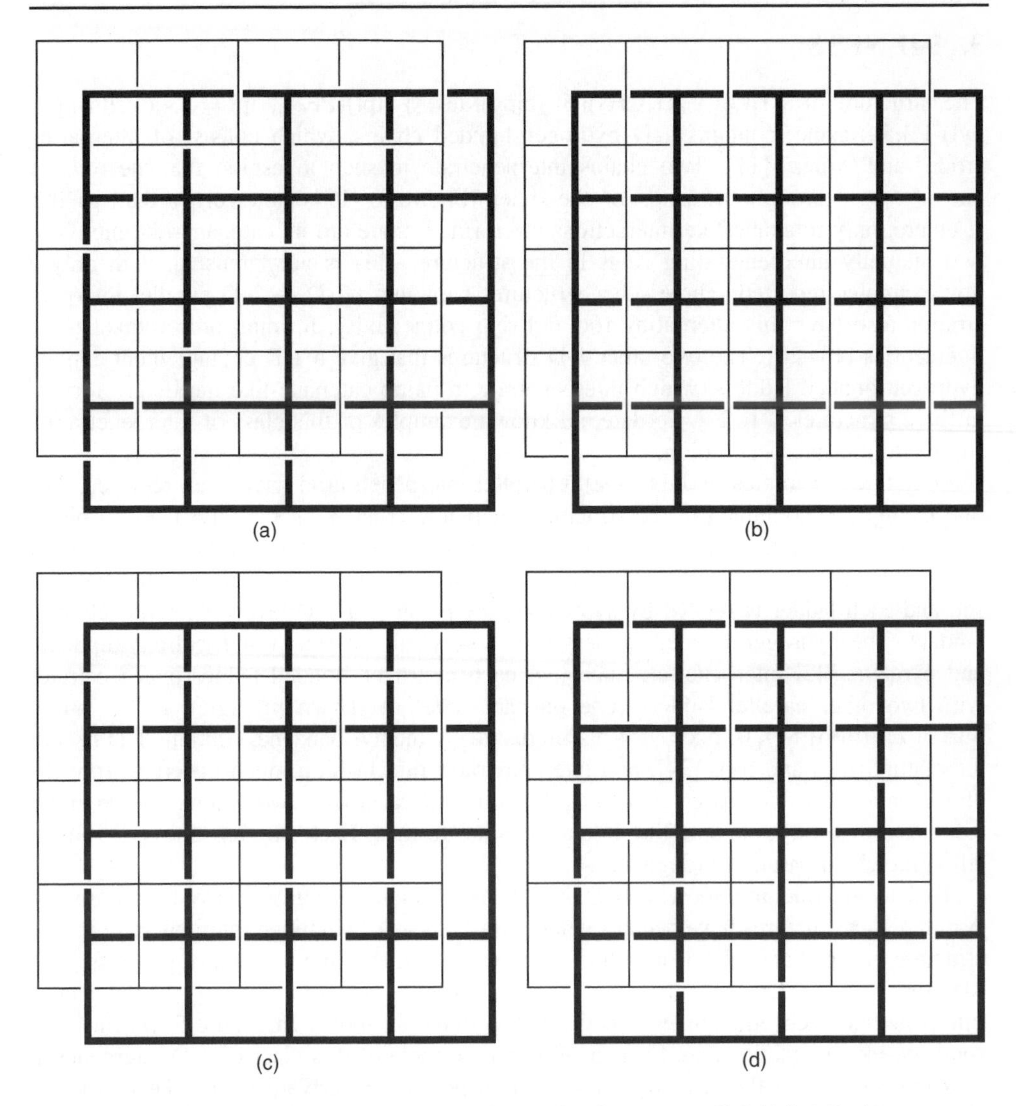

Figure 2. Four different interpenetration modes for two parallel (4,4) 2-D sheets.

interpenetration, as shown in Figure 2. Therefore, these topological aspects must also be explained in a complete structural description; almost always through a diagram.

Structures with 3-D nets can also display different modes of interpenetration. While these may not be as clear cut as the parallel/inclined and dimensional increase aspects of 1-D and 2-D nets, they are nonetheless important, and are akin to the different modes for 2-D (4,4) sheets shown in Figure 2. However, different modes are less common for 3-D nets than they are for 2-D nets. Some particularly striking examples are seen for structures with diamond networks, and are discussed below, as are representative examples for each of the major types of interpenetration.

3. 1-D NETS

The structure of $\{(HO_2CC_6H_4CH_2)(Bu_2bipy)Me_2Pt\}_3\{tpt\}(PF_6)_3$, tpt = 2,4,6-tri(4-pyridyl)-1,3,5-triazine, contains 1-D hydrogen-bonded chains, which consist of alternating "rods" and "rings" [17]. Two chains interpenetrate in such a fashion that the rods of one net pass through the rings of the other (Figure 3). It is noteworthy that, in this structure, only rotaxane-like interactions are formed; there are no catenane-like motifs of two mutually interpenetrating rings in the structure. This is very unusual, with only a few examples reported. Three other structures showing a 1-D → 1-D parallel interpenetration also have this alternating rod and ring connectivity, forming only rotaxane-like interactions [18–20]. The two other 1-D structures that give a 1-D entanglement contain hydrogen-bonded ladders, which interpenetrate in pairs; catenane-like motifs are formed in these structures [21, 22]. To date, all known examples of this class of interpenetration show twofold interpenetration.

Only three examples of 1-D → 2-D parallel interpenetration have been reported. The first example is contained in the structures of $[Cu_2L_3(MeCN)_2](X)_2 \cdot solv$, L = 1,4-bis(4-pyridyl)butadiyne, X = PF_6^-, BF_4^- [23]. The structures contain 1-D ladders. Each ladder interpenetrates with four others, two on either side. As the ladders are parallel but offset, and each ladder is related by symmetry, the result is a 2-D layer of interpenetrating ladders. The hydrogen-bonded ladders in a co-crystal formed by 4,4′-sulfonyldiphenol and pyrazine [22] interpenetrate such that each hydrogen-bonded ladder interpenetrates with two other parallel ladders (one on each side) as shown in Figure 4. The structure of $Zn_2(bix)_3(SO_4)_2$, bix = 1,4-bis(imidazol-1-ylmethyl)benzene, contains 1-D nets of alternating rings and rods [24], and interpenetrates in a fashion much different from the aforementioned ladder-based structures. Each net interacts with two others, one on either side, such that only rotaxane-like interactions are formed. Each ring has one rod from an adjoining chain passing through it.

To date, no structures showing 1-D → 3-D parallel interpenetration have been reported. A number of structures have been reported with 1-D → 2-D inclined interpenetration. The structure of $Ag_2(bix)_3(NO_3)_2$ contains 1-D nets with alternating rod and ring motifs [25], like the Zn/SO_4 structure discussed above. However, in $Ag_2(bix)_3(NO_3)_2$ the chains interpenetrate each other in two inclined directions, as shown in Figure 5. Again, only rotaxane-like interactions are formed. Another example of this class of 1-D interpenetration is contained in the structure of a calcium porphyrin derivative [26]. This structure contains chains comprising loops that interpenetrate in two inclined directions. Each ring contains a node of another net, with only rotaxane-like interactions resulting. By contrast, a series of coordination polymers containing the ligand 1-(1-imidazolyl)-4-(imidazol-1-ylmethyl)benzene show 1-D → 2-D inclined interpenetration in which catenane-like interactions are formed [27].

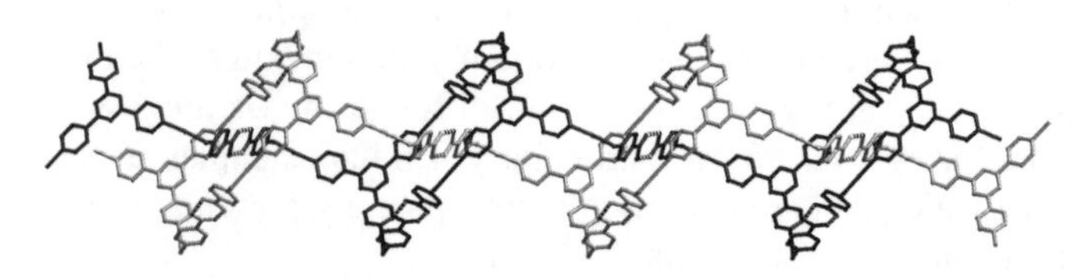

Figure 3. 1-D → 1-D parallel interpenetration in the structure of $\{(HO_2CC_6H_4CH_2)(Bu_2bipy)$-$Me_2Pt\}_3\{tpt\}(PF_6)_3$ [17]. Striped bonds represent hydrogen bonds.

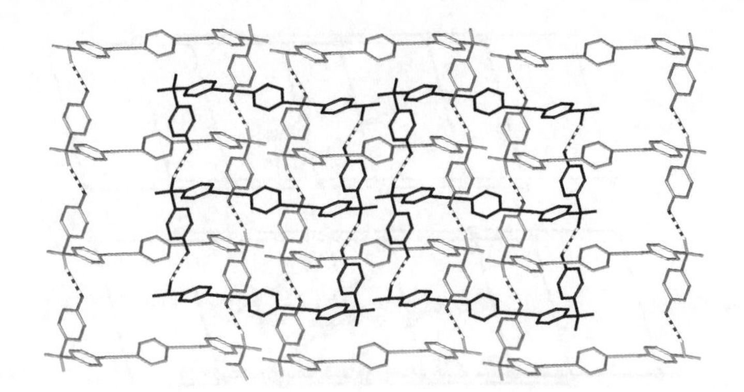

Figure 4. Hydrogen-bonded ladders in a 4,4′-sulfonyldiphenol–pyrazine co-crystal showing 1-D → 2-D interpenetration [22]. Striped bonds represent hydrogen bonds.

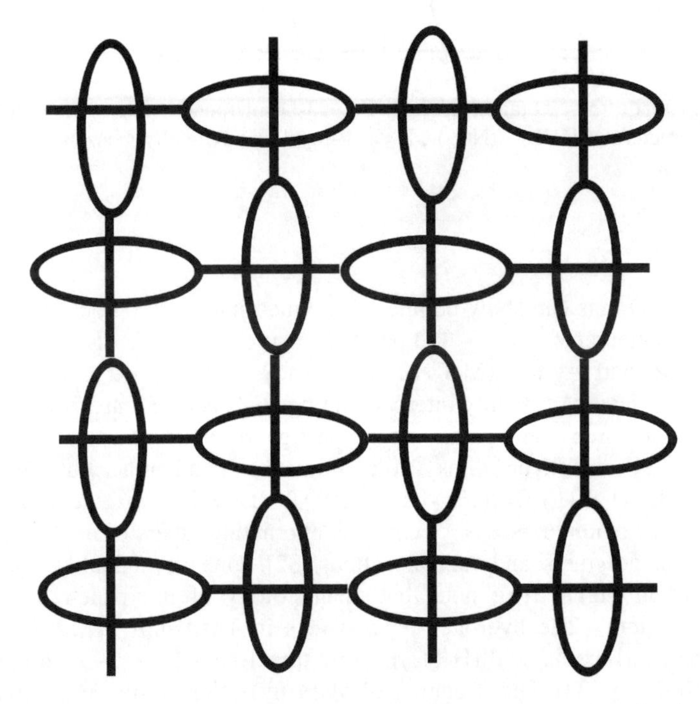

Figure 5. Schematic representation of the 1-D → 2-D interpenetration in the structure of $Ag_2(bix)_3(NO_3)_2$ [25].

1-D → 3-D inclined interpenetration is shown by a number of structures, all of which contain ladderlike nets. In the structure of Cu_2(ip)(bipy), ip = isophthalate, the interpenetration is threefold – that is, each ladder window is penetrated by two other ladders [28]. In the structure of $[CdL_{1.5}](NO_3)_2$, L = 1,4-bis(4-pyridylmethyl)benzene, the interpenetration is fivefold [29] (Figure 6), as it is for $Cd_2(nbpy4)_3(NO_3)_4$, nbpy4 = N, N'-bis-(4-pyridinylmethylene)-1,5-naphthalenediamine [30].

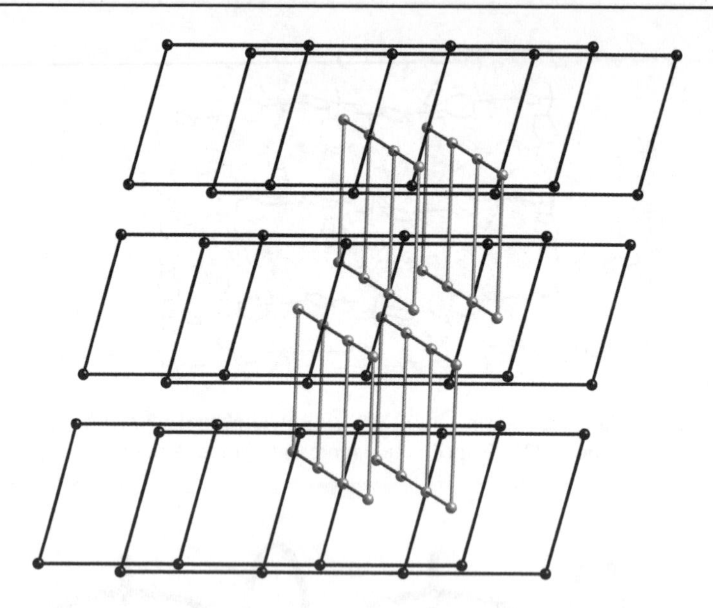

Figure 6. Schematic representation of the 1-D → 3-D interpenetration of coordination polymer ladders in the structure of $[CdL_{1.5}](NO_3)_2$, L = 1,4-bis(4-pyridylmethyl)benzene; spheres represent the Cd nodes [29].

4. 2-D NETS

Like 1-D nets, 2-D nets can show parallel or inclined interpenetration, resulting in either 2-D or 3-D entanglements. 2-D → 2-D parallel interpenetration is displayed by the structures of Ag(tcm) and Ag(tcm)(MeCN), tcm = tricyanomethanide, $C(CN)_3^-$ [31]. Both structures contain layers of doubly interpenetrating (6,3) nets (Figure 7(a)) in which each Ag coordinates to three tcm ligands, and each ligand coordinates to three Ag cations. In Ag(tcm)(MeCN), the silver atoms also coordinate to pendant acetonitrile ligands. The structure of θ-$(BEDT\text{-}TTF)_2Cu_2(CN)(dca)_2$, BEDT-TTF = bis(ethylenedithio)tetrathiafulvalene, dca = dicyanamide, $N(CN)_2^-$ contains alternating layers of doubly interpenetrating coordination polymers and radical cations [32]. One such doubly interpenetrating layer is shown in Figure 7(b); note that the topology of interpenetration is different from that of Ag(tcm). The hydrogen-bonded nets in TMA·bipy, TMA = trimesic acid (1,3,5-benzenetricarboxylic acid (H_3btc)), show threefold 2-D → 2-D parallel interpenetration of (6,3) sheets [33]. The structures of $M_2(azpy)_3(NO_3)_4\cdot solv$, M = Co and Cd, and azpy = *trans*-4,4′-azobis(pyridine), contain triply interpenetrating (6,3) nets that have a herringbone grid geometry rather than the hexagonal grid geometry of the three previous examples [34]. The record is held by one polymorph of $Ag(TEB)(CF_3SO_3)$, TEB = 1,3,5-tris(4-ethynylbenzonitrile)benzene, in which discrete layers of sixfold interpenetrating (6,3) nets are formed [35].

Another, less commonly observed topology for 2-D nets with three-connecting nodes is 4.8^2. A small number of structures with interpenetrating nets of this topology have been reported. The first was the structure of $Cd_2(NO_3)_4(dpt)_2(MeCN)$, dpt = 2,4-bis(4-pyridyl)-1,3,5-triazine [36]. It contains two interpenetrating 4.8^2 networks (Figure 8), as do two later structures containing trigonal imidazole-donor ligands [37, 38]. Hittorf's phosphorus also contains doubly interpenetrating sheets based on nets containing three-connecting nodes; in this case the topology is $8^2.10$ [39].

Some of the different interpenetration topologies for 2-D → 2-D parallel interpenetration of (4,4) nets were shown in Figure 2. We have already discussed the twofold interpenetration of Cu(tcm)(bipy) (Figure 1(b)), in which the networks interpenetrate with the topology shown in Figure 2(c). The hydrogen-bonded nets formed by 4,4′-sulfonyldiphenol interpenetrate with the same topology [40], as do the coordination nets in α-Cu(dca)(bpee) [41]. By comparison, the two (4,4) nets in $[Cd(4\text{-pic})_2\{Ag(CN)_2\}_2](4\text{-pic})$, 4-pic = 4-methylpyridine [42], interpenetrate with the topology shown in Figure 2(d), and the two hydrogen-bonded nets in $[Re_4(CO)_{12}(OH)_4]\cdot 2bipy\cdot 2MeOH$ have the same topology as Figure 2(b) [43]. The topology of interpenetration shown in Figure 2(a) is shown by the structure of $Cd_2(bpp)_4(NO_3)_4(H_2O)$, bpp = 1,3-bis(4-pyridyl)propane [44]. Three nets interpenetrate in the acentric structure of CdL_2, L = 3-[2-(4-pyridyl)ethenyl] benzoate, which shows very significant nonlinear optical properties [45]. The highest degree of 2-D → 2-D parallel interpenetration for (4,4) nets reported to date is fivefold, shown by the structure of $CoL_2(NCS)_2$, L = 1,2-bis(4-pyridinecarboxamido)ethane [46].

Figure 7. Two different modes of 2-D → 2-D parallel interpenetration of (6,3) sheets, represented by the structures of (a) Ag(tcm) [31] and (b) the anionic coordination polymer layers in θ-$(BEDT\text{-}TTF)_2Cu_2(CN)(dca)_2$ [32].

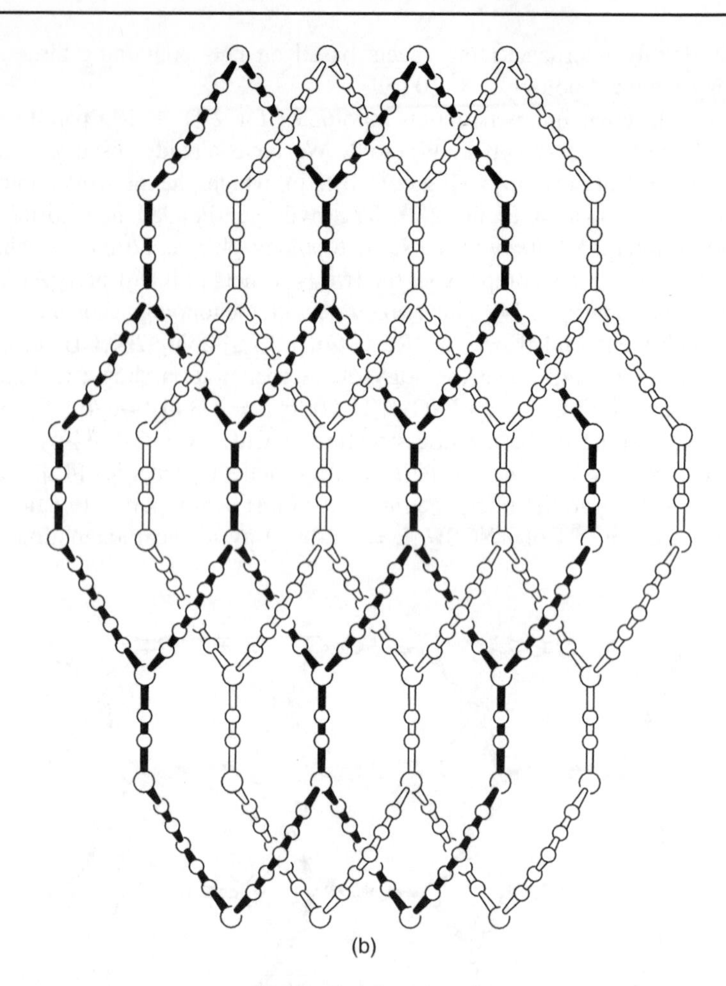

Figure 7. (*continued*)

The fascinating compound $Zn_4L_8 \cdot HL \cdot H_2O$, L = 3-[2-(4-pyridyl)ethenyl]benzoate, contains three different types of layers, two of which are single sheets, while the third consists of two parallel interpenetrating (4,4) nets [47]. Similarly, in the structures of $Ag(bpp)_2(XF_6)$, X = P and As, layers of single (4,4) sheets and doubly interpenetrating (4,4) sheets alternate [48]. The structures of $M(bipy)(azpy)_2(NCS)_2 \cdot H_2O$, M = Mn and Co, display a remarkable structure in which triple helices defined by hydrogen bonds are cross-linked by M-bipy-M bridges, generating three independent (4,4) nets (Figure 9(a)) [49]. The mode of interpenetration is unprecedented and very different from those shown in Figure 2. Another structure with an unusual mode of interpenetration is $[Mn(p\text{-XBP4})_3](ClO_4)_2$, *p*-XBP4 = *N*, *N′*-*p*-phenylenedimethylenebis(pyridin-4-one) [50]. The four-connecting nodes (Mn atoms) are bridged in one direction by $Mn(p\text{-XBP4})_2Mn$ rings, and in the other direction by Mn(*p*-XBP4)Mn rods. Two sheets interpenetrate such that the rods pass through the rings in a rotaxane-like fashion (Figure 9(b)).

Structures showing 2-D → 3-D parallel interpenetration typically contain individual networks that are either highly undulating or have some degree of "thickness". The

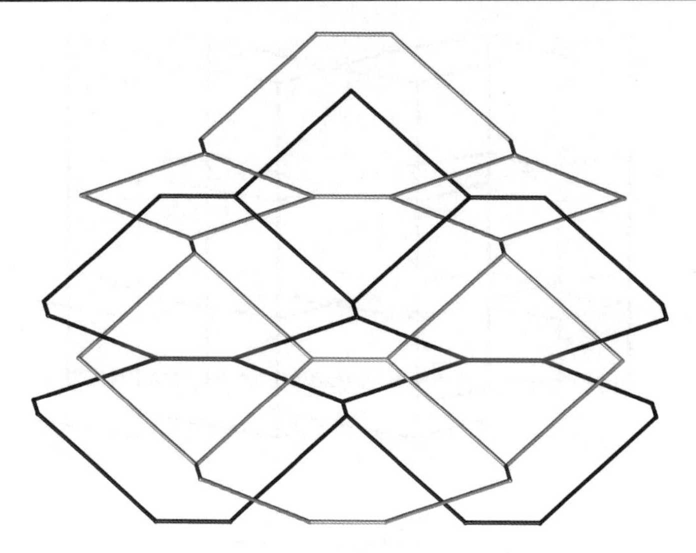

Figure 8. Schematic representation of the interpenetration of two 4.8^2 nets in the structure of $Cd_2(NO_3)_4(dpt)_2(MeCN)$ [36].

structure of α-trimesic acid is an example of the former [51]. The structure contains highly undulating hydrogen-bonded hexagonal (6,3) sheets. Each window of each sheet has three other networks passing through it. However, each sheet actually interpenetrates with nine other sheets (Figure 10). Interestingly, three of these sheets have coincident mean planes with the first sheet (as seen for 2-D → 2-D parallel interpenetration), while three have mean planes parallel but "above" that of the first sheet, while the final three are "below" the first sheet. Another structure in which each sheet interpenetrates with both coplanar and offset parallel sheets is $AgL(CF_3SO_3)\cdot 0.5H_2O$, L = 1,3,5-tris(4-cyano-phenoxymethyl)-2,4,6-trimethylbenzene [52]. In this structure, layers of doubly interpenetrating 2-D → 2-D (6,3) sheets interpenetrate with other parallel but offset such layers, giving the overall 3-D entanglement. Each (6,3) net thus interpenetrates with two others, one coplanar and one in the next double layer. The structure of $[Cu(bpee)_{1.5}(PPh_3)]PF_6\cdot 1\cdot 5CH_2Cl_2$ is much simpler, with each corrugated (6,3) sheet interpenetrating with one above and one below [53]. In $Cu_3(bipy)_2(pydc)_2\cdot 4H_2O$, pydc = pyridine-2,4-dicarboxylate, each undulating sheet interpenetrates four others, two on either side [54].

2-D → 3-D parallel interpenetration of "thick" layers is highlighted by the structure of $Cu_4(dca)_4(bipy)_3(MeCN)_2$ [41]. The 2-D net in this structure consists of two parallel (6,3) sheets tied together by bipy bridges. Each net interpenetrates with four others, two above and two below (Figure 11). The structure is similar to that of $Ag_3[Si(p\text{-}C_6H_4CN)_4]_2(PF_6)_3\cdot 1\cdot 6THF\cdot 0.5C_6H_6\cdot 2CH_2Cl_2$ [55]. The "thickness" in the 2-D nets in $[Co_5(bpe)_9(H_2O)_8(SO_4)_4](SO_4)\cdot 14H_2O$, bpe = 1,2-bis(4-pyridyl)ethane, is even more pronounced, with each consisting of *five* layers linked by bpe bridges [56]. Each net is penetrated by four others.

In contrast to the α polymorph discussed above, the γ polymorph of trimesic acid shows 2-D inclined interpenetration [51]. Each window of each hydrogen-bonded (6,3) sheet is penetrated by three other sheets (Figure 12). The structures of Cu(bipy)X, X = Cl, Br and I, contain (6,3) sheets (with Cu_2X_2 nodes) in which each window is again penetrated by

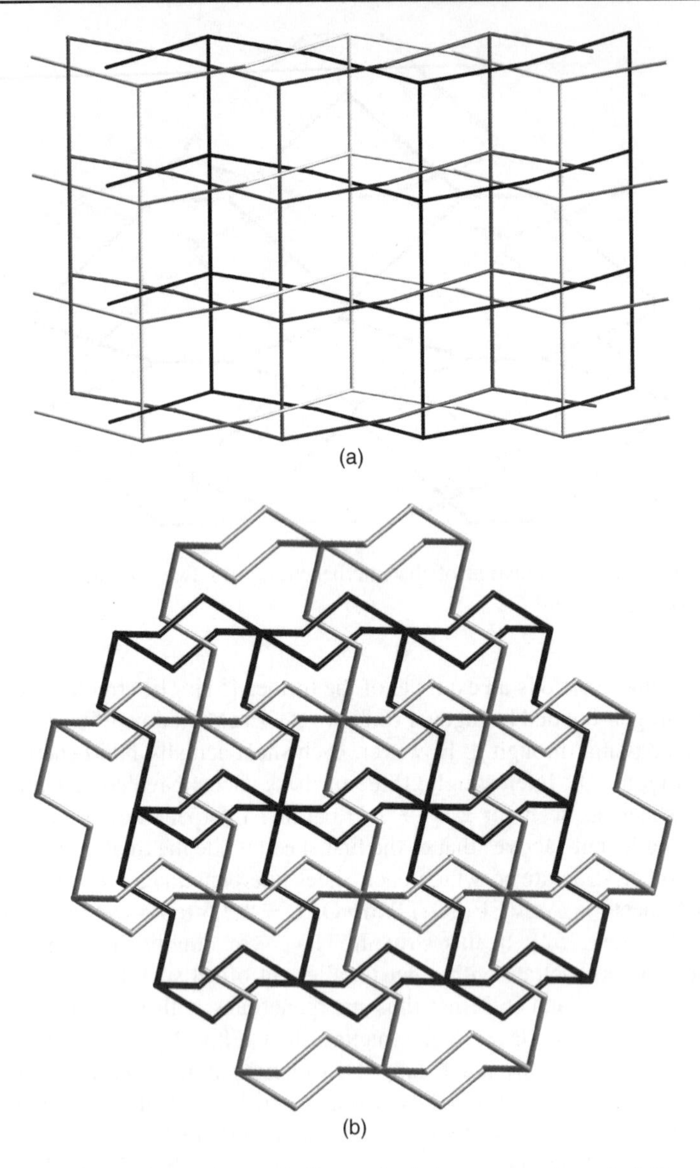

Figure 9. Schematic representations of the unusual interpenetration of 2-D → 2-D parallel (4,4) nets in the structures of (a) $M(bipy)(azpy)_2(NCS)_2{\cdot}H_2O$, M = Mn and Co [49] and (b) $[Mn(p\text{-}XBP4)_3](ClO_4)_2$ [50].

three other inclined sheets [57]. The compound $(rad)_2[Mn_2\{Cu(opba)\}_3]$, rad^+ = 2-(4-*N*-methylpyridinium)-4,4,5,5-tetramethylimidazolin-1-oxyl-3-oxide and opba = *o*-phenylenebis(oxamate) contains two sets of interpenetrating (6,3) sheets inclined at 73° to each other [58]. In each layer, every window is penetrated by a rod from one inclined layer. However, the interpenetrating networks are cross-linked by weak Cu···O interactions between the sheets and the radical rad^+ cations. The compound shows long-range magnetic ordering below 22.5 K.

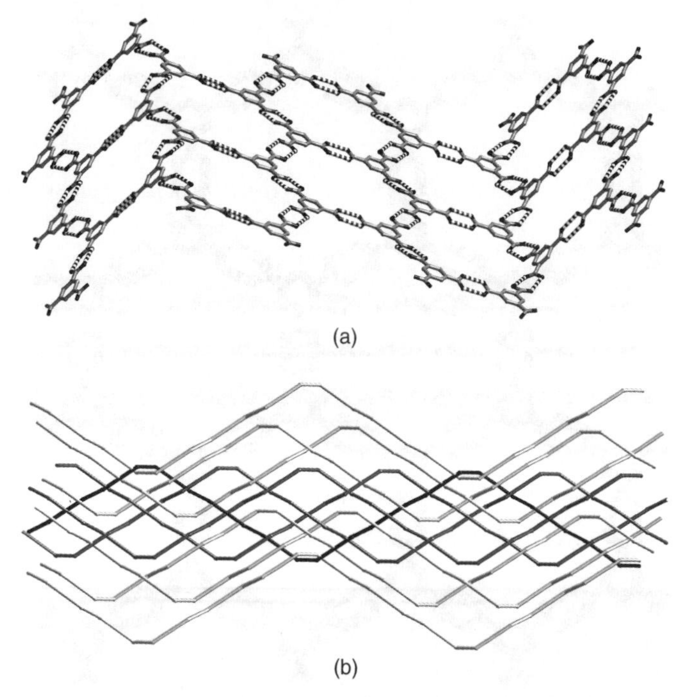

Figure 10. (a) One undulating hydrogen-bonded (6,3) sheet in the structure of α-trimesic acid [51]. Hydrogen bonds are shown by the striped bonds, and (b) the interpenetration of nine other such sheets through a single net (viewed side-on), generating 2-D → 3-D parallel interpenetration.

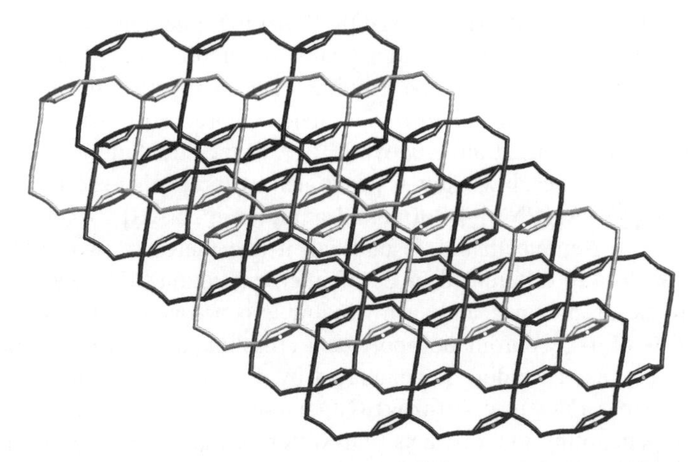

Figure 11. 2-D → 3-D parallel interpenetration of thick 2-D layers (viewed side-on) in the structure of $Cu_4(dca)_4(bipy)_3(MeCN)_2$ (bipy ligands represented schematically) [41].

Zaworotko has outlined three topologically different modes for inclined interpenetration of (4,4) sheets, which he labelled diagonal/diagonal, parallel/parallel, and parallel/diagonal [59]. In the parallel mode, the window of one sheet is penetrated by a rod of the other, whereas in the diagonal mode the window of the given sheet contains a node of the interpenetrating sheet. The (4,4) sheets in the structures of $[M(bipy)_2(H_2O)_2](SiF_6)$,

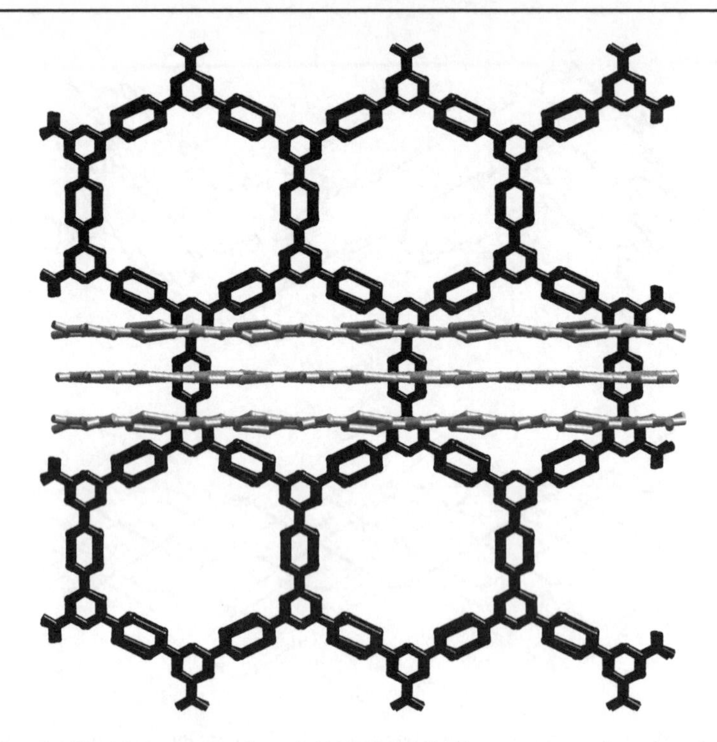

Figure 12. Inclined interpenetration of (6,3) sheets in the structure of γ-trimesic acid [51].

M = Zn, Cd and Cu, and $[Cd(bipy)_2(H_2O)(OH)](PF_6)$ interpenetrate with the diagonal/diagonal mode [60], as shown schematically in Figure 13(a). Each sheet window contains the metal node of one interpenetrating sheet. The structure of $Co(dca)_2(bipy)\cdot 0.5H_2O\cdot 0.5MeOH$ shows the parallel/parallel mode (Figure 13(b)); each window is penetrated by a rod of an interpenetrating network [61]. In the structure of $CdL_2(NO_3)_2$, L = 1,2-bis(4-pyridyl)hexane, each window is penetrated by three other sheets [44]. The structure of $[Cd(4\text{-ampy})_2\{Ag(CN)_2\}_2][Cd(mea)(4\text{-ampy})\{Ag(CN)_2\}_2]_2$, 4-ampy = 4-aminopyridine and mea = 2-aminoethanol, is unusual in that there are two different types of (4,4) sheets that display different degrees of interpenetration [62]. One type is defined by a 2-D coordination polymer, and each window is penetrated by two nets of the other type that consists of 1-D coordination polymers crosslinked by hydrogen bonds. In this second type of net, each window is penetrated by only one net of the other type. The structure of $[Ni(azpy)_2(NO_3)_2]_2[Ni_2(azpy)_3(NO_3)_4]\cdot 4CH_2Cl_2$ is even more remarkable as it contains interpenetrating inclined nets with different topologies – (6,3) nets, in which each window is penetrated by two interpenetrating nets, and (4,4) sheets, in which each window is penetrated by one interpenetrating net [63].

While most examples of 2-D inclined interpenetration involve two inclined sets of interpenetrating sheets, often perpendicular or close to it, a few examples have been reported with three or four inclined stacks of sheets. Both the hydrogen-bonded nets of $Pt(HL)_2L_2\cdot 2H_2O$, HL = 4-pyridylcarboxylic acid [64] and the coordination polymer nets of $Fe(bpb)_2(NCS)_2\cdot 0.5H_2O$, bpb = 1,4-bis(4-pyridyl)butadiene [65] define (4,4) sheets that interpenetrate in three mutually perpendicular directions. In each case, the windows

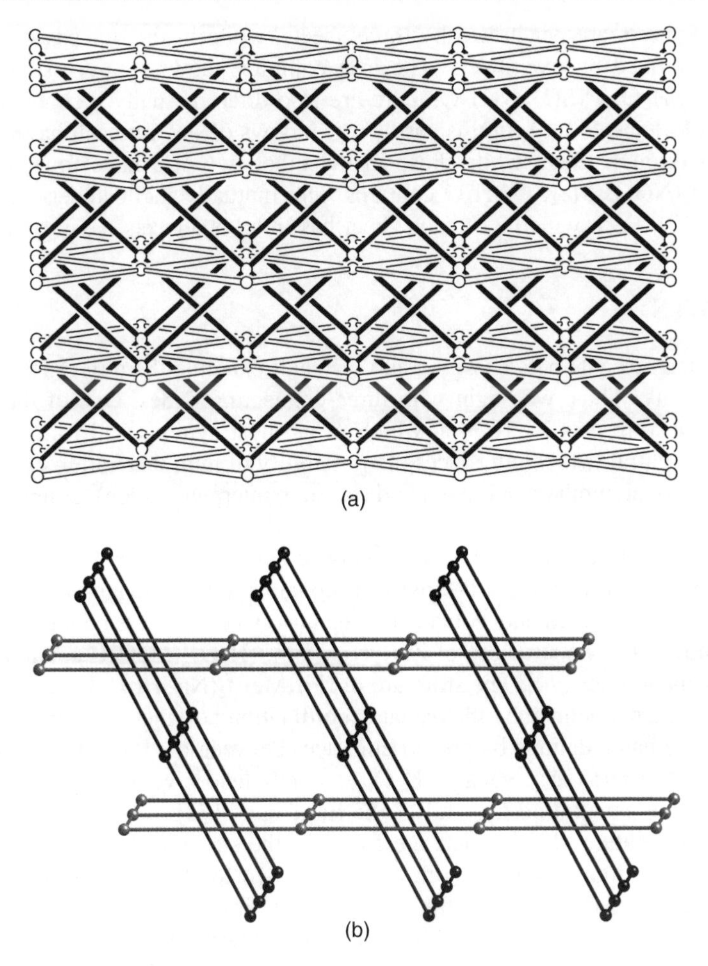

Figure 13. Schematic representations of inclined interpenetration of (4,4) 2-D sheets showing (a) diagonal/diagonal topology [59] for $[M(bipy)_2(H_2O)_2](SiF_6)$, M = Zn and Cd, Cu [60], and (b) parallel/parallel topology [59] for $Co(dca)_2(bipy)\cdot 0.5H_2O\cdot 0.5MeOH$ [61].

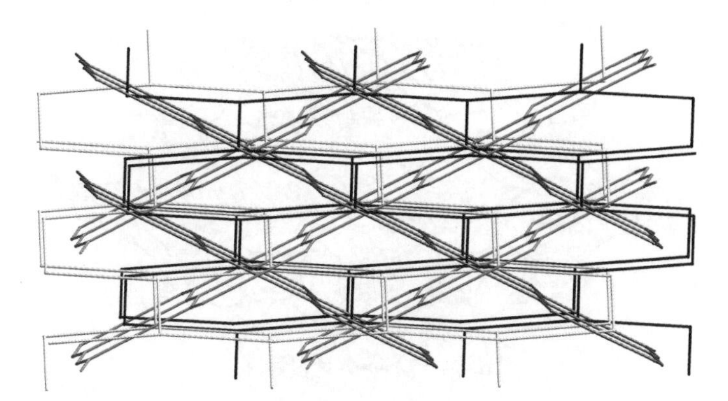

Figure 14. Schematic representation of the interpenetration of four mutually inclined sets of (6,3) 2-D sheets in the structure of $Co_2(azpy)_3(NO_3)_4\cdot Me_2CO\cdot 2H_2O$ [67].

in one set of sheets are penetrated by three sheets, while in the other two sets the windows are penetrated by five sheets. The iron complex shows spin-crossover properties. In $[Ni_6(bpe)_{10}(H_2O)_{16}](SO_4)_6 \cdot xH_2O$, there are also three mutually inclined sets of interpenetrating (4,4) sheets [66]. In this case, the windows of one set each have two inclined nets passing through them, while those in the other two sets have four. The structure of $Co_2(azpy)_3(NO_3)_4 \cdot Me_2CO \cdot 2H_2O$ contains *four* mutually inclined sets of (6,3) sheets (Figure 14) [67]. Each window of each sheet has three other nets passing though it.

5. 3-D NETS

There are numerous 3-D nets possible, and we shall examine them in order of increasing node connectivity. Thus, we begin with three-connecting nodes. One of the simplest of the three-connecting nets is (10,3)-a, which in its most symmetrical form is cubic. It is also inherently chiral, which introduces the possibility of interpenetration of nets of all the same hand, of equal numbers of either hand (racemic interpenetration), or unequal numbers of either hand. The later case has yet to be reported. Cyanamide (H_2NCN) contains two interpenetrating hydrogen-bonded nets of opposite hand (Figure 15) [68]. The chloroform solvate of 3,3′,5,5′-tetramethyl-4,4′-bipyrazole contains four hydrogen-bonded nets (two of either hand), while the methanol solvate and the unsolvated γ phase contain six nets (three of either hand) [69]. The structure of $Zn(tpt)_{2/3}(SiF_6)(H_2O)_2(MeOH)$ contains eight nets, four of each enantiomer [70]. The structure of $(Ph_3MeP)_2[NaCr(ox)_3]$, ox = oxalate, contains one (10,3)-a net defined by the oxalate coordination polymer, and an interpenetrating one, of opposite hand, defined by phenyl embraces between the Ph_3MeP^+ cations [71]. In contrast, $Ni_3(btc)_2(py)_6(eg)_6 \cdot {\sim}3eg \cdot {\sim}4H_2O$, eg = ethylene glycol, and related structures, contains four interpenetrating nets, all of the same hand [72].

Two interpenetrating (10,3)-b nets are found in the structures of $Zn_3(tpt)_2X_6$, X = Cl and I [3, 73]. The iodine structures have been shown to shrink or swell depending on the passage of guests in and out of the structure. $Cu(bipy)_{3/2} \cdot NO_3 \cdot 1{\cdot}25H_2O$ [74] and one polymorph of $Ag(TEB)(CF_3SO_3)$ [75] contain six interpenetrating (10,3)-b nets. Four interpenetrating (10,3)-d nets are seen in the structures of $Co(H_2biim)_3 \cdot 0.8DMF \cdot 0.5H_2O$, H_2biim = biimidazole [76] and $CoL_{1.5}(NO_3)_2 \cdot H_2O$, L = 1,4-bis(3-pyridyl)-2,3-diaza-1,3-butadiene [77]. $NaTi_2(PS_4)_3$ contains two interpenetrating (8,3)-c nets [78], while

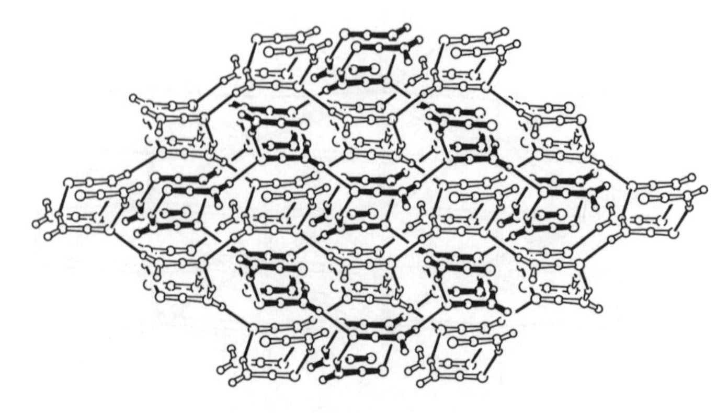

Figure 15. Two hydrogen-bonded (10,3)-a nets in the structure of cyanamide (H_2NCN) [68]. Hydrogen bonds are represented by the thin lines.

$[Ph_3PCH_2Ph][Cd(tp)Cl]$, tp = terephthalate, contains two nets with 8^2.10-a (or LiGe) topology [79].

The largest class of interpenetrating nets are those with the 4-connected diamond topology. A classic example is found in the structures of $M(CN)_2$, M = Cd and Zn [1, 80, 81], which contain two interpenetrating coordination nets (Figure 16). The maximum number of interpenetrating nets observed to date are 11, seen in the hydrogen-bonded structure of the 1:2 co-crystal of tetrakis[4-(3-hydroxyphenyl)phenyl]methane and benzoquinone [82]. The maximum for a coordination polymer, namely $[Ag(ddn)_2]NO_3$, ddn = 1,12-dodecanedinitrile, is 10 [83]. One particularly unusual example is the structure of $K_2[PdSe_{10}]$, which contains two interpenetrating nets [84]. The nets are chemically different – in one net, the Pd nodes are connected by Se_4^{2-} links, in the other they are connected by Se_6^{2-}. Despite the different compositions, the Pd· · · Pd distance is the same in both nets, as it must be; the Se_6^{2-} link is more twisted than the Se_4^{2-} link.

Nearly all structures with interpenetrating diamond nets show the same mode of interpenetration, which we shall call the "normal" mode. A feature of this mode is the relationship between adamantane-like cages from the interpenetrating nets. The two related adamantane units for any two interpenetrating nets are defined by the passing of a rod from one net through a hexagonal window of an adamantane cavity of the other net [41]. This rod is then used to define the second adamantane cavity. In the structure of β-Cu(dca)(bpee), the adamantane cavities of the five crystallographically identical interpenetrating diamondoid networks, defined as above, display different relationships to each other [41]. Two of the adamantane units interact with the first in the normal fashion, whereas the other two do not (Figure 17). Abnormal interpenetration is also

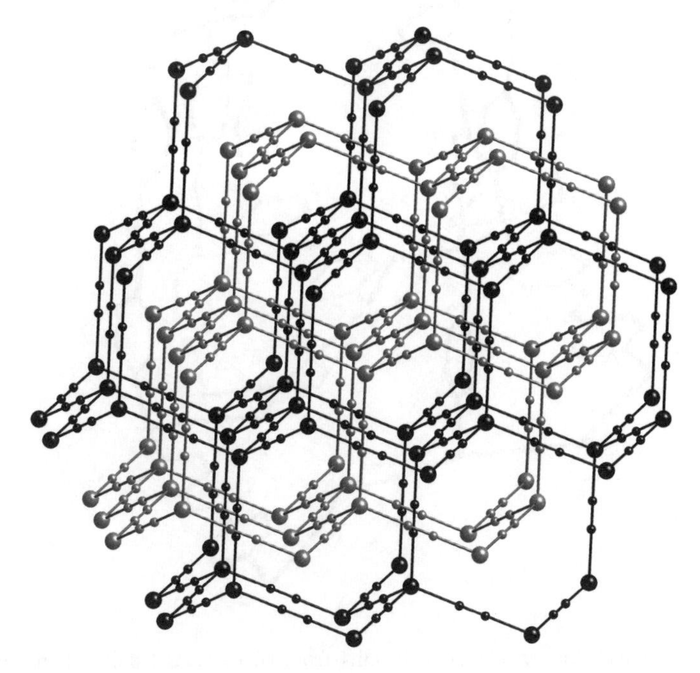

Figure 16. Two interpenetrating diamondoid nets in the structure of $M(CN)_2$, M = Zn and Cd [1, 80, 81].

observed between the five interpenetrating hydrogen-bonded diamond nets in the structures of adamantane-1,3,5,7-tetracarboxylic acid and methanetetrapropionic acid [85]. In these structures, any two nets interpenetrate in the normal fashion; only when a third net is considered does the unusual interpenetration become apparent. Abnormal interpenetration in the structure of Cd(imidazole-4-acrylate)$_2$·1·7H_2O is likely directed by hydrogen-bonding interactions between the four interpenetrating diamondoid coordination polymer nets [86]. Abnormal modes of interpenetration are also seen in the structures of CuL_2PF_6, L = 2,7-diazapyrene (3 nets) [83, 87], Ni(fum)(bpe), fum = fumarate (5 nets) [88], $(NH_4)(Me_2NH_2)(CdL_2)$, L = 3,3′-azodibenzoate (6 nets) [89], Ag(3,3′-DCPA)$_2$$ClO_4$·$H_2O$, 3,3′-DCPA = 3,3′-dicyanodiphenylacetylene (8 nets) [83, 90], and [Ag(ddn)$_2$]XF_6, X = P and As (8 nets) [83].

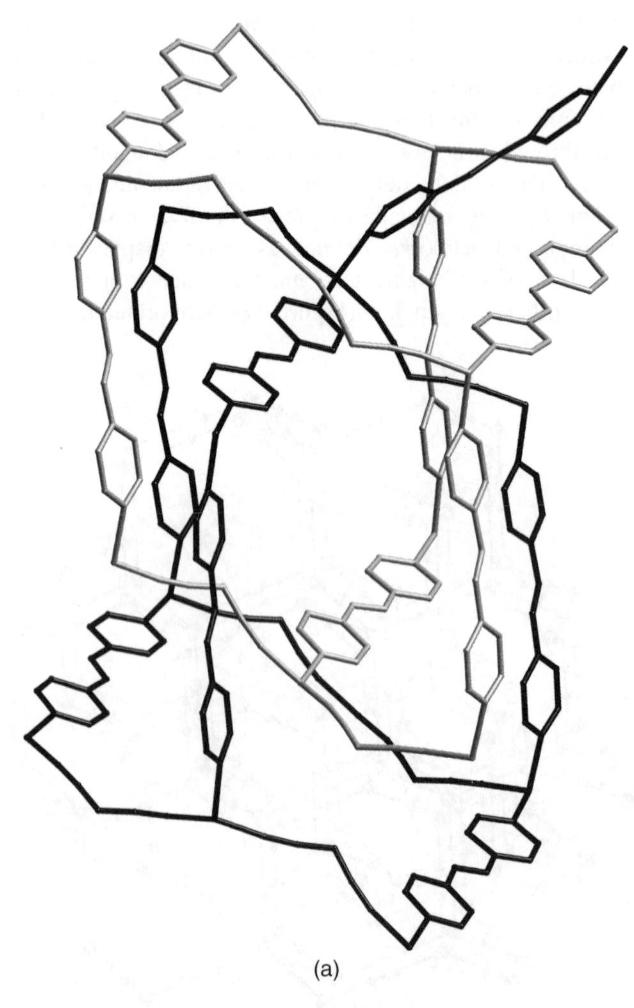

(a)

Figure 17. Interpenetration of adamantane units from two selected pairs of the five interpenetrating diamondoid nets in the structure of β-Cu(dca)(bpe) showing (a) a normal relationship and (b) an abnormal relationship [41].

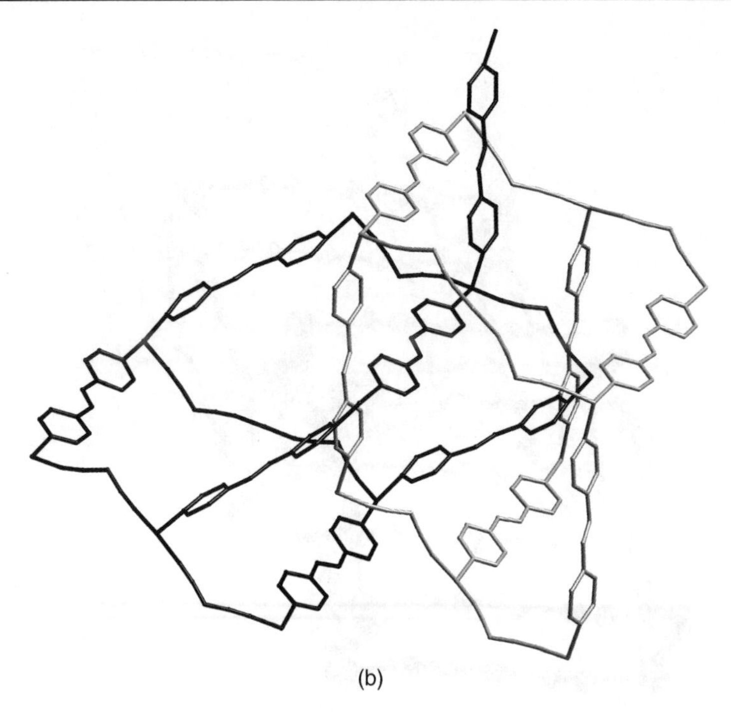

Figure 17. *(continued)*

Other 4-connected nets are less common. Six interpenetrating nets with the chiral quartz topology (all of the same hand) are observed in the structures of $M[Au(CN)_2]_3$, M = Zn and Co [91]. The structure of $Cu(tcp)CuBF_4{\cdot}17C_6H_5NO_2$, Cu(tcp) = 5,10,15,20-tetrakis(4-cyanophenyl)-21H,23H-prophine copper(II), contains two PtS nets that interpenetrate in an asymmetric fashion [92]. $Cu(bipy)_2(CF_3SO_3)_2$ contains two interpenetrating $4^2.8^4$ nets [93], while two $4^2.6^3.8$ nets are present in Cu(SCN)(bpe) [94]. $[Ag(sebn)_2]XF_6$, sebn = sebaconitrile, and X = P and As contains four $SrAl_2$ nets [95], while the 1:1 co-crystal of cyanuric acid and biuret [96] and the coordination polymer $[Cu(bpe)(H_2O)(SO_4)]$ [97] both contain two $CdSO_4$ nets. Interpenetration of NbO nets occurs for $FeL_2[Ag(CN)_2]_2{\cdot}2/3H_2O$, L = 3-cyanopyridine (three nets) [98], and for $Cu_2(OMe)_2L_2{\cdot}0.69H_2O$, HL = 9-acridinecarboxylic acid, in which the two interpenetrating nets contain $Cu_2(\mu\text{-}OMe)_2$ dimers acting as nodes [99].

Five-connected nets are rare, but the structures of $Cu(bipy)_{1.5}Cr_2O_7{\cdot}H_2O$ [100], Cu_2(2,5-dimethylpyrazine)$(dca)_4$ [101] and $Zn_2(OH)$(btc)(pipe), pipe = piperazine [102], all contain two interpenetrating nets with five-connecting nodes. In the latter case, the nodes are $Zn_2(OH)(O_2CR)$ clusters.

The second most common form of interpenetration is that of 6-connected α-Po networks, although none to date have been reported with more than three interpenetrating nets. The compounds Mn(N, N'-butylenebisimidazole)$_2(X)_2$, X = BF_4, ClO_4, ClO_4/PF_6 and ClO_4/AsF_6, contain two interpenetrating α-Po nets [103], as do the structures of α-M(dca)$_2$(pyrazine), M = Mn, Fe, Co, Ni, Cu and Zn [104], and $(Me_3Sn)_3Rh(SCN)_6$ [105]. Despite the formation of two interpenetrating nets in the structure of $Tb_2(ADB)_3[Me_2SO]_4{\cdot}$

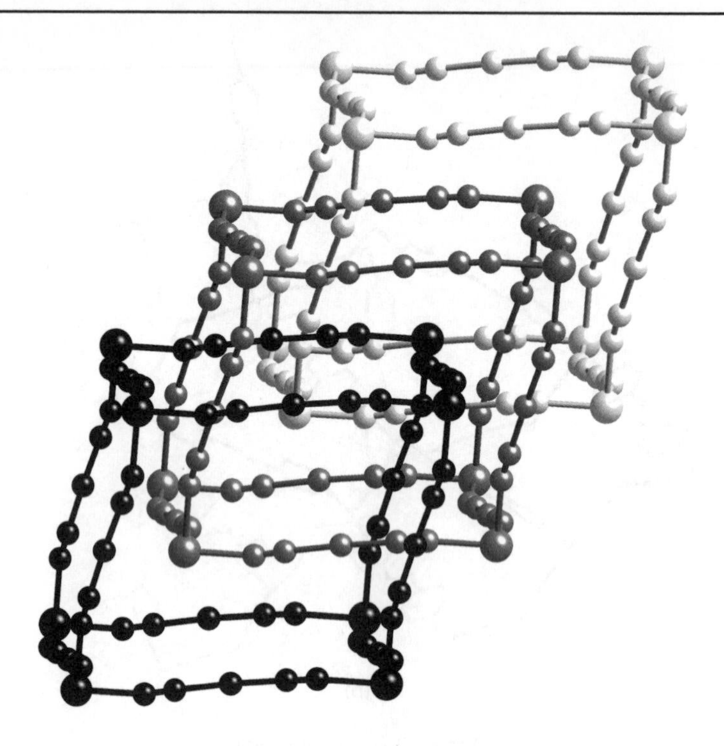

Figure 18. Three interpenetrating α-Po nets in the structure of $Rb[Cd\{Ag(CN)_2\}_3]$ [107].

$16Me_2SO$, ADB = 4,4′-azodibenzoate, the free volume is still 71% of the crystal volume [106]. The fact that more than two interpenetrating networks were not formed is ascribed to the presence of bulky $Tb_2(O_2CR)_6$ clusters acting as the nodes. Three interpenetrating nets are shown by a series of related compounds typified by the structure of $Rb[Cd\{Ag(CN)_2\}_3]$ (Figure 18) [107]. Three interpenetrating nets are also shown by the coordination polymers $Zn(bib)_3(BF_4)_2$, bib = bis(imidazole)butyne, $Co(bib)_3(NO_3)_2$, and $Cd(bix)_3(X)_3$, X = ClO_4 and NO_3 [108].

Like diamondoid nets, α-Po nets show a "normal" mode of interpenetration, and only one example of abnormal interpenetration has been reported. In the normal mode of interpenetration, each cubelike cavity catenates with eight such cavities in the interpenetrating network. In the structure of $[Mn(bpe)(H_2O)_4](ClO_4)_2(bpe)_4$, which contains nets defined by both coordination and hydrogen bonds, a cubic cavity of one net catenates with ten cavities of the other net, as shown in Figure 19 [109].

The structure of $Eu[Ag(CN)_2]_3{\cdot}3H_2O$ represents a rare example of interpenetrating (threefold) 6-connected nets with topology other than α-Po [110]. The structures of $M_3(bpdc)_3(bipy){\cdot}solv$, M = Co and Zn and bpdc = biphenyldicarboxylate, contain two interpenetrating 8-connected nets, with $M_3(O_2CR)_6$ clusters acting as nodes [111].

Thus far, we have only mentioned nets containing nodes of only one degree of connectivity. However, a number of simple nets are possible that contain nodes of two or more degrees of connectivity. A fundamental 3,4-connected net is Pt_3O_4; two such nets that interpenetrate asymmetrically are found in the structure of $Cu_3(BTB)_2(H_2O)_3{\cdot}9DMF{\cdot}2H_2O$, H_3BTB = 4,4′,4″-benzene-1,3,5-triyl-tribenzoic acid [112]. Two 3,4-connected nets are also found in

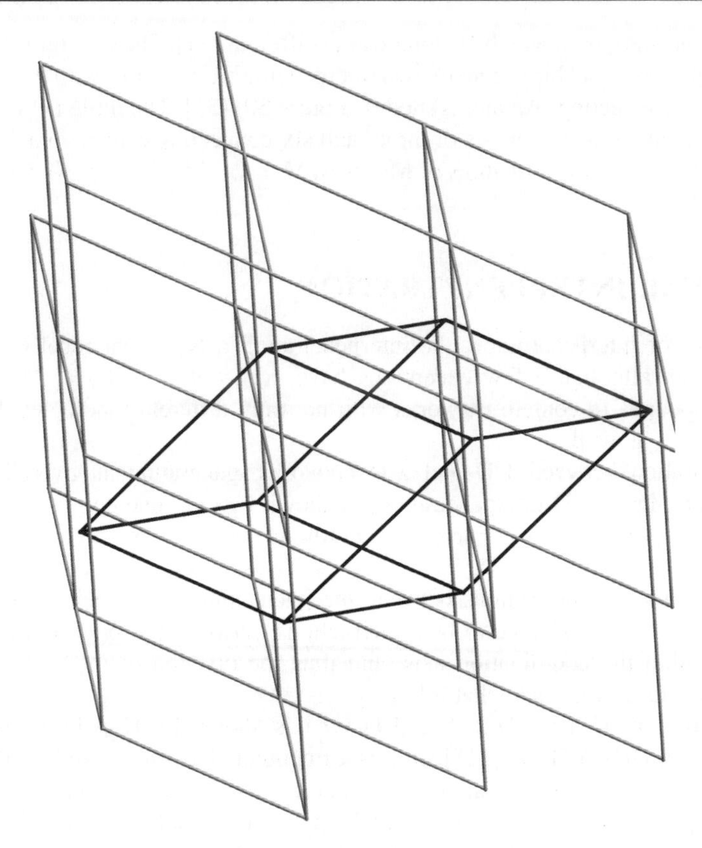

Figure 19. An abnormal interpenetration mode between α-Po nets in the structure of [Mn(bpe)-$(H_2O)_4$]$(ClO_4)_2$(bpe)$_4$ [109], represented schematically. The central cage catenates with ten analogous cages of the second net (rather than the usual eight).

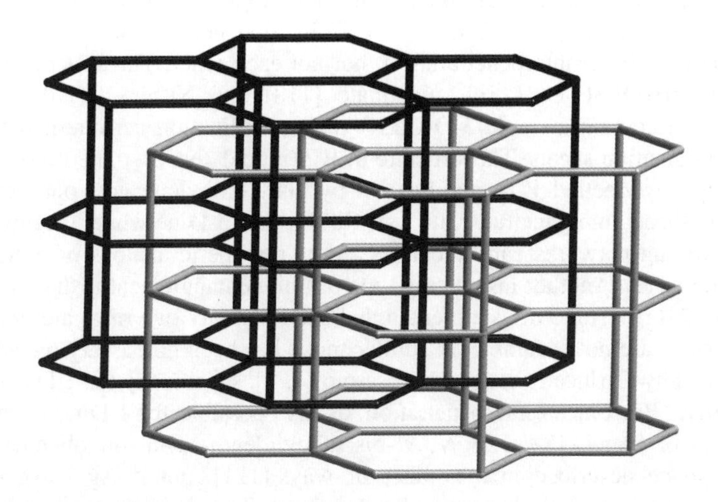

Figure 20. Schematic representation of the two interpenetrating 3,5-connected nets in the structure of Ag(tcm)(dabco) [31].

Ag(tcm)(phenazine)$_{1/2}$, in which the topology is different [31]. The structures of Ag(tcm)L, L = dabco, pyrazine and bipy, contain two interpenetrating networks with three-connecting (tcm) and five-connecting (Ag atoms) nodes (Figure 20) [31]. The rutile net is another fundamental binodal net as it consists of three- and six-connecting centers in a 2:1 ratio. Two such nets are found in the structures of $M(tcm)_2$, M = Cr, Mn, Fe, Co, Ni, Cu, Zn, Cd and Hg [113].

6. UNUSUAL INTERPENETRATION

In most cases of interpenetration, the interpenetrating nets are chemically and topologically identical, although a few exceptions have been outlined above. Thus, it is very rare for compounds to contain networks with not only different topologies, but different dimensionalities as well.

Interpenetration between 1-D and 2-D networks, generating an overall 2-D entanglement (1-D/2-D $\rightarrow$ 2-D interpenetration) is shown by only one compound, $Co(bipy)_{2.5}(NO_3)_2{\cdot}2C_{14}H_{10}$ [114]. In this structure, 1-D railroad coordination polymers interpenetrate with 2-D (4,4) sheets of phenanthrene molecules. The sheets are defined by intermolecular aromatic edge-to-face interactions between the phenanthrene molecules. However, there are also edge-to-face and face-to-face interactions between the phenanthrene and bipy molecules within the coordination nets, and thus the division of the structure into two types of separate nets is somewhat arbitrary.

Interpenetration of the 1-D/2-D $\rightarrow$ 3-D type is shown by $[Cu_5(bpp)_8(SO_4)_4(EtOH)(H_2O)_5](SO_4){\cdot}EtOH{\cdot}25{\cdot}5H_2O$ [115]. In this compound, 1-D chains and 2-D (4,4) sheets interpenetrate in an inclined fashion, such that each sheet window has a rod from one chain passing through it, and each chain loop has two sheets passing through it; see Figure 21.

1-D/3-D interpenetration is shown by the structure of $[Co(bix)_2(H_2O)_2](SO_4){\cdot}7H_2O$, which contains 3-D $CdSO_4$ nets interpenetrating with 1-D chains [116]. The structure of $M(L)F_2{\cdot}14H_2O$, M = Cd and Zn and L = hexakis(imidazol-1-ylmethyl)benzene, shows 2-D/3-D interpenetration [117]. The α-Po coordination net has two different 2-D (6,3) hydrogen-bonded sheets interpenetrating it, but not each other. The structures of $[Cu_2L_4{\cdot}3H_2O][Cu_4L_4{\cdot}2H_2O]{\cdot}3H_2O$, L = isonicotinate [118] and $Ni(bipy)_2(H_2PO_4)_2{\cdot}C_4H_9OH{\cdot}H_2O$ [119] both contain a single 3-D $CdSO_4$-like coordination net interpenetrating with 2-D (4,4) coordination sheets. The structure of $[Co(mppe)_2(NCS)_2]{\cdot}2[Co(mppe)_2(NCS)_2]{\cdot}5MeOH$, mppe = 1-methyl-1′-(4-pyridyl)-2-4-pyrimidyl)ethylene, goes one step further – the 2-D (4,4) sheets interpenetrate with two $CdSO_4$-like 3-D networks (Figure 22) [120].

Interpenetrating networks can be considered as polymeric analogs of molecular catenanes and rotaxanes. Another fascinating type of finite entanglement is that of Borromean rings (Figure 23(a)). Three rings are entangled such that any two rings are *not* catenated, and yet the rings are not separable. A similar motif has been noted very recently in polymeric networks by Carlucci, Ciani and Proserpio [7, 8], specifically for 2-D nets, although they show that "Borromean interpenetration" is also possible for 1-D nets. For example, the structures of $[Ag_2L_3]X_2$, L = *N*, *N*′-bis(salicylidene)-1,4-diaminobutane, X = NO_3 and ClO_4 can be described in a number of ways [121], but if Ag$\cdots$Ag interactions are ignored, then 2-D (6,3) sheets can be discerned. These sheets are highly puckered, and interpenetrate in a 2-D $\rightarrow$ 3-D parallel fashion. However, close inspection of the interpenetration (Figure 23(b)), reveals that although the layers are inseparable, any two

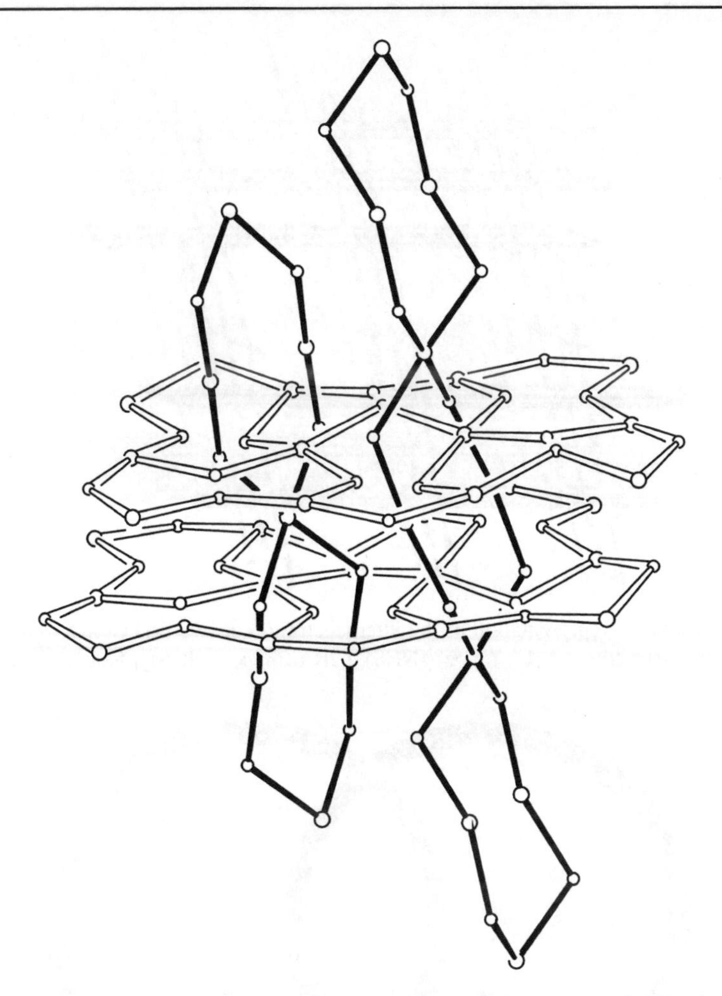

Figure 21. Schematic representation of the 1-D/2-D → 3-D interpenetration in the structure of $[Cu_5(bpp)_8(SO_4)_4(EtOH)(H_2O)_5](SO_4)\cdot EtOH\cdot 25\cdot 5H_2O$ [115].

adjacent layers are not interpenetrating. The structure of $[Cu(tmeda)_2\{Ag(CN)_2\}_3]ClO_4$, tmeda = N, N, N', N'-tetramethylethylenediamine, displays a different type of interpenetration, 2-D → 2-D parallel interpenetration [122]. Nonetheless, while the structure contains layers of three entangled (6,3) sheets (Figure 23(c)), no two sheets in this layer interpenetrate – it is only the combination of all three that makes them inseparable. A number of other cases of Borromean interpenetration have also been identified [7, 8].

7. CONSEQUENCES OF INTERPENETRATION

There is no doubt that interpenetration is a fascinating topological subject. But it can also produce new materials with interesting physical properties. While the presence of interpenetrating networks obviously reduces the microporosity of that which would otherwise be observed for a single net, it does not necessarily preclude porosity. For example,

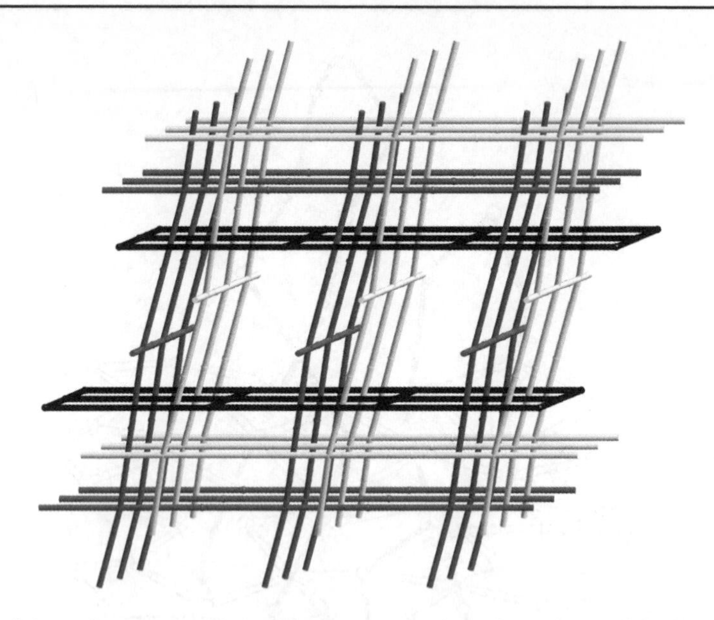

Figure 22. Schematic representation of the interpenetration of two 3-D $CdSO_4$-like nets and 2-D (4,4) sheets in the structure of $[Co(mppe)_2(NCS)_2]\cdot 2[Co(mppe)_2(NCS)_2]\cdot 5MeOH$ [120].

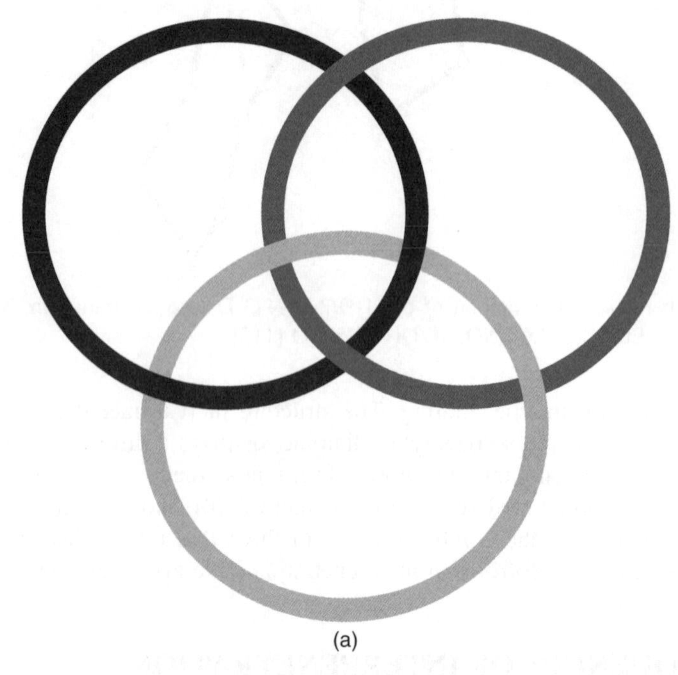

(a)

Figure 23. (a) Three Borromean rings, (b) schematic representation of the Borromean interpenetration [7, 8] in the 2-D → 3-D parallel interpenetration of (6,3) sheets in the structure of $[Ag_2L_3]X_2$, L = *N*, *N'*-bis(salicylidene)-1,4-diaminobutane, X = NO_3 and ClO_4 [121], and (c) schematic representation of the Borromean interpenetration in the 2-D → 2-D parallel interpenetration of (6,3) sheets in the structure of $[Cu(tmeda)_2\{Ag(CN)_2\}_3]ClO_4$ [122].

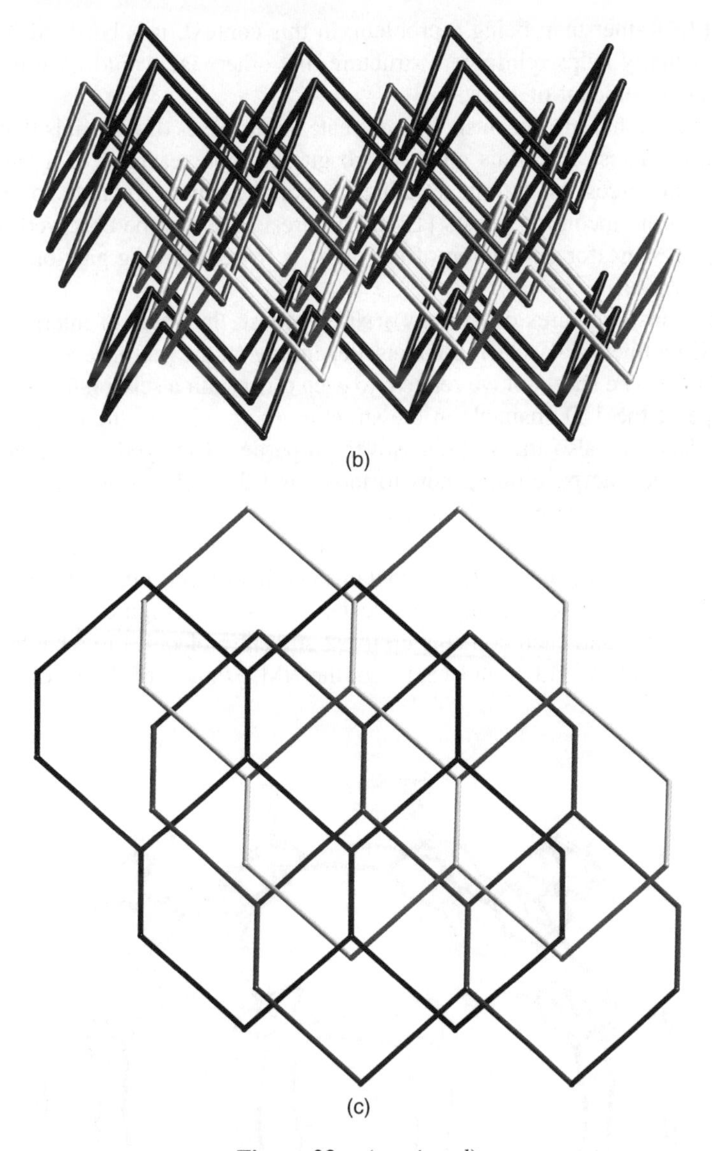

(b)

(c)

Figure 23. (*continued*)

the structure of $Cu_3(tpt)_4(ClO_4)_3$ contains two interpenetrating 3,4-connected networks, which can also be related to the α-Po net if vertex-sharing $[Cu_6(tpt)_4]^{6+}$ cages within the network are considered nodes [123]. The frameworks, however, only account for approximately 65% of the crystal volume, the rest being occupied by highly disordered anions in essentially liquid solvent. Interpenetration can even be beneficial for the microporosity of a material. Yaghi *et al.* have reported a structure containing two interpenetrating Pt_3O_4 networks that show reversible adsorption of large amounts of gases and organic solvents [112]. Rather than being a problem, in this context, it is believed that the interpenetration actually helps reinforce a structure that otherwise would have been expected to collapse upon removal of guests.

Interpenetration has also been shown to create new classes of materials that have "flexible" structures. These materials only absorb guests upon reaching of a "gate-opening" pressure, which forces the interpenetrating networks to move relative to each other in order to absorb the incoming guests [124]. Hysteresis can also be observed, whereby the gate-closing pressure for desorption of the guest is lower than the gate-opening pressure for guest absorption.

In microporous structures of $Fe(azpy)_2(NCS)_2$.guest, the angle of interpenetration between the 2-D inclined (4,4) sheets changes, depending on the presence and nature of guest molecules [125]. The sheets move relative to each other with a scissorslike motion, changing the shape of the 1-D channels in the structure. Not only are the channel dimensions guest dependant, but also the spin-crossover properties displayed by the Fe(II) centers. The ability of the interpenetrating nets to move and flex relative to each other is likely an important factor in the guest dependant magnetic properties.

One compound in which interpenetration has fascinating consequences is $Zn_3(tpt)_2(CN)_3(NO_3)_3 \cdot$solv [126]. The structure contains two interpenetrating networks, which, due to the interpenetration, create completely sealed off cavities (Figure 24). The cavities are sizable (>1290 Å^3) and each contains up to 22 solvent molecules in what are essentially isolated liquid droplets. Although broadened, the NMR signals of these solvents can even

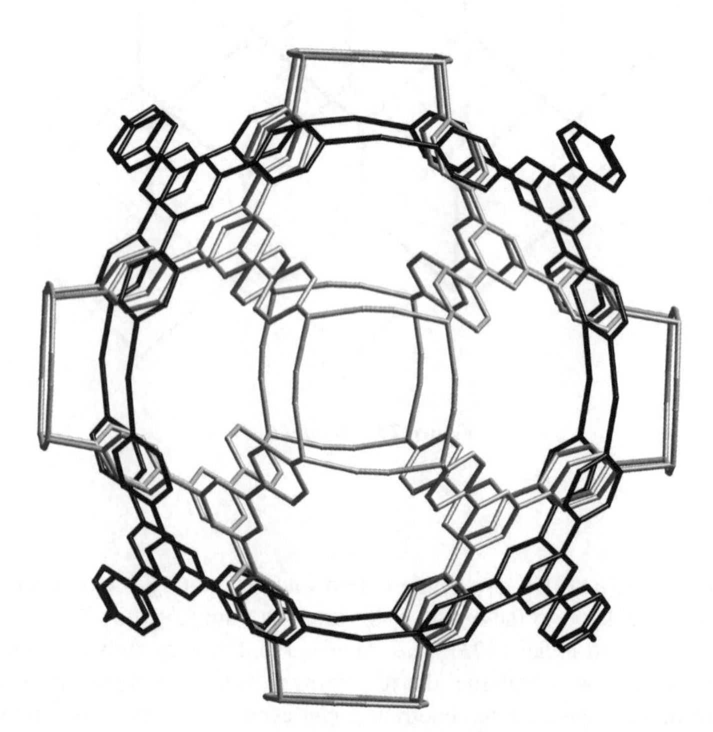

Figure 24. A single enclosed cavity formed by the two interpenetrating networks in the structure of $Zn_3(tpt)_2(CN)_3(NO_3)_3 \cdot$solv [126].

be observed by placing the solid crystals in an ordinary NMR instrument, confirming their liquidity.

Another consequence of interpenetration is the placement of the molecular components in close proximity with each other. This is particularly important in the structure of $Cu(dcnqi)_2$, dcnqi = dicyanoquinodiimine, and its derivatives [127]. These compounds contain seven interpenetrating diamondoid networks in which Cu(I) cations act as nodes. The interpenetration is such that the dcnqi links line up in π-bonded stacks that pass through the structure. It is these stacks that are responsible for the metal-like electrical conductivity.

8. SELF-PENETRATION

As mentioned earlier, interpenetrating networks can be usefully considered as polymeric analogs of 0D molecular catenanes and rotaxanes [4, 128]. So the question naturally arises as to what a polymeric analog of a 0D molecular knot would look like. For many 3-D networks, knotlike motifs can be defined, provided enough nodes are incorporated into the pathway that defines the knot. Thus, a limiting condition is required to make the network–knot analogy meaningful.

A useful mathematical tool commonly used in the analysis of networks is the "shortest circuit". The shortest circuit is the pathway containing the least number of nodes that leaves a given node via a given link, and returns to the node via a different given link. Thus, shortest circuits can be defined for each pair of links from each type of node. These shortest circuits can then be used to define the network. For example, the 2-D (6,3) notation, used throughout this article, means that the network has three-connecting nodes, and all the shortest circuits are 6-gons (i.e. contain six nodes). This can also be used in the Schläfli notation for this net, 6^3. In this case, the superscripted 3 refers not to the node connectivity, but to the fact that there are three shortest circuits that are 6-gons, one for each of the three different possible pairs of links from the node which can be used to define a shortest circuit.

An important consequence of this definition is the fact that all shortest circuits are not of the same size – they depend on the particular node and particular pairs of links chosen. Thus, for the rutile (TiO_2) network discussed earlier, which contains three-connecting and six-connecting centers, the shortest circuits are 4-, 6- and 8-gons, and the overall Schläfli symbol for this net is $(4^2.6^{10}.8^3)(4.6^2)_2$. Therefore, we define a self-penetrating network, a polymeric analog of a molecular knot, as a single network in which its smallest circuits are penetrated by links in the same network [4, 129].

The relationship between single networks, interpenetrating networks and self-penetrating networks is best illustrated by example. The structures of α-$M(dca)_2$, M = Cr, Mn, Fe, Co, Ni and Cu, contains single networks with the aforementioned rutile topology (Figure 25(a)) [130]. The metal atoms are octahedral and six-connecting and the dca ligands are three-connecting. The three-connecting tcm ligand is slightly larger than the dca ligand, so in the structures of $M(tcm)_2$, M = Cr, Mn, Fe, Co, Ni, Cu, Zn, Cd and Hg, two interpenetrating rutile networks are formed (Figure 25(b)) [113]. However, when both ligands are mixed, as in the compounds M(dca)(tcm), M = Co, Ni and Cu, a single, self-penetrating network with topology very close to, but different from, rutile is formed (indeed, it has the same Schläfli symbol); see Figure 25(c) [131]. As for the doubly interpenetrating rutile nets of $M(tcm)_2$, the six- and eight-membered shortest circuits are penetrated but not the four-membered shortest circuits. Thus, the self-penetrating rutile-related network for the dca/tcm compounds is a structural compromise between the single rutile networks of the dca compounds and the doubly interpenetrating rutile networks of the tcm compounds. Interestingly, the magnetic properties of the three M(dca)(tcm) compounds are also a compromise. Thus, for M = Co, Ni and Cu, the

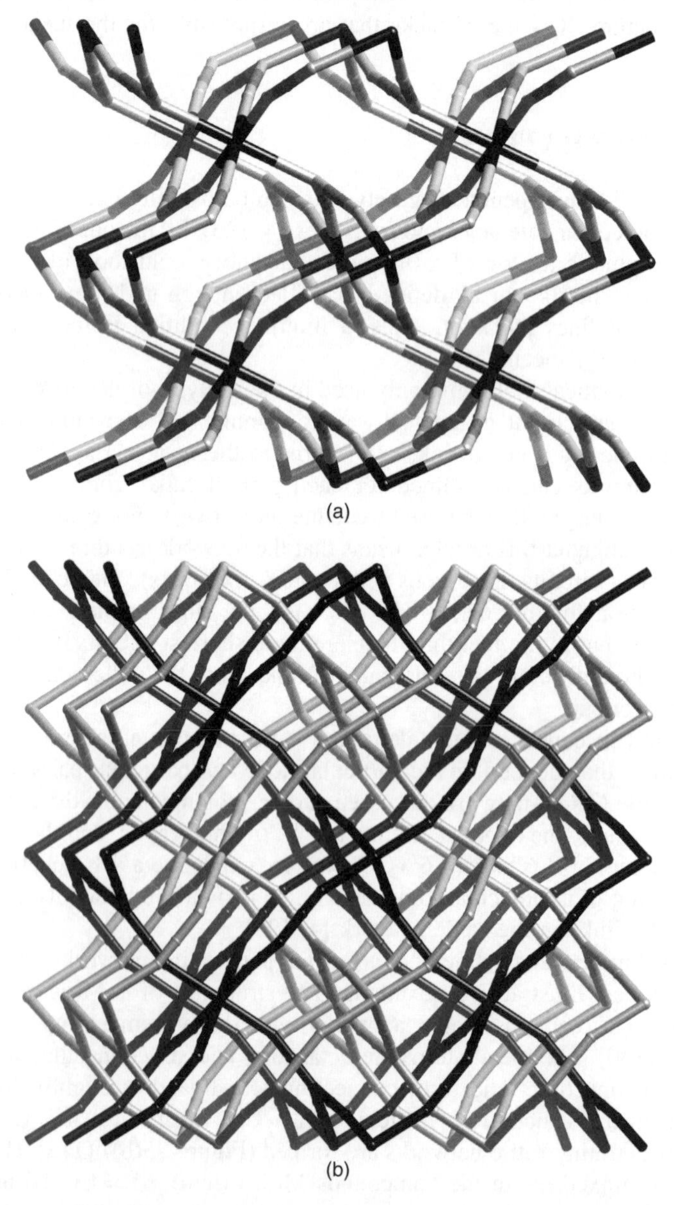

Figure 25. (a) A single rutile net in the structures of α-M(dca)$_2$ [130]; (b) two interpenetrating rutile nets in the structures of M(tcm)$_2$ [113] and (c) the single self-penetrating net (represented schematically) with the same Schläfli symbol as rutile found in the structures of M(dca)(tcm) [131].

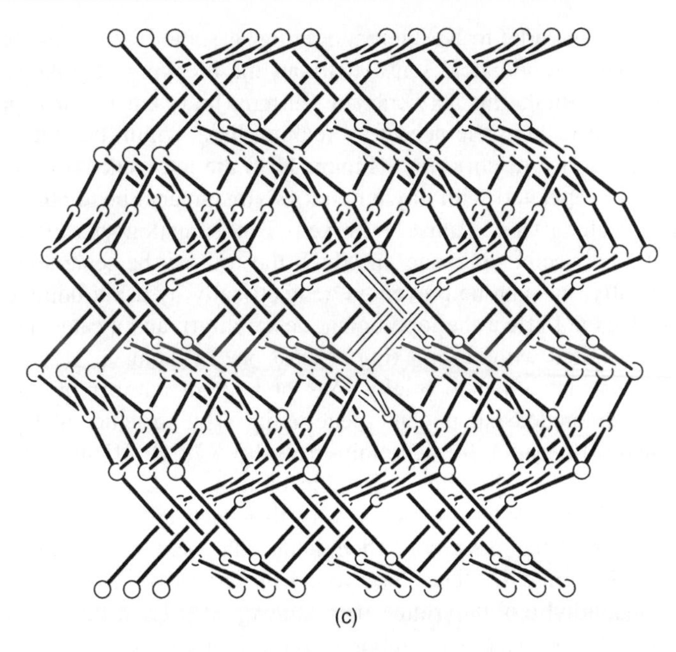

Figure 25. (*continued*)

α-M(dca)$_2$ complexes show long-range magnetic ordering, the M(tcm)$_2$ compounds do not, and the M(dca)(tcm) compounds do show ordering, but at lower temperatures than the α-M(dca)$_2$ analog [131].

A number of other more fundamental nets have also been shown to be self penetrating. These include the (12,3) net, as exemplified by the structure of Ni(tpt)(NO$_3$)$_2$ [129], the (8,4) net, as contained in the structure of α-Ni(pyara)$_2$(H$_2$O)$_2$, pyaraH = 3-(3-pyridyl)-acrylic acid [132], and the coesite net, as seen in [Ag(2-ethpyz)$_2$]SbF$_6$, 2-ethpyz = 2-ethylpyrazine [133]. Other reported self-penetrating networks include the structures of Cd$_2$(4,4′-pytz)$_3$(μ-NO$_3$)(NO$_3$)$_3$(MeOH), 4,4′-pytz = 3,6-bis(pyridin-4-yl)-1,2,4,5-tetrazine [134], [Zn$_3$(OH)$_3$(bpp)$_3$](NO$_3$)$_3\cdot$8$\cdot$67H$_2$O [135], and Zn(OAc)$_2$(bpe)$\cdot$2H$_2$O [136].

In contrast, the structure of [Co$_2$(bpe)$_3$(SO$_4$)$_2$(MeOH)$_2$].xSolv cannot be classed as self penetrating, despite its containing catenated 6-membered rings, because the shortest circuits are four- and five-membered rings [66]. The structure of Cu(HCO$_2$)$_2$(bipy) also contains catenated six-membered rings but is again not self penetrating as the shortest circuits are five-membered rings [137].

9. ENTANGLED BUT NOT INTERPENETRATING

We have discussed a number of fascinating interpenetrating (and self-penetrating) networks here, but we have perhaps not addressed a fundamental question: *how do we define interpenetration?* This is an important question because many structures can be complicated, and interpenetration is not the only form of entanglement that can occur. We have, for example, already mentioned the interdigitation of 2-D networks in the structure of Cu(tcm)(hmt) (Figure 1(a)).

For two or more networks to be interpenetrating, it should be impossible to separate them (in an imaginary, topological sense) without the breaking of links within the networks. Thus, the interdigitating networks in Cu(tcm)(hmt) are not interpenetrating, as they can be separated without the need for breaking links within the nets. However, the interpenetrating sheets in Cu(tcm)(bipy) (Figure 1(b)) are interpenetrating – the nets cannot be topologically separated with any amount of (imaginary) bond stretching, bending or other distortions alone; bonds must be broken. This is analogous to the difference, on a molecular scale, between a rotaxane, in which the rod can be separated from the ring (at least topologically, this can be prevented geometrically by using bulky end groups on the rod, but this does not affect the topological description), and a catenane, in which the rings cannot be separated without breaking at least one of them.

Nonetheless, it is instructive to briefly look at a small number of other fascinating structures that are entangled but not interpenetrating. The structure of HgI_2L, L = 2,6-bis(4-pyridinylmethyl)-benzo[1,2-c:4,5-c′]dipyrrole-1,3,5,7(2H, 6H)-tetrone contains 1-D chains that "weave" in two directions, creating clothlike 2-D layers (Figure 26) [138]. In each layer, the chains weave in such a way that each chain passes under two perpendicular chains, over the next two, then under the next two again, and so forth. Fivefold helices that form 1-D "tubes" are found in $Ni(acac)_2L{\cdot}x MeCN{\cdot}y H_2O$, L = C_2-symmetric 1,1-binaphthyl-6,6′-bipyridine derivatives [139]. Each tube is entangled with four adjacent such tubes. Nine such tubular single helices are entangled in the structure of $Cd(bpea)(phen)_2$, H_2bpea = biphenylethene-4,4′-dicarboxylic acid, phen = 1,10-phenanthroline [140].

The compounds $[CuL(solv)(NO_3)_2][CuL_{1.5}(NO_3)_2].2solv$, L = 1,4-bis(4′-pyridylethynyl)benzene, solv = MeOH and EtOH, contain 1-D ladders and 1-D chains [141]. The chains pass through the windows of the ladders in an inclined fashion. Again, it

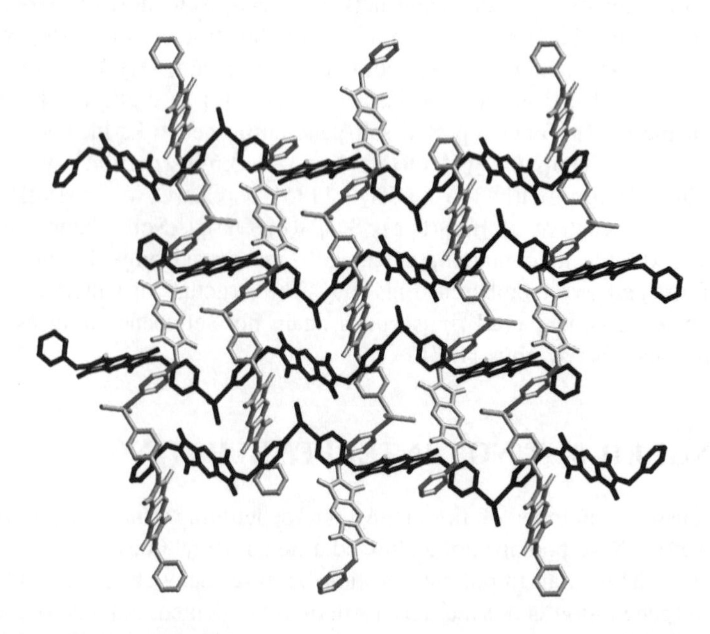

Figure 26. Interwoven, noninterpenetrating 1-D chains in the structure of HgI_2L, L = 2,6-bis(4-pyridinylmethyl)-benzo[1,2-c:4,5-c′]dipyrrole-1,3,5,7(2H, 6H)-tetrone [138].

should be stressed that this is not an interpenetrating system, as while the ladders are penetrated by the chains, the chains are not penetrated by the ladders, and thus the two motifs could be topologically separated without bond breakage. A similar motif is also seen in the structure of $[CuL_3(NO_3)_2]\cdot2[CuL_2(H_2O)(NO_3)](NO_3)_2\cdot EtOH$, L = 1,6-bis(4-pyridyl)hexane, in which 1-D nets pass through windows of other, inclined 1-D nets, but are not penetrated themselves [142]. In the structure of $[Ag(bpp)][Ag_2(bpp)_2(ox)]NO_3$, 1-D chains pass in an inclined fashion through the windows of 2-D (6,3) sheets [143]. In $[Mn(dca)_2(H_2O)_2]\cdot H_2O$, the 1-D nets pass though (4,4) 2-D sheets, although in this case there is extensive hydrogen-bonding linking the two motifs [144]. In these two examples, the 1-D chains and the 2-D nets are inclined, and thus, each 1-D chain passes through an infinite number of sheets. In the structure of $[bpeH][La(NO_3)_4(H_2O)(bpe)]$, 1-D hydrogen-bonded chains run through the windows of corrugated (4,4) sheets such that the chains are parallel to the sheets, resulting in a "Chinese blinds" motif [145]. Each chain thus penetrates only one sheet.

Kim and coworkers have reported a series of structures in which cucurbituril "beads" are threaded onto ligands in a rotaxane-like fashion [146]. These rotaxane ligands are then used in the construction of coordination polymers. In one fascinating example, the coordination polymers formed *do* interpenetrate, showing 2-D inclined interpenetration of (6,3) sheets [147]. Loeb and Hoffart have also recently reported a series of 3-D coordination polymers that contain rotaxane-based ligands [148]. These 3-D networks are, in fact, also doubly interpenetrating – four compounds contain two interpenetrating α-Po nets, while another contains two interpenetrating 6-connected nets with more complicated and very unusual topology.

We discussed earlier the interdigitation of 2-D sheets in the structure of Cu(tcm)(hmt). In $[Cd_2(bpea)(pt)(phen)_2][Cd(pt)(phen)]\cdot2H_2O$, pt = phthalate, layers of 1-D chains alternate with 2-D (4,4) sheets; the pendant phen ligands of the 1-D chains penetrate the windows of the 2-D sheets [140]. The structure of $Ag_2(bpethy)_5(BF_4)_2$, bpethy = 1,2-bis(4-pyridyl)ethyne, shows interdigitation of 1-D "railroad" polymers, that is, ladders with pendant ligands projecting from either side [149]. The 1-D chains lie in a parallel fashion to form 2-D planes; each railroad window is penetrated by rods from four adjoining nets. The structures of $[M_2(bipy)_3(H_2O)(phba)_2](NO_3)_2\cdot4H_2O$, M = Cu and Co, phba = 4-hydroxybenzoate have similar railroad 1-D nets. However, in this case, adjoining railroads are inclined to each other (although their directions of propagation remain parallel) [150]. This results in an overall 3-D motif. In this case, each railroad window is penetrated by rods from two adjoining nets. Finally, the beautiful structure of $[Cu_2(bpdc)_2(phen)_2(H_2O)]_2\cdot2H_2O$ contains extensively interdigitated molecular squares [151].

10. CONCLUSIONS

A full analysis of the way networks interpenetrate is an important component in the true understanding of network structures. While its effect on microporosity is obvious, the phenomenon also has the potential to affect other properties, such as nonlinear optical, magnetic or electronic properties. Thus, a true crystal engineer should focus not only on the network topology, but the interpenetration topology and its consequences as well.

I have described here only an illustrative selection of the many interpenetrating structures now known; a full list is given on the interpenetration website [152]. By nature, this website gives an extensive but not comprehensive list; there are undoubtedly interpenetrating structures reported that have yet to be listed. However, promising moves have been made toward a systematic search of the CSD and ICSD structural databases for such structures [9]. These techniques, and the continued research of crystal engineers around the world, will undoubtedly reveal many fascinating new modes of interpenetration.

REFERENCES

1. B. F. Hoskins and R. Robson, *J. Am. Chem. Soc.*, **112**, 1546–1554 (1990).
2. M. O'Keeffe, M. Eddaoudi, H. Li, T. Reineke and O. M. Yaghi, *J. Solid State Chem.*, **152**, 3–20 (2000).
3. S. R. Batten and R. Robson, *Angew. Chem., Int. Ed. Engl.*, **37**, 1460–1494 (1998).
4. S. R. Batten and R. Robson, in *Molecular Catenanes, Rotaxanes and Knots, A Journey Through the World of Molecular Topology* (Eds. J.-P. Sauvage and C. Dietrich-Buchecker), Wiley-VCH, Weinheim, 77–105, 1999.
5. S. R. Batten, *CrystEngComm*, **3**, 67–73 (2001).
6. S. R. Batten, in *Encyclopedia of Supramolecular Chemistry* (Eds. J. L. Atwood and J. W. Steed), Marcel Dekker, New York, 735–741, 2004.
7. L. Carlucci, G. Ciani and D. M. Proserpio, *CrystEngComm*, **5**, 269–279 (2003).
8. L. Carlucci, G. Ciani and D. M. Proerpio, *Coord. Chem. Rev.*, **246**, 247–289 (2003).
9. V. A. Blatov, L. Carlucci, G. Ciani and D. M. Proserpio, *CrystEngComm*, **6**, 377–395 (2004).
10. J. S. Miller, *Adv. Mater.*, **13**, 525–527 (2001).
11. S. T. Hyde, A.-K. Larsson, T. Di Matteo, S. Ramsden and V. Robins, *Aust. J. Chem.*, **56**, 981–1000 (2003).
12. (a) O. D. Friedrichs, M. O'Keefe and O. M. Yaghi, *Solid State Sci.*, **5**, 73–78 (2003); (b) O. D. Friedrichs, M. O'Keeffe and O. M. Yaghi, *Acta Crystallogr., Sect. A*, **59**, 22–27 (2003).
13. S. R. Batten, B. F. Hoskins and R. Robson, *Chem. Eur. J.*, **6**, 156–161 (2000).
14. O. D. Friedrichs, M. O'Keeffe and O. M. Yaghi, *Acta Crystallogr., Sect. A*, **59**, 515–525 (2003).
15. M. O'Keeffe and B. G. Hyde, *Crystal Structures. I. Patterns and Symmetry*, Mineralogical Society of America Monograph, Mineralogical Society of America, Washington, 1996.
16. (a) A. F. Wells, *Three-Dimensional Nets and Polyhedra*, Wiley-Interscience, New York, 1977; (b) A. F. Wells, *Further Studies of Three-Dimensional Nets, ACA Monograph No. 8*, American Crystallographic Association, 1979.
17. C. S. A. Fraser, M. C. Jennings and R. J. Puddephatt, *Chem. Commun.*, 1310–1311 (2001).
18. C. J. Kuehl, F. M. Tabellion, A. M. Arif and P. J. Stang, *Organometallics*, **20**, 1956–1959 (2001).
19. Z. Wang, Y. Cheng, C. Liao and C. Yan, *CrystEngComm*, **3**, 237–242 (2001).
20. J. Fan, W.-Y. Sun, T. Okamura, Y.-Q. Zheng, B. Sui, W.-X. Tang and N. Ueyama, *Cryst. Growth Des.*, **4**, 579–584 (2004).
21. H.-F. Zhu, J. Fan, T. Okamura, W.-Y. Sun and N. Ueyama, *Chem. Lett.*, 898–899 (2002).
22. G. Ferguson, C. Glidewell, R. M. Gregson and E. S. Lavender, *Acta Crystallogr., Sect. B*, **55**, 573–590 (1999).
23. (a) A. J. Blake, N. R. Champness, A. Khlobystov, D. A. Lemenovskii, W.-S. Li and M. Schröder, *Chem. Commun.*, 2027–2028 (1997); (b) M. Maekawa, H. Konaka, Y. Suenaga, T. Kuroda-Sowa and M. Munakata, *J. Chem. Soc., Dalton Trans.*, 4160–4166 (2000).
24. L. Carlucci, G. Ciani and D. M. Proserpio, *Cryst. Growth Des.*, **5**, 37–39 (2005).
25. B. F. Hoskins, R. Robson and D. A. Slizys, *J. Am. Chem. Soc.*, **119**, 2952–2953 (1997).
26. M. E. Kosal, J. H. Chou and K. S. Suslick, *J. Porph. Phthal.*, **6**, 377–381 (2002).

27. (a) H.-F. Zhu, W. Zhao, T. Okamura, J. Fan, W.-Y. Sun and N. Ueyama, *New J. Chem.*, **28**, 1010–1018 (2004); (b) H. F. Zhu, J. Fan, T. Okamura, W.-Y. Sun and N. Ueyama, *Cryst. Growth Des.*, **5**, 289–294 (2005).
28. J. Tao, X. Yin, R. Huang and L. Zheng, *Inorg. Chem. Commun.*, **5**, 1000–1002 (2002).
29. (a) M. Fujita, Y. J. Kwon, O. Sasaki, K. Yamaguchi and K. Ogura, *J. Am. Chem. Soc.*, **117**, 7287–7288 (1995); (b) M. Fujita, O. Sasaki, K.-Y. Watanabe, K. Ogura and K. Yamaguchi, *New J. Chem.*, **22**, 189–191 (1998).
30. C.-Y. Su, A. M. Goforth, M. D. Smith and H.-C. zur Loye, *Chem. Commun.*, 2158–2159 (2004).
31. (a) J. Konnert and D. Britton, *Inorg. Chem.*, **5**, 1193–1196 (1966); (b) S. R. Batten, B. F. Hoskins and R. Robson, *New J. Chem.*, **22**, 173–175 (1998); (c) B. F. Abrahams, S. R. Batten, B. F. Hoskins and R. Robson, *Inorg. Chem.*, **42**, 2654–2664 (2003).
32. (a) G. Saito, H. Yamochi, T. Nakamura, T. Komatsu, N. Matsukawa, T. Inoue, H. Ito, T. Ishiguro, M. Kusunoki, K. Sakaguchi and T. Mori, *Synth. Met.*, **55–57**, 2883–2890 (1995); (b) T. Komatsu, H. Sato, N. Matsukawa, T. Nakamura, H. Yamochi, G. Saito, M. Kusunoki, K. Sakaguchi and S. Kagoshima, *Synth. Met.*, **70**, 779–780 (1995); (c) T. Komatsu, H. Sato, T. Nakamura, N. Matsukawa, H. Yamochi, G. Saito, M. Kusunoki, K. Sakaguchi and S. Kagoshima, *Bull. Chem. Soc. Jpn.*, **68**, 2233–2244 (1995).
33. C. V. K. Sharma and M. J. Zaworotko, *Chem. Commun.*, 2655–2656 (1996).
34. (a) M. A. Withersby, A. J. Blake, N. R. Champness, P. A. Cooke, P. Hubberstey and M. Schröder, *New J. Chem.*, **23**, 573–575 (1999); (b) M. A. Withersby, A. J. Blake, N. R. Champness, P. A. Cooke, P. Hubberstey, A. L. Realf, S. J. Teat and M. Schröder, *J. Chem. Soc., Dalton Trans.*, 3261–3268 (2000).
35. D. Venkataraman, S. Lee, J. S. Moore, P. Zhang, K. A. Hirsch, G. B. Gardner, A. C. Covey and C. L. Prentice, *Chem. Mater.*, **8**, 2030–2040 (1996).
36. S. A. Barnett, A. J. Blake, N. R. Champness, J. E. B. Nicolson and C. Wilson, *J. Chem. Soc., Dalton Trans.*, 567–573 (2001).
37. S.-Y. Wan, J. Fan, T. Okamura, H.-F. Zhu, X.-M. Ouyang, W.-Y. Sun and N. Ueyama, *Chem. Commun.*, 2520–2521 (2002).
38. J. Fan, W.-Y. Sun, T. Okamura, W.-X. Tang and N. Ueyama, *Inorg. Chem.*, **42**, 3168–3175 (2003).
39. V. H. Thurn and H. Krebs, *Acta Crystallogr., Sect. B*, **25**, 125–135 (1969).
40. (a) C. Glidewell and G. Ferguson, *Acta Crystallogr., Sect. C*, **52**, 2528–2530 (1996); (b) C. Davies, R. F. Langler, C. V. K. Sharma and M. J. Zaworotko, *Chem. Commun.*, 567–568 (1997).
41. S. R. Batten, A. R. Harris, P. Jensen, K. S. Murray and A. Ziebell, *J. Chem. Soc., Dalton Trans.*, 3829–3836 (2000).
42. T. Soma and T. Iwamoto, *Chem. Lett.*, 821–824 (1994).
43. S. B. Copp, S. Subramanian and M. J. Zaworotko, *Angew. Chem., Int. Ed. Engl.*, **32**, 706–709 (1993).
44. M. J. Plater, M. R. St, J. Foreman, T. Gelbrich, S. J. Coles and M. B. Hursthouse, *J. Chem. Soc., Dalton Trans.*, 3065–3073 (2000).
45. (a) W. Lin, O. R. Evans, R.-G. Xiong and Z. Wang, *J. Am. Chem. Soc.*, **120**, 13272–13273 (1998); (b) O. R. Evans and W. Lin, *Chem. Mater.*, **13**, 3009–3017 (2001).
46. S. U. Son, B. Y. Kim, C. H. Choi, S. W. Lee, Y. S. Kim and Y. K. Chung, *Chem. Commun.*, 2528–2529 (2003).
47. O. R. Evans and W. Lin, *Chem. Mater.*, **13**, 3009–3013 (2001).
48. L. Carlucci, G. Ciani, D. M. Proserpio and S. Rizzato, *CrystEngComm*, **4**(22): 121–129 (2002).
49. (a) B. Li, G. Yin, H. Cao, Y. Liu and Z. Xu, *Inorg. Chem. Commun.*, **4**, 451–453 (2001); (b) B. Li, H. Liu, Y. Xu, J. Chen and Z. Xu, *Chem. Lett.*, 902–903 (2001).

50. D. M. L. Goodgame, S. Menzer, A. M. Smith and D. J. Williams, *Angew. Chem., Int. Ed. Engl.*, **34**, 574–575 (1995).
51. (a) D. J. Duchamp and R. E. Marsh, *Acta Crystallogr., Sect. B*, **25**, 5–19 (1969); (b) F. H. Herbstein, M. Kapon and G. M. Reisner, *Acta Crystallogr., Sect. B*, **41**, 348–354 (1985); (c) F. H. Herbstein, *Isr. J. Chem.*, **6**, IVp–Vp (1968); (c) J. E. Davies, P. Finocchiaro and F. H. Herbstein, in *Inclusion Compounds*, Vol. 2, Chap. 11 (Eds. J. L. Atwood, J. E. D. Davies and D. D. MacNicol), Academic Press, London, 1984; (d) F. H. Herbstein, *Top. Curr. Chem.*, **140**, 107–139 (1987); (e) F. H. Herbstein, M. Kapon and G. M. Reisner, *Proc. R. Soc. London, Ser. A*, **376**, 301–318 (1981).
52. S. Banfi, L. Carlucci, E. Caruso, G. Ciani and D. M. Proserpio, *Cryst. Growth Des.*, **4**, 29–32 (2004).
53. J. M. Knaust, S. Lopez and S. W. Keller, *Inorg. Chim. Acta*, **324**, 81–89 (2001).
54. X.-M. Zhang and X.-M. Chen, *Eur. J. Inorg. Chem.*, 413–417 (2003).
55. F.-Q. Liu and T. D. Tilley, *Inorg. Chem.*, **36**, 5090–5096 (1997).
56. L. Carlucci, G. Ciani, D. M. Proserpio and S. Rizzato, *Chem. Commun.*, 1319–1320 (2000).
57. (a) O. M. Yaghi and G. Li, *Angew. Chem., Int. Ed. Engl.*, **34**, 207–209 (1995); (b) S. R. Batten, J. C. Jeffery and M. D. Ward, *Inorg. Chim. Acta*, **292**, 231–237 (1999); (c) J. Y. Lu, B. R. Cabrera, R.-J. Wang and J. Li, *Inorg. Chem.*, **38**, 4608–4611 (1999); (d) A. J. Blake, N. R. Brooks, N. R. Champness, P. A. Cooke, M. Crew, A. M. Deveson, L. R. Hanton, P. Hubberstey, D. Fenske and M. Schröder, *Cryst. Eng.*, **2**, 181–195 (1999).
58. (a) H. O. Stumpf, L. Ouahab, Y. Pei, D. Grandjean and O. Kahn, *Science*, **261**, 447–449 (1993); (b) H. O. Stumpf, L. Ouahab, Y. Pei, P. Bergerat and O. Kahn, *J. Am. Chem. Soc.*, **116**, 3866–3874 (1994); (c) M. G. F. Vaz, L. M. M. Pinheiro, H. O. Stumpf, A. F. C. Alcantara, S. Golhen, L. Ouahab, O. Cador, C. Mathoniere and O. Kahn, *Chem. Eur. J.*, **5**, 1486–1495 (1999); (d) M. G. F. Vaz, H. O. Stumpf, N. L. Speziali, C. Mathoniere and O. Cador, *Polyhedron*, **20**, 1761–1769 (2001); (e) M. A. Novak, M. G. F. Vaz, N. L. Speziali, W. V. Costa and H. O. Stumpf, *Polyhedron*, **22**, 2391–2394 (2003).
59. M. J. Zaworotko, *Chem. Commun.*, 1–9 (2001).
60. (a) R. W. Gable, B. F. Hoskins and R. Robson, *J. Chem. Soc., Chem. Commun.*, 1677 (1990); (b) R. Robson, B. F. Abrahams, S. R. Batten, R. W. Gable, B. F. Hoskins and J. Liu, in *Supramolecular Architecture* (Ed. T. Bein), *ACS Symposium Series 499*, American Chemical Society, Washington, 256–273, 1992.
61. P. Jensen, S. R. Batten, B. Moubaraki and K. S. Murray, *J. Chem. Soc., Dalton Trans.*, 3712–3722 (2002).
62. T. Soma and T. Iwamoto, *Acta Crystallogr., Sect. C*, **52**, 1200–1203 (1996).
63. L. Carlucci, G. Ciani and D. M. Proserpio, *New J. Chem.*, **22**, 1319–1321 (1998).
64. C. B. Äakeröy, A. M. Beatty and D. S. Leinen, *Angew. Chem., Int. Ed. Engl.*, **38**, 1815–1819 (1999).
65. N. Moliner, C. Munoz, S. Letard, X. Solans, N. Menendez, A. Goujon, F. Varret and J. A. Real, *Inorg. Chem.*, **39**, 5390–5393 (2000).
66. L. Carlucci, G. Ciani, D. M. Proserpio and S. Rizzato, *CrystEngComm*, **5**, 190–199 (2003).
67. M. Kondo, M. Shimamura, S. Noro, S. Minakoshi, A. Asami, K. Seki and S. Kitagawa, *Chem. Mater.*, **12**, 1288–1299 (2000).
68. (a) M. A. Brook, R. Faggiani, C. J. L. Lock and D. Seebach, *Acta Crystallogr., Sect. C*, **44**, 1981–1984 (1988); (b) B. H. Torrie, R. von Dreele and A. C. Larson, *Mol. Phys.*, **76**, 405–410 (1992); (c) C. L. Christ, *Acta Crystallogr.*, **4**, 77–77 (1951).
69. I. Boldog, E. B. Rusanov, J. Sieler, S. Blaurock and K. V. Domasevitch, *Chem. Commun.*, 740–741 (2003).
70. B. F. Abrahams, S. R. Batten, H. Hamit, B. F. Hoskins and R. Robson, *Chem. Commun.*, 1313–1314 (1996).
71. V. M. Russell, D. C. Craig, M. L. Scudder and I. G. Dance, *CrystEngComm*, **2**, 16–23 (2000).

72. (a) C. J. Kepert and M. J. Rosseinsky, *Chem. Commun.*, 31–32 (1998); (b) C. J. Kepert, T. J. Prior and M. J. Rosseinsky, *J. Am. Chem. Soc.*, **122**, 5158–5168 (2000); (c) T. J. Prior and M. J. Rosseinsky, *Inorg. Chem.*, **42**, 1564–1575 (2003).
73. (a) K. Biradha and M. Fujita, *Angew. Chem., Int. Ed. Engl.*, **41**, 3392–3395 (2002); (b) O. Ohmori, M. Kawano, M. Fujita, *J. Am. Chem. Soc.*, **126**, 16292–16293 (2004).
74. O. M. Yaghi and H. Li, *J. Am. Chem. Soc.*, **117**, 10401–10402 (1995).
75. (a) G. B. Gardner, D. Venkataraman, J. S. Moore and S. Lee, *Nature*, **374**, 792–795 (1995); (b) G. B. Gardner, Y.-H. Kiang, S. Lee, A. Asgaonkar and D. Venkataraman, *J. Am. Chem. Soc.*, **118**, 6946–6953 (1996).
76. (a) L. Ohrstrom, K. Larsson, S. Borg and S. T. Norberg, *Chem. Eur. J.*, **7**, 4805–4810 (2001); (b) K. Larsson and L. Ohrstrom, *CrystEngComm*, **5**, 222–225 (2003).
77. Y.-B. Dong, M. D. Smith and H.-C. zur Loye, *Inorg. Chem.*, **39**, 4927–4935 (2000).
78. X. Cieren, J. Angenault, J.-C. Courtier, S. Jaulmes, M. Quarton and F. Robert, *J. Solid State Chem.*, **121**, 230–235 (1996).
79. J.-C. Dai, X.-T. Wu, Z.-Y. Fu, S.-M. Hu, W.-X. Du, C.-P. Cui, L.-M. Wu, H.-H. Zhang and R.-Q. Sun, *Chem. Commun.*, 12–13 (2002).
80. H. S. Zhdanov, *C. R. Acad. Sci. URSS*, **31**, 352–354 (1941).
81. E. Shugam and H. S. Zhdanov, *Acta Physiochim. URSS*, **20**, 247–252 (1945).
82. D. S. Reddy, T. Dewa, K. Endo and Y. Aoyama, *Angew. Chem., Int. Ed. Engl.*, **39**, 4266–4286 (2000).
83. L. Carlucci, G. Ciani, D. M. Proserpio and S. Rizzato, *Chem. Eur. J.*, **8**, 1520–1526 (2002).
84. K. W. Kim and M. G. Kanatzidis, *J. Am. Chem. Soc.*, **114**, 4878–4883 (1992).
85. (a) O. Ermer, *J. Am. Chem. Soc.*, **110**, 3747–3754 (1988); (b) O. Ermer, A. Kusch and C. Robke, *Helv. Chim. Acta*, **86**, 922–929 (2003).
86. Y.-H. Liu, H.-C. Wu, H.-M. Lin, W.-H. Hou and K.-L. Lu, *Chem. Commun.*, 60–61 (2003).
87. A. J. Blake, N. R. Champness, A. N. Khlobystov, D. A. Lemenovskii, W.-S. Li and M. Schröder, *Chem. Commun.*, 1339–1340 (1997).
88. S. Konar, E. Zangrando, M. G. B. Drew, J. Ribas and N. R. Chaudhuri, *Dalton Trans.*, 260–266 (2004).
89. Z.-F. Chen, R.-G. Xiong, B. F. Abrahams, X.-Z. You and C.-M. Che, *J. Chem. Soc., Dalton Trans.*, 2453–2455 (2001).
90. K. A. Hirsch, S. R. Wilson and J. S. Moore, *Inorg. Chem.*, **36**, 2960–2968 (1997).
91. (a) B. F. Hoskins, R. Robson and N. V. Y. Scarlett, *Angew. Chem., Int. Ed. Engl.*, **34**, 1203–1204 (1995); (b) S. C. Abrahams, L. E. Zyontz and J. L. Bernstein, *J. Chem. Phys.*, **76**, 5458–5462 (1982).
92. B. F. Abrahams, B. F. Hoskins, D. M. Michail and R. Robson, *Nature*, **369**, 727–729 (1994).
93. L. Carlucci, N. Cozzi, G. Ciani, M. Moret, D. M. Proserpio and S. Rizzato, *Chem. Commun.*, 1354–1355 (2002).
94. Q.-M. Wang, G.-C. Guo and T. C. W. Mak, *Chem. Commun.*, 1849–1850 (1999).
95. L. Carlucci, G. Ciani, P. Macchi, D. M. Proserpio and S. Rizzato, *Chem. Eur. J.*, **5**, 237–243 (1999).
96. N. M. Stainton, K. D. M. Harris and R. A. Howie, *J. Chem. Soc., Chem. Commun.*, 1781–1784 (1991).
97. D. Hagrman, R. P. Hammond, R. Haushalter and J. Zubieta, *Chem. Mater.*, **10**, 2091–2100 (1998).
98. A. Galet, V. Niel, M. C. Munoz and J. A. Real, *J. Am. Chem. Soc.*, **125**, 14224–14225 (2003).
99. X.-H. Bu, M.-L. Tong, H.-C. Chang, S. Kitagawa and S. R. Batten, *Angew. Chem., Int. Ed. Engl.*, **43**, 192–195 (2004).
100. L. Pan, N. Ching, X. Huang and J. Li, *Chem. Commun.*, 1064–1065 (2001).
101. W.-F. Yeung, S. Gao, W.-T. Wong and T.-C. Lau, *New J. Chem.*, **26**, 523–525 (2002).
102. Z. Shi, G. Li, L. Wang, L. Gao, X. Chen, J. Hua and S. Feng, *Cryst. Growth Des.*, **4**, 25–27 (2004).

103. (a) P. C. M. Duncan, D. M. L. Goodgame, S. Menzer and D. J. Williams, *Chem. Commun.*, 2127–2128 (1996); (b) L. Ballester, I. Baxter, P. C. M. Duncan, D. M. L. Goodgame, D. A. Grachvogel and D. J. Williams, *Polyhedron*, **17**, 3613–3623 (1998).
104. (a) P. Jensen, S. R. Batten, G. D. Fallon, D. C. R. Hockless, B. Moubaraki, K. S. Murray and R. Robson, *J. Solid State Chem.*, **145**, 387–393 (1999); (b) P. Jensen, S. R. Batten, B. Moubaraki and K. S. Murray, *J. Solid State Chem.*, **159**, 352–361 (2001); (c) J. L. Manson, C. D. Incarvito, A. L. Rheingold and J. S. Miller, *J. Chem. Soc., Dalton Trans.*, 3705–3706 (1998); (d) J. L. Manson, Q. Huang, J. W. Lynn, H.-J. Koo, M.-H. Whangbo, R. Bateman, T. Otsuka, N. Wada, D. N. Argyriou and J. S. Miller, *J. Am. Chem. Soc.*, **123**, 162–172 (2001); (e) C. M. Brown and J. L. Manson, *J. Am. Chem. Soc.*, **124**, 12600–12605 (2002).
105. E. Siebel and R. D. Fischer, *Chem. Eur. J.*, **3**, 1987–1991 (1997).
106. T. M. Reineke, M. Eddaoudi, D. Moler, M. O'Keeffe and O. M. Yaghi, *J. Am. Chem. Soc.*, **122**, 4843–4844 (2000).
107. (a) B. F. Hoskins, R. Robson and N. V. Y. Scarlett, *J. Chem. Soc., Chem. Commun.*, 2025–2026 (1994); (b) S. C. Abrahams, J. L. Bernstein and R. Liminga, *J. Chem. Phys.*, **73**, 4585–2128 (1980); (c) U. Geiser and J. A. Schlueter, *Acta Crystallogr., Sect. C*, **59**, i21–i23 (2003); (d) W. Dong, L.-N. Zhu, Y.-Q. Sun, M. Liang, Z.-Q. Liu, D.-Z. Liao, Z.-H. Jiang, S.-P. Yan and P. Cheng, *Chem. Commun.*, 2544–2545 (2003); (e) M. Zabel, S. Kuhnel and K.-J. Range, *Acta Crystallogr., Sect. C*, **45**, 1619–1621 (1989); (f) L. Pauling and P. Pauling, *Proc. Natl. Acad. Sci. U.S.A.*, **60**, 362–367 (1968); (g) A. Ludi and H. U. Gudel, *Helv. Chim. Acta*, **51**, 1762–1765 (1968); (h) H. U. Gudel, A. Ludi and P. Fischer, *J. Chem. Phys.*, **56**, 674–675 (1972); (i) H. U. Gudel, A. Ludi, P. Fischer and W. Halg, *J. Chem. Phys.*, **53**, 1917–1923 (1970); (j) H. U. Gudel, A. Ludi and H. Burki, *Helv. Chim. Acta*, **51**, 1383–1389 (1968); (k) A. Ludi, H. U. Gudel and V. Dvorak, *Helv. Chim. Acta*, **50**, 2035–2039 (1967); (l) R. Haser, C. E. de Broin and M. Pierrot, *Acta Crystallogr., Sect. B*, **28**, 2530–2537 (1972); (m) A. D. Kirk, H. L. Schlafer and A. Ludi, *Can. J. Chem.*, **48**, 1065–1072 (1970).
108. B. F. Abrahams, B. F. Hoskins, R. Robson and D. A. Slizys, *CrystEngComm*, **4**, 478–482 (2002).
109. C. S. Hong, S.-K. Son, Y. S. Lee, M.-J. Jun and Y. Do, *Inorg. Chem.*, **38**, 5602–5601 (1999).
110. Z. Assefa, R. J. Staples and J. P. Fackler Jr., *Acta Crystallogr., Sect. C*, **51**, 2527–2529 (1995).
111. (a) Q. Fang, X. Shi, G. Wu, G. Tian, G. Zhu, R. Wang and S. Qiu, *J. Solid State Chem.*, **176**, 1–4 (2003); (b) L. Pan, H. Liu, X. Lei, X. Huang, D. H. Olson, N. J. Turro and J. Li, *Angew. Chem., Int. Ed. Engl.*, **42**, 542–546 (2003).
112. B. Chen, M. Eddaoudi, S. T. Hyde, M. O'Keeffe and O. M. Yaghi, *Science*, **291**, 1021–1023 (2001).
113. (a) S. R. Batten, B. F. Hoskins and R. Robson, *J. Chem. Soc., Chem. Commun.*, 445–446 (1991); (b) S. R. Batten, B. F. Hoskins, B. Moubaraki, K. S. Murray and R. Robson, *J. Chem. Soc., Dalton Trans.*, 2977–2986 (1999); (c) J. L. Manson, C. Campana and J. S. Miller, *Chem. Commun.*, 251–252 (1998); (d) H. Hoshino, K. Iida, T. Kawamoto and T. Mori, *Inorg. Chem.*, **38**, 4229–4232 (1999); (e) J. L. Manson, E. Ressouche and J. S. Miller, *Inorg. Chem.*, **39**, 1135–1141 (2000).
114. K. V. Domasevitch, G. D. Enright, B. Moulton and M. J. Zaworotko, *J. Solid State Chem.*, **152**, 280–285 (2000).
115. L. Carlucci, G. Ciani, M. Moret, D. M. Proserpio and S. Rizzato, *Angew. Chem., Int. Ed. Engl.*, **39**, 1506–1510 (2000).
116. L. Carlucci, G. Ciani and D. M. Proserpio, *Chem. Commun.*, 380–381 (2004).
117. B. F. Hoskins, R. Robson and D. A. Slizys, *Angew. Chem., Int. Ed. Engl.*, **36**, 2752–2755 (1997).
118. J. Y. Lu and A. M. Babb, *Chem. Commun.*, 821–822 (2001).
119. Y.-C. Jiang, Y.-C. Lai, S.-L. Wang and K.-H. Lii, *Inorg. Chem.*, **40**, 5320–5321 (2001).
120. D. M. Shin, I. S. Lee, Y. K. Chung and M. S. Lah, *Chem. Commun.*, 1036–1037 (2003).

121. M. L. Tong, X.-M. Chen, B.-H. Ye and L.-N. Ji, *Angew. Chem., Int. Ed. Engl.*, **38**, 2237–2240 (1999).
122. D. B. Leznoff, B.-Y. Xue, R. J. Batchelor, F. W. B. Einstein and B. O. Patrick, *Inorg. Chem.*, **40**, 6026–6034 (2001).
123. B. F. Abrahams, S. R. Batten, H. Hamit, B. F. Hoskins and R. Robson, *Angew. Chem., Int. Ed. Engl.*, **35**, 1690–1692 (1996).
124. R. Kitaura, K. Seki, G. Akiyama and S. Kitagawa, *Angew. Chem., Int. Ed. Engl.*, **42**, 428–431 (2003).
125. G. J. Halder, C. J. Kepert, B. Moubaraki, K. S. Murray and J. D. Cashion, *Science*, **298**, 1762–1765 (2002).
126. S. R. Batten, B. F. Hoskins and R. Robson, *J. Am. Chem. Soc.*, **117**, 5385–1765 (1995).
127. (a) O. Ermer, *Adv. Mater.*, **3**, 608–611 (1991); (b) K. Sinzger, S. Hunig, M. Jopp, D. Bauer, W. Bietsch, J. U. von Schutz, H. C. Wulf, R. K. Kremer, T. Metzenthin, R. Bau, S. I. Khan, A. Lindbaum, C. L. Lengauer and E. Tillmanns, *J. Am. Chem. Soc.*, **115**, 7696–7705 (1993); (c) A. Aumuller, P. Erk, G. Klebe, S. Hunig, J. U. von Schutz and H.-P. Werner, *Angew. Chem., Int. Ed. Engl.*, **25**, 740–741 (1986); (d) R. Kato, H. Kobayashi, A. Kobayashi, T. Mori and H. Inokuchi, *Chem. Lett.*, 1579–1582 (1987); (e) R. Kato, H. Kobayashi and A. Kobayashi, *J. Am. Chem. Soc.*, **111**, 5224–5232 (1989); (f) A. Kobayashi, R. Kato, H. Kobayashi, T. Mori and H. Inokuchi, *Solid State Commun.*, **64**, 45–51 (1987); (g) S. Hunig, A. Aumuller, P. Erk, H. Meixner, J. U. von Schutz, H.-J. Gross, U. Langohr, H.-P. Werner, H. C. Wolf, C. Burschka, G. Klebe, K. Peters and H. G. v. Schnering, *Synth. Met.*, **27**, B181–B188 (1988); (h) R. Kato, H. Kobayashi, A. Kobayashi, T. Mori and H. Inokuchi, *Synth. Met.*, **27**, B263–B268 (1988); (i) A. Kobayashi, T. Mori, H. Inokuchi, R. Kato and H. Kobayashi, *Synth. Met.*, **27**, B275–B280 (1988); (j) A. Aumuller, P. Erk, S. Hunig, J.-U. von Schutz, H. P. Werner, H. C. Wolf and G. Klebe, *Mol. Cryst. Liq. Cryst. Sci. Technol.*, **156**, 215–221 (1988).
128. J.-P. Sauvage and C. Dietrich-Buchecker, *Molecular Catenanes, Rotaxanes and Knots, A Journey Through the World of Molecular Topology*, Wiley-VCH, Weinheim, 1999.
129. B. F. Abrahams, S. R. Batten, M. J. Grannas, H. Hamit, B. F. Hoskins and R. Robson, *Angew. Chem., Int. Ed. Engl.*, **38**, 1475–1477 (1999).
130. (a) S. R. Batten, P. Jensen, B. Moubaraki, K. S. Murray and R. Robson, *Chem. Commun.*, 439–440 (1998); (b) M. Kurmoo and C. J. Kepert, *New J. Chem.*, **22**, 1515–1524 (1998); (c) J. L. Manson, C. R. Kmety, Q.-Z. Huang, J. W. Lynn, G. M. Bendele, S. Pagola, P. W. Stephens, L. M. Liable-Sands, A. L. Rheingold, A. J. Epstein and J. S. Miller, *Chem. Mater.*, 1998, **10**, 2552–2560 (1998).
131. P. Jensen, D. J. Price, S. R. Batten, B. Moubaraki and K. S. Murray, *Chem. Eur. J.*, **6**, 3186–3195 (2000).
132. M.-L. Tong, X.-M. Chen and S. R. Batten, *J. Am. Chem. Soc.*, **125**, 16170–16171 (2003).
133. L. Carlucci, G. Ciani, D. M. Proserpio and S. Rizzato, *J. Chem. Soc., Dalton Trans.*, 3821–3827 (2000).
134. M. A. Withersby, A. J. Blake, N. R. Champness, P. A. Cooke, P. Hubberstey and M. Schröder, *J. Am. Chem. Soc.*, **122**, 4044–4046 (2000).
135. M. J. Plater, M. R. St, J. Foreman, T. Gelbrich and M. B. Hursthouse, *J. Chem. Soc., Dalton Trans.*, 1995–2000 (2000).
136. M. T. Ng, T. C. Deivaraj, W. T. Klooster, G. J. McIntyre and J. J. Vittal, *Chem. Eur. J.*, **10**, 5853–5859 (2004).
137. J. L. Manson, J. G. Lecher, J. Gu, U. Geiser, J. A. Schlueter, R. Henning, X. Wang, A. J. Schultz, H.-J. Koo and M.-H. Whangbo, *Dalton Trans.*, 2905–2911 (2003).
138. Y.-H. Li, C.-Y. Su, A. M. Goforth, K. D. Shimizu, K. D. Gray, M. D. Smith and H.-C. zur Loye, *Chem. Commun.*, 1630–1631 (2003).
139. Y. Cui, S. J. Lee and W. Lin, *J. Am. Chem. Soc.*, **125**, 6014–6015 (2003).

140. X.-L. Wang, C. Qin, E.-B. Wang, L. Xu, Z.-M. Su and C.-W. Hu, *Angew. Chem., Int. Ed. Engl.*, **43**, 5036–5040 (2004).
141. M. B. Zaman, M. D. Smith and H.-C. zur Loye, *Chem. Commun.*, 2256–2257 (2001).
142. M. J. Plater, M. R. St, J. Foreman, T. Gelbrich and M. B. Hursthouse, *Cryst. Eng.*, **4**, 319–328 (2001).
143. M.-L. Tong, Y.-M. Wu, J. Ru, X.-M. Chen, H.-C. Chang and S. Kitagawa, *Inorg. Chem.*, **41**, 4846–4848 (2002).
144. (a) K. S. Murray, S. R. Batten, B. Moubaraki, D. J. Price and R. Robson, *Mol. Cryst. Liq. Cryst. Sci. Technol.*, **335**, 313–322 (1999); (b) S. R. Batten, P. Jensen, C. J. Kepert, M. Kurmoo, B. Moubaraki, K. S. Murray and D. J. Price, *J. Chem. Soc., Dalton Trans.*, 2987–2997 (1999).
145. C. V. K. Sharma and R. D. Rogers, *Chem. Commun.*, 83–84 (1999).
146. K. Kim, *Chem. Soc. Rev.*, **31**, 96–107 (2002).
147. (a) D. Whang and K. Kim, *J. Am. Chem. Soc.*, **119**, 451–452 (1997); (b) K.-M. Park, D. Whang, E. Lee, J. Heo and K. Kim, *Chem. Eur. J.*, **8**, 498–508 (2002).
148. D. J. Hoffart and S. J. Loeb, *Angew. Chem., Int. Ed. Engl.*, **44**, 901–904 (2005).
149. L. Carlucci, G. Ciani and D. M. Proserpio, *Chem. Commun.*, 449–450 (1999).
150. M.-L. Tong, H.-J. Chen and X.-M. Chen, *Inorg. Chem.*, **39**, 2235–2238 (2000).
151. G.-F. Liu, B.-H. Ye, Y.-H. Ling and X.-M. Chen, *Chem. Commun.*, 1442–1443 (2002).
152. Known Examples of Interpenetration. http://web.chem.monash.edu.au/Department/Staff/Batten/Intptn.htm (2005).

9

Architecture and Functional Engineering Based on Paddlewheel Dinuclear Tetracarboxylate Building Blocks

SUSUMU KITAGAWA and SHUHEI FURUKAWA
Department of Synthetic Chemistry and Biological Chemistry, Graduate School of Engineering, Kyoto University, Katsura 615-8510, Japan.

1. INTRODUCTION

1.1. Paddlewheel Dinuclear Building Block

Remarkable progress has been made in the area of molecular inorganic–organic hybrid compounds. The design, synthesis and characterization of infinite one-, two-, and three-dimensional (1D, 2D, and 3D) networks has been an area of rapid growth and of great current interest for both their structural and topological novelty as well as for their potential applications as functional materials [1–6]. The main strategy for the assembly of target architectures by rational methods is a molecular building–block approach; organic molecules with functional groups (such as pyridinyl and carboxylate groups) and metal ions and/or metal clusters with bonds and angles regulated for their coordination are used as sides and nodes of architectures (Figure 1). The key to success is the design of molecular building blocks that direct the formation of the desired architectural, chemical and physical properties of the resulting assemblies.

Among the numerous examples of building blocks, a family of paddlewheel dinuclear tetracarboxylate complexes [7], $[M^{II}_2(O_2CR)_4]$, is an excellent candidate for controlling not only entire network structures but also function, because of the following reasons (Figure 2): (1) the paddlewheel units constructed from two divalent metal ions (M^{II}) and

Frontiers in Crystal Engineering. Edited by Edward R.T. Tiekink and Jagadese J. Vittal

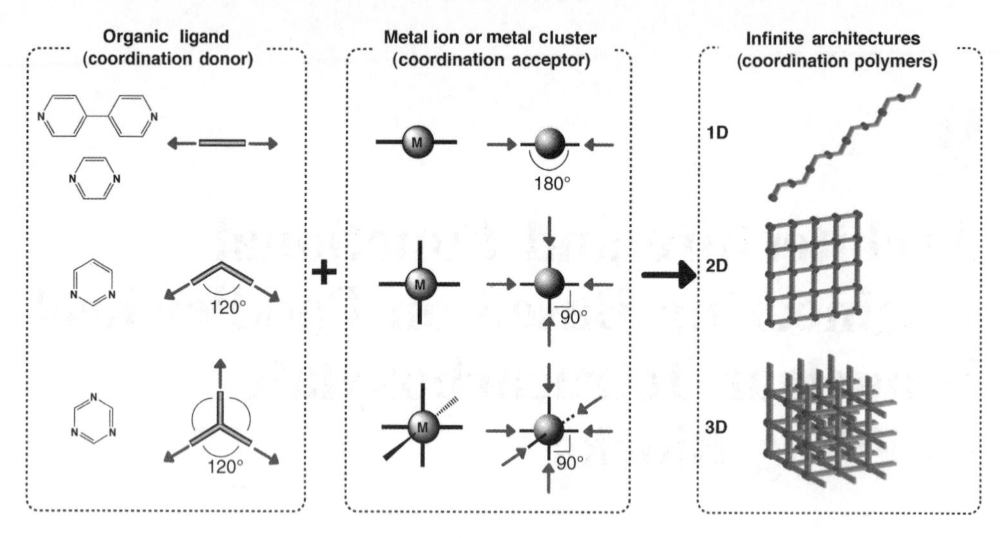

Figure 1. Schematic representation of a molecular building–block approach for infinite (1D, 2D or 3D) architectures.

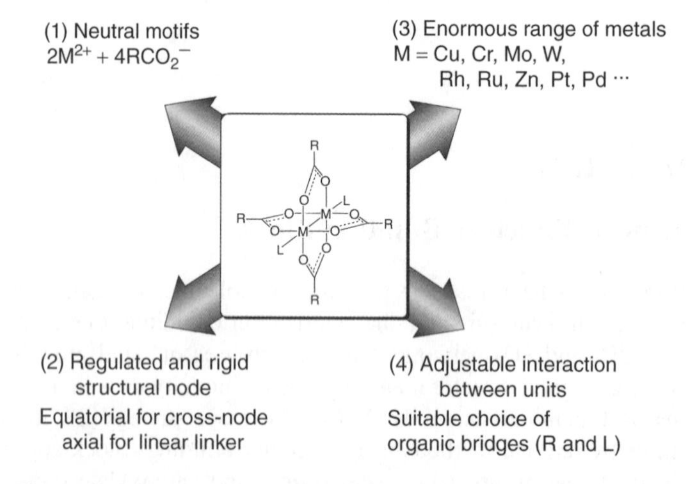

Figure 2. Characteristics of paddlewheel dinuclear building blocks.

four uninegative carboxylate ligands (RCO_2^-) can be used as neutral structural motifs; (2) the regulated angle of the dinuclear environment allows itself to use the rigid structural node such as equatorial positions for the cross nodes and axial positions for the linkers; (3) a large range of metal elements is potentially available to form homologous structures, each of which provides unique spectroscopic and magnetic properties due to the electronic configurations of the chosen metal center and metal–metal interaction; (4) by suitable choice of both equatorial- and axial-bridging linkers, the nature and degree of interaction between adjacent paddlewheel units can be finely controlled. These advantages allows the replacement and introduction of various metal ions into the dinuclear cores with retention of the architectures, which in turn provides opportunities to control not only the assembled structures but also their physical properties.

1.2. High-dimensional Frameworks and Their Functions

Comprehensive and thorough studies of paddlewheel units have been conducted since the first crystallographic reports of their copper and chromium acetate complexes, $[M^{II}_2(O_2CCH_3)_4(H_2O)_2]$ ($M^{II} = Cu^{II}$ [8] and Cr^{II} [9]), in 1953. During the subsequent five decades, a wide variety of infinite architectures have been synthesized and structurally characterized. While a number of 1-D polymeric chain complexes have been known for a long time, 2-D or 3-D high-dimensional structure were only first reported in 1993 [10]. In recent years, much attention has been paid to evaluating the physical properties of these high-dimensional architectures, such as magnetism and conductivity, and more recently, porous properties have been found for several complexes. Depending on the nature of the metal–metal bonds (metal–metal interactions), magnetic and conductive properties are attributed to the metal center and its oxidation state. As a result, copper, rhodium and ruthenium centers have been employed frequently. On the other hand, rigid building blocks containing dicopper or dizinc cores allow for the formation of porous frameworks. In this chapter, we focus upon a description of 2-D and 3-D coordination architectures based on Cu_2, Rh_2, Ru_2 and Zn_2 cores, and their functions.

1.3. Electronic Structures of Dinuclear Cores

When the physical properties of dinuclear cores are discussed, the characteristic electronic structures due to metal–metal interactions are all important (Figure 3) [7]. When two metal atoms approach each other, only five nonzero overlaps between the pairs of d orbitals on the two atoms are possible owing to symmetry considerations. Four of the five d orbitals, that is, d_z^2, d_{xz}, d_{yz} and d_{xy}, give rise to σ, π (doubly degenerate), and δ-bonding orbitals and their antibonding counterparts. The remaining pair of $d_{x^2-y^2}$ orbitals interacts primarily with the set of four ligands, and makes a strong contribution to metal–ligand bonding.

From this consideration, a dicopper complex has no metal–metal bond as both bonding and antibonding orbitals are fully occupied. However, there is a weak coupling of the unpaired electrons, one on each copper(II) center, giving rise to a singlet ground state with a triplet state lying only a few kilojoules per mole above it; the latter state is thus appreciably populated at normal temperatures and the complexes are paramagnetic.

In dirhodium complexes, 14 electrons are used to fill the orbitals; thus, their complexes are diamagnetic. The Rh^{II}_2 cores can undergo redox reactions to give mixed valence and paramagnetic, $Rh_2^{I,II}$ and $Rh_2^{II,III}$ species, with retention of structure.

The diruthenium core has a striking feature owing to its characteristic electronic structures. In terms of the simple picture of metal–metal bonding orbitals, 12 electrons have been contributed, and so there is an unpaired electrons in each of the degenerate π^* orbitals of the Ru^{II}_2 core. Therefore, the series of diruthenium complexes, $[Ru^{II}_2(O_2CR)_4]$, is paramagnetic ($S = 1$). This core also undergoes redox reactions to give oxidative analogs, $Ru_2^{II,III}$ and Ru^{III}_2. It should be noted that $[Ru_2^{II,III}(O_2CR)_4]^+$ appears to have π^* and δ^* orbitals that are close in energy or accidentally triply degenerate. Therefore, one electron

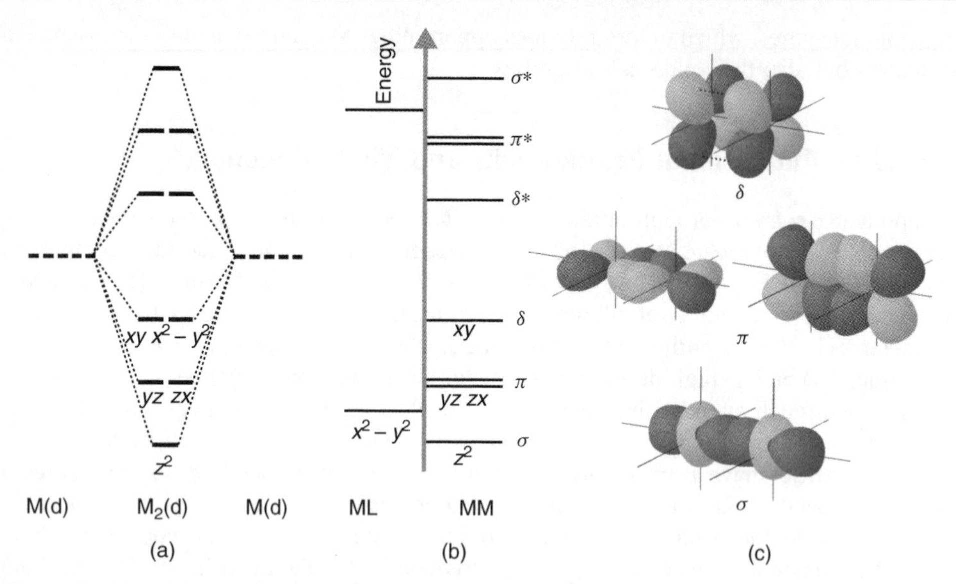

Figure 3. Energy-level diagrams of metal–metal interaction between two transition metal ions (a), and metal–ligand bonding and metal–metal bonds (b). Schematic representation of orbitals describing metal–metal interactions (c).

Table 1. Electronic configurations and bond orders of selected metal–metal bonded complexes

Compounds	Electronic configuration	Bond order
$[Cu^{II}_2(O_2CCH_3)_4(H_2O)_2]$	$\sigma^2\pi^4\delta^2\delta^{*2}\pi^{*4}\sigma^{*2}$	0
$[Rh^{II}_2(O_2CCH_3)_4(H_2O)_2]$	$\sigma^2\pi^4\delta^2\delta^{*2}\pi^{*4}$	1
$[Ru^{II}_2(O_2CCH_3)_4(H_2O)_2]$	$\sigma^2\pi^4\delta^2\delta^{*2}\pi^{*2}$	2
$[Ru_2^{II,III}(O_2CCH_3)_4Cl]$	$\sigma^2\pi^4\delta^2\delta^{*1}\pi^{*2}$	2.5
$[Mo^{II}_2(O_2CCH_3)_4]$	$\sigma^2\pi^4\delta^2\delta^{*0}\pi^{*0}$	4

is removed from δ^* orbital with oxidation from $[Ru^{II}_2(O_2CR)_4]$ to $[Ru_2^{II,III}(O_2CR)_4]^+$, affording a high-spin paramagnetic state ($S = 3/2$). The electronic configuration and bond orders of the type of paddlewheel complexes are summarized in Table 1.

2. SYNTHETIC STRATEGY

2.1. *In situ* Building Blocks and Preorganized Building Blocks

General lack of control over the character of frameworks produced from the traditional synthetic methods is directly related to the fact that the starting entities do not maintain their structure during the reaction, leading to poor correlation between reactants and products. The design and synthesis of an infinite network can be realized by starting with *preorganized* and rigid molecular building blocks that will maintain their structural integrity throughout the construction process. Alternatively, the use of well-defined conditions that lead to the formation of such building blocks *in situ* is an equally viable approach to the design of infinite structures. From this viewpoint, the generation processes of paddlewheel building blocks are divided into two categories: *preorganized* building blocks and *in situ* building blocks, which are strongly dependent on the reaction kinetics of the substitution reaction at the equatorial and axial positions of dinuclear cores.

In the case of Cu_2 and Zn_2 cores, paddlewheel cores are generated *in situ* by the reaction of metal sources with organic carboxylates under ambient conditions, since both the equatorial and axial positions of the cores are labile for substitution reactions. Therefore, paddlewheel building blocks are used as cross-linked nodes or octahedral-linked nodes (Figure 4(b)). These *in situ* building blocks are known as *secondary building units* [11].

By contrast, equatorial carboxylates of Rh_2 and Ru_2 cores are inert for substitution but the axial positions of these can be easily exchanged, allowing these positions to function as linear linkers (Figure 4(a)). In this regard, it is the *preorganized* paddlewheel carboxylate complex that is useful for the construction of infinite networks. Further, there have been no reports of structurally characterized extended networks containing Rh_2 and Ru_2 cores linked by equatorial positions.

2.2. Preparation of Preorganized Building Blocks (Rh_2 and Ru_2)

The versatile set of carboxylate derivatives available as *preorganized* building blocks allows one to control both assembled structures as well as the electronic properties of the dinuclear cores. The wide variety of $[Rh^{II}_2(O_2CR)_4]$ complexes are known due to the simple procedure of synthesis, namely, the reduction of $RhCl_3{\cdot}3H_2O$ with acetic acid in alcoholic solution leading to $[Rh^{II}_2(O_2CCH_3)_4]$, which can be followed by a simple carboxylate exchange reaction with desired carboxylic acid (eq. 1) [7].

$$Rh^{III}Cl_3{\cdot}3H_2O + CH_3CO_2H \rightarrow [Rh^{II}_2(O_2CCH_3)_4] \quad (1)$$

The Ru_2 building blocks are of great importance owing to their characteristic magnetic properties of the Ru_2 core ($S = 1$ for the neutral $[Ru^{II}_2(O_2CR)_4]$ complexes and $S = 3/2$ for the cationic $[Ru_2{}^{II,III}(O_2CR)_4]^+$ complexes) [12]. However, this chemistry has principally focused on the cationic form rather than the neutral form because of synthetic difficulties in the preparation of the neutral complexes. The well-accepted synthesis of

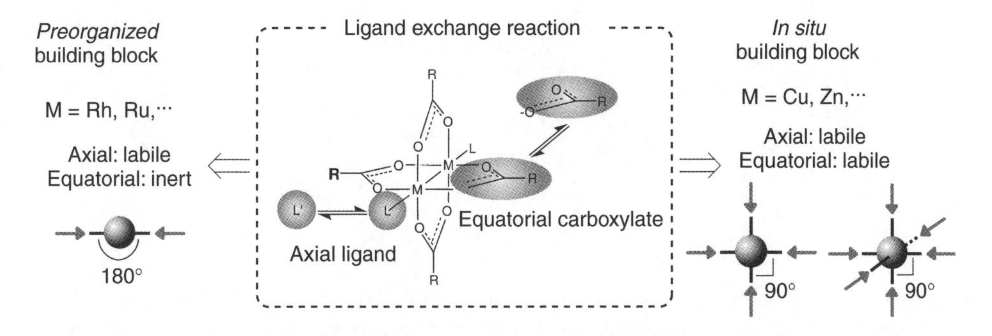

Figure 4. Definition of *preorganized* building block (left) and *in situ* building block (right).

the cationic complexes is the reaction of $RuCl_3{\cdot}3H_2O$ with acetic acid giving rise to $[Ru_2^{II,III}(O_2CCH_3)_4Cl]$, which is the starting material for the carboxylate substitution (eq. 2) [13]:

$$Ru^{III}Cl_3{\cdot}3H_2O + CH_3CO_2H \rightarrow [Ru_2^{II,III}(O_2CCH_3)_4Cl] \qquad (2)$$

On the other hand, the neutral complex has been usually synthesized via a "ruthenium blue solution" [14] prepared by the reduction of $RuCl_3{\cdot}3H_2O$ by H_2 in the presence of excess alkali metal carboxylate [15]. The reaction proceeds through a very air-sensitive neutral intermediate, $[Ru^{II}_2(O_2CH)_4]$ or $[Ru^{II}_2(O_2CCH_3)_4]$, followed by a substitution reaction with the desired carboxylates (eq. 3):

$$Ru^{III}Cl_3{\cdot}3H_2O \xrightarrow{H_2} \text{"Ru blue solution"} \xrightarrow{CH_3CO_2H} [Ru^{II}_2(O_2CCH_3)_4] \qquad (3)$$

Moreover, Ru^{II}_2 carboxylate complexes are generally air sensitive, while Rh_2^{II} and $Ru_2^{II,III}$ complexes are air stable. This fact restricts the production of a variety of infinite networks containing Ru^{II}_2 units. However, it is worth noting that a recent report describes a facile one-pot synthesis of Ru^{II}_2 complexes [16]. In this procedure, a simple reaction of $[Ru_2^{II,III}(O_2CCH_3)_4(THF)_2](BF_4)$, the desired carboxylic acid, and *N*,*N*-dimethylaniline (used as the solvent) produces the desired neutral diruthenium complex in good yield as a consequence of a one-electron reduction followed by a carboxylate exchange reaction.

2.3. Coordination Mode of Axial Positions

Whereas the equatorial positions of the cores are occupied by four carboxylates, axial positions are coordinated by ligands frequently containing N– or O–donating atoms. Therefore, a wide variety of ligands can be used as organic linkers (Figure 5(a)). In addition, mutual coordination between metal ions and the oxygen atoms of the carboxylate ligands can be generated due to the strong preference of axial coordination, which affords an extended molecular system as shown in Figure 5(b).

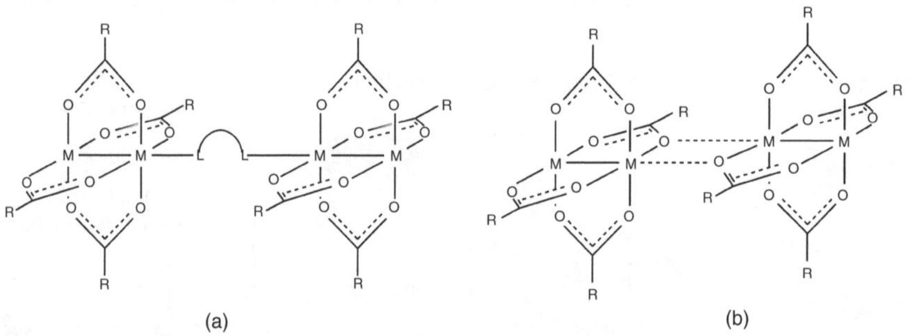

(a) (b)

Figure 5. Coordination mode of axial positions: (a) the axial coordination by organic linkers containing N– or O–donating atoms and (b) the mutual coordination between metal ions and oxygen atoms of coordinated carboxylates.

3. ARCHITECTURE ENGINEERING BASED ON PREORGANIZED BUILDING BLOCKS

3.1. High-dimensional (2D and 3D) Architectures

In the past ten years or so, the design and synthesis of infinite coordination networks has been a central theme in the realm of crystal engineering. Paddlewheel dinuclear complexes have been used as a *preorganized* building block or as a rigid linear linker, leading to a wide variety of 1-D architectures such as linear and zigzag chains with suitable organic linkers. On the other hand, higher-dimensional 2-D and 3-D networks have received relatively little attention. In order to construct 2-D and 3-D coordination networks, bridging ligands providing the nodes for networks should have three or higher connectable sites.

In 1993, the first such example became available, namely $[\{Rh_2(O_2CCF_3)_4\}_2(TCNE)]$ (**1**) [10], TCNE is tetracyanoethene, in which all the four cyanide groups are involved in coordination to create a 2-D sheet structure. It is worth noting that this complex is also the first example of an infinite structure containing TCNE, which functions as a μ_4-η^4 bridging ligand. The assembly of $[Cu_2(O_2CCH_3)_4]$ with an organic three-connected node, 2,4,6-tri(4-pyridyl)-1,3,5-triazine (TPT), leads to $[\{Cu_2(O_2CCH_3)_4\}_3(TPT)_2]$ (**2**) [17], which is a 2-D honeycomblike array with large hexagonal cavities (approximately 34–37 Å) as shown in Figure 6. As an alternative to interpenetration between 2-D networks, large cavities are accommodated by TPT nodes from the sheets on opposites sides, which are attracted together to form $\pi \cdots \pi$ association inside the hole. When a porphyrin scaffold is used as the four-connected node, a 1-D zigzag chain (**3**) and a 2-D sheet structure (**4**) are generated with $[Cu_2(O_2CCH_3)_4]$ [18]. These structures are obtained from the same molecular components but different solvent on synthesis.

Alternatively, an approach using metalloligands has been investigated to control the assembly of building blocks that possess second coordination comprising exo-oriented donor atoms. The simple metal cyanide, $[Co(CN)_6]^{3-}$, is used as multiconnecting linker,

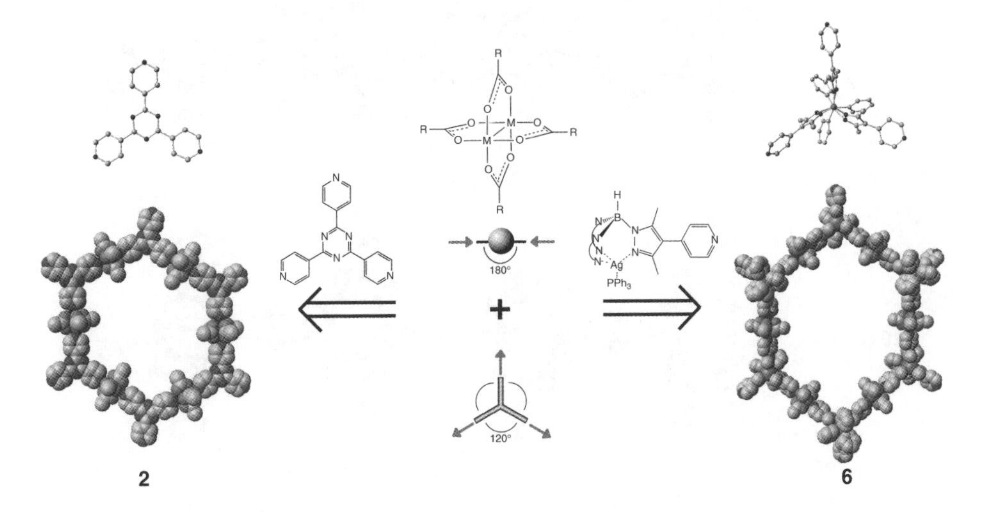

Figure 6. Crystal structures of **2** (left) and **6** (right). Hydrogen atoms are omitted for clarity.

which gives a 2-D gridlike network (**5**) with $[Rh_2(O_2CCH_3)_4]$ [19]. Each $[Co(CN)_6]^{3-}$ unit offers the four in-plane cyanide groups for connection to the Rh_2 building blocks and the two remaining apical cyanides are bound to K^+ counter cations. A neutral silver(I) complex with hydrotris(4-(4-pyridyl)-pyrazolyl)borate and PPh_3 acts as a tripodal metalloligand owing to the exodentate pyridyl groups. Figure 6 illustrates that the reaction of the metalloligand with $[Rh_2(O_2CCH_3)_4]$ forms a 2-D honeycomblike network (**6**) [20]. Owing to the intrinsic deviation of the metalloligand from a flat trigonal orientation, the infinite coordination structure appears as a corrugated layer architecture, in which the silver centers are located at the vertices of a chairlike structure.

3.2. Architectures with High Affinity at Axial Positions of $[Rh_2(O_2CCF_3)_4]$

Among the many kinds of paddlewheel complexes, the dirhodium(II) trifluoroacetate, $[Rh_2(O_2CCF_3)_4]$, is an intriguing molecular building block for infinite coordination assemblies because of the strong preference of axial coordination arising from the strong electron-withdrawing effect of the four CF_3 groups [21, 22]. In other words, this complex is a very strong bifunctional Lewis acid and so forms complexes with relatively weak Lewis bases.

The interaction of molecular iodine, I_2, with $[Rh_2(O_2CCF_3)_4]$ using a solid/vapor deposition ("solvent-less") technique gave a zigzag chain polymer $[\{Rh_2(O_2CCF_3)_4\}I_2]\cdot I_2$ (**7**) [23] formed by alternating $[Rh_2(O_2CCF_3)_4]$ and I_2 molecules and having an unprecedented bidentate bridging coordination mode of molecular diiodine as represented in Figure 7. Another iodine molecule is clathrated, by just filling voids in the structure without any close contacts with the neighboring atoms. Another interesting synthetic aspect of this

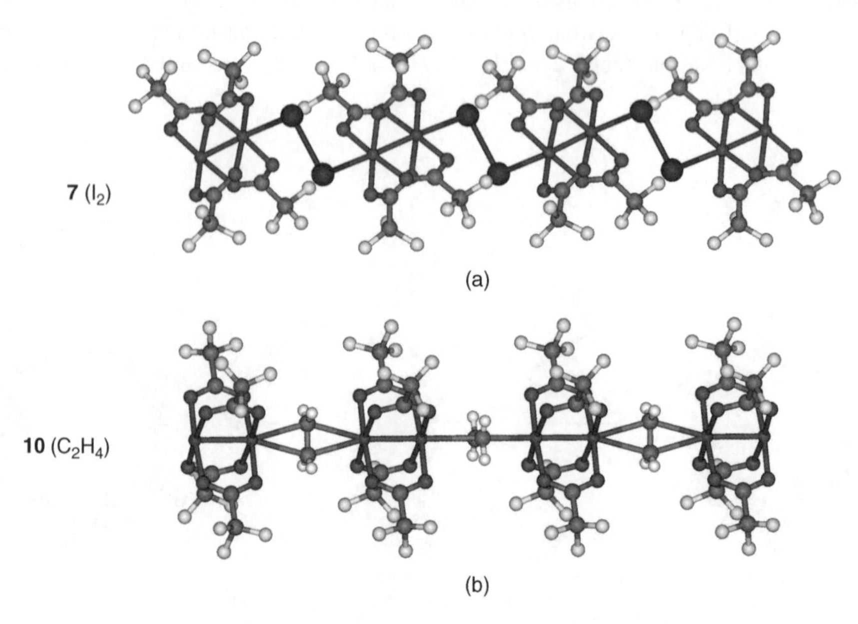

Figure 7. Crystal structures of **7** (a) and **10** (b).

result is the resistance of the $[Rh_2(O_2CCF_3)_4]$ unit toward oxidation by I_2 at temperatures above 150 °C.

When $[Rh_2(O_2CCF_3)_4]$ was exposed to vapors of elemental sulphur, with different ratios of Rh_2 to sulfur in the initial mixtures, two sulfur adducts of $[Rh_2(O_2CCF_3)_4]$ were isolated as shown in Figure 8: $[\{Rh_2(O_2CCF_3)_4\}_m(S_8)_n]$ ($m : n$ = 1:1 and 3:2) [24]. The structure of $[\{Rh_2(O_2CCF_3)_4\}(S_8)]$ (**8**) comprises $[Rh_2(O_2CCF_3)_4]$ units with the eight-membered S_8 rings 1,3-attached to their axial positions to form a 1-D zigzag chain. The structure of $[\{Rh_2(O_2CCF_3)_4\}_3(S_8)_2]$ (**9**) is particularly interesting as each S_8 molecule is 1,3,6-coordinated to three dirhodium complexes, while each Rh_2 unit binds to two different S_8 rings, which gives a pseudo-2-D architectures comprising octagon units ($4Rh_2$ and $4S_8$ blocks) forming a ribbon-type extended structure. Alternatively, the 2-D architecture can be considered as being built of two polymeric chains of the monoadduct $[\{Rh_2(O_2CCF_3)_4\}(S_8)]$ stapled together by an additional $[Rh_2(O_2CCF_3)_4]$ molecule lying between each pair of facing S_8 rings, which brings the total composition to an $[Rh_2(O_2CCF_3)_4]$:(S_8) ratio of 3:2.

The "solvent-less" synthetic approach also allows the formation of the infinite organometallic architectures supported by $Rh\cdots\pi$ interactions. This study was thoroughly elaborated upon by employing various types of π-ligands, ranging from the smallest ligand, such as ethylene, to planar and nonplanar Polycyclic Aromatic Hydrocarbon rings (PAHs).

The cosublimation of $[Rh_2(O_2CCF_3)_4]$ with 1,2-diiodoethane, ICH_2CH_2I, afforded the 1-D chain structure $[Rh_2(O_2CCF_3)_4(C_2H_4)]$ (**10**) in which ethylene molecules bridge two Rh_2 units through $Rh\cdots\pi$ bonds as shown in Figure 7 [25]. Similar 1-D structures with alternating arrangements of $[Rh_2(O_2CCF_3)_4]$ and π-ligand were isolated with benzene [26], *p*-xylene [26] and hexamethylbenzene [27] instead of ethylene. In all complexes, an off-center complexation takes place with opposite edges of the ligand approaching the axial sites of the neighboring Rh_2 units in $\eta^2(1,2)$:$\eta^2(4,5)$ bridging fashion. In the case of diphenylacetylene, containing both aromatic rings and alkyne carbon–carbon triple bond, the different conditions of sublimation led to isolation of both a discrete bis-adduct

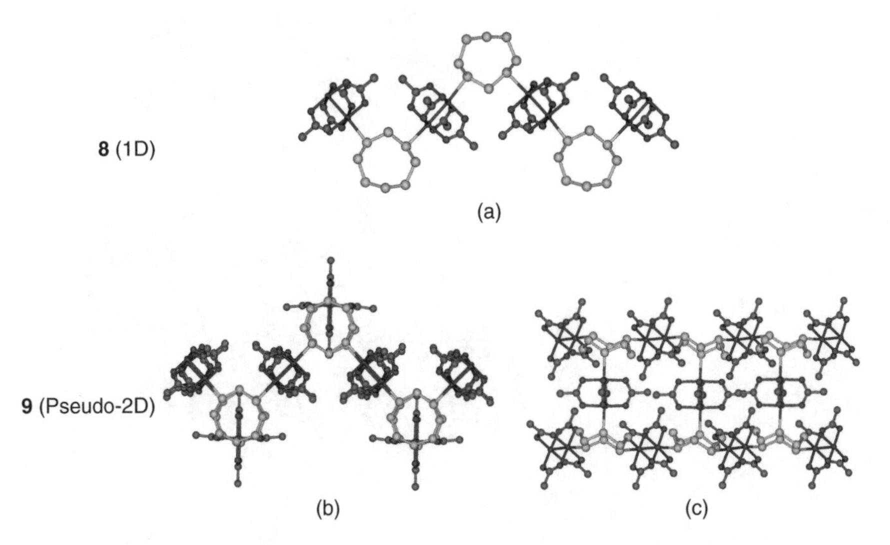

Figure 8. Crystal structures of **8** and **9**: (a) 1-D zigzag chain structure of **8**; (b) a side view and (c) a top view of pseudo-2-D ladderlike structure of **9**. Fluoride atoms are omitted for clarity.

and a 1-D chain complex [28]. In the former, two ligands coordinate the Rh_2 core through two alkyne carbon atoms. The latter complex is of particular interest as both arene and alkyne functions of the ligand are involved in coordination to the Rh_2 units, so as to form a 1-D chain structure.

Complexes with planar PAHs such as triphenylene have revealed a variety of structural motifs ranging from discrete molecules to extended 1-D chains and 2-D networks [29]. Most PAHs gave 1-D chain structures with an alternating arrangement as found for the monoaromatic rings. Each ligand exhibits a bidentate-bridging coordination with two rhodium atoms attached to opposite sides of the aromatic plane, that is, in a one-up, one-down fashion. One of the more remarkable structures is the 3:2 adduct formed with triphenylene, that is, $[\{Rh_2(O_2CCF_3)_4\}_3(\text{triphenylene})_2]$ (**11**) [29]. Each ligand exhibits a tridentate-bridging coordination using three of the C–C bonds, each from a different aromatic ring, affording a pseudo-2-D ladder structure in which formally two infinite chains of monoadducts are stapled together by additional $[Rh_2(O_2CCF_3)_4]$ units as illustrated in Figure 9. In this array, an elemental unit can be considered as an octagon formed by four dirhodium complexes and four triphenylene molecules, which is reminiscent of the structure of $[\{Rh_2(O_2CCF_3)_4\}_3(S_8)_2]$ (Figure 8) [24]. Chrysene also gives a 2-D network with a 3:1 composition (**12**) (Figure 9) [29]. Each chrysene molecule acts as in a tetradentate mode, bridging to four metal centers of four different Rh_2 cores, two on each side of the aromatic plane. Two C–C bonds on each of the outer aromatic rings of chrysene are used in the "up-and-down" fashion. Actually, for such a coordination mode as observed for chrysene, a 2:1 ratio would be enough to create such a brick-wall structural motif. Two other dirhodium complexes axially bind only one aromatic ligand each but are linked

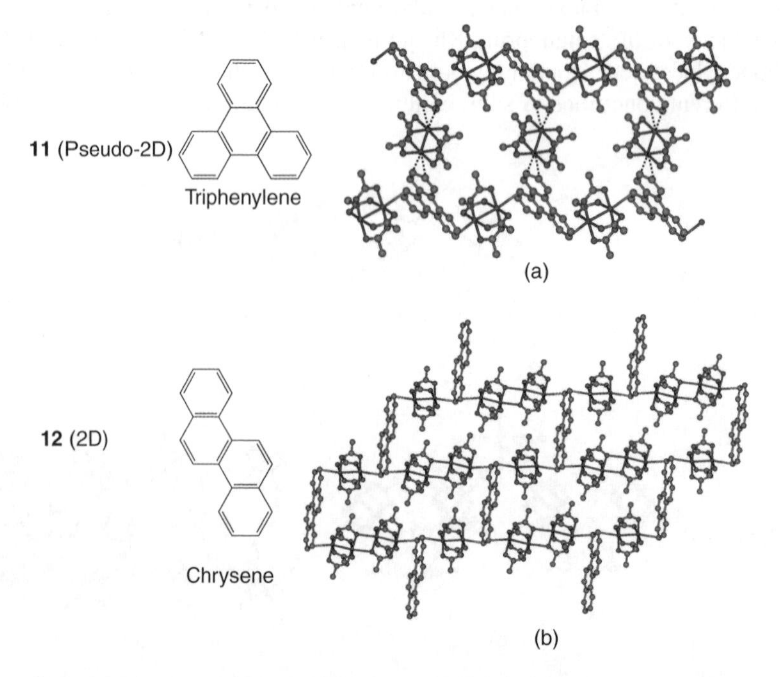

Figure 9. Crystal structures of **11** and **12**: a pseudo-2-D ladderlike structure of **11** (a) and a 2-D brick-wall structure of **12** (b). Hydrogen atoms and fluoride atoms are omitted for clarity.

together at their other ends through the mutual axial coordination of carboxylic oxygen atoms. It is worth noting that the site-selectivity of η^2-metal coordination in PAHs is correlated with the π-bond order of the C=C bonds or the π-electron density distribution, as determined by simple Hückel calculations.

The use of nonplanar PAH ligands such as corannulene and hemifullerene leads to fascinating architectures due to the convex (exo) and concave (endo) surfaces of these bowl-shaped molecules. Two products, $[\{Rh_2(O_2CCF_3)_4\}_m(\text{corannulene})_n]$ ($m : n = 1{:}1$ and 3:2), were isolated by the control of the reagent ratio in the solid state as shown in Figure 10 [30]. The complex with a 1:1 composition is a 1-D zigzag chain (**13**) comprising alternating Rh_2 units and corannulene molecules; one dirhodium complex approaches the ligand from the convex side of the ligand while the other binds from the concave side. The adduct with a 3:2 composition has an infinite 2-D layered network (**14**) comprising giant hexagonal cells constructed out of six Rh_2 units and six corannulene molecules. Each corannulene molecule is coordinated to three Rh_2 units: two from the convex side and one from the concave side. Remarkably, only rim carbon atoms of the ligand are involved in coordination.

With hemifullerene, only one product, $[\{Rh_2(O_2CCF_3)_4\}_3(\text{hemifullerene})]$ (**15**) [31], was isolated in spite of attempts to change the product composition by varying the reaction temperature and the ratio of reagents. Although the authors reported the framework as a

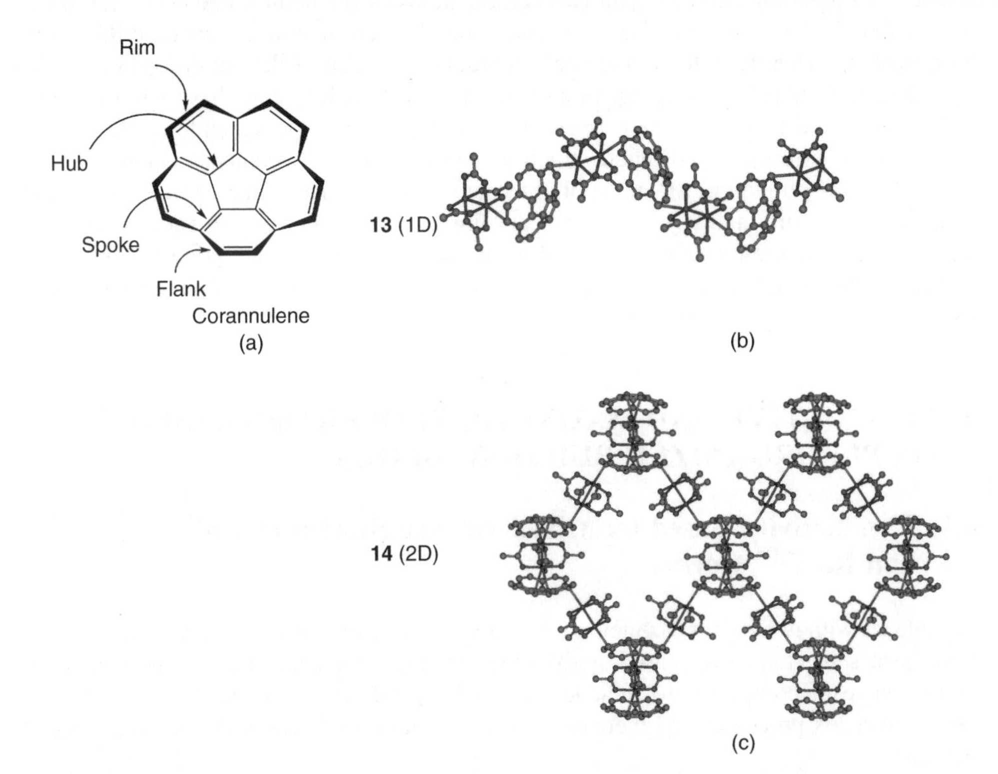

Figure 10. (a) Schematic representation of corannulene. Crystal structures of **13** and **14**: (b) a 1-D chain structure of **13** and (c) a 2-D layered structure of **14**. Hydrogen atoms and fluoride atoms are omitted for clarity.

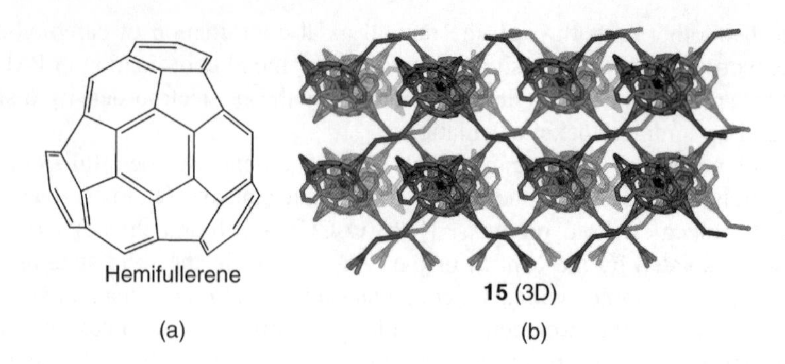

Figure 11. (a) Schematic representation of hemifullerene. (b) Crystal structure of a doubly interpenetrated 3-D framework of **15**. Trifluoroacetate groups and hydrogen atoms are omitted for clarity.

2-D layer, reinspection of crystal structure indicates the double interpenetrated 3-D framework (Figure 11). This complex has an infinite 3-D framework built upon tetrabridged coordination; three rhodium centers approach the hemifullerene ligand from the convex side and one is bound to the concave side. Whereas the three exo-coordinated rhodium centers form bonding contacts with two carbon atoms of the hemifullerene (η^2 coordination mode), the coordination of the sole endo-bound rhodium atom is best described as in an η^1 fashion. Therefore, the tetradentate coordination mode of the hemifullerene can be described as η^2:η^2:η^2:η^1-bridging. In addition, two of four Rh_2 units have hemifullerene molecules bound to both of their open axial positions; one bridges endo/endo, and the other exo/exo, affording a 1-D chain structure with alternating Rh_2 units and hemifullerene molecules. The remaining Rh_2 units bind to the convex surface of the ligand at one axial position of the core, and two such units are linked at the other end to form a dimer of dimers core structure through the mutual axial coordination of carboxylic oxygen atoms. Moreover, this dimer of dimers connects the 1-D chain motifs positioned obliquely, giving rise to two types of interpenetrated 3-D framework.

4. CONDUCTIVE AND MAGNETIC PROPERTIES BASED ON PREORGANIZED BUILDING BLOCKS

4.1. Conductivity Based on Mixed-valence System of $Rh^{II}{}_2$ and $Rh_2{}^{II,III}$ Cores

Complexes with two or more centers of mixed valency in similar or identical surroundings have been subjected to increased study. One direction of recent effort is the expansion of the range of complexes to include those with paddlewheel motifs because of their enhanced redox processes and increased internal degrees of freedom due to metal–metal bonding.

To date, two examples of mixed-valence infinite architectures have been reported based on paddlewheel building blocks, namely, $[Rh_2(acam)_4]^{0/+}$ (Hacam = acetamide): a 2-D honeycomb network for $[\{Rh_2(acam)_4\}_3(\mu_3\text{-}Cl)_2]$ (**16**) [32] and a 3-D diamondoid network

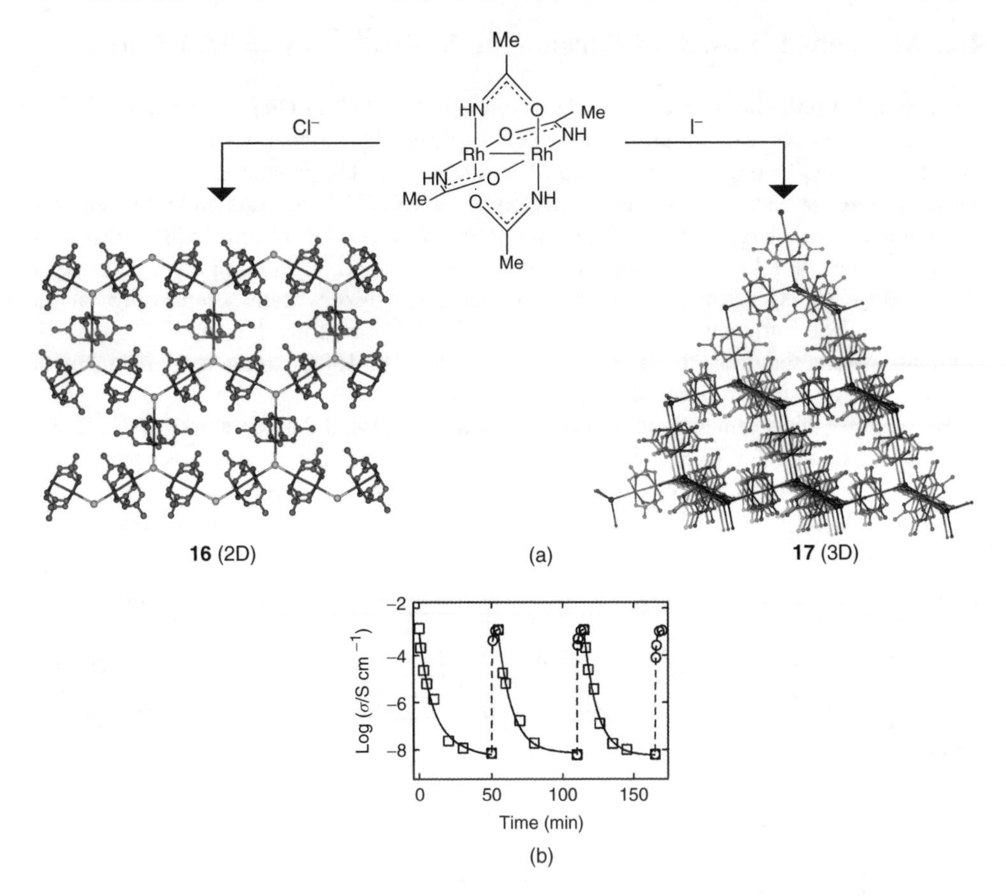

Figure 12. (a) Crystal structures of **16** and **17**: A 2-D honeycomb network of **16** (left) and a 3-D diamondoid framework of **17** (right). Guest water molecules and hydrogen atoms are omitted for clarity. (b) Oscillation in electrical conductivity of a pellet of **17**·6H_2O arising from cycles of dehydration under vacuum at room temperature (squares and solid lines) and rehydration in a moisture-saturated argon flow at room temperature (circles and broken lines). Reprinted with permission from [33]. Copyright 2004, American Chemical Society.

for [{Rh_2(acam)$_4$}$_2$(μ_4-I)] (**17**), as shown in Figure 12 [33]. Complex **16** comprises three pairs of crystallographically independent dinuclear units, six crystallographically equivalent Cl^- ions at the corners and two crystallographically independent water molecules with intralayer hydrogen bonding. From the examination of bond distances, the neutral ${Rh_2}^{4+}$ state is pinned at one core and the cationic radical ${Rh_2}^{5+}$ state at the other two sites, which arises from their hydrogen bonds with water molecules. The observed low electrical conductivity (2×10^{-7} S cm^{-1}) reflects the localization of the oxidation states. On the other hand, **17** is constructed from one crystallographically equivalent Rh_2 core and an iodide, and has water molecules that are hydrogen bonded to the 3-D network. The electrical conductivity of this complex, 1.4×10^{-3} S cm^{-1}, was much higher than that of the honeycomb network. It is of interest that the conductivity decreased to 7.0×10^{-9} S cm^{-1} upon dehydration of the complex and oscillated over a range of 10^5 during the dehydration-rehydration cycles.

4.2. Magnetism Based on Paramagnetic $Ru_2^{II,III}$ ($S = 3/2$) Cores

The cationic diruthenium tetracarboxylate complex, $[Ru_2^{II,III}(O_2CR)_4]^+$, has a $\sigma^2\pi^4\delta^2\delta^{*1}\pi^{*2}$ valence electronic configuration in which the three highest-energy electrons are shared over the essentially degenerate π^* and δ^* orbitals [34]. The magnetic properties are of particular interest because of the unusually high spin ($S = 3/2$) for a second row coordination complex with large zero-field splitting, ZFS, ($D = +63 \pm 11$ cm^{-1}) [35]. Whereas, a large number of 1-D chain complexes have been reported, most complexes of the type $[Ru_2^{II,III}(O_2CR)_4Cl]$ form chloride-bridged, linear or zigzag chains, depending on the nature of R. The intrachain interaction between $Ru_2^{II,III}$ cores involves antiferromagnetic coupling. When the Ru–Cl–Ru bond angle is close to 180°, strong antiferromagnetic interactions are usually observed [36].

By contrast, high-dimensional structures such as 2-D and 3-D networks are still rare (only one example for each). These high-dimensional structures are constructed from $Ru_2^{II,III}$ cores and hexacyanometallate (Figure 13). In the 2-D layered array of $[\{Ru_2^{II,III}(O_2C(CH_3)_3)_4\}(H_2O)M^{III}(CN)_6]$, M = Fe (**18**) and Co(**19**) [37], the M atom, located on a

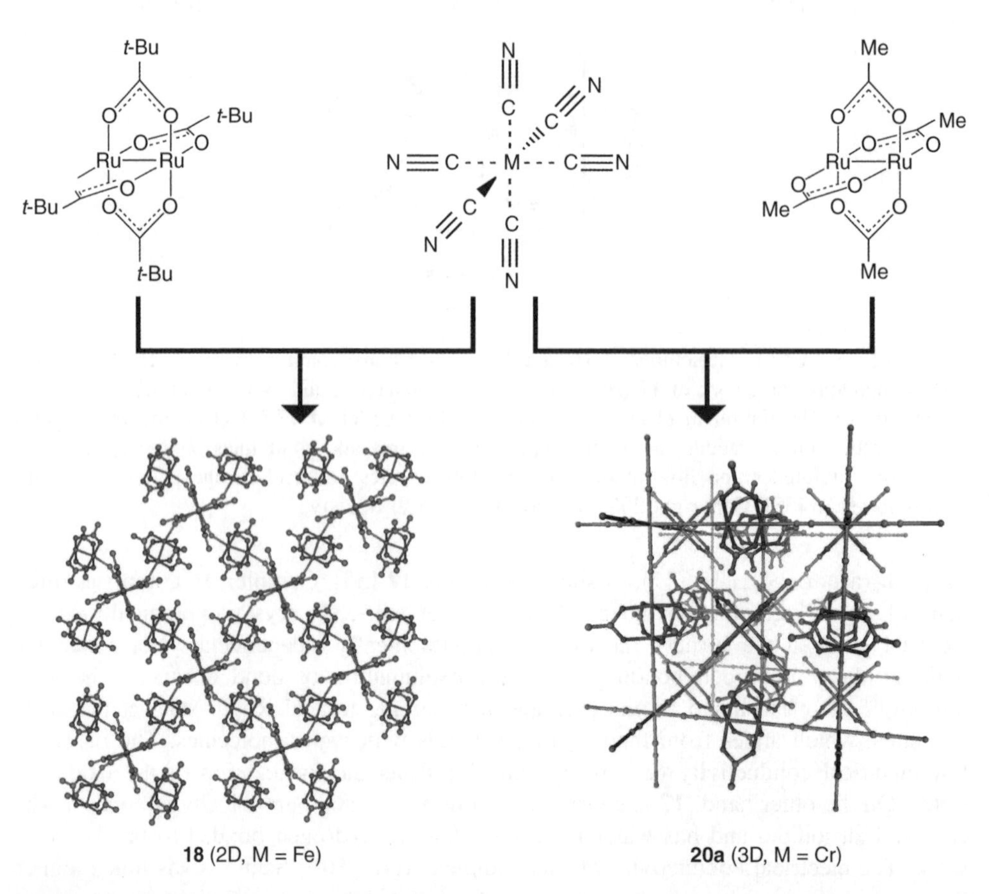

Figure 13. Crystal structures of **18** and **20a**: a 2-D layered structure of **18** (left) and a double interpenetrated 3-D body-centered cubic framework of **20a** (right). Methyl groups of *tert*-butyl group and hydrogen atoms are omitted for clarity.

fourfold axis, uses four cyano groups within the *ab* plane to connect the Ru_2 dimer units, affording a 2-D sheet made up of 28-membered [–M–CN–Ru_2–NC–]$_4$ rings, while one of the remaining two *trans*-cyano ligands is terminal along the *c* direction and the other coordinates to the ruthenium atom of Ru_2 dimer units with the opposite axial site capped by a water molecule. The temperature dependence of magnetic susceptibility for **18** shows that the magnetic moment gradually decreases with decreasing temperature and reaches a minimum value at around 20 K. Upon further cooling, the magnetic moment increases abruptly, reaching a maximum value at 8 K, and then decreases. This type of variation suggests ferrimagnetic behavior with a $Ru_2^{II,III}$-Fe^{III} antiferromagnetic interaction. Above 100 K, the magnetic susceptibility data can be fitted to the Curie–Weiss expression with $\theta = -34.8$ K, which is consistent with the presence of an antiferromagnetic interaction. The effective magnetic moment of the Co^{III} analog shows a monotonic decrease with lowering of temperature, and above 10 K can be fitted with $\theta = -23.0$ K. However, the zero-field splitting contribution was not considered.

The assemblies of $[Ru_2^{II,III}(O_2CCH_3)_4]^+$ with $[M^{III}(CN)_6]^{3-}$, M = Cr (**20**), Fe (**21**) and Mn (**22**), afford molecule-based magnets [38]. These complexes were prepared under two different conditions, that is, aqueous and nonaqueous (acetonitrile) routes. Unfortunately, only one complex $[Ru_2^{II,III}(O_2CCH_3)_4]_3[Cr^{III}(CN)_6]$ (**20a**), prepared from aqueous solution, was structurally characterized by the Rietveld analysis of the synchrotron powder diffraction data, which provides the double interpenetrated 3-D body centered cubic structure with [–M–CN–Ru_2–NC–] linkages as shown in Figure 13. Molecular modeling reveals that a second interpenetrating lattice would be unexpected due to the large steric crowding between the two lattices when the carboxylate group is bigger than the acetate, for example, propionate. Therefore, the assemblies of the analogous tetrapivalate complex, $[Ru_2^{II,III}(O_2CC(CH_3)_3)_4]^+$, afforded 2-D arrays, as mentioned above. On the other hand, the nonaqueous route gave $[Ru_2^{II,III}(O_2CCH_3)_4]_3[Cr^{III}(CN)_6]\cdot x(CH_3CN)$ (**20b**), as an amorphous solid. As no void exists in **20b** allowing accommodation of acetonitrile in the crystal structure and in consideration of its low density (1.90 $g\,cm^{-3}$) compared with the interpenetrating lattice, that is, 2.11 $g\,cm^{-3}$, the structure of **20b** does not correspond to that of **20a** obtained from the aqueous method. Both crystalline **20a** and amorphous **20b** phases of $[Ru_2^{II,III}(O_2CCH_3)_4]_3[Cr^{III}(CN)_6]$ are molecular-based ferrimagnets with $T_c = 33$ K and $\theta = -40$ K for **20a** and $T_c = 34.5$ K and $\theta = -70$ K for **20b** (Figures 14(a) and 14(b)). While the field dependence of the magnetization for **20a** showed an unusual constricted hysteresis loop, which has been attributed to metamagnetism caused by canted spins, **20b** did not have this constricted hysteresis loop but contained a classical hysteresis loop (Figure 14(c)). The superexchange mechanisms between the $Ru_2^{II,III}$ cores and Cr^{III} ions are considered as antiferromagnetic coupling between the $d\pi^*$ orbitals of the $Ru_2^{II,III}$ cores and d orbitals Cr^{III} through $p\pi^*$ orbitals of cyanide rather than due to the antiferromagnetic coupling between orthogonally aligned $d\delta^*$ orbitals of the $Ru_2^{II,III}$ cores to the d orbitals of Cr^{III} (Figure 14(d)).

4.3. Magnetism Based on Paramagnetic Ru_2^{II} ($S = 1$) Cores

As mentioned in the section describing the preparation of building blocks (Section 2.2), very little work has been performed on Ru^{II}_2 complexes, in part, owing to their air-sensitive nature, let alone the difficulty of their synthesis. To date, only three extended structures are known: one 1-D chain complex and two 2-D layer complexes.

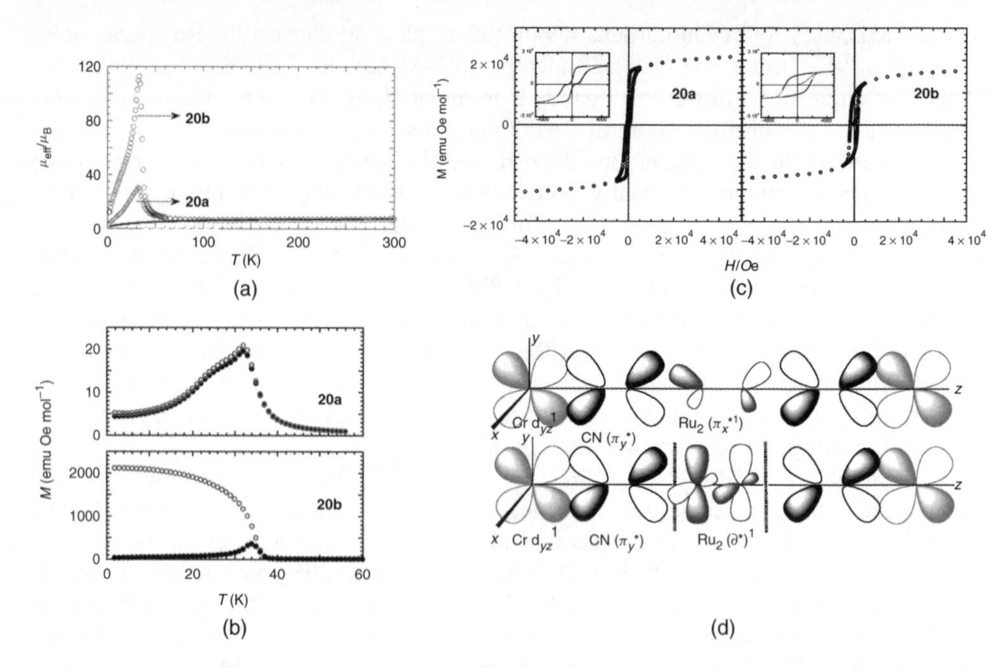

Figure 14. (a) $\mu_{\text{eff}}(T)$ data; (b) field cooled (FC, ○) and zero field cooled (ZFC, ●) $M(T)$ data, and (c) $M(H)$ data at 2 K taken from ±50,000 Oe for **20a** and **20b**. Inserts highlight the region from ±5000 Oe. (d) Nearest neighbor overlap in the xz plane for a ⋯Cr–CN–Ru_2–NC–Cr⋯ segment (45° relative rotation): depicting the nonorthogonal overlap of the Cr^{III} d_{yz}, CN π_y*, and Ru_2 π_y* orbitals (top) and depicting the nonbonding orthogonal overlap of the Cr^{III} d_{yz}, CN π_y*, and Ru_2 δ* orbitals (bottom). Reprinted with permission from [38]. Copyright 2004 American Chemical Society.

The 1-D $Ru^{II}{}_2$ polymers shown in Figure 15, [$Ru^{II}{}_2(O_2CCF_3)_4$(phenazine)] (**23**) [39], displays relatively linear Ru–Ru–N bond angles, albeit with some canting away from linearity (15°) by phenazine units with the Ru–N⋯N angles being about 165°. Magnetic measurements revealed a decrease in χT as the temperature is lowered, again, primarily due to the ZFS. The behavior and fitting process, using the molecular field approximation model [40], indicates weak antiferromagnetic interactions between $Ru^{II}{}_2$ cores with the fitted parameters being $zJ = -3.0$ cm^{-1} and D = 277 cm^{-1}. The weak antiferromagnetic coupling was ascribed to the less than ideal orbital overlap caused by the aforementioned canting angle.

In 2-D layer structure incorporating [$Ru_2^{II}(O_2CCF_3)_4$] units and 7,7,8,8-tetracyanoquinodimethane (TCNQ) bridges, [{$Ru^{II}{}_2(O_2CCF_3)_4$}$_2$(TCNQ)]·3(toluene) (**24**) [41], all four cyano groups of TCNQ are coordinated to independent Ru_2 molecules, affording a distorted hexagonal network as shown in Figure 15. Interestingly, **24** appears to show significant electronic delocalization. Evidence for this delocalization came from the IR and visible spectra of the complex, which show characteristic features of a partially reduced TCNQ with concomitant oxidation of one or more of the $Ru^{II}{}_2$ units to the $Ru_2{}^{II,III}$ form due to significant metal to ligand π back-donation. Unfortunately, the temperature dependence of the magnetic susceptibility and the theoretical fitting for this complex were not reported.

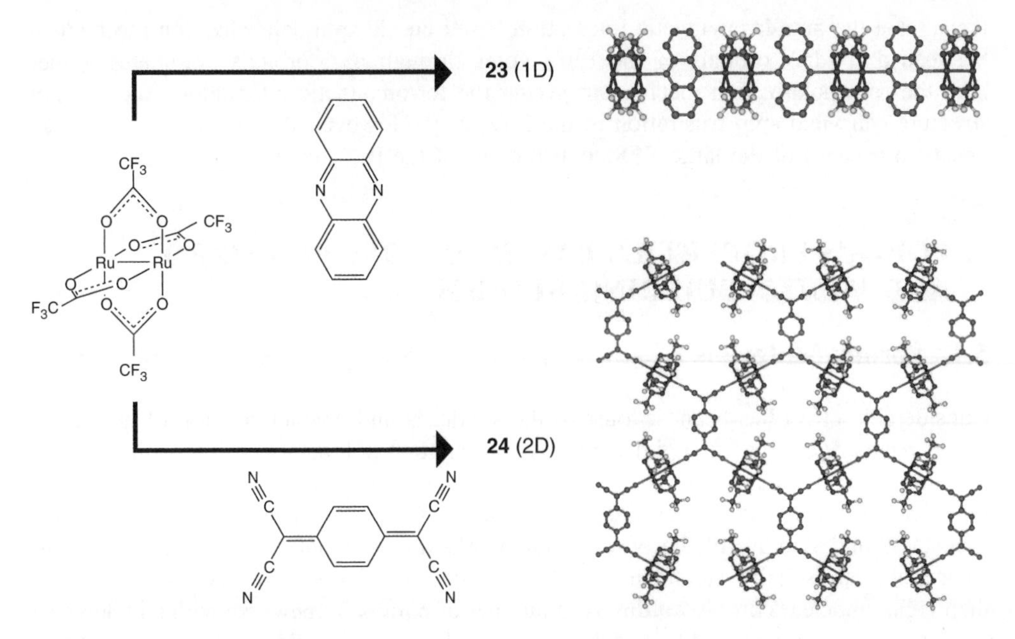

Figure 15. Crystal structures of **23** and **24**: a 1-D chain structure of **23** (top) and a 2-D honeycomb network of **24** (bottom). Hydrogen atoms are omitted for clarity.

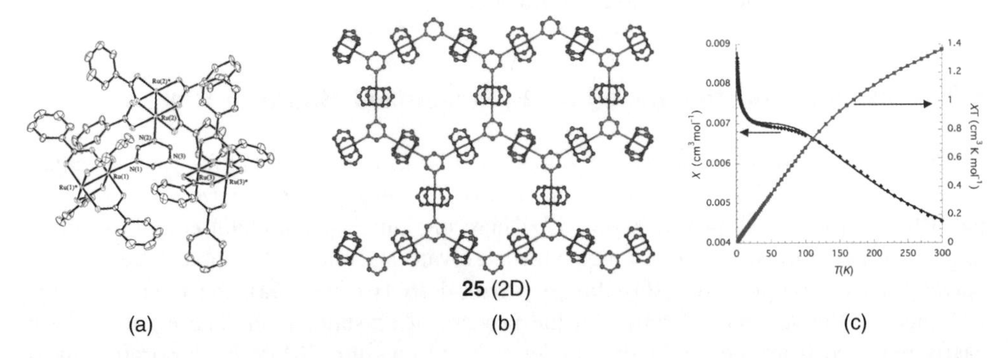

Figure 16. (a) Structure of one triazine molecule coordinated to three independent $[Ru^{II}(O_2CPh)_4]$ units in **25** (ORTEP representation). (b) A 2-D honeycomb structure of **25**. Phenyl groups are omitted for clarity. (c) $\chi(T)$ and $\chi T(T)$ data for **25**. The solid line represents the theoretical fit. Reprinted with permission from [42]. Copyright The Royal Society of Chemistry 2005.

The other 2-D layered structure, $[\{Ru^{II}_2(O_2CPh)_4\}(\text{triazine})]$ (**25**) [42], has the flawless hexagonal honeycomb network shown in Figure 16(b), where all three nitrogen atoms of triazine are coordinated to independent Ru_2 units (Figure 16(a)). In this structure, the paramagnetic centers are located on the midpoint of sides of honeycomb (6,3) network, which is interpreted as the magnetic 2-D Kagomé lattice. The overall magnetic behavior is very similar to that of the 1-D chain complex as mentioned above (Figure 16(c)). The theoretical fitting was also performed by the same equation used in the 1-D chain complex, which provides the values $zJ = -2.2\ \text{cm}^{-1}$ and $D = 254\ \text{cm}^{-1}$. The negative zJ value means that the antiferromagnetic interaction based on the spin delocalization mechanism between $d\pi^*$–$d\pi^*$ orbitals of the Ru^{II}_2 cores through $p\pi^*$ orbitals dominates, rather than the spin polarization mechanism giving the ferromagnetic interaction. Such a spin structure can yield spin frustration in the 2-D array. However, this has not been clearly observed because of the large ZFS contributions of the Ru^{II}_2 cores.

5. POROUS PROPERTIES BASED ON PREORGANIZED AND *IN SITU* BUILDING BLOCKS

5.1. Porous Materials

Considerable effort has been devoted to the synthesis and characterization of new crystalline nanosized porous materials, such as coordination polymers and inorganic zeolites, because of their versatile applicability to gas storage, molecular sieves, size- or shape-selective catalysis, and ion exchange. For successful performance of these functions, robustness and stability of the porous frameworks are essential. The paddlewheel dinuclear motifs are suitable candidates for the construction of robust frameworks because of their rigid dinuclear core. Accordingly, a number of porous frameworks with paddlewheel motifs have been reported. Most of these are based on *in situ* building blocks with copper and zinc ions that allow access to the high-dimensional frameworks [43]. On the other hand, there are few reports of the porous materials with *preorganized* building blocks, but recently, notable advances have been made.

5.2. Porous Properties Based on Preorganized Building Blocks

There is only one series of porous coordination polymers, or absorbent for gas molecules, with the *preorganized* building blocks, that is, $[M_2(O_2CPh)_4(pyrazine)]$, M = Cu (**26**) [44] and Rh (**27**) [45], as shown in Figure 17. These one-dimensional chain complexes crystallized in two forms: one as a diacetonitrile solvate and the other, solvent-free. In the solvated chain complex, $[Cu_2(O_2CPh)_4(pyrazine)]\cdot 2(CH_3CN)$ (**26a**), the micropores are occupied by the solvent molecules in the process of crystallization. These solvents are easily removed from the micropores under reduced pressure. While N_2 absorption measurements show that three N_2 molecules are absorbed per $[Cu_2(O_2CPh)_4(pyrazine)]$ unit, the absorbed amount of water molecules is very small compared with those of N_2 [44]. This indicates that the micropores are extremely hydrophobic.

The solvent-free complexes were synthesized as $[M_2(O_2CPh)_4(pyrazine)]$, M = Cu (**26b**) and Rh (**27b**). The crystal packing of these complexes feature very narrow void space and these appear to expand to allow adsorption of external gas molecules. Surprisingly, this solid also demonstrates the ability to act as an absorbent for gas [45]. As shown in Figure 17, in **26a** N_2 molecules are free to enter the space with spontaneous structural changes in the progress of gas adsorption, whereas water molecules are not adsorbed. It is noteworthy that these nonporous complexes include gas molecules such as CO_2 [46] and O_2 [47] at low temperature; the structures of these clathrates have been established by x-ray crystallography. These experiments provide evidence that the adsorption of gas molecules into the nonporous coordination polymer causes a structural transformation.

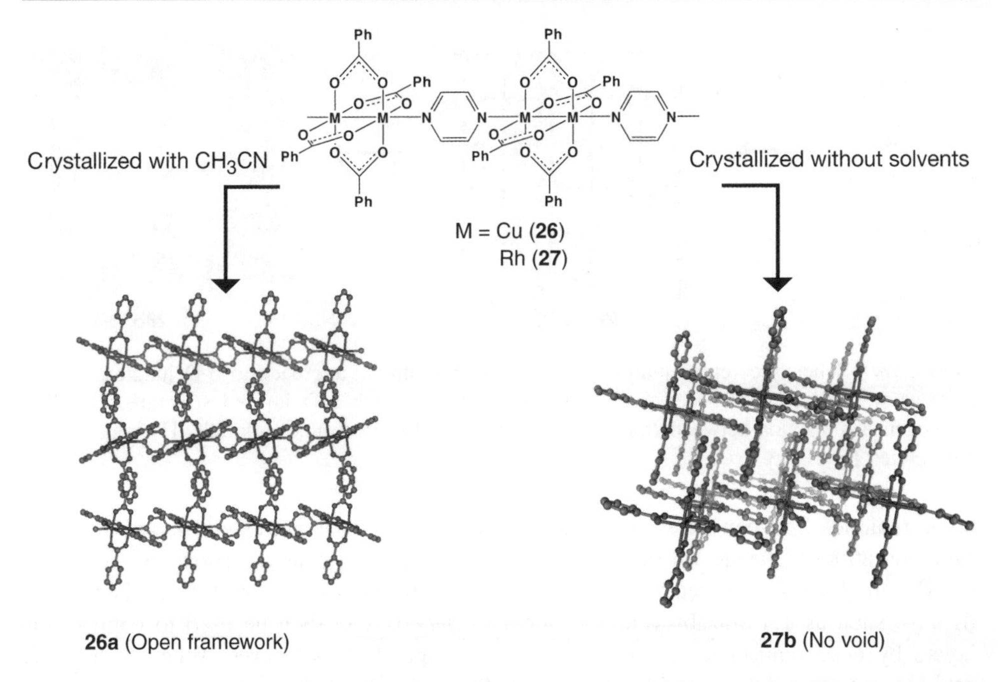

Figure 17. Crystal structures of **26a** and **27b**: a open framework constructed from 1-D chain motifs of $[Cu^{II}_2(O_2CPh)_4(prz)]$ with CH_3CN as guest solvent molecules of **26a** (left) and crystal packing of 1-D chain motifs of $[Rh^{II}_2(O_2CPh)_4(prz)]$ without void spaces of **27b** (right). Guest CH_3CN molecules and hydrogen atoms are omitted for clarity.

5.3. Porous Properties Based on *In Situ* Building Blocks

Self-assembly of transition metal ions and multifunctional organic ligands represents a successful paradigm for the single-step synthesis of infinite molecular structures. The *in situ* building blocks containing copper(II) and zinc(II) ions are included into this category of single-step synthesis. Especially, $[Cu_2(O_2CR)_4]$ units give rise to numerous examples of infinite coordination architectures due to its simple preparation compared with the *pre-organized* building blocks incorporating ruthenium and rhodium ions: the simple mixing of copper(II) sources with organic carboxylates under conditions of ambient temperature and pressure.

In recent progress in the design and synthesis of porous coordination polymers, *in situ* building blocks of the paddlewheel dinuclear cores have been shown to be one of the most significant and essential motifs for the construction of stable and robust porous frameworks, which show permanent porosity without any guest molecules in the pores. Thus, among the series of Metal-Organic Framework structures (MOFs), several porous and nonporous architectures feature paddlewheel cores [43]. In addition, a series of coordination polymers based on nanosized polyhedral building blocks (square and triangular) have been constructed from the paddlewheel motifs with nonlinear dicarboxylates [48].

The simplest strategy for the construction of porous frameworks is to use the paddlewheel motifs as the rigid octahedral nodes linked by linear organic linkers in both equatorial and axial positions, which gives rise to the rectangular-latticed frameworks. This strategy has been successfully realized by the rational synthetic method to insert

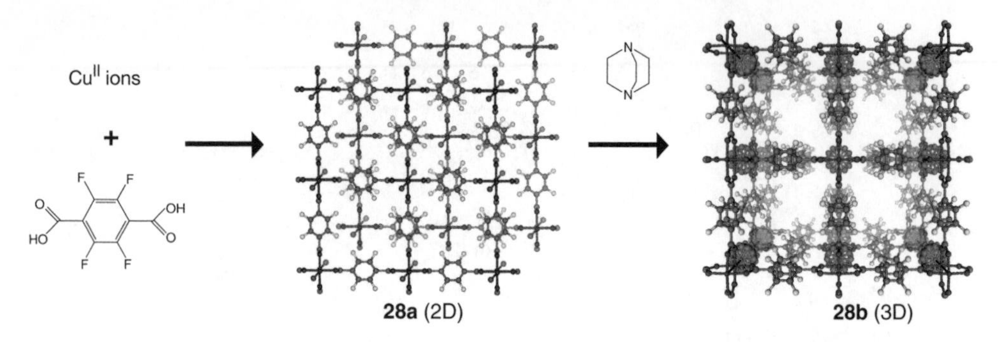

Figure 18. Schematic representation of the rational synthetic approach to 3-D jungle-gym like framework; a reaction between Cu^{II} ions and tfbdc affords a 2-D layered structure, **28a**. The heterogeneous reaction between **28a** and dabco gives a 3-D open framework, **28b**. Hydrogen atoms are omitted for clarity.

pillar moieties between layers of two-dimensional layered complexes (Figure 18) [49]. Thus, tetrafluorobenzene-1,4-dicarboxylate (tfbdc), as the linear dicarboxylate, and copper(II) ions form a 2-D square-grid layer with dimensions of 10.9×10.8 Å^2, **28a**. Pillar ligands, such as 1,4-diazabicyclo[2,2,2]octane (dabco), can then be used to connect the layers by coordination to the axial positions of the dicopper cores. Whereas the 2-D grid structure was determined by x-ray crystallography, the heterogeneous pillar insertion reaction resulted in the degradation of the crystal transparence. Therefore, synchrotron x-ray powder diffraction measurements were performed and the structure was refined by Rietveld analysis. The crystal structure of this framework, [$Cu_2(tfbdc)_2(dabco)$] (**28b**), shows that 3-D interconnected channels exist and their dimensions are approximately 6.3×6.3 Å^2 and 3.5×4.7 Å^2. The calculated solvent-accessible pore volumes and surface areas are 0.54 cm^3 g^{-1} (55% of unit cell volume) and 2020 m^2 g^{-1}, respectively (using a sphere with diameter of 3.0 Å).

This 3-D scaffold is highly suitable for rational modification because it is easy to replace tfbdc with other linear dicarboxylate ligands such as 1,4-benzenedicarboxylate (bdc), 2,3-difluorobenzene-1,4-dicarboxylate (dfbdc), 1,4-naphthalenedicarboxylate (ndc), and 2,3-dimethoxy-1,4-dicarboxylate (dmdbc) [50]. All analogous complexes have adsorption properties for gas molecules such as nitrogen and argon. Moreover, not only can the bridging dicarboxylate linker be substituted but also the pillar ligand can be replaced with other $N \cap N$ bridging ligands. The heterogeneous reaction between the 2-D grid motif, [$Cu_2(bdc)_2$], and the 4,4′-bipyridyl (4,4′-bpy) as the pillar affords the 3-D framework, [$Cu_2(bdc)_2(4,4\text{-bpy})$] (**29**) [51]. The results of powder x-ray diffraction and, molecular mechanics and dynamics simulations indicate **29** did not have the simple 3-D jungle-gym framework but has the mutually interpenetrating structure of two 3-D frameworks as represented in Figures 19(a) and 19(b). The insertion of 4,4′-bpy, which has a longer bridging span than dabco, creates a large void space associated with one framework, into which the other framework is allowed to occupy. However, this doubly interpenetrating framework still has 1-D cylindrical channels with 3.4×3.4 Å^2 windows. The adsorption property for methane is very striking. Below a specific pressure, the amount of methane adsorbed in the micropores increases only slightly from zero, and above that specific pressure the amount adsorbed increases suddenly. There is a hysteresis loop between the adsorption and desorption (Figure 19(c)). Such an adsorption behavior of methane is typical of an

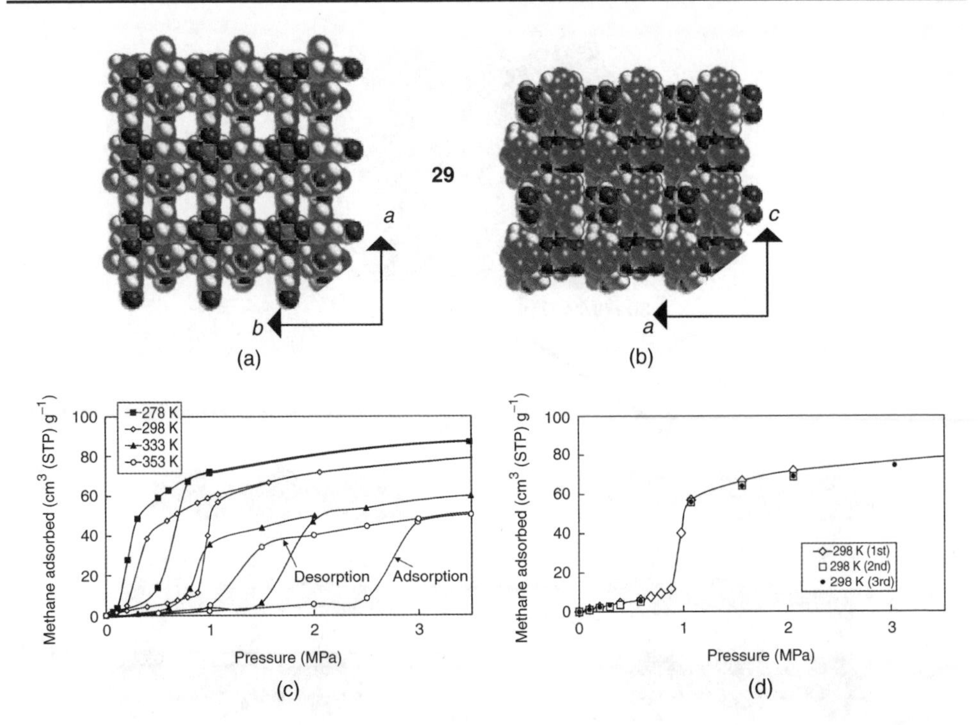

Figure 19. Space-filling model of the crystal structure of **29**: (a) a view down the *c*-axis and (b) a view down the *b*-axis. The structure model was optimized by Molecular Mechanics (MM) with the program package Cerius 2. High-pressure methane adsorption–desorption isotherms: (c) temperature dependence and (d) cyclic characteristics of methane adsorption. Reprinted with permission from [51]. Copyright The Royal Society of Chemistry 2002.

adsorption process in which a guest-induced phase transition takes place [52]. The second and third isotherms of the repeated methane adsorption behavior are the same as that of the first isotherm (Figure 19(d)), indicating that the pore structure of **29**, after adsorption, reverts to the original structure before adsorption by methane desorption. These phenomena are interpreted in the following terms: when the pressure of methane reaches a certain pressure (referred to as a gate-opening pressure) [53], a guest-induced phase transition takes place and the interpenetrating structures is changed locally so as to enlarge the pore size of the channel. A similar adsorption isotherm was also seen in the case of methanol adsorption. The powder x-ray diffraction measurements for **29**, recorded before and after methanol adsorption, strongly suggest that the 3-D framework is retained but only that the style of the interpenetration is changed by adsorption.

The analogous structure for **28b** was also prepared by the paddlewheel dizinc unit, $[Zn_2(bdc)_2(dabco)]$ (**30**), which also has permanent porosity and gas adsorption properties [54]. However, its framework rigidity is quite different from the dicopper analogs (Figure 20). In an as-synthesized complex, **30a**, containing guest molecules, that is, four DMF and a half H_2O, the bdc linker is unusually bent, which results in severe twisting of the paddlewheel cores from an ideal square grid. When the guest solvent molecules are removed from pores, the bridging bdc ligand is now linear (**30b**), affording a perfect 2-D square grid as well as **28b**, with an increase of the cell volume from 1091.8(4) to

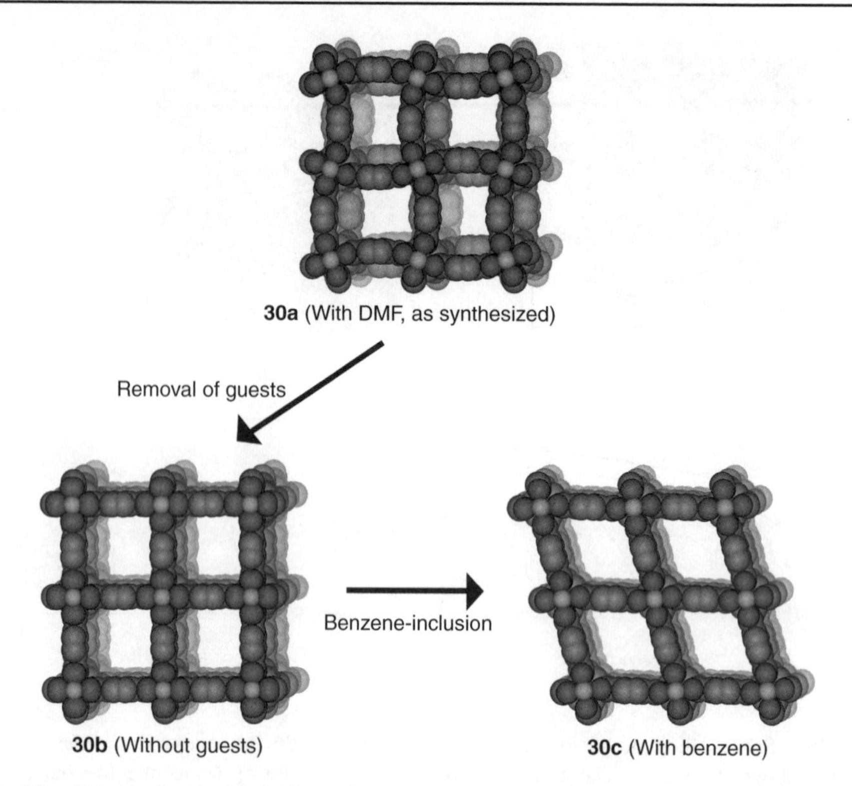

Figure 20. Schematic representation of the structural change of the 3-D framework, **30a** (as synthesized compound including 4DMF and 1/2H_2O as guests): the removal of guests affords **30b**, and the introduction of benzene molecules to pores gives **30c**. All guests molecules and hydrogen atoms are omitted for clarity.

1147.6(3) Å^3. Moreover, the inclusion of benzene leads to a different mode of framework distortion; the perfect 2-D square grid changes to a 2-D rhombic grid (**30c**) with the shrinkage of volumes (1114.2(2) Å).

6. CONCLUSION AND OUTLOOK

In this perspective, we provide evidence that the paddlewheel dinuclear tetracarboxylate motifs are intriguing building blocks for the construction of stable frameworks with high dimensionality as well as having the possibility of being exploited as functional electronic materials based on the magnetism and conductivity arising from the characteristic metal–metal interactions. The former functions such as porous properties are especially applicable to the *in situ* building blocks (i.e. containing Cu_2 and Zn_2 cores) and the latter to the *preorganized* building blocks (i.e. containing Ru_2 and Rh_2 cores). At present, these properties have been independently developed.

From this standpoint, we are now developing the second stage of the chemistry of the paddlewheel dinuclear units; the first stage should be recognized as chemistry of the dinuclear core itself, especially, the metal–metal bonding properties, and the second stage is the extended framework development summarized in this review.

The third stage is envisaged as the unification of the independently developed areas: functions of framework solids themselves, such as porous properties, and functions of its electronic structure, such as magnetic and conductive properties. It is of importance that metal ions inside the framework are freely replaced with other metal ions. Especially, the integration of the paramagnetic paddlewheel cores such as $Rh_2^{II,III}$, $Ru_2^{II,III}$ and Ru^{II}_2 cores into stable and high-dimensional porous frameworks offers much promise, not only for multifunctional framework materials, such as molecular-based porous magnets or conductors, but also of novel hybridized functions with guest molecules, for instance, the charge-transfer property between the host framework and the guest molecules (gas or organic molecules). As mentioned above, there have been no reports of structurally characterized extended networks with Rh_2 and Ru_2 cores linked via equatorial positions. Therefore, the development of appropriate synthetic methodologies is essential to realize these novel functions.

REFERENCES

1. S. Kitagawa, R. Kitaura and S.-I. Noro, *Angew. Chem., Int. Ed.*, **43**, 2334–2375 (2004).
2. B. Moulton and M. J. Zaworotko, *Chem. Rev.*, **101**, 1629–1658 (2001).
3. S. R. Batten and R. Robson, *Angew. Chem., Int. Ed.*, **37**, 1460–1494 (1998).
4. A. J. Blake, N. R. Champness, P. Hubberstey, W.-S. Li, M. A. Withersby and M. Schröder, *Coord. Chem. Rev.*, **183**, 117–138 (1999).
5. M. W. Hosseini, *CrystEngComm*, **6**, 318–322 (2004).
6. L. Carlucci, G. Ciani and D. M. Proserpio, *Coord. Chem. Rev.*, **246**, 247–289 (2003).
7. F. A. Cotton and R. A. Walton, *Multiple Bonds between Metal Atoms*, 2nd ed., Clarendon Press, Oxford, 1993.
8. J. N. van Niekerk, *Acta Cryst.*, **6**, 277–232 (1953).
9. J. N. van Niekerk, F. R. L. Schoeng and J. F. de Wet, *Acta Cryst.*, **6**, 501–504 (1953).
10. F. A. Cotton and Y. Kim, *J. Am. Chem. Soc.*, **115**, 8511–8512 (1993).
11. M. Eddaoudi, D. B. Moler, H. Li, B. Chen, T. M. Reineke, M. O'Keeffe and O. M. Yaghi, *Acc. Chem. Res.*, **34**, 319–330 (2001).
12. M. A. S. Aquino, *Coord. Chem. Rev.*, **170**, 141–202 (1998).
13. T. A. Stephenson and G. Wilkinson, *J. Inorg. Nucl. Chem.*, **28**, 2285–2291 (1966).
14. D. Rose and G. Wilkinson, *J. Chem. Soc. A*, 1791–1795 (1970).
15. A. J. Lindsey, G. Wilkinson, M. Motevalli and M. B. Hursthouse, *J. Chem. Soc., Dalton Trans.*, 2321–2326 (1985).
16. S. Furukawa and S. Kitagawa, *Inorg. Chem.*, **43**, 6464–6472 (2004).
17. S. R. Batten, B. F. Hoskins, B. Moubaraki, K. S. Murray and R. Robson, *Chem. Commun.*, 1095–1096 (2000).
18. B. Zimmer, V. Bulach, M. W. Hosseini, A. De Cian and N. Kyritsakas, *Eur. J. Inorg. Chem.*, 3079–3082 (2002).
19. J. Lu, W. T. A. Harrison and A. J. Jacobson, *Chem. Commun.*, 399–400 (1996).
20. K.-T. Youm, S. Hub, Y. J. Park, S. Park, M.-G. Choi and M.-J. Jun, *Chem. Commun.*, 2384–2385 (2004).
21. F. A. Cotton, E. V. Dikarev and X. Feng, *Inorg. Chem. Acta.*, **237**, 19–26 (1995).
22. F. A. Cotton, E. V. Dikarev and S.-E. Stiriba, *Inorg. Chem.*, **38**, 4877–4881 (1999).
23. F. A. Cotton, E. V. Dikarev and M. A. Petrukhina, *Angew. Chem., Int. Ed.*, **39**, 2362–2364 (2000).
24. F. A. Cotton, E. V. Dikarev and M. A. Petrukhina, *Angew. Chem., Int. Ed.*, **40**, 1521–1523 (2001).

25. F. A. Cotton, E. V. Dikarev, M. A. Petrukhina and R. E. Taylor, *J. Am. Chem. Soc.*, **123**, 5831–5832 (2001).
26. F. A. Cotton, E. V. Dikarev, M. A. Petrukhina and S.-E. Stiriba, *Polyhedron*, **19**, 1829–1835 (2000).
27. F. A. Cotton, E. V. Dikarev and S.-E. Stiriba, *Organometallics*, **18**, 2724–2726 (1999).
28. F. A. Cotton, E. V. Dikarev, M. A. Petrukhina and S.-E. Stiriba, *Organometallics*, **19**, 1402–1405 (2000).
29. F. A. Cotton, E. V. Dikarev and M. A. Petrukhina, *J. Am. Chem. Soc.*, **123**, 11655–11663 (2001).
30. M. A. Petrukhina, K. W. Andreini, J. Mack and L. T. Scott, *Angew. Chem., Int. Ed.*, **42**, 3375–3379 (2003).
31. M. A. Petrukhina, K. W. Andreini, L. Peng and L. T. Scott, *Angew. Chem., Int. Ed.*, **43**, 5477–5481 (2004).
32. Y. Takazaki, Z. Yang, M. Ebihara, K. Inoue and T. Kawamura, *Chem. Lett.*, **32**, 120–121 (2003).
33. Y. Fuma, M. Ebihara, S. Kutsumizu and T. Kawamura, *J. Am. Chem. Soc.*, **126**, 12238–12239 (2004).
34. M. C. Barral, S. Herrero, R. Jiménez-Aparicio, M. R. Torres and F. A. Urbanos, *Angew. Chem., Int. Ed.*, **44**, 305–307 (2005).
35. R. Jiménez-Aparicio, F. A. Urbanos and J. M. Arrieta, *Inorg. Chem.*, **40**, 613–619 (2001).
36. F. D. Cukiernik, D. Luneau, J.-C. Marchon and P. Maldivi, *Inorg. Chem.*, **37**, 3698–3704 (1998).
37. D. Yoshioka, M. Mikuriya and M. Handa, *Chem. Lett.*, **31**, 1044–1045 (2002).
38. T. E. Vos, Y. Liao, W. W. Shum, J.-H. Her, P. W. Stephens, W. M. Reiff and J. S. Miller, *J. Am. Chem. Soc.*, **126**, 11630–11639 (2004).
39. H. Miyasaka, R. Clérac, C. S. Campos-Fernández and K. R. Dunbar, *J. Chem. Soc., Dalton Trans.*, 858–861 (2001).
40. C. J. O'Connor, *Prog. Inorg. Chem.*, **29**, 203–283 (1982).
41. H. Miyasaka, C. S. Campos-Fernández, R. Clérac and K. R. Dunbar, *Angew. Chem., Int. Ed.*, **39**, 3831–3835 (2000).
42. S. Furukawa, M. Ohba and S. Kitagawa, *Chem. Commun.*, 865–867 (2005).
43. O. M. Yaghi, M. O'Keeffe, N. W. Ockwig, H. K. Chae, M. Eddaoudi and J. Kim, *Nature*, **423**, 705–714 (2003).
44. R. Nukada, W. Mori, S. Takamizawa, M. Mikuriya, M. Handa and H. Naono, *Chem. Lett.*, **28**, 367–368 (1999).
45. S. Takamizawa, T. Hiroki, E.-I. Nakata, K. Mochizuki and W. Mori, *Chem. Lett.*, **31**, 1208–1209 (2002).
46. S. Takamizawa, E.-I. Nakata, H. Yokoyama, K. Mochizuki and W. Mori, *Angew. Chem., Int. Ed.*, **42**, 4331–4334 (2003).
47. S. Takamizawa, E.-I. Nakata and T. Saito, *Angew. Chem., Int. Ed.*, **43**, 1368–1371 (2004).
48. H. Abourahma, G. J. Bodwell, J. Lu, B. Moulton, I. R. Pottice, R. B. Walsh and M. J. Zaworotko, *Cryst. Growth Des.*, **3**, 513–519 (2003).
49. R. Kitaura, F. Iwahori, R. Matsuda, S. Kitagawa, Y. Kubota, M. Takata and T. C. Kobayashi, *Inorg. Chem.*, **43**, 6522–6524 (2004).
50. R. Matsuda and S. Kitagawa, *unpublished work*.
51. K. Seki, *Phys. Chem. Chem. Phys.*, **4**, 1968–1971 (2002).
52. R. Kitaura, K. Seki, G. Akiyama and S. Kitagawa, *Angew. Chem., Int. Ed.*, **42**, 428–431 (2003).
53. D. Li and K. Kaneko, *Chem. Phys. Lett.*, **335**, 50–56 (2001).
54. D. N. Dybtsev, H. Chun and K. Kim, *Angew. Chem., Int. Ed. Engl.*, **43**, 5033–5036 (2004).

10

Supramolecular Interactions in Directing and Sustaining Coordination Molecular Architectures

XIAO-MING CHEN and MING-LIANG TONG
School of Chemistry and Chemical Engineering, Sun Yat-Sen University, Guangzhou 510275, P. R. China.

1. INTRODUCTION

Since the pioneering reports of Hoskins, Robson and coworkers in 1990 [1], crystal engineering of metal complexes, especially coordination polymers, has attracted great interest. Many supramolecular architectures based on metal–ligand interactions have, in fact, been designed for purely aesthetic reasons. The intense interest in this field, given impetus by synthetic and theoretical chemists, crystallographers and materials scientists, has resulted in not only beautiful and diversified structures, but also potential applications such as electronic, magnetic, optical, absorbent and catalytic materials [2].

Compared to inorganic species, coordination molecular architectures built upon molecular building blocks hold great promise for processing, flexibility, structural diversity and geometrical control, such as the size, shape and symmetry. Coordination molecular architectures can be assembled with metal ions in different geometries (e.g. tetrahedral, square-planar and octahedral) and various multifunctional ligands, such as linear or angular exo-bidentate, planar or tripodal tridentate, into multidimensional frameworks. It is obvious that the coordination behavior of metal ions as well as structures and coordination behavior of the ligands play a fundamental role in controlling the structures of coordination molecular architectures, which can exhibit a wide range of interesting zero-dimensional (0-D), 1-D, 2-D or 3-D structural features.

Frontiers in Crystal Engineering. Edited by Edward R.T. Tiekink and Jagadese J. Vittal

Hydrogen-bonding and aromatic–aromatic $\pi \cdots \pi$ stacking interactions have significant directionality and reliability compared with the weaker electrostatic interactions and van der Waals forces. The strength of moderate hydrogen bonds falls within a range of 15–40 kJ mol^{-1}. The $\pi \cdots \pi$ interactions are usually weaker in both directionality and strength compared to that of hydrogen bonding. Thus, typical $\pi \cdots \pi$ interactions have strength of approximately 10 kJ mol^{-1} [3]. Therefore, the strength of each of hydrogen bonding and $\pi \cdots \pi$ interactions are between the range of covalent bonding and van der Waals forces. In fact, hydrogen bonding is the most utilized of non-covalent interactions and as such is used as a powerful organizing force in the designing of solids [4]. Since they are directional and selective, hydrogen bonds are somewhat similar to metal–ligand coordination bonds in both directing (via molecular recognition and geometrical control) and sustaining the molecular aggregates [5]. The $\pi \cdots \pi$ interaction plays a similar role in the formation of molecular aggregates as aromatic rings can only be arranged into three possible ways: perfectly face-to-face, offset (or slipped) face-to-face and edge-to-face, of which the first orientation is rarely observed in coordination polymers [6].

Besides the above-mentioned supramolecular interactions, other weak intermolecular interactions, such as metallophilic interactions between closed-shell d^{10} metal ions (e.g. copper(I), silver(I) and gold(I)), can also have significant influence on their polymeric/supramolecular structures. Usually, the strongest metallophilic interaction is aurophilicity, which was estimated to be comparable in strength to the usual hydrogen bonding [7]. Theoretical studies on dimeric models of linear two-coordinated complexes indicate that metallophilicity decreases in the order gold(I) > silver(I) > copper(I) [8]. Even though they are weak, such interactions may be exploited in assembling interesting coordination architectures. It is noteworthy that, so far, no review focusing mainly on the role of metallophilic interactions in the construction of coordination molecular architectures has yet been published.

In this review, we summarize a number of interesting examples in which hydrogen-bonding, $\pi \cdots \pi$ and metal–metal interactions play an important role in the construction of coordination supramolecular architectures. As several relevant reviews have been published, we will focus on the more recent and representative cases.

2. MOLECULAR ARCHITECTURES ASSEMBLED BY HYDROGEN-BONDING INTERACTIONS

As stated previously, hydrogen bonds have been widely exploited in assembling coordination molecular architectures. Such architectures can be generated from 0-D, 1-D and 2-D species. Several reviews have been published to describe hydrogen-bonded networks of coordination complexes (see also Chapters 11 and 12 in this book) [9]. Therefore, we will only select a few recent examples involving aquometal centers, carboxylic acid groups, CH$\cdots$X bonding sites and imidazoles, as well as those of biological relevance.

2.1. Metal Complexes with Aquometal Centers

A typical class of hydrogen-bonded supramolecular architecture has been well established involving hydrogen bonds between aquometal centers and 4,4′-bipyridine (4,4′-bpy), as either ligands or solvated molecules. Such systems, especially those of 4,4′-bpy-bridged linear metal complexes extending though hydrogen bonds formed between the aqua ligands and solvated 4,4′-bpy molecules into networks [10], as well as those of 4,4′-bpy analogs and 4,4′-bpy-N,N'-dioxides [11], have been reviewed recently [12]. Of course, discrete metal-4,4′-bpy complexes can also form extensive hydrogen-bonded architectures, as exemplified by $[Mn(4,4'\text{-bpy})_2(H_2O)_4](ClO_4)_2\cdot(4,4'\text{-bpy})_4$ (**1**), which features a hydrogen-bonded 3-D molecular network with triangular channels hosting the ClO_4^- anions [13].

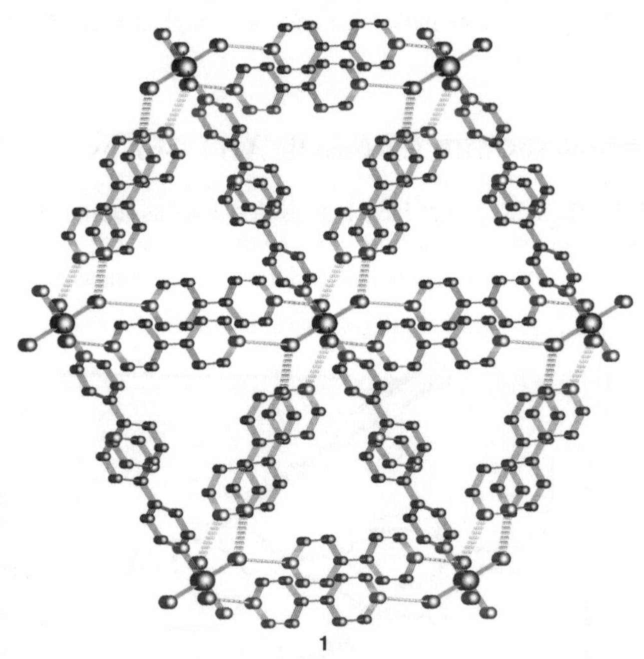

1

Top view showing the triangular channels in **1**.
Hydrogen bonds are represented by broken lines.

Aside from the formation of 0-D and 1-D complexes, it is well known that 4,4′-bpy can be readily used to generate 2-D square grids, which may be stacked in an offset fashion in the solid state in the absence of further structure-controlling factors, resulting in the formation of smaller channels [14]. To obtain larger channels and/or to observe host–guest supramolecular interactions, guest molecules may be utilized. Actually, the 2-D square-grid networks $[M(4,4'\text{-bpy})_2(H_2O)_2](ClO_4)_2\cdot(2,4'\text{-bpy})_2\cdot H_2O$, M = Cd(II) and Zn(II) (**2**) [15], and related 2-D grids [14] exhibit interesting packing fashions, hosting inorganic anions, neutral or cationic organic molecules in the channels. Upon anchoring of the 2,4′-bpy guests by hydrogen bonding ($O\cdots N = 2.77$ Å) between the aqua ligands and the 4-pyridyl ends in **2**, the superposition of such square grids creates large square channels, approximately 11.5×11.5 Å, as shown in Figure 1.

An interesting 2-D hydrogen-bonded honeycomb network was revealed in $(4,4'\text{-}H_2bpy)_2[Ni(NCS)_4(H_2O)_2](NO_3)_2$ (**3**) ($4,4'\text{-}H_2bpy$ = 4,4′-bipyridinium) [16], in which the aqua ligands in the octahedral $[Ni(NCS)_4(H_2O)_2]^{2-}$, nitrate and $[4,4'\text{-}H_2bpy]^{2+}$ ions are involved in the formation of hydrogen-bonded $R^{10}{}_{12}(64)$ rings and approximate 24×20 Å subunits (Figure 2(a)). Another unusual feature of **3** is that such nets are triply interwoven to generate "chicken-coop" sheets; similar interpenetration phenomena have been rarely documented in the hydrogen-bonded networks [5, 17, 18].

In a porous 3-D hydrogen-bonded $[Co(OH_2)_6][H_2(TC\text{-}TTF)]\cdot 2H_2O$ (**4**) ($H_2(TC\text{-}TTF)^{2-}$ = doubly deprotonated tetra(carboxyl)tetrathiafulvalene dianion), extensive hydrogen bonds exist between eight of the twelve protons in the hexaaquometal cation and six carboxy oxygen atoms [19]. Slipped stacks of the TTF anion and hydrogen-bonded chains of $[Co(H_2O)_6]^{2+}$ define 1-D channels that are filled with disordered water molecules (Figure 2(b)). It is notable that such microporous framework remained stable upon removal of the guest water molecules and has a high degree of selectivity toward guest sorption.

2.2. Metal Complexes with Carboxylic Acid Groups

Hydrogen-bonded carboxylic groups have been widely used as supramolecular synthons in crystal engineering of organic architectures [5] and in the same way have also been widely utilized in the construction of coordination architectures. A simple example to illustrate

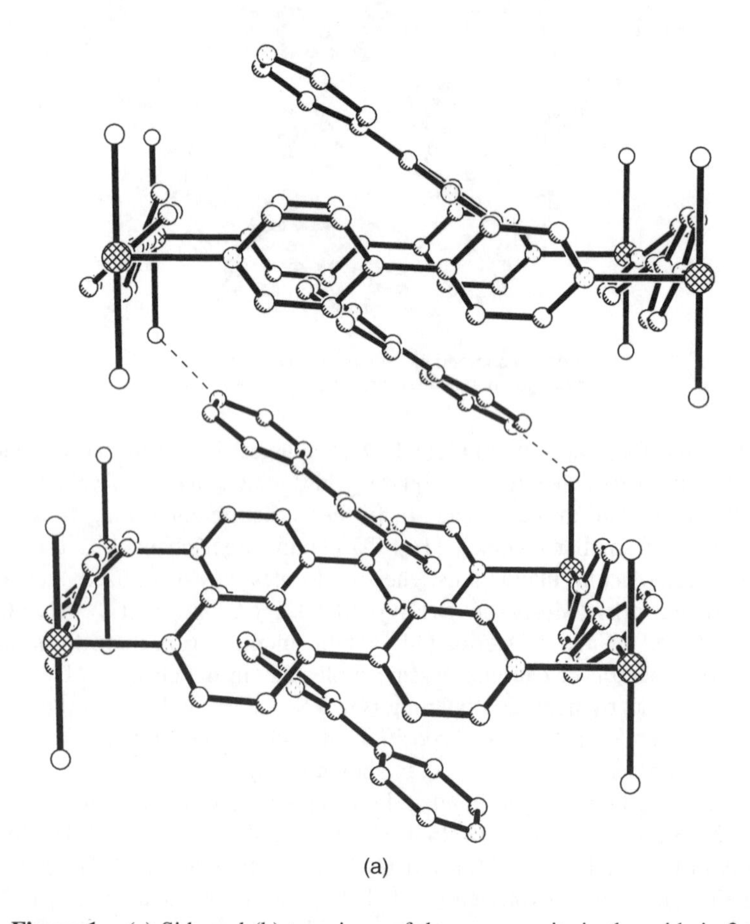

(a)

Figure 1. (a) Side and (b) top views of the square units in the grids in **2**.

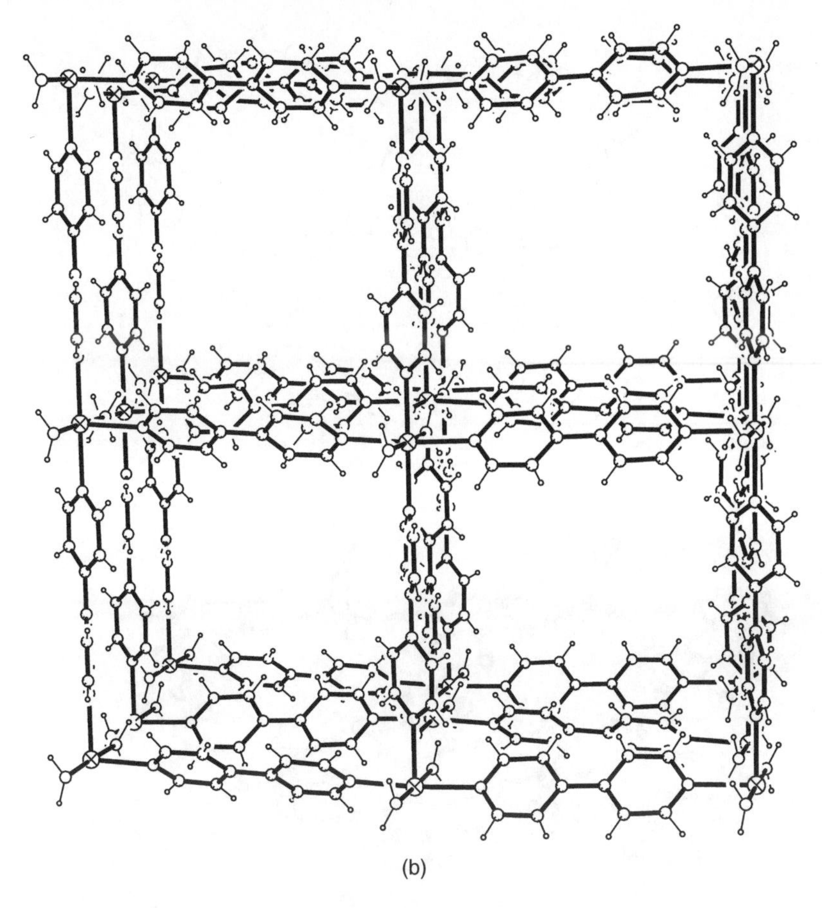

(b)

Figure 1. *(continued)*

this strategy is realized in $[Cd(SCN)_2(Hna)_2][na]$ (Hna = nicotinic acid) (**5**) [20], in which the two trans positions around the six-coordinated cadmium(II) are occupied by two N-coordinated na ligands, forming double hydrogen bonds (O$\cdots$O 2.508(3) Å) between the uncoordinated carboxyl groups from adjacent chains to generate layers (Figure 3(a)). Meanwhile, another na ligand forms a head-to-head hydrogen bond with the guest na molecule. Structural extension of the host framework has also been successfully achieved by the use of a longer analog of na, or by the use of fumaric acid, to insert into the double hydrogen-bonded carboxylic dimer, leading to the formation of new 2-D networks that host large aromatic guests [21].

In a square-planar platinum(II) complex, $[Pt(ina)_2(Hina)_2]$ (**6**) (Hina = isonicotinic acid), singly hydrogen-bonded carboxylic dimers creates a square grid (Figure 3(b)) [18]. The nets interpenetrate in three mutually perpendicular planes and the remaining cavities (*ca.* 5 Å in diameter) are occupied by water molecules. One advantage of using neutral complexes is that, since no counterions are present, there is more space available for guest molecules.

The viability of hydrogen–bond-directed assembly is also demonstrated by modified porphyrin complexes, which are of particular interest for their photoelectronic properties.

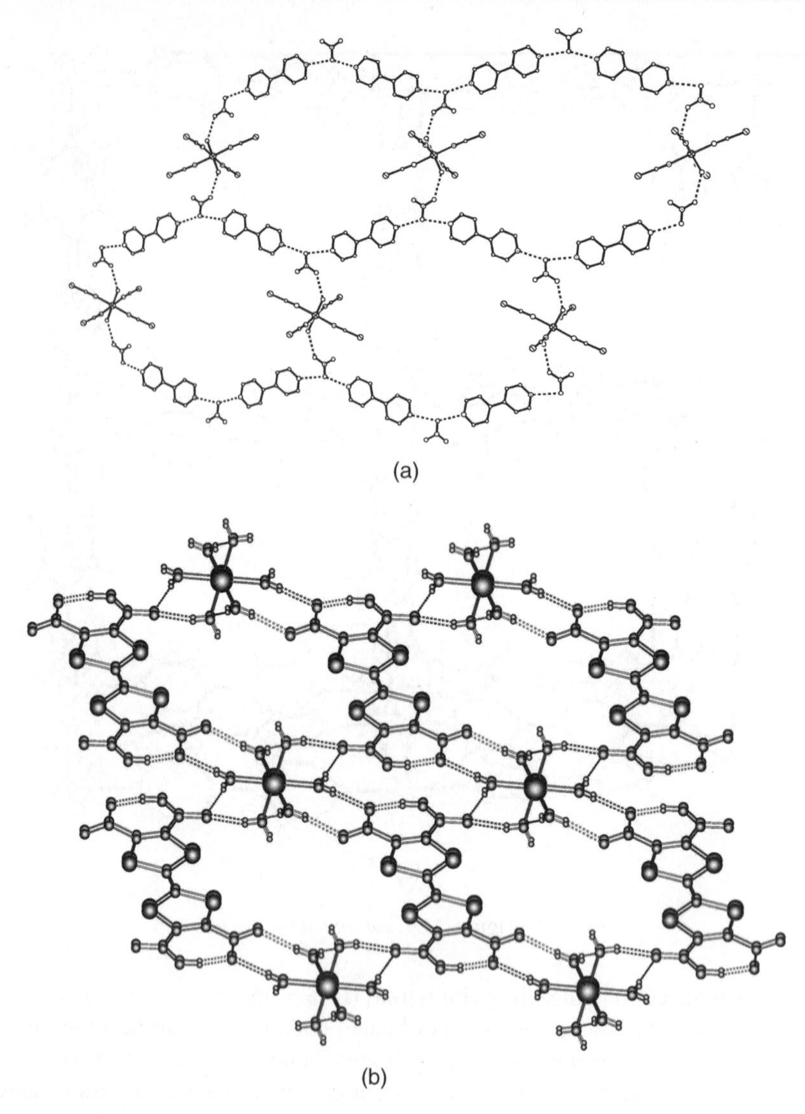

Figure 2. (a) The hydrogen-bonded triple interpenetration 2-D honeycomb network in **3** and (b) hydrogen-bonded 3-D network in **4**.

Carboxylic acid–substituted porphyrin complexes can be assembled to generate flat, gridlike 2-D networks through acid–acid hydrogen-bond interactions, as exemplified by aquazinc tetra(4-carboxyphenyl)porphyrin (aqua-ZnTCPP) (**7**) [22] illustrated in Figure 4(a). Related compounds have also been investigated [23].

The isostructural complexes $[M(Hdcbpy)_3]\cdot nH_2O$ (**8**) [24] (M = Co(III) or Rh(III), and H_2dcbpy = 5,5′-dicarboxy-2,2′-bipyridyl) exhibit two interpenetrating, homochiral rhombohedral networks (Figure 4(b)), linked by very strong carboxylic acid–carboxylate hydrogen bonds (*ca.* O· · · O 2.46 Å), in which each complex acts as a node for six strong hydrogen bonds, having pseudo-3-fold symmetry.

2.3. Metal Complexes with Imidazole Groups

Imidazole, as a simple molecule and having a histidine residue, participates in ligating metal centers in metalloenzymes and exhibits a strong tendency to form hydrogen bonds with an acceptor, for example, in the carboxylate-histidine-zinc triad systems observed in the catalytic processes of many zinc enzymes [25]. Such triads can be used in constructing 3-D architectures based on infinite [Zn(Him)$_2$(dicarboxylate)] chains and hydrogen bonding [26].

More interesting examples of coordination molecular architectures involve polyimidazole ligands. Chiral metal complexes of achiral tripodal ligands were used to assemble homochiral layers [27]. The chiral complexes have the general formula [M(H$_3$L)]X$_n$·*x*(sol) (M = Co(III) or Fe(II); X = ClO_4^- or NO_3^-; $n = 2$ or 3; sol = solvent; H$_3$L = tris[2-(((imidazol-4-yl)methylidene)amino)ethyl]amine or analogous ligands). Each [M(H$_3$L)]$^{n+}$ cation (**9**) induces the chirality of clockwise (C) and anticlockwise (A) enantiomers due to the screw coordination arrangement of the achiral tripodal ligand around the M^{III}

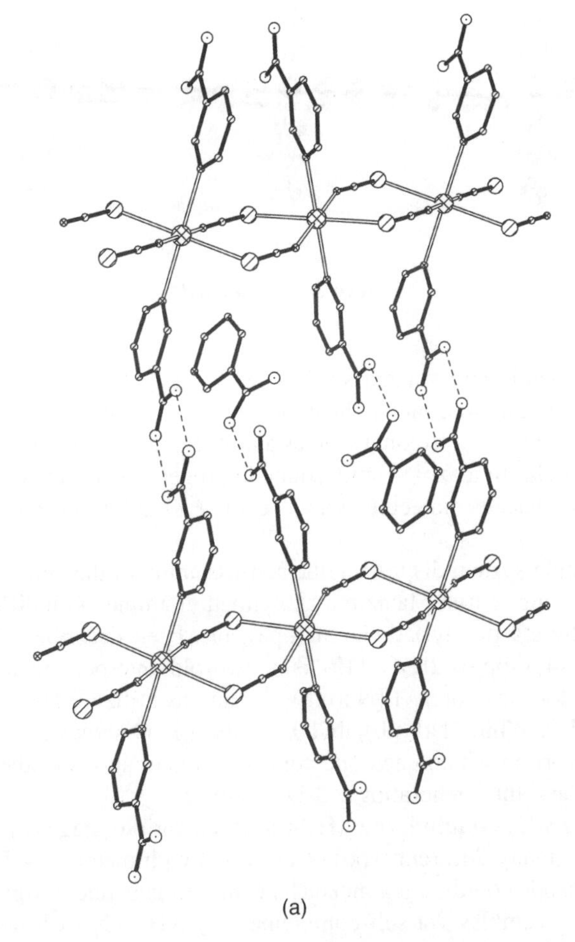

(a)

Figure 3. (a) A structural unit in **5** and (b) single hydrogen-bonded square grid in **6**.

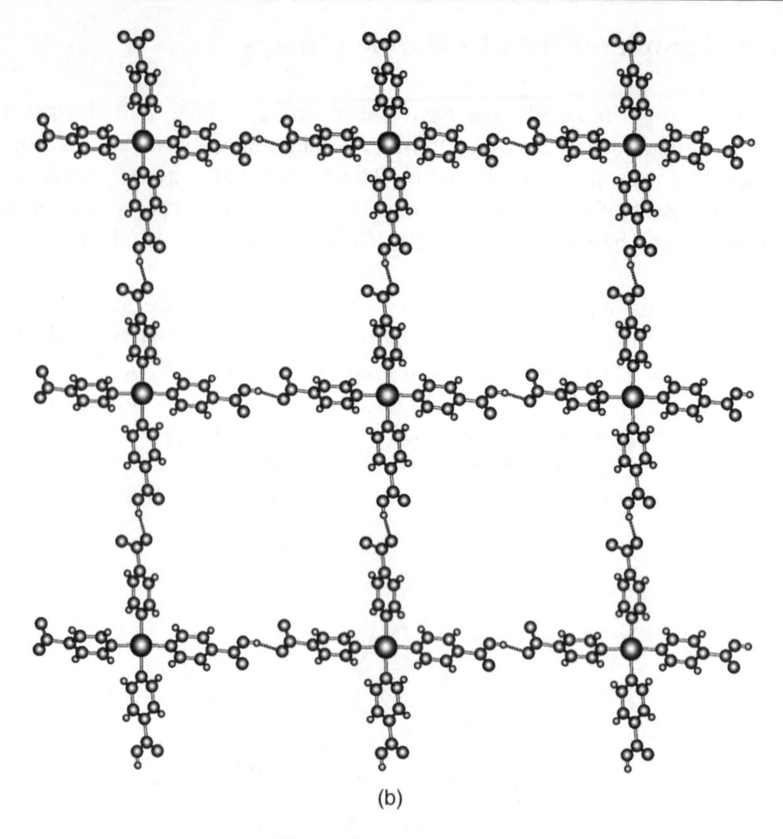

(b)

Figure 3. (*continued*)

ion. The formally hemi-deprotonated species $[M(H_{1.5}L)]^{1.5+}$, which functions as a self-complementary chiral building block, generates an extended homochiral layer comprising a hexanuclear structure with a trigonal void as a unit (Figure 5(a)). The 2-D structure arises from the intermolecular imidazole–imidazolate hydrogen bonds between $[M(H_3L)]^{3+}$ and $[M(L)]^0$, in which adjacent molecules with the same chirality are arrayed in an up-and-down fashion.

Another interesting system displaying the self-assembly of the chiral building block Δ- or Λ-$[Nd(ntb)_2]^{3+}$, ntb = tris(2-benzimidazolylmethyl)amine, with different spacers 4,4′-bpy and *trans*-1,2-bis(4-pyridyl)ethylene) (bpe), has been described [28]. For instance, $[Nd(ntb)_2][ClO_4]_3{\cdot}(4,4'\text{-bpy})_3{\cdot}2H_2O$ (**10**) is a twofold interpenetrated 3-D network in which the spacers 4,4′-bpy connect $[Nd(ntb)_2]^{3+}$ cations utilizing NH···N bonding interactions (Figure 5(b)). While $[Nd(ntb)_2][ClO_4]_3{\cdot}(bpe)_3{\cdot}H_2O$ shows a topologically similar but achiral framework in which each bpe connects either cations of the same chirality or a pair of enantiomers, thus generating a 3-D racemate.

Also, interestingly, 2,2′-biimidazole (H_2bim) in its various stages of deprotonation has been used to create many different types of networks with metal ions. For example, three bidentate Hbim ligands coordinate a nickel(II) center to generate a trigonal $[Ni(Hbim)_3]^-$ anion (**11**), which assembles via self-complementary NH···N hydrogen bonds, furnishing a 2-D hexagonal or "honeycomb" motif with encapsulated K^+ ions (Figure 6) [29].

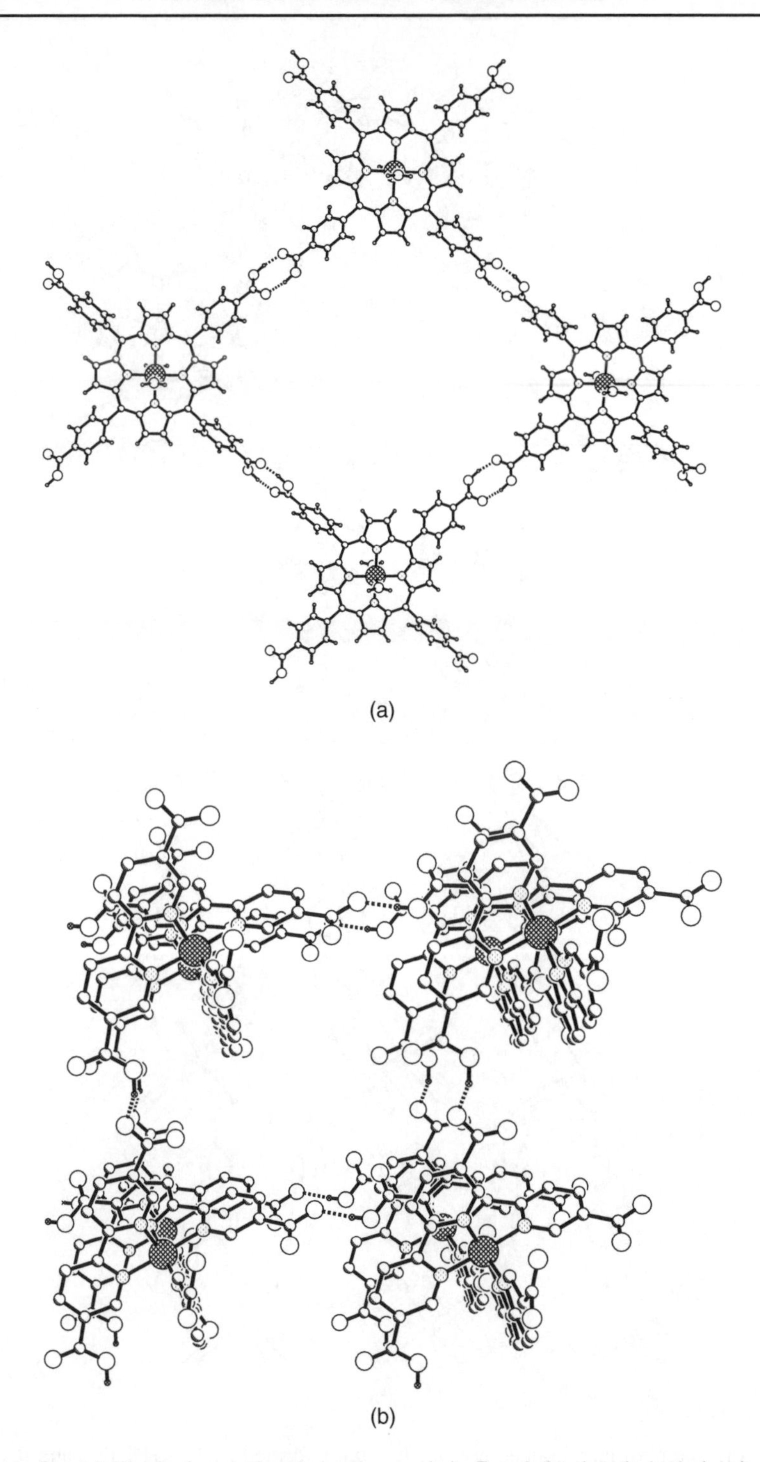

Figure 4. (a) The hydrogen-bonded square unit in **7** and (b) rhombohedral unit in **8**.

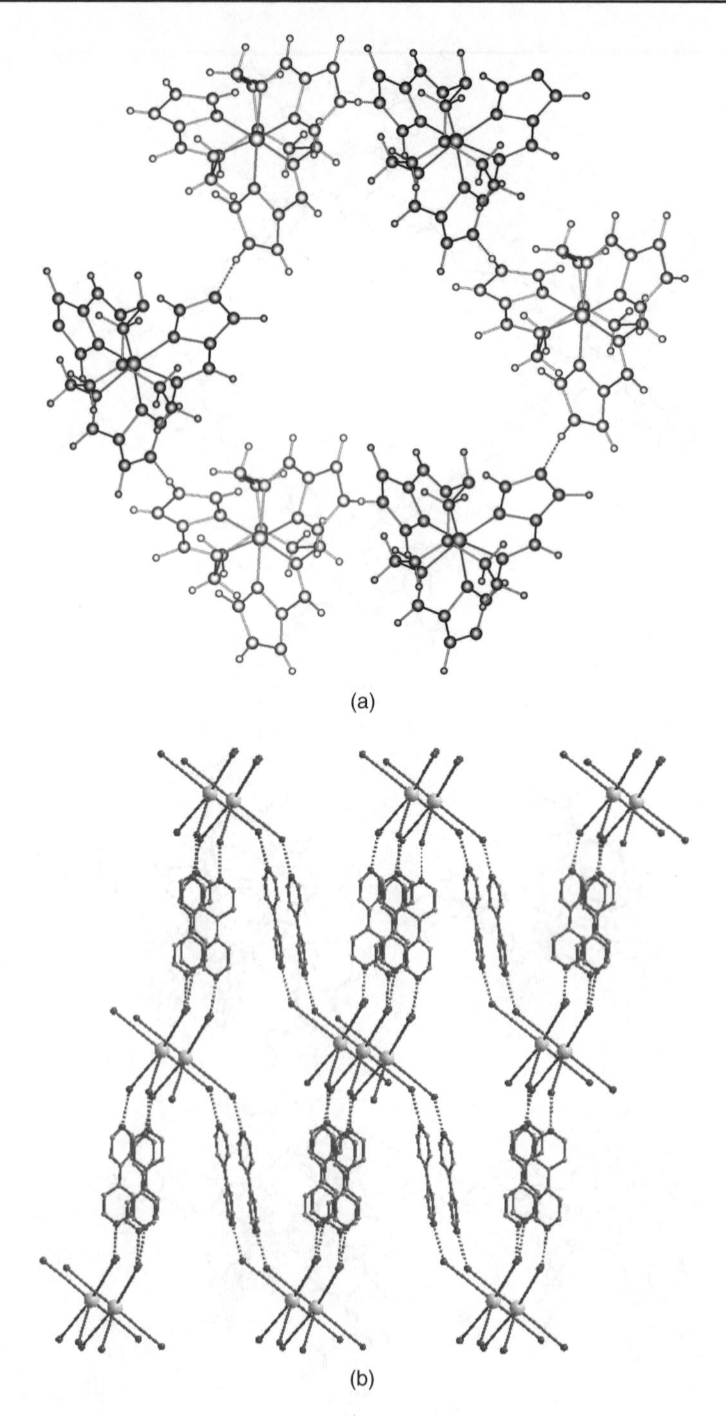

Figure 5. (a) The hydrogen-bonded hexanuclear units formed by $[Co(H_3L)]^{3+}$ and $[Co(L)]^0$ in **9** and (b) a single achiral 3-D network in **10**. For clarity, each 2-benzimidazolylmethyl arm of the ntb ligand in (b) is represented by a long rod joining each NH group to the Ln^{3+} cation.

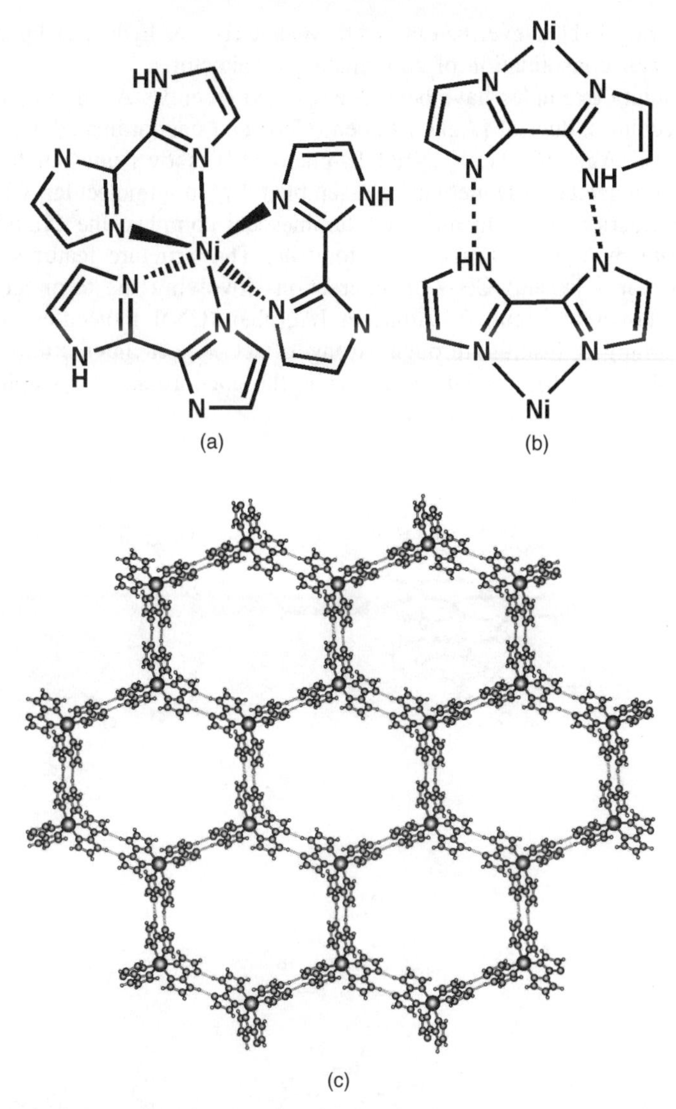

Figure 6. Chemical structures of (a) coordination geometry and (b) hydrogen bonding in **11** and (c) the 2-D anionic "honeycomb" network.

Using larger cations, such as tetrabutylammonium, hydrogen-bonded zigzag chains are produced instead. Further incorporation of hydrogen-bond acceptor groups, such as sulfate and carbonate as spacers, yielded different molecular architectures, which exhibit porous structures upon stacking of 2-D networks [30].

2.4. Metal Complexes with CH$\cdots$X Bonding Sites

Although nonclassic CH$\cdots$X (X = O, N, π, etc.) hydrogen bonding is relatively weak compared with conventional hydrogen-bonding interactions, these have now been well acknowledged as being important in crystal engineering, especially in the realm of organic crystal engineering [31]. Nevertheless, such weak CH···X hydrogen bonding are also recognized in their consolidation of coordination architectures.

A few intriguing examples have been documented recently. A solvothermally synthesized 1:1 molecular adduct of [Ag(1,10-phen)(CN)] and uncoordinated 1,10-phenanthroline (1,10-phen), [Ag(1,10-phen)(CN)](1,10-phen) (**12**), shows unusual threefold interpenetrating, 4-connected 3-D networks assembled by supramolecular CH···N bonds and $\pi \cdots \pi$ interactions [32]. In this photoluminescent complex, the cyanide group was generated *in situ* by C–C cleavage of acetonitrile. The structure features very unusual CH···N bonds, $\pi \cdots \pi$ and CH···π interactions involving the terminal cyanide and 1,10-phen as shown in Figure 7. Adjacent [Ag(phen)(CN)] molecules are stacked in an inversely arranged fashion through strong offset $\pi \cdots \pi$ interactions (face-to-face distance *ca.* 3.40 Å) to form a 1-D array, while the uncoordinated 1,10-phen molecules

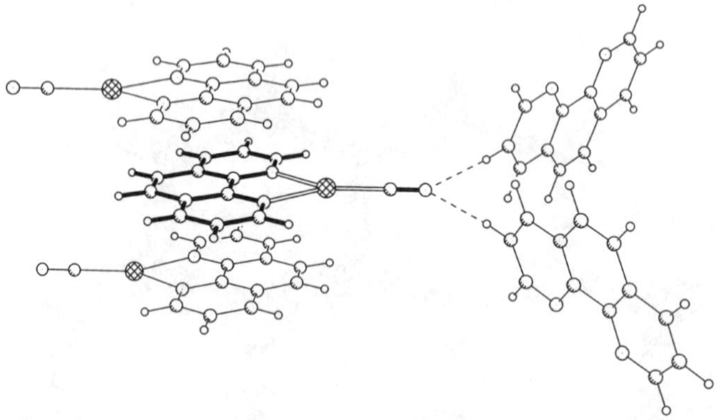

(a)

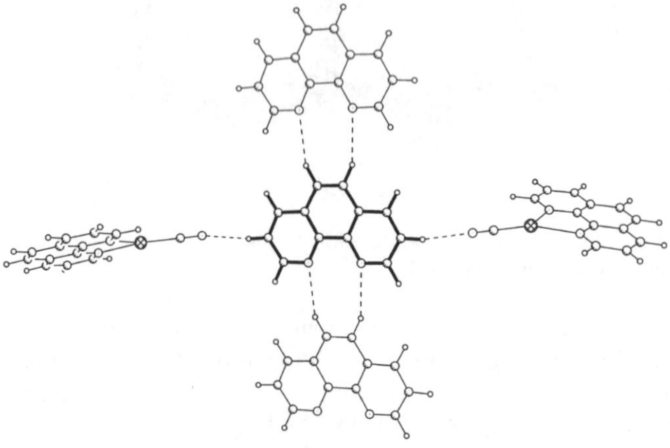

(b)

Figure 7. (a) The tetrahedral node defined by two [Ag(1,10-phen)(CN)] species through $\pi \cdots \pi$ interactions and two uncoordinated 1,10-phen groups through the CH···N bonds; (b) square-planar node defined by an uncoordinated 1,10-phen group connected with two [Ag(phen)(CN)] and two uncoordinated 1,10-phen through CH···N bonds, (c) 1,10-phen-CH···π interactions and (d) the crystal packing pattern of **12** viewed along the direction of the *a*-axis.

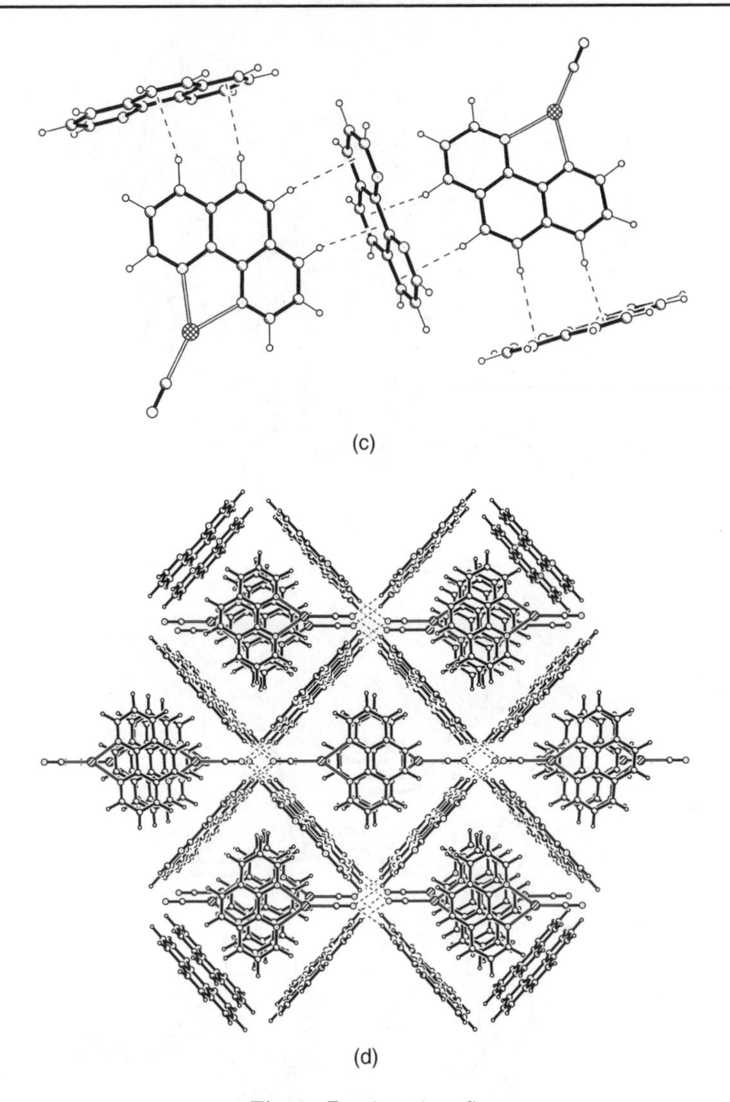

Figure 7. *(continued)*

are interconnected to give extended ribbons with relatively long CH···N(phen) contacts (C···N 3.61 Å). The arrays and the ribbons are extended to a 3-D structure by strong CH···N(CN) interactions (C···N 3.41 Å). Such coordination molecular architectures consolidated by supramolecular $\pi \cdots \pi$, CH···N and CH···π interactions are rarely observed, although a few purely organic species exhibit these interesting interactions [33].

In [Zn(2,2′-bpy)(1,4-bdc)](2,2′-bpy) (bdc = benzenedicarboxylate) (**13**), each 1,4-bdc group acts in a bis-bidentate mode to bridge two zinc(II) ions, resulting in [Zn(2,2′-bpy)(1,4-bdc)] zigzag chains with the 2,2′-bpy ligands on either side [34]. The chains are self-assembled into undulating layers via CH···O interactions (3.49 Å) formed between the phenyl-CH groups and the coordinated carboxy oxygen atoms (Figure 8(a)). Furthermore, all lateral 2,2′-bpy ligands from adjacent layers are intercalated in a zipperlike

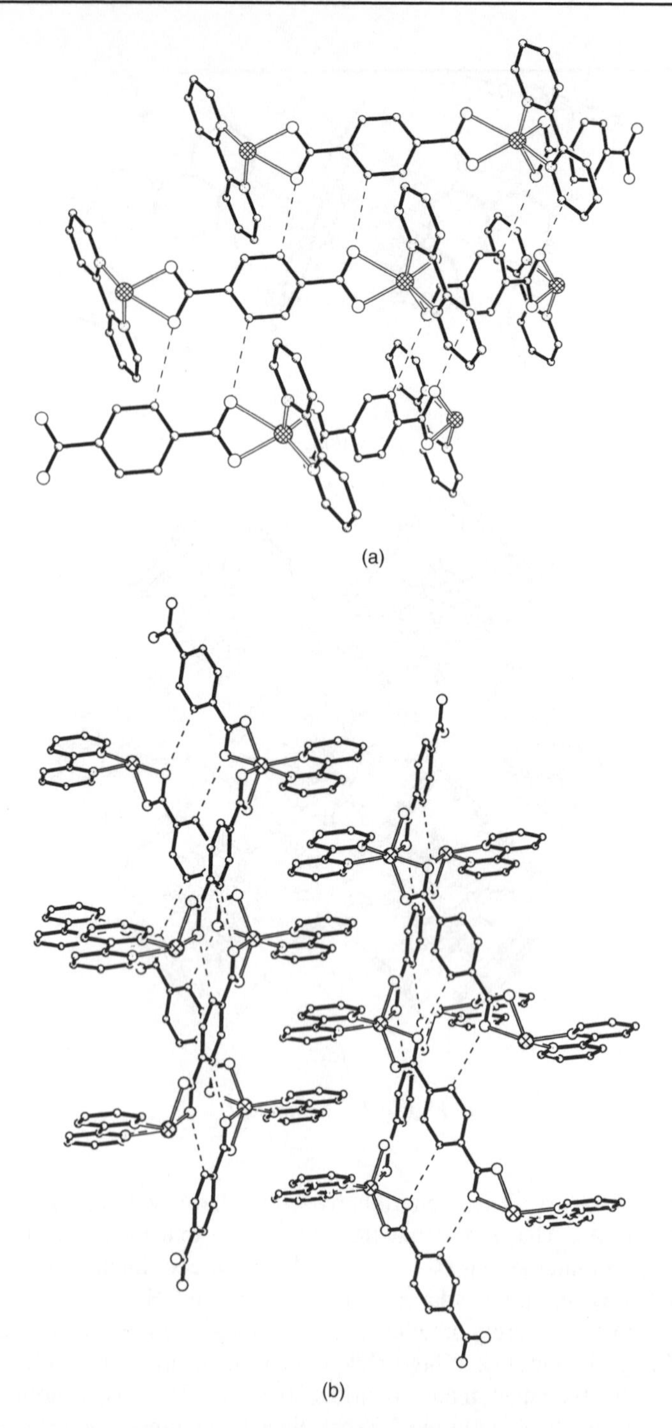

Figure 8. 3-D structure constructed by the zigzag chains with CH···O hydrogen bonds and $\pi \cdots \pi$ interactions in **13**: (a) a top view of a hydrogen-bonded layers; (b) a side view of two layers exhibiting the intercalation of the lateral bpy ligands and (c) the 3-D network viewed along the *c*-axis.

(c)

Figure 8. *(continued)*

fashion via $\pi \cdots \pi$ interactions between the 2,2′-bpy groups (Figure 8(b)), in combination with CH···O interactions between the 2,2′-bpy and carboxy groups, to furnish 3-D supramolecular network with 1-D nanosized saddlelike channels (Figure 8(c)). The guest bpy molecules reside in the channels and exhibit parallel, offset intermolecular $\pi \cdots \pi$ interactions (face-to-face distance 3.5 Å) between them.

Usually, control of reaction conditions is important in constructing coordination polymers, especially for hydrothermal reactions. For example, a series of zigzag chains with the general formula $[Ni(1,4\text{-bdc})L](H_2O)_n \cdot m(1,4\text{-}H_2bdc)$ ($n = 0$, 1; $m = 0$, 0.5, 0.75; L = 1,10-phen or 2,2′-bpy) have been synthesized under different hydrothermal reactions [35]. The $[Ni(1,4\text{-bdc})(2,2'\text{-bpy})] \cdot (1,4\text{-}H_2bdc)_{0.75}$ complex (**14**) has a framework that is very similar to that of **13**, but the guest 1,4-H_2bdc molecules in **14** are hydrogen bonded to each other, along with the formation of extensive supramolecular interactions with the chains.

It is quite obvious that supramolecular interactions may impart unusual properties to the generated frameworks owing to the flexibility of such interactions. Even weak CH···X interactions hold the promise of new properties, as exemplified by a neutral, laminated metal-organic framework (MOF) structures, $[Fe(pydc)(4,4'\text{-bpy})] \cdot H_2O$ (**15**) (H_2pydc = 2,5-dicarboxypyridine) [36]. In **15**, octahedral iron(II) ions are bridged by pydc and 4,4′-bpy ligands into a 2-D square grid with large cavities (atom–atom distance *ca.* 11.5 × 8.8 Å) (Figure 9(a)). Interlayer CH···O bonds (C···O 3.00–3.41 Å) are found between the pyridyl-CH groups and uncoordinated carboxy oxygen atoms at the edges of adjacent layers (Figure 9(b)). Such hydrogen bonds contribute to the layer alignment to give the 2-D porosity: the offset stacking of such layers results in smaller channels (atom–atom distance 4.5 × 11.5 Å) perpendicular to the layers, and gives rise to additional channels (atom–atom distance *ca.* 5.4 × 7.8 Å) parallel to the layers. Therefore, MOF **15** still possesses channels having 18.6% volume of the unit cell, which are occupied by the lattice water molecules.

The MOF of **15** is very robust to heat and common organic solvents. It was observed that it shrank into the guest-free framework under appropriate heating in N_2 or vacuum treatment. Anhydrous **15** could readily take up ethanol vapor to furnish an ethanol-solvated crystal structure. Also, **15** could be transformed directly from the parent single crystals into methanol- or ethanol-solvated single crystals when soaked in the corresponding solutions with concomitant cell volume changes.

The above example, along with the more interesting example of flexible frameworks featuring $\pi \cdots \pi$ interactions (see **32** in Section 3.3), may serve as the prototypes of flexible porous metal-organic materials.

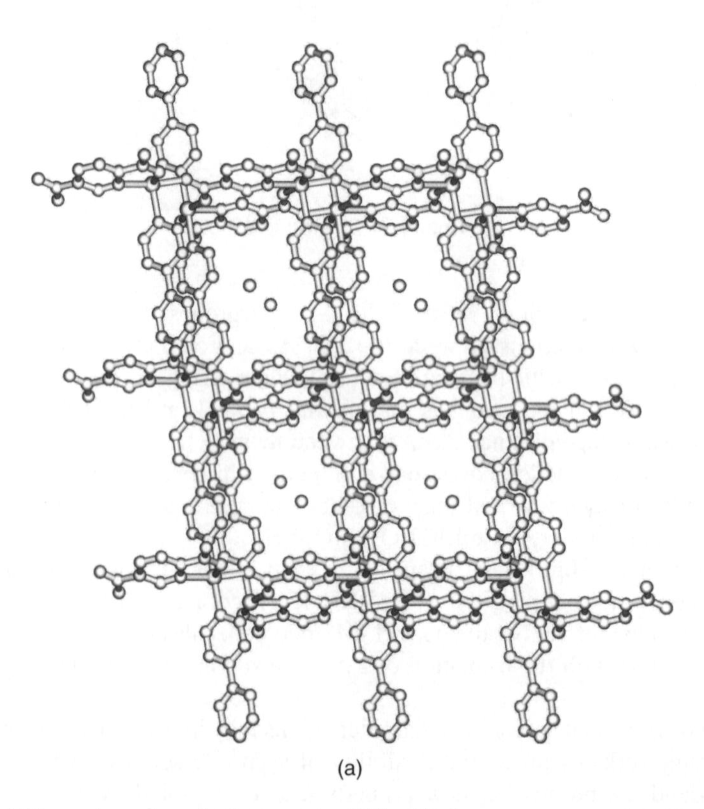

Figure 9. (a) The rectangular-grid framework of **15** with the guest water molecules and (b) offset stacking of the layers.

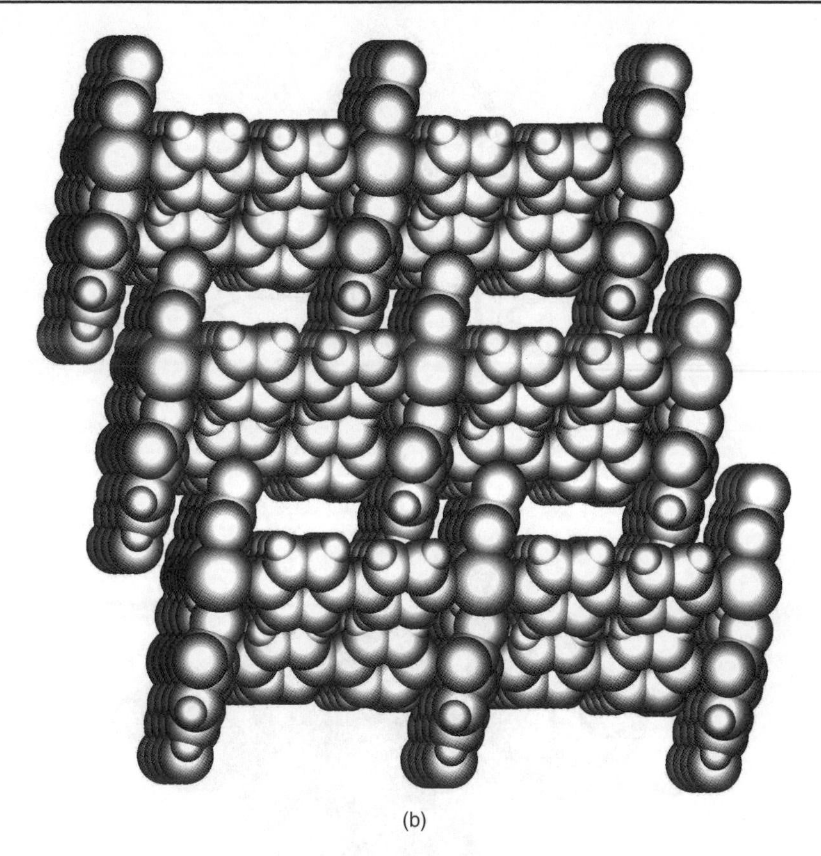

(b)

Figure 9. *(continued)*

2.5. Metal Complexes of Biological Relevance

Obviously, hydrogen-bonding interactions are critically important in biology, as evidenced by the hydrogen bonds in DNA, proteins and other systems. Consequently, investigations on hydrogen bonding operating in model complexes related to biological systems have been the subject of considerable and ongoing interest [37].

Stimulated by their biological existence, helical structures, including metal helicates, have attracted significant interest in supramolecular chemistry for many years. So far, many of the multistrand helices have been formed through metal ligation with twisted organic bridging ligands as strands. Multistrand helices or helicates assembled solely by supramolecular interactions are still rare. Recently, a very simple but efficient strategy for the construction of ligand strands has been documented for the dinuclear double helicates $[Co(L)(HL)Cl]_2$ (**16**) (HL = 6-methylpyridine-2-methanol or analogous ligands) by employing hydrogen bonding as a construction element for the ligand strands [38]. As shown in Figure 10, a pair of ligands are singly hydrogen-bonded via their oxygen atoms to form a bis-bidentate ligand strand. Two such strands wrap two cobalt(II) cations to generate a dimeric double-strand helicate. A chloride anion completes the coordination

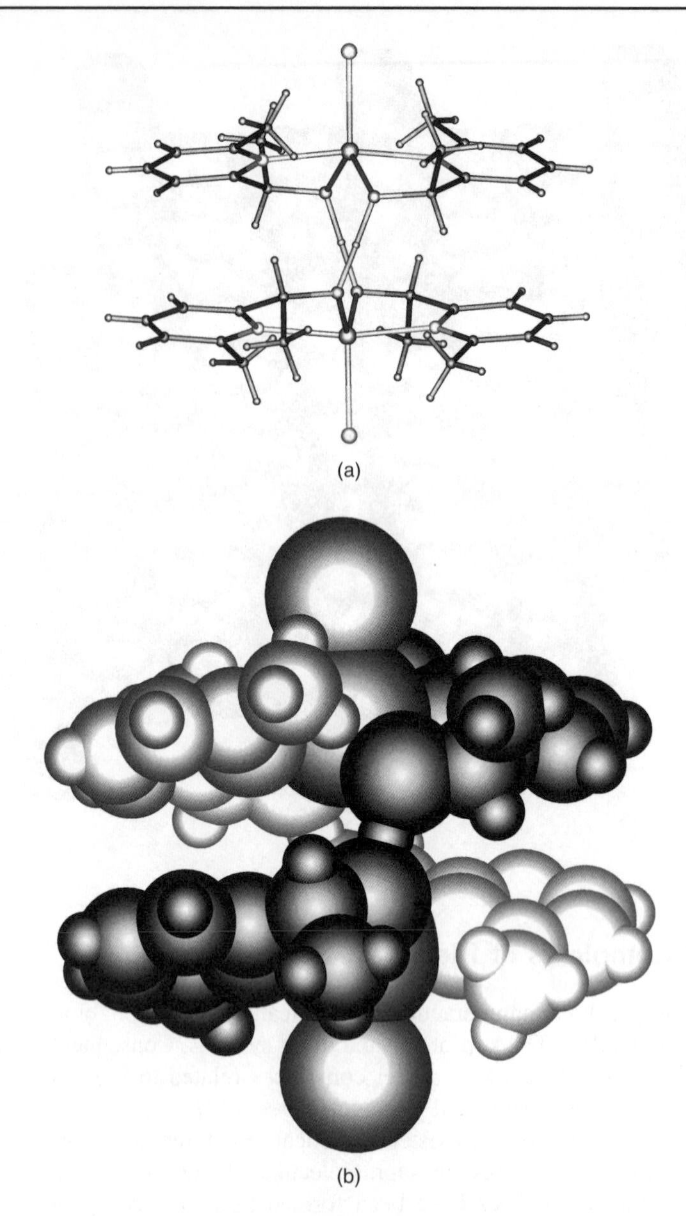

Figure 10. (a) Ball-and-stick and (b) space-filling representations of the *P* enantiomer of helicate **16**.

sphere of each pentacoordinate cobalt(II) center. Other interesting multistrand helices will be discussed later in Sections 3.2 and 4.2.

Also worthy of mention is the recent report of the self-assembly of [Mn(quinoline-2-carboxylate)$_2$(H_2O)$_2$] with adenine, in which the adenine molecules form simultaneously the Watson–Crick and Hoogsteen types of hydrogen bonding [39]. A cationic quaternary platinum(III) complex of 2-amino-9-(4-hydroxy-5-hydroxymethyl-tetrahydro-furan-2-yl)-1,9-dihydropurin-6-one, 1-[2-(acridin-9-ylamino)ethyl]-1,3-dimethylthiourea and ethylenediamine (en) involving a Pt–acridine interaction is a member of a new class of potent cytotoxic agents that associate with DNA through a combination of monofunctional metallation of nucleobase-nitrogen and intercalation of the acridine chromophore of the sulfur-bound nonleaving group into the DNA base stack [40]. The structure features a rare type of mispairing between the base pair formed by the two platinum-modified guanines, which is sandwiched between the acridine ligands.

3. MOLECULAR ARCHITECTURES ASSEMBLED VIA $\pi \cdots \pi$ INTERACTIONS

In the solid state, $\pi \cdots \pi$ interactions are widely observed in the construction of supramolecular organic and metal-organic structures. Indeed, in combination with coordination bonds, $\pi \cdots \pi$ interactions can be employed to construct very interesting coordination architectures, which may not only be of aesthetic interest but also have many potential applications in materials science, since the weakly directional interactions may be used as a tool in the design of flexible and even smart materials.

3.1. 0-D Metal Complexes

Although thus far supramolecular architectures assembled from discrete metal complexes through $\pi \cdots \pi$ interactions are somewhat limited, as an important type of supramolecular force, such interactions in principle can be used to control the process of molecular recognition and self-assembly, leading to fascinating molecular architectures.

For example, reaction of (17*E*)-3-((*E*)-3-((pyridine-2-yl)methyleneamino)benzyl-*N*-(pyridine-2-yl)methylene)benzeneamine (L) with $[Cu(MeCN)_4]BF_4$ yielded an unusual arc-shaped helicate $[Cu_2(L)_2](BF_4)_2$ (**17**) owing to the curvature of the bridging L ligands; consequently, four arc-shaped helicates aggregate to form an uncommon cyclic array held together by $\pi \cdots \pi$ interactions between the pyridyl ends of the L ligands [41]. The chirality of each helicate alternates around the central axis so that the overall structure is achiral.

A large, neutral molecule $[Cu_2(bpa)_2(1{,}10\text{-phen})_2(H_2O)]_2 \cdot 2H_2O$ (**18**) (bpa = biphenyl-4,4′-dicarboxylate) possessing a centrosymmetric tetranuclear rhombic core was constructed with long, bridging bpa ligands [42]. As illustrated in Figure 11(a), each $[Cu(1{,}10\text{-phen})]^{2+}$ unit at a corner is bridged by two bpa ligands, one end being monodentate and the other being chelating. The rhombic cavity has a 15.1 Å edge and the two diagonals are 16.8 and 25.1 Å. The two lateral 1,10-phen ligands at the acute corners penetrate oppositely into the rhombic cavity of the third rhombus to generate a molecular node, featuring a strong $\pi \cdots \pi$ interaction between them (face-to-face distance 3.46 Å). Such interpenetration of adjacent rhombi gives rise to 2-D arrays.

A porous architecture assembled from discrete molecular hexagons by $\pi \cdots \pi$ interactions was found in $[Cu_6(\text{5-nitro-1,3- benzenedicarboxylate})_6(DMSO)_6(MeOH)_3(H_2O)_9](MeOH)_6$ (**19**), featuring effective inner and outer diameters of *ca.* 8 Å and 31.4 Å, respectively [43], which are arranged into flattened layers, and are further packed in an ABCABC fashion to generate 1-D hourglass-shaped channels (effective diameter 8 Å), as shown in Figure 11(b). The interlayer separations are 3.34 Å, indicative of strong face-to-face $\pi \cdots \pi$ interactions.

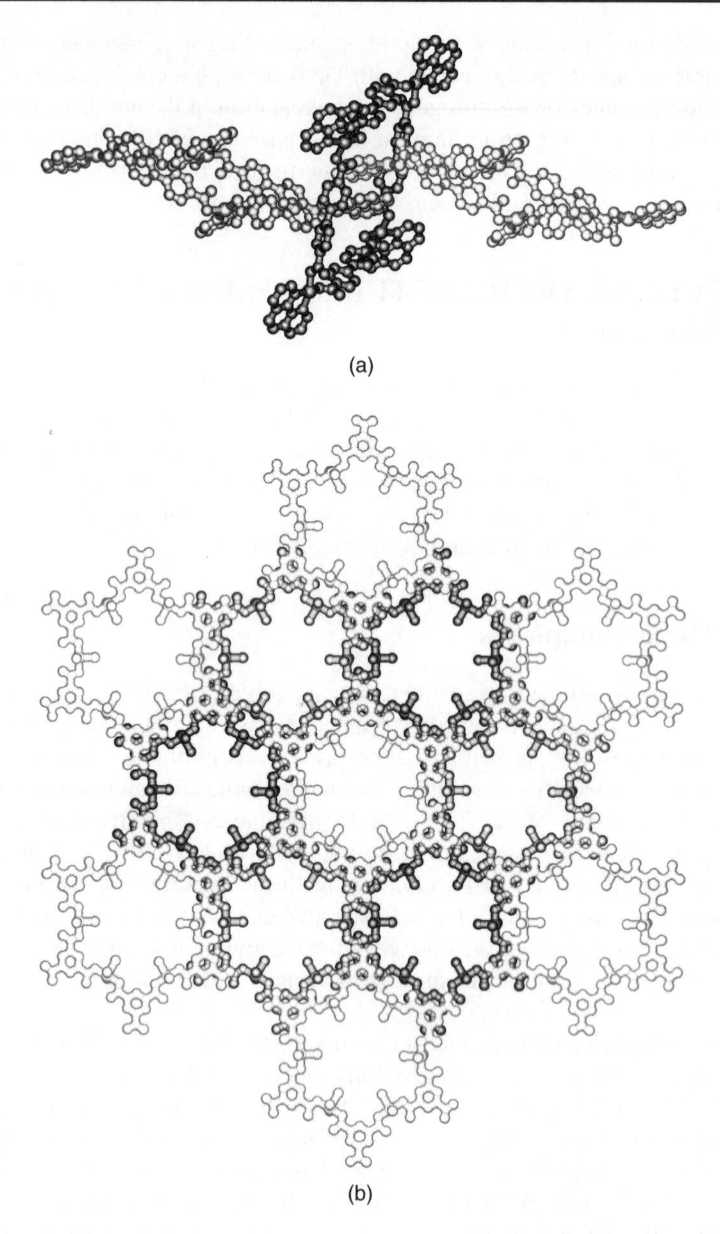

Figure 11. (a) The penetration in **18** and (b) packing of the large hexagonal molecules in **19**.

Another intriguing $\pi \cdots \pi$ stacked architecture assembled from discrete metal complexes is found in the crystal structure of $[Ln_2(dnba)_4(DMF)_8][Mo_6O_{19}]$ (dnba = 3,5-dinitrobenzoate) [44]. This is constructed by offset $\pi \cdots \pi$ interactions between the aromatic groups of tetracarboxylate-bridged $[Ln_2(dnba)_4(DMF)_8]^{2+}$ dimers (3.4–3.5 Å) to encapsulate the large polyoxometalate anion. It is also noteworthy to mention that, in agreement with other literature reports [45], the structure of this complex implies that nitro-substituted phenyl groups have a strong tendency to form face-to-face $\pi \cdots \pi$ interactions. Another example is found in a triple helicate Fe_2L_3 (H_2L = $(C_6H_4OH)CH{=}N{-}N{=}CH(C_6H_4OH)$, which assembles via face-to-face $\pi \cdots \pi$ and CH$\cdots\pi$ interactions into 3-D frameworks with channels to include different organic guests [46].

3.2. 1-D Coordination Polymers

Bidentate chelate ligands, such as en, 2,2′-bpy and 1,10-phen, are strong ligands that invariably reduce the number of metal coordination sites for other ligands. Therefore, mixed linear dicarboxylate and chelate ligands can form linear or zigzag chains [47]. In particular, by using suitably bent or V-shaped organic bridges to improve the helicity as well as length of the pitch of polymeric chains, the aromatic chelates may provide $\pi \cdots \pi$ interaction sites that may lead to double-strand helices or zippers, which are still rare in the literature.

Hydrothermal synthesis using appropriate precursors has afforded infinite, neutral single-strand helical polymers $[Cu(1,3\text{-bdc})(2,2'\text{-bpy})]\cdot 2H_2O$ (**20**) [48] and $[Cu_2(1,3\text{-bdc})_2(1,10\text{-phen})_2(H_2O)]$ (**21**) [49]. More fascinatingly, [Cu(oba)(1,10-phen)] (**22**) with a bent and longer 4,4′-oxy-bis(benzoate) (oba) ligand, displays an unusual double-strand helical structure (Figure 12(a)) [48], in which the copper(II) center is four coordinated by two oxygen atoms from two oba ligands and two 1,10-phen-nitrogen atoms. The 1,10-phen ligands are extended in a parallel fashion on both sides of the single-strand chain at a face-to-face distance of 6.84 Å to give a spatial arrangement suitable for aromatic intercalation; thus, each pair of independent, homochiral chains intertwine with each other into a double-strand helix of C_2 symmetry though interstrand face-to-face $\pi \cdots \pi$ interactions (3.42 Å) between the 1,10-phen ligands within the double-strand helix. Unfortunately, in terms of a crystal engineering target, both left- and right-handed helices coexist in the crystal. An analogous, unique double-strand helix (**39**) assembled by ligand-unsupported argentophilic interactions will be described in Section 4.2.

In contrast, a more flexible and longer dicarboxylate, ethylenedi(4-oxybenzoate) (eoba) ligand, allowed for the formation of a zipperlike, double-strand chain in [Cu(eoba)(1,10-phen)] (**23**) (Figure 12(b)) [48], in which the 1,10-phen ligands are decorated at one side of this chain in a slanted fashion. They recognize those from another chain through offset face-to-face $\pi \cdots \pi$ interactions (*ca.* 3.34 Å); the adjacent Cu$\cdots$Cu separation is 14.94 Å.

When using different metal ions, different molecular architectures may be derived. For example, cadmium(II) or zinc(II) ions are bridged by 1,3-bdc ligands into ribbons in [M(1,3-bdc)(1,10-phen)], M = Cd or Zn [50], which are further packed through intercalation of the lateral 1,10-phen ligands via $\pi \cdots \pi$ interactions in a zipperlike fashion into 2-D arrays.

To a certain extent, the above observations imply that the most critical factors for the pairing of single-strand helical coordination chains into double-strand helices brought about by $\pi \cdots \pi$ interactions lies in the coordination behavior of the metal ion, the length and flexibility of the dicarboxylate ligands as well as the suitability of bpy-like ligands for $\pi \cdots \pi$ interactions. So far, crystal structures of homochiral coordination chains assembled by achiral bridging ligands are extremely limited, and thus remain a challenging topic [51].

On the other hand, such homochiral chains can be more easily constructed with chiral bridging ligands; an interesting example was recently established with a chiral ligand, 2,2′-dimethoxy-1,1′-binaphthyl-3,3′-bis(4-vinylpyridine) (L), which forms homochiral chains

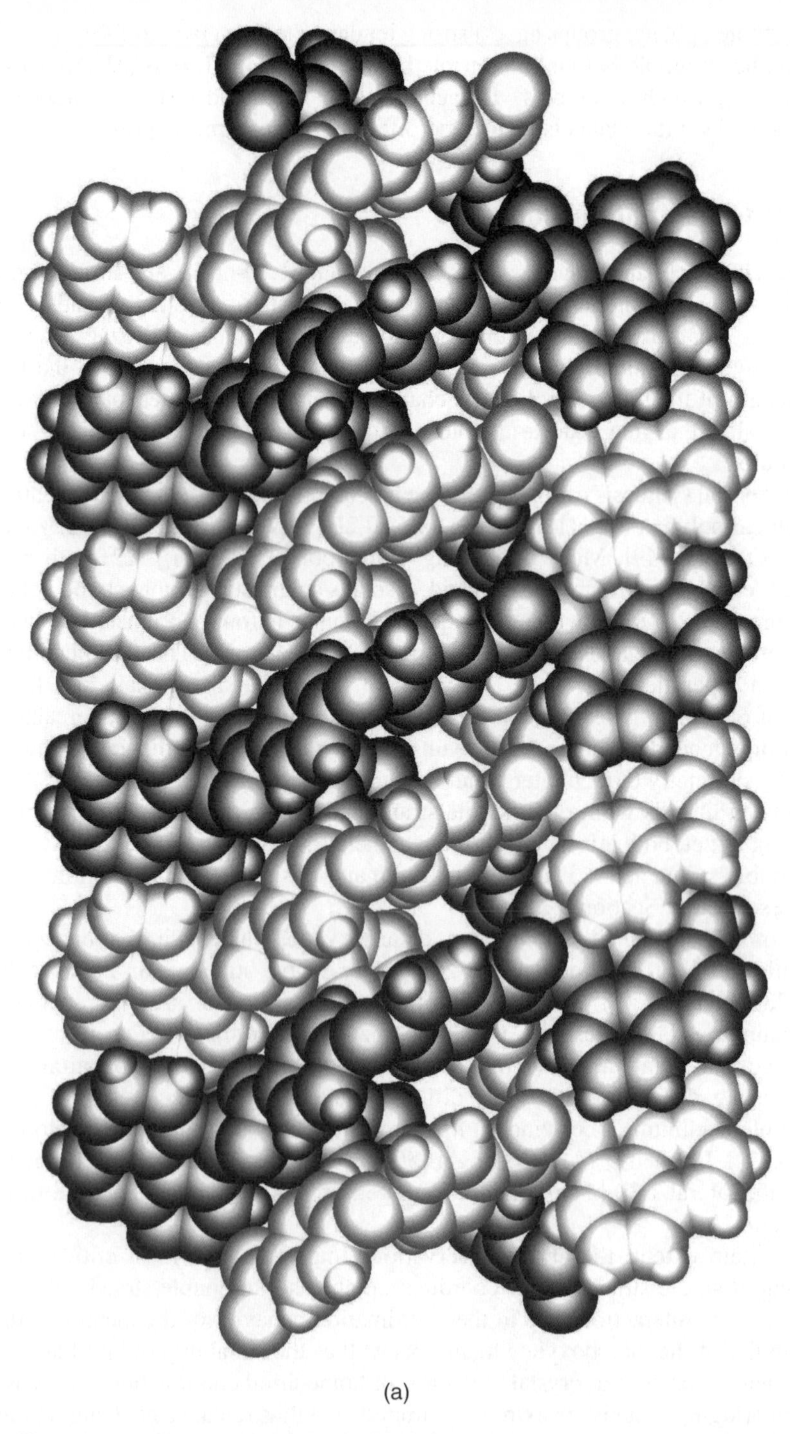

(a)

Figure 12. (a) A double-stranded helix in **22** and (b) a zipperlike double-strand chain in **23**.

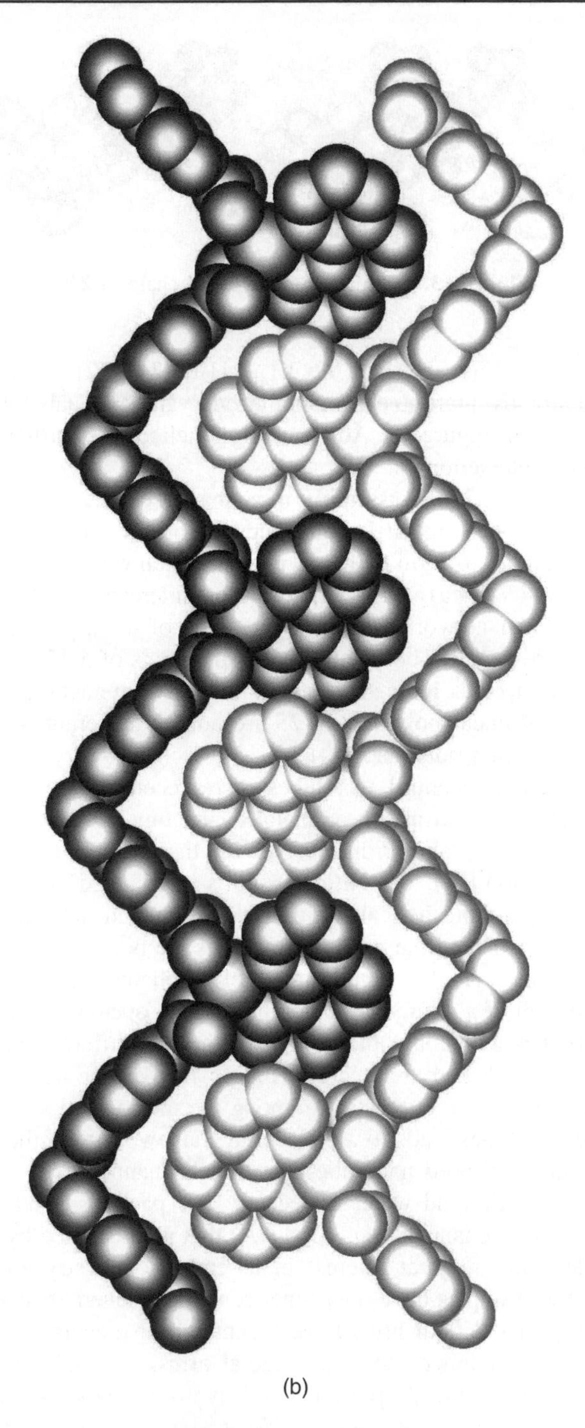

(b)

Figure 12. *(continued)*

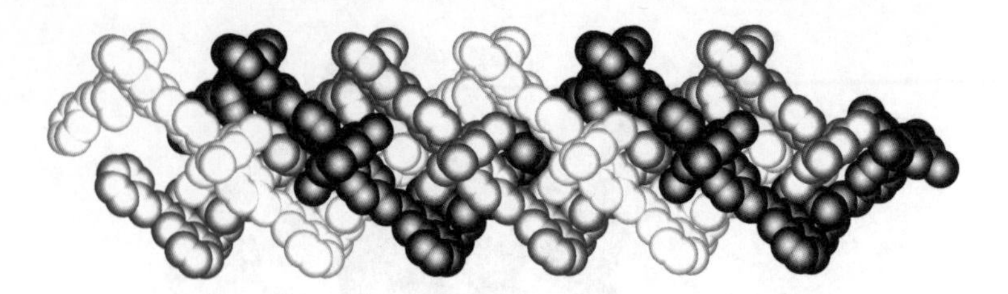

Figure 13. A triple-strand helical chain in **24**.

with $Ni(acac)_2$, namely, $[Ni(acac)_2(L)]\cdot(MeOH)_{1.5}\cdot H_2O$ (**24**) [52]. In the solid state, three left-handed helical strands intertwine one another via van der Waals interactions to form a triple helix as shown in Figure 13. Adjacent triple helices are further assembled into a 2-D array via $\pi\cdots\pi$ interactions.

Another interesting case is the assembly of S-shaped chains into 3-D porous molecular networks. The 4,7-phen ligand is rigid and capable of binding to transition metal fragments with 120° angles to form S-shaped chains, which was demonstrated in $[Cu(4,7\text{-}phen)(H_2O)_3](ClO_4)_2\cdot(4,7\text{-}phen)_2$ (**25**) [53]. The most interesting feature of **25** is the 3-D honeycomblike molecular network formed by stacking of the adjacent triple chains in a parallel fashion (Figure 14(a)). The face-to-face separations of 3.45–3.76 Å between the interchain 4,7-phen molecules indicate significant $\pi\cdots\pi$ interactions. The arrangement and stacking fashion of linear polymers in **25** are notably different from those found in the known 1-D coordination polymers [54].

Periodically ordered homochiral nanotubes were recently assembled via interlocking quintuple helices formed from linear metal-connecting nodes of $Ni(acac)_2$ and C_2-symmetric 1,1-binaphthyl-6,6′-bipyridines (L) with the general formula of $[Ni(acac)_2(L)]\cdot x$S (**26**) (S = solvents) [55]. The $Ni(acac)_2$ units are bridged by the binaphthyl backbones of L to form an infinite helical chain. The left-handed helix is generated around the crystallographic 4_1-axis and the naphthyl moieties are pointing away from the helical axis to generate a hollow cylinder. Five infinite helical chains are associated in a parallel fashion to form the wall of a tetragonal nanotube with an opening of *ca.* 2 × 2 nm. Each helix further intertwines with four other helices from four different nanotubes to give a periodically ordered interlocked architecture (Figure 14(b)). The framework is stabilized by two types of strong $\pi\cdots\pi$ interactions among the intertwined vinylnaphthyl groups. Interlocking of the nanotubes leads to a 3-D chiral framework with the eclipsing of nanotube corners. Partially eclipsed nanotubes have open channels (*ca.* 17 × 17 Å), which are filled with guest MeCN and water molecules. The packing of adjacent tubes in **26** also leads to smaller open channels (*ca.* 7 × 11 Å) that are occupied by water molecules.

Molecular ladders are of great interest in assembly chemistry and are well documented [2b–d]. Most examples have only inner rungs, but several exceptions comprising the same inner rungs and lateral arms have recently been documented [56]. Two such examples having different inner rungs and lateral arms, $[M_2(4,4'\text{-}bpy)_3(H_2O)_2(phba)_2](NO_3)_2\cdot 4H_2O$ (M = Cu(II) and Co(II); phba = 4-hydroxybenzoate), have been reported [57], in which 4,4′-bpy ligands are bridges while η^2-phba ligands are lateral arms. Such lateral arms are threaded into the $[M_4(4,4'\text{-}bpy)_4]$ squares of adjacent ladders, with each square penetrated by two phba arms belonging to two different ladders, in a fashion opposite to that observed in the pseudorotaxane mentioned in Section 3.1 [42]. Such

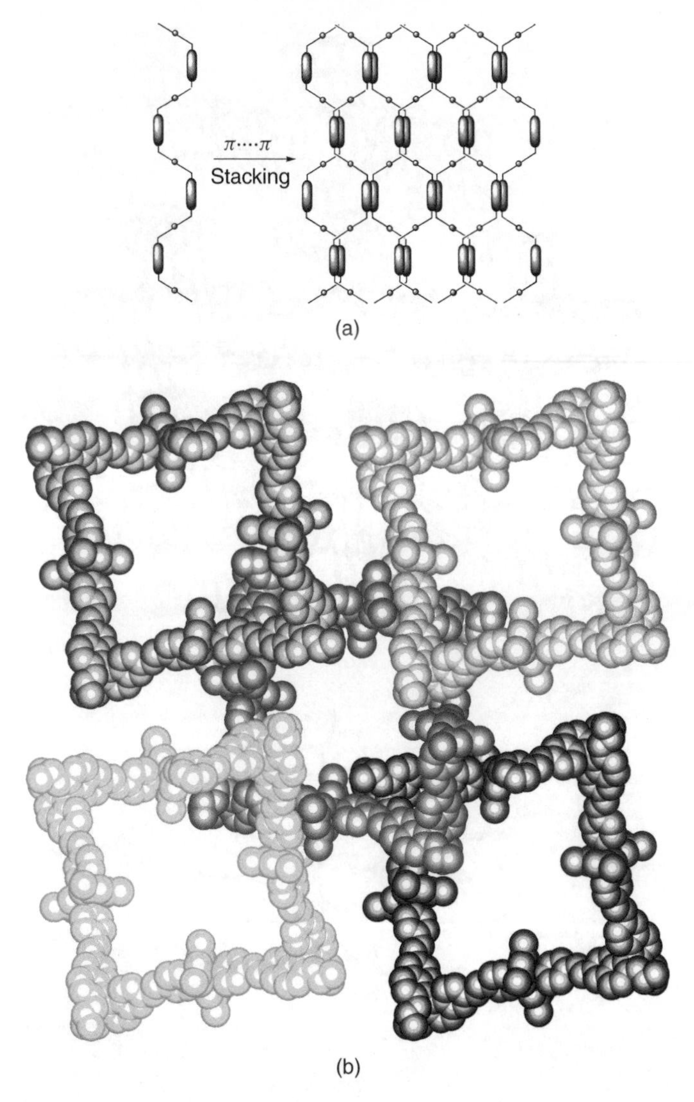

Figure 14. (a) Self-assembly of S-shaped chains into 3-D porous molecular network in **25** and (b) a space-filling model showing the interlocking of adjacent helical chains in **26**.

multiple threading results in a 3-D network with 1-D channels filled with the nitrate and water guests.

Intermolecular aromatic interactions have been documented in a series of silver(I) complexes with flexible linear ligands 1,3-bis(8-thioquinolyl)propane and 1,4-bis(8-thioquinolyl)butane (btqb) with different propylene and butylene linkers between the quinoline ends [58]. For example, in $[Ag_2(btqb)(CF_3SO_3)(CH_3CN)](CF_3SO_3)$ (**27**), the tetrahedral silver(I) atoms are bridged by the sulfur atoms to form cyclic Ag_4S_4 subunits. Such subunits are connected through the butylene linkers into columnlike chains in which the quinoline groups extend into four directions. These chains then recognize each other via $\pi \cdots \pi$ interactions to form a 3-D framework as illustrated in Figure 15(a).

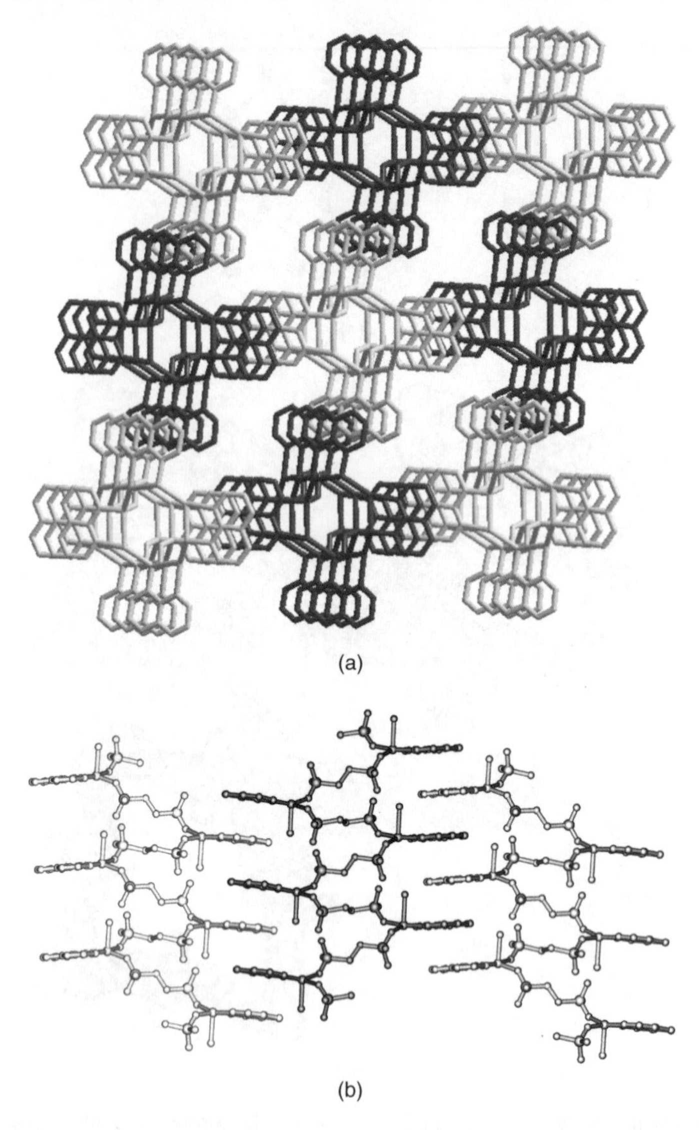

Figure 15. (a) The 3-D network formed by the columnlike chains via four directional aromatic stacking interactions in **27** and (b) the corrugated $\pi \cdots \pi$ stacked layer in **28**.

Introduction of 1,10-phen units into the metal-organophosphate phases creates a zigzag chain [Cu(edp)(1,10-phen)] (**28**) (edp = ethylenediphosphate), which represents a new supramolecular machine based on a 1D copper(II) organophosphonate system as the prototype of electrochemical artificial muscles [59]. The zigzag backbone of **28** comprises phosphonate groups that may serve as carriers of protons or metal ions, while the lateral 1,10-phen plays an important role in forming a $\pi \cdots \pi$ intercalated layer structure between adjacent chains (Figure 15(b)), which makes the layer conducting. Like natural muscle fibers, the corrugated sheets are composed of mats of chain bundles joined by $\pi \cdots \pi$ interactions. Each copper(II) atom is in a square-pyramidal geometry and is chelated by a 1,10-phen, two oxygen atoms from two ethylenediphosphate groups and one aqua ligand. The triggering signals of the actuator in **28** could be envisaged to be an electrochemical signal that initiates a redox process, converting Cu(II) to Cu(I). Upon insertion of Li^+ ions into this material, electrochemical studies show an actuator response detectable by cyclic voltammetry, with concomitant changes in the unit-cell volume. This system presents important implications for the understanding of the actuating mechanism of natural molecular motors and the design of new ionic-switched electrochemical artificial muscle systems.

3.3. 2-D Coordination Polymers

It is apparent that $\pi \cdots \pi$ interactions are not restricted to arranging 0-D and 1-D species into higher-dimensional structures. Indeed, a number of recent examples also demonstrate that some regular layers can recognize each other through $\pi \cdots \pi$ interactions. A typical example was reported in a silver(I)-hexamethylenetetramine (hmt) system [45]. In this case, the lateral aromatic monocarboxylates in hexagonal layers of $[Ag(\mu_3\text{-hmt})L]\cdot nH_2O$ (**29**), L = 4-nitrobenzoate or analogous ligands, could intercalate between neighboring layers in a zipperlike fashion to generate 3-D arrays with micropores. Such a $\pi \cdots \pi$ intercalation strategy can also be applied to assemble inorganic layers into 3-D architectures, as exemplified by a mixed zinc-vanadium phosphate $[Zn(1,10\text{-phen})Zn(VO)(PO_4)_2]$ (**30**) (Figure 16) [60], in which the zipperlike intercalation is very strong (face-to-face separation *ca.* 3.3 Å). Analogous intercalation phenomena have also been observed [61].

For assembly of rigid double layers into 3-D microporous architectures via $\pi \cdots \pi$ interactions, a case in point was demonstrated in $[Cu_2(4,4'\text{-bpy})_5(H_2O)_4]\cdot\text{anions}\cdot 2H_2O\cdot 4EtOH$ (**31**) (anions = $4PF_6^-$ or $2PF_6^-$ and $2ClO_4^-$) [62]. There are three distinct types of coordination modes for the 4,4′-bpy ligands. One is an infinite bridging mode linking the copper(II) atoms into linear chains in two directions, being inclined to each other by *ca.*

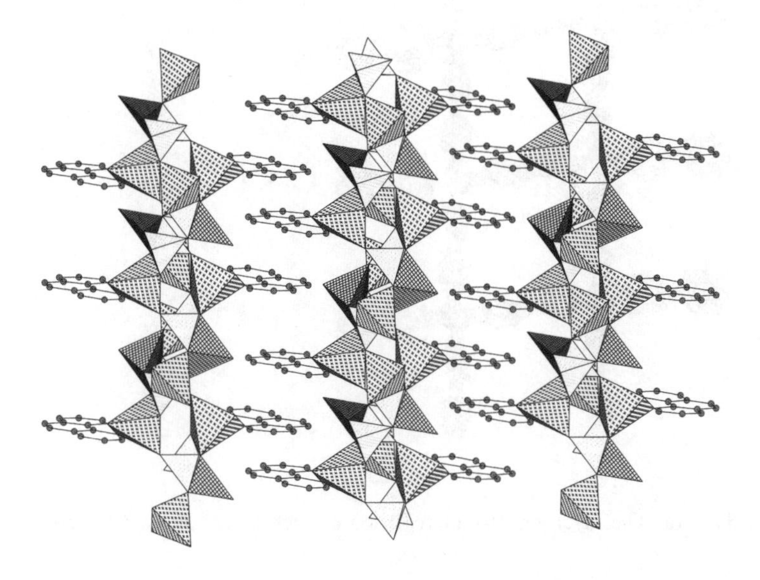

Figure 16. The interlayer $\pi \cdots \pi$ intercalation in **30**.

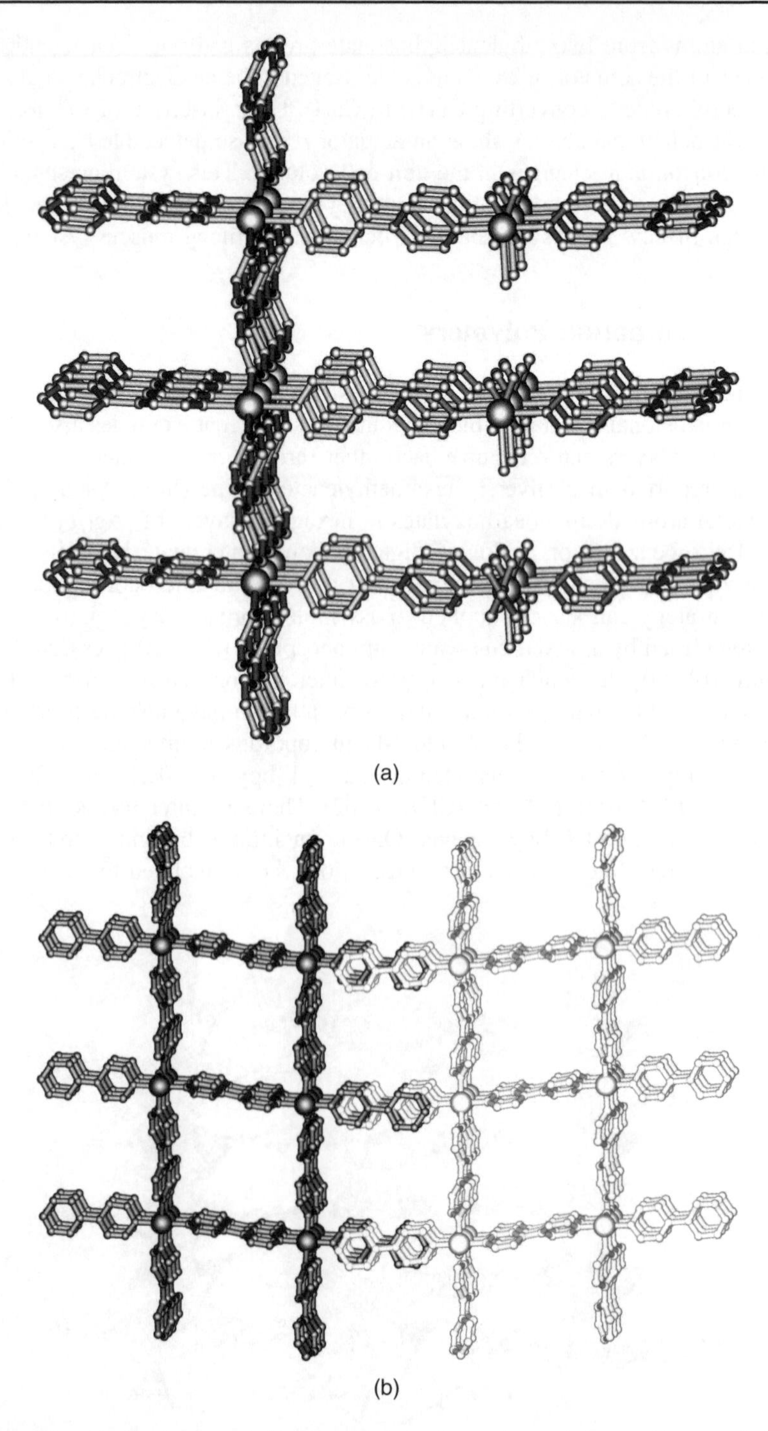

Figure 17. (a) The layer and (b) intercalation of such layers into a 3-D network in **31**.

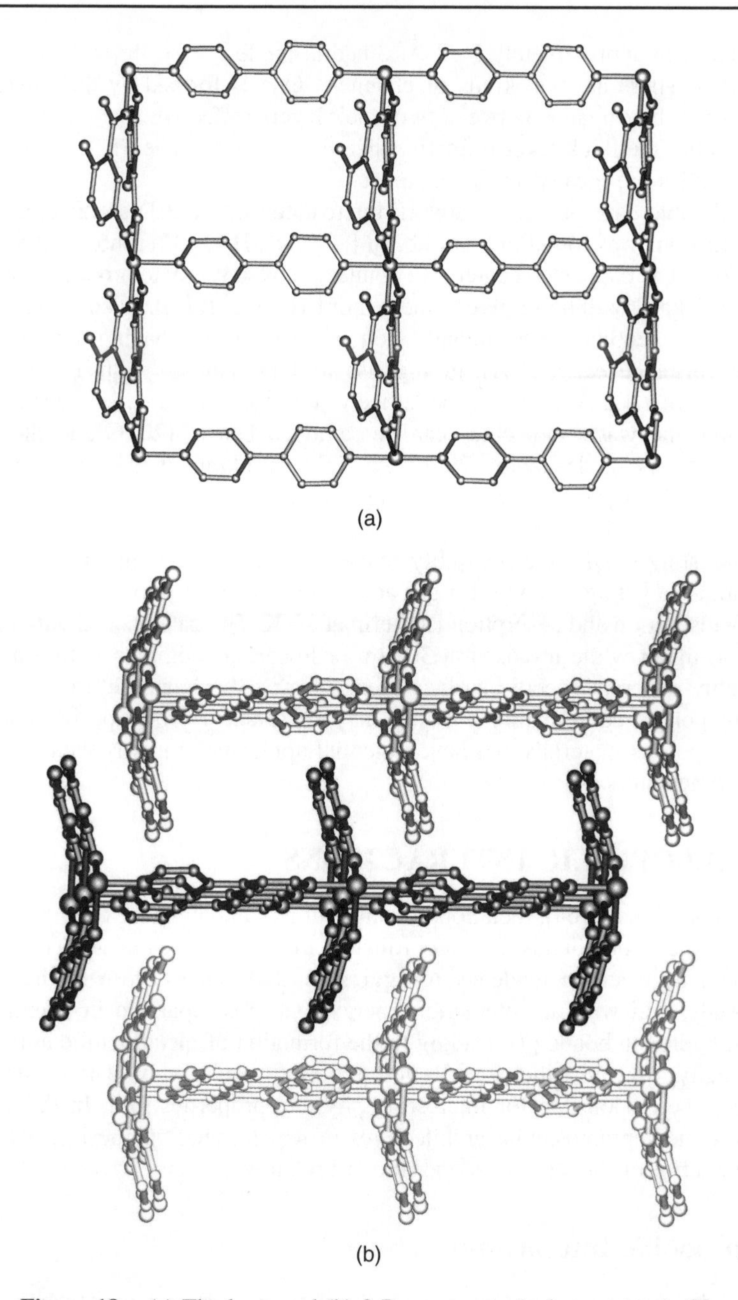

Figure 18. (a) The layer and (b) 3-D $\pi \cdots \pi$ stacked structures in **32**.

44° (slanted from bottom to top or perpendicular to the paper in Figure 17(a)). The second type bridges the two chains thus formed into a thick layer. The third coordination mode is as a terminal ligand, which is in a trans position to the second kind of 4,4′-bpy ligand. These three modes of the 4,4′-bpy afford an unprecedented 2-D structure. Each thick layer assembles by $\pi \cdots \pi$ interactions between the terminal 4,4′-bpy groups (shortest interring

C···C distance is approximately 3.37 Å, dihedral angle 30°) to form a porous network (Figure 17(b)). There are two kinds of channels. One is formed by the intercalation of the terminal 4,4′-bpy ligands between two thick layers (effective size *ca.* 3 × 7 Å) while another is within the thick layer (effective size *ca.* 6 × 6 Å). The PF_6^- anions and guest solvent molecules are located in the channels.

A more flexible case of a 2-D network intercalated into a 3-D architecture built with $\pi \cdots \pi$ interactions was found in $[Cu(dhbc)_2(4,4'\text{-bpy})]\cdot H_2O$ (**32**) (dhbc = 2,5-dihydroxybenzoate) [63]. The copper(II) centers are connected by 4,4′-bpy to produce straight chains that are interlinked by dhbc to give a sheet motif (Figure 18(a)). Such sheets are packed with the nearest neighbor dhbc ligands from adjacent layers having significant $\pi \cdots \pi$ interactions (distance *ca.* 3.44 Å), giving rise to 1-D channels with a cross section of effective size 3.6 × 4.2 Å, in which the lattice water molecules are located (Figure 18(b)). Upon heating, the water molecules can be removed below 120 °C and the anhydrous framework remains stable up to 170 °C. A slight shrinking of the layer gap (effective size *ca.* 3.0 × 4.2 Å) attributable to a gliding motion of the π-stacked rings was observed with powder x-ray diffraction.

Very interestingly, with the flexibility rendered by the $\pi \cdots \pi$ interactions, anhydrous **32** demonstrates a gating behavior for gas adsorption and desorption by the N_2 (molecular size 3.3 Å) adsorption and desorption isotherm at 77 K. The gate opened with the pressure at 50 atm or higher while it closed at 30 atm or lower, concomitant with a hysteresis as well as highly different specific surface area: 24 $m^2\,g^{-1}$ versus 320 $m^2\,g^{-1}$. Therefore, this kind of porous coordination polymer may represent a prototype of smart, flexible and dynamic porous materials that have potential applications in gas separation, sensors, switches and actuators.

4. METALLOPHILIC INTERACTIONS

As reviewed previously [64], metallophilic attractions can take place between closed-shell metal ions in their compounds. Low-coordinated metal ions such as gold(I), silver(I) and copper(I) display a strong tendency to aggregate at distances close to the sum of the van der Waals radii with an interaction energy that is comparable in strength to weak and modest hydrogen bonds [7], leading to the formation of metallophilic attractions, that is, aurophilicity, argentophilicity and cuprophilicity, respectively. These attractions are considered to be responsible for interesting physical properties [65]. In this section, we summarize some supramolecular architectures in which metallophilic interactions play a role in the control of the structures and dimensionalities.

4.1. Cuprophilic Interactions

Supramolecular architectures featuring cuprophilicity are not well documented. An extended 2-D network $[Cu(bpp)]BF_4$ (**33**) (bpp = 1,3-bis(4-pyridyl)propane) was solvothermally produced by use of an ionic liquid (1-butyl-3-methylimidazolium)(BF_4) as solvent [66]. Each copper(I) center in **33** is ligated to two nitrogen atoms from different bpp ligands to generate wavelike chains. Adjacent chains are extended into a wavelike brick-wall layer of (6,3) topology via weak interchain Cu···Cu interactions (3.002(2) Å), which exceed the van der Waals radius sum of two copper atoms (2.8 Å) [67]. The structure is similar to a silver(I) complex (see **38** in Section 4.2).

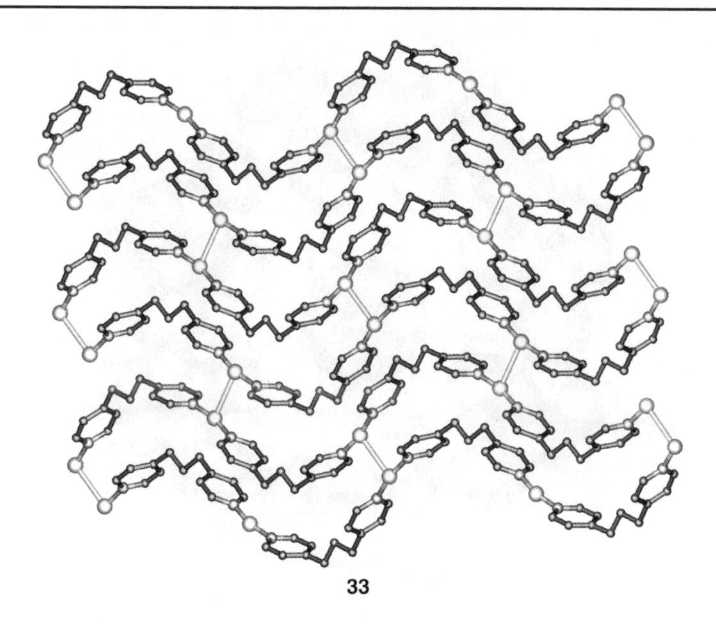

33

Simple imidazolate derivatives, as bent exo-bidentate ligands, can be used to prepare an organically cornered uniform molecular octagon, $[Cu_8(mim)_8]\cdot$(toluene) (**34**) (mim = 2-methylimidazolate), via a hydrothermal approach [68]. The neutral, flattened octagon of **34** comprises eight copper(I) centers, eight mim anions and a guest molecule, such as toluene, in the cavity (Figure 19(a)). The octagons further stack to a thick layer via short intermolecular Cu$\cdots$Cu contacts (2.786(1)–2.879(1) Å) (Figure 19(b)).

Furthermore, with step-by-step increase of temperature and amount of aqueous ammonia, reduction of Cu^{II} to Cu^{I} to different extents was controlled to generate several mixed-valent copper complexes. For example, Cu^{II} ions (Cu(1) and Cu(2)) in $[Cu_2(im)_3]$ (**35**) [69] are coordinated in highly flattened tetrahedral geometries, and Cu^{I} ions (Cu(3) and Cu(4)) are in linear coordination environments. Short contacts exist between copper(I) atoms (2.7293(6) and 2.8144(5) Å) unsupported by direct ligand bridges, extending into linear chains. The whole structure of **35** is a novel, uninodal 4-connected 3-D topological (4.8^5) net where the copper(II) centers are treated as nodes and, im, as well as im-Cu(I)-im, fragments as linkers, in which some of the eight-membered circuits $Cu(II)_8Cu(I)_4(im)_{12}$ are catenated by other eight-membered circuits (Figure 20). Thus, this arrangement is a rare example of an unusual self-entanglement net [2e].

Another interesting copper(I) complex, [Cu(im)] (**36**), was produced from the reaction of $Cu(O_2CCH_3)_2\cdot 4H_2O$ with im ligands in 3-methyl-1-butanol under solvothermal conditions at 140° [70]. The copper(I) centers in **36** are bridged by im groups in a trans fashion to furnish infinite chains, which run in two approximately perpendicular directions and intersect with interchain Cu$\cdots$Cu interactions (*ca.* 2.8 Å), forming the 2-D sheet of a typical $(8^2 10)$ net (Figure 21).

4.2. Argentophilic Interactions

A great number of silver(I) complexes with short Ag$\cdots$Ag separations have been structurally characterized in which the Ag$\cdots$Ag separations are markedly shorter than the

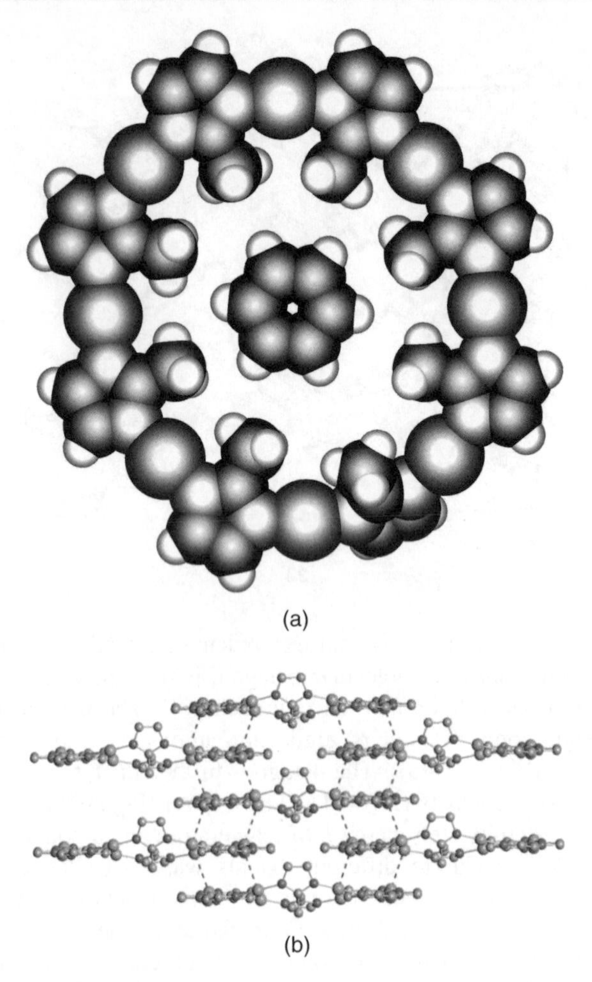

Figure 19. (a) The structure of $[Cu_8(mim)_8]$·toluene and (b) 2-D network of **34** linked by Cu···Cu contacts.

sum of the van der Waals radii of two silver atoms (3.44 Å) [66], indicating the significance of the Ag···Ag interactions thus formed. Many examples of short Ag···Ag distances are ligand supported, with the bridging or capping ligands being, for example, bis(phosphino)methane and carboxylate. Because of their weak nature, the most important structural evidence for argentophilicity should be found in ligand-unsupported Ag···Ag interactions and such a phenomenon was reported in a 1-D carboxylate-bridged complex more than one decade ago [71]; this was verified by resonance Raman spectra [64]. In this section, we display the assembly of supramolecular architecture via Ag···Ag interactions as well as via silver–ligand coordination bonds. It is noted that the structural motifs depend on the anion, solvent and ligand fashion as well as on the reaction conditions [72].

As a linear ligand, 4,4′-bpy has been widely employed to assemble with silver(I) into 1-D or 2-D motifs. The hydrothermally synthesized $[Ag(4,4'\text{-bpy})](NO_3)$ (**37**) comprises polymeric chains further assembled into a 3-D arrangement via ligand-unsupported

Ag$\cdots$Ag interactions (2.97 Å) as shown in Figure 22(a) [73]. The overall structure consists of triple interpenetrated networks having 6 × 23 Å channels, where the nitrate ions reside. This material showed reversible exchange of MoO_4^-, BF_4^- and SO_4^{2-} anions.

Using H_2PO_4 as the source of anion, a similar structural unit was obtained but with distinct crystal packing. In [Ag(4,4′-bpy)](H_2PO_4)·(H_3PO_4), each pair of $[Ag(4,4'\text{-bpy})]_\infty$ chains is linked into a molecular ladder by ligand-supported Ag$\cdots$Ag interactions (3.29 Å) [74]. When the strongly coordinating acetate was introduced, the analogous ladder [Ag(4,4′-bpy)(μ-$MeCO_2$)]·$3H_2O$ was derived, in which the acetate bridges adjacent silver(I) centers, thereby reducing the effective positive charge at each silver(I) atom with concomitant shortening of the Ag$\cdots$Ag distance to 3.12 Å [74]. By contrast, [Ag(4,4′-bpy)](NO_2) [75] and [Ag(4,4′-bpy)](BF_4)·(MeCN)(H_2O) [76] are 1-D polymeric chains without Ag$\cdots$Ag interactions.

An angular ligand 2,4′-bpy, a 4,4′-bpy analog, has also been investigated in this context, offering [Ag(2,4′-bpy)]ClO_4 (**38**), which comprises 2,4′-bpy-bridged helical chains [77]. These chains further self-assemble into a 2-D network via weak ligand-unsupported Ag$\cdots$Ag interactions (3.1526(6) Å) (Figure 22(b)), which is structurally similar to the copper(I) polymer (**33**) [66].

A more interesting example was found in [Ag(bpp)](CF_3SO_3) (**39**) [78], in which the flexible exo-bidentate bpp ligands bind to metal atoms forming right- and left-handed

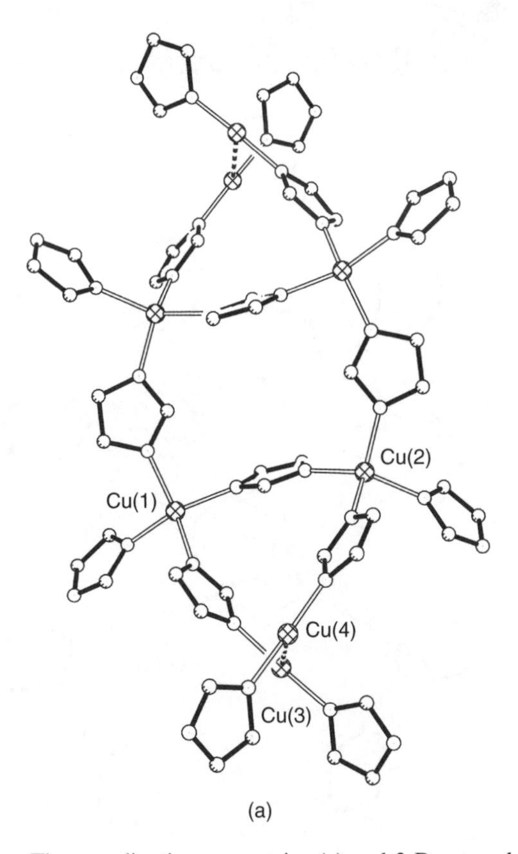

Figure 20. The coordination geometries (a) and 3-D network (b) in **35**.

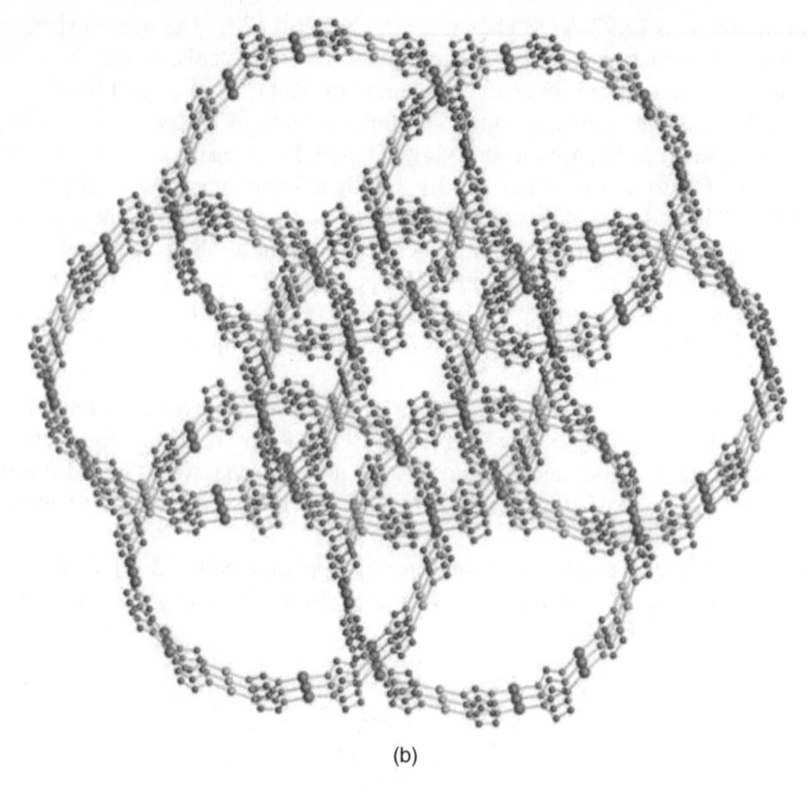

(b)

Figure 20. (*continued*)

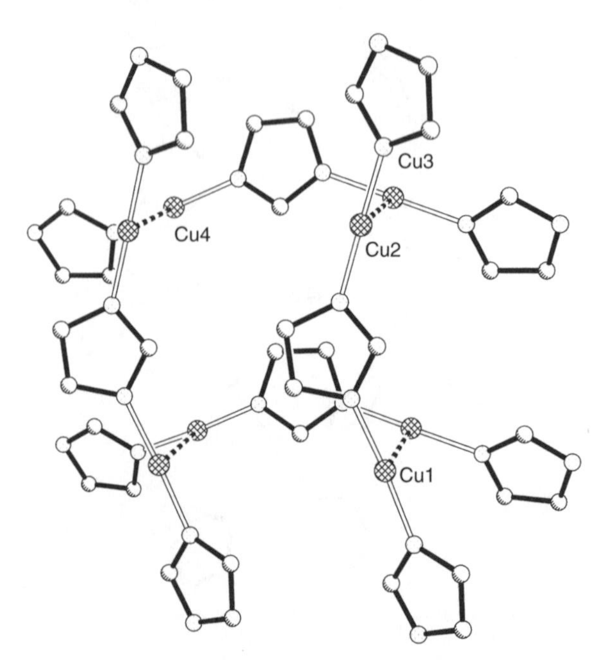

Figure 21. The layer of **36** viewed along the *c*-direction.

strands that further intertwine via Ag$\cdots$Ag interactions at 3.09 Å into an infinite double helix (Figure 22(c)). Such double-strand helix formed under direction of the Ag$\cdots$Ag interactions is very rare, and is analogous to **22**, generated via $\pi\cdots\pi$ interactions [48].

A bent ligand, *N*-(4-pyridinylmethyl)-4-pyridinecarboxamide (ppa), was employed to generate an interesting 3-D network, of formula $[Ag_2(ppa)_2(ox)]\cdot 9H_2O$ (**40**) [79] (ox = oxalate), constructed by infinite $[Ag(ppa)]_n$ (6,3) layers pillared by ox ligands, in which adjacent ppa-bridged zigzag chains are linked into corrugated layers by ligand-unsupported Ag$\cdots$Ag contacts (3.202(1) Å) and offset $\pi\cdots\pi$ interactions (Figure 23(a)). The lattice

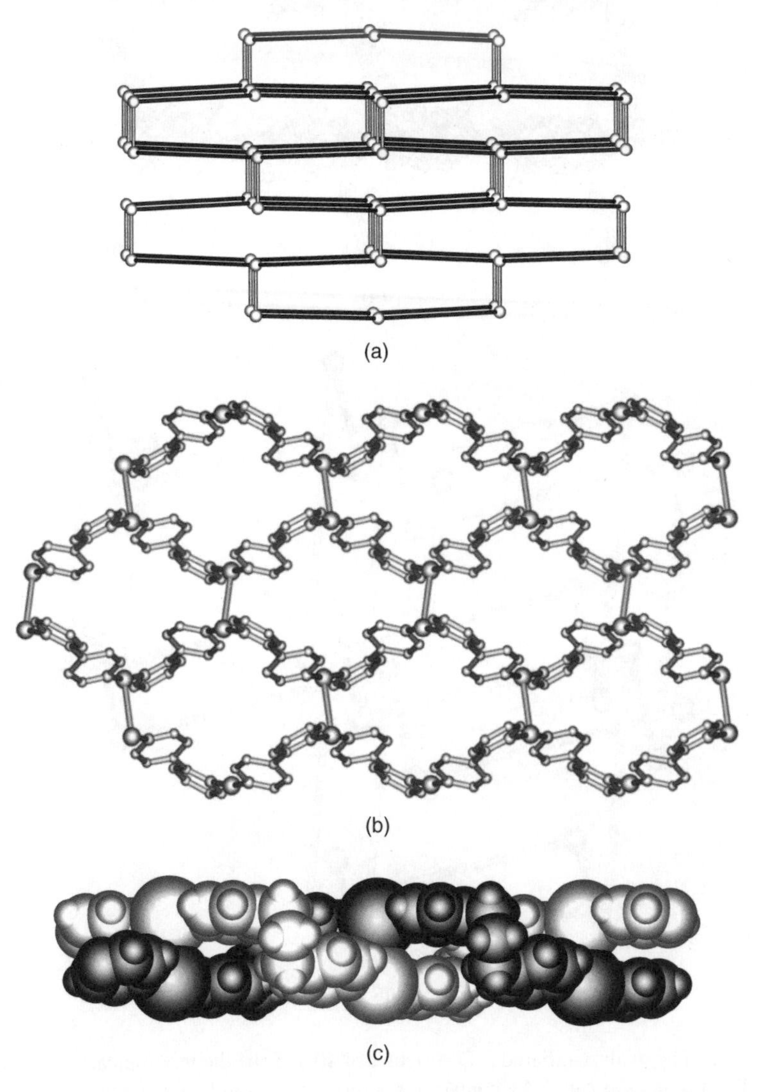

Figure 22. (a) A single 3-D framework of **37**; the 4,4′-bpy ligands and Ag$\cdots$Ag interactions are presented as long and short bars, respectively, (b) the 2-D network constructed by helical chains with Ag$\cdots$Ag interactions in **38** and (c) the space-filling representation of the double helix in **39**.

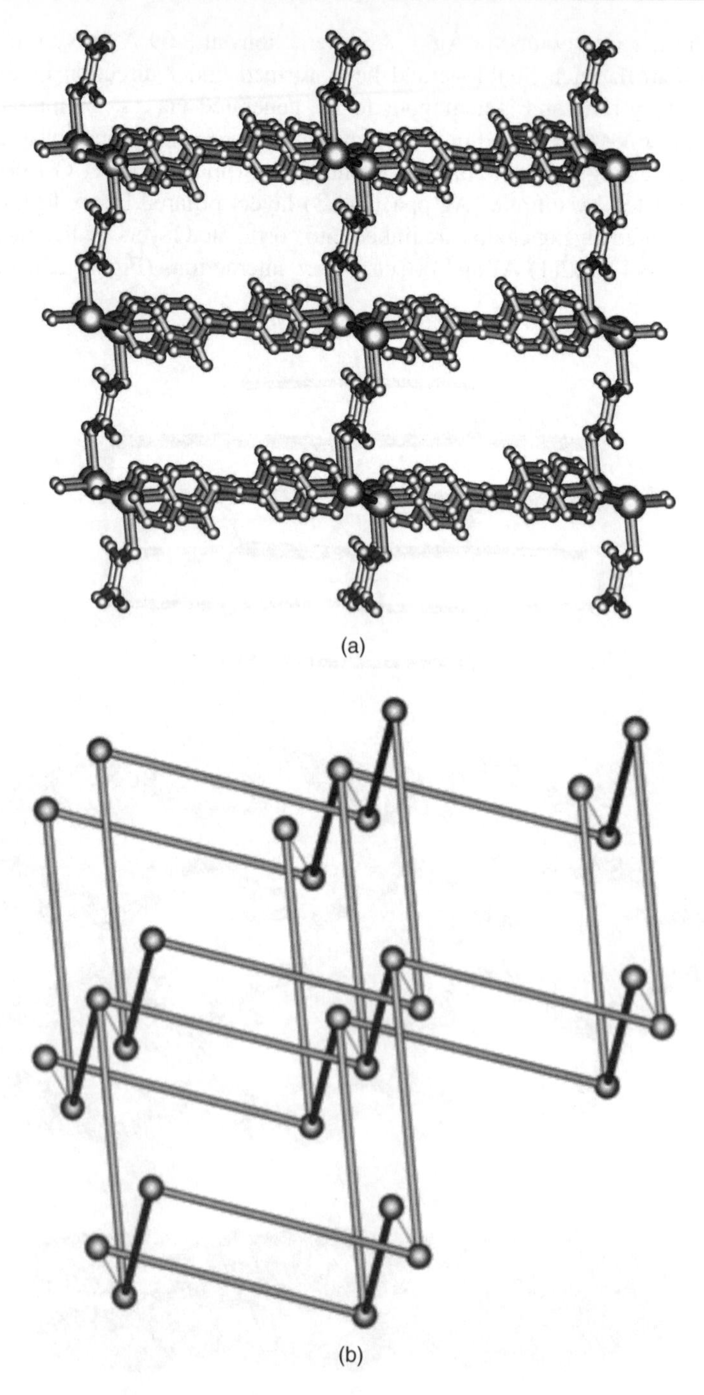

Figure 23. (a) The oxalate-pillared 3-D structure of **40** and (b) the topological net of **41**, where the bridging L ligands and the Ag$\cdots$Ag contacts are presented as solid bars and thin lines, respectively.

water molecules are situated at the rectangular channels (8.1 × 11.9 Å) and stabilized by hydrogen bonds.

Moreover, reaction of a flexible bis-monodentate Schiff base spacer ligand, 1,4-bis(salicylaldeneamino)butane (H_2L), with $AgNO_3$ or $AgClO_4$ offered $[Ag_2(H_2L)_3]X_2$ (**41**), X = NO_3^- or ClO_4^-, which exhibit highly undulated (6,3) layers, with six-membered rings of 3-connected silver(I) atoms (Figure 23(b)) [80]. The silver(I) centers have inter-layer ligand-unsupported Ag···Ag interactions (*ca.* 2.94 Å). Different topologies can be described for **41** depending on which links are included in the topological model. A 4-connected self-penetrating 3-D array results if these Ag···Ag contacts are taken into account, or an overall 6-connected *a*-Po net results if the disilver units were considered as a node. More interestingly, if the Ag···Ag interactions are neglected, an interlinked net via Borromean links results [81].

As shown in Figure 24, an interesting silver(I) wire, $[Ag_2(tren(mim)_3)](NO_3)_2 \cdot H_2O$ (**42**), was obtained by self-assembly of the tripodal Schiff base (tren(mim)$_3$ = tris{2-[2-(1-methyl)imidazolyl]methyliminoethyl}amine) with $AgNO_3$, exhibiting Ag···Ag interactions [82]. Analogous spiral arrays were also demonstrated in silver(I) complexes of 4′-thiomethyl-2,2′:6′,2′-terpyridine (ttpy), $[Ag_5(ttpy)_5(MeCN)_3]X_5$ (X = ClO_4^- and BF_4^-) [83].

A triple interpenetrated 3-D spin-crossover polymer $\{Fe(3\text{-}CNpy)_2[Ag(CN)_2]_2\} \cdot 2/3H_2O$ (**43**) with tunable argentophilic interactions was assembled by Fe(II), $[Ag(CN)_2]^-$ and 3-cyanopyridine [84]. Both crystal structures of the high-spin (HS) and low-spin (LS) forms of **43** adopt the trigonal *P*-3 space group. The equatorial bond lengths of the elongated octahedral iron(II) centers, defined by the nitrogen atoms of four $[Ag(CN)_2]^-$ groups, are shorter than axial bond lengths formed with 3-CNpy (LS). The bis-monodentate $[Ag(CN)_2]^-$ bridges and iron(II) centers are assembled to form a $\{[Fe[Ag(CN)_2]_2]\}$ 4-connected 3-D NbO net (Figure 25), which is interpenetrated by two other identical but independent nets. Interestingly, changeable internet close Ag···Ag contacts (3.256(2) vs 3.1593(6) Å) in the HS and LS forms, as well as spin-crossover phenomenon, was observed by crystallography and accompanying magnetochemical study.

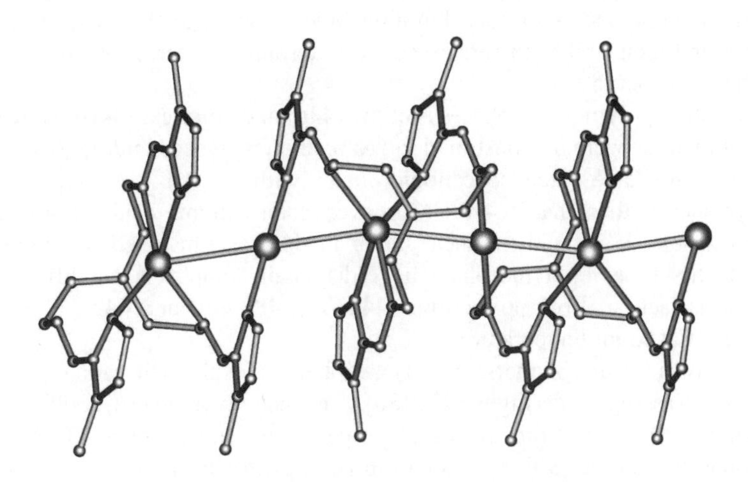

Figure 24. Structure of the chain in **42**.

Figure 25. The three interlocked networks of NbO topology in **43**. The argentophilic interactions are presented by thin lines.

4.3. Aurophilic Interactions

Self-assembly in gold(I) chemistry has also attracted particular interest in the design and synthesis of hybrid organic–inorganic molecular materials. This is because gold(I) centers are versatile as components of such building blocks owing to their tendency for linear coordination and their ability to form secondary aurophilic interactions that can be used to direct the self-assembly.

Two interesting polymorphic phases (yellow (**44a**) and colorless (**44b**)) showing strong aurophilic attractions were demonstrated in $[Au(cyclohexyl\ isocyanide)]_2](PF_6)$ (**44**) [85]. Both **44a** and **44b** have extended cationic chains with the Au···Au separation in **44a** significantly shorter than that in **44b**. Moreover, each cationic chain in **44b** is slightly kinked into a helix compared to linear chain in **44a**, as illustrated in Figure 26. This represents the first case of polymorphism in gold(I) chain complexes that differ in the mode of aurophilic attraction. Both polymorphs **44a** (λ_{max}480 nm) and **44b** ($\lambda_{max} = 424$ nm) are luminescent at room temperature.

Very recently, a linear polymeric gold(I) 4-diphenylphosphanylbenzoate $[Au(4\text{-}Ph_2PC_6H_4CO_2)]$ (**45**) was reported (Figure 27) [86]. The helicity in these double chains (see the following text) arises from the usual propeller-like arrangement of phenyl groups at the phosphorus centers, which associate in pairs primarily through Au···Au contacts (3.1643(9) Å) to give a racemic *P*,*M* double-chain structure.

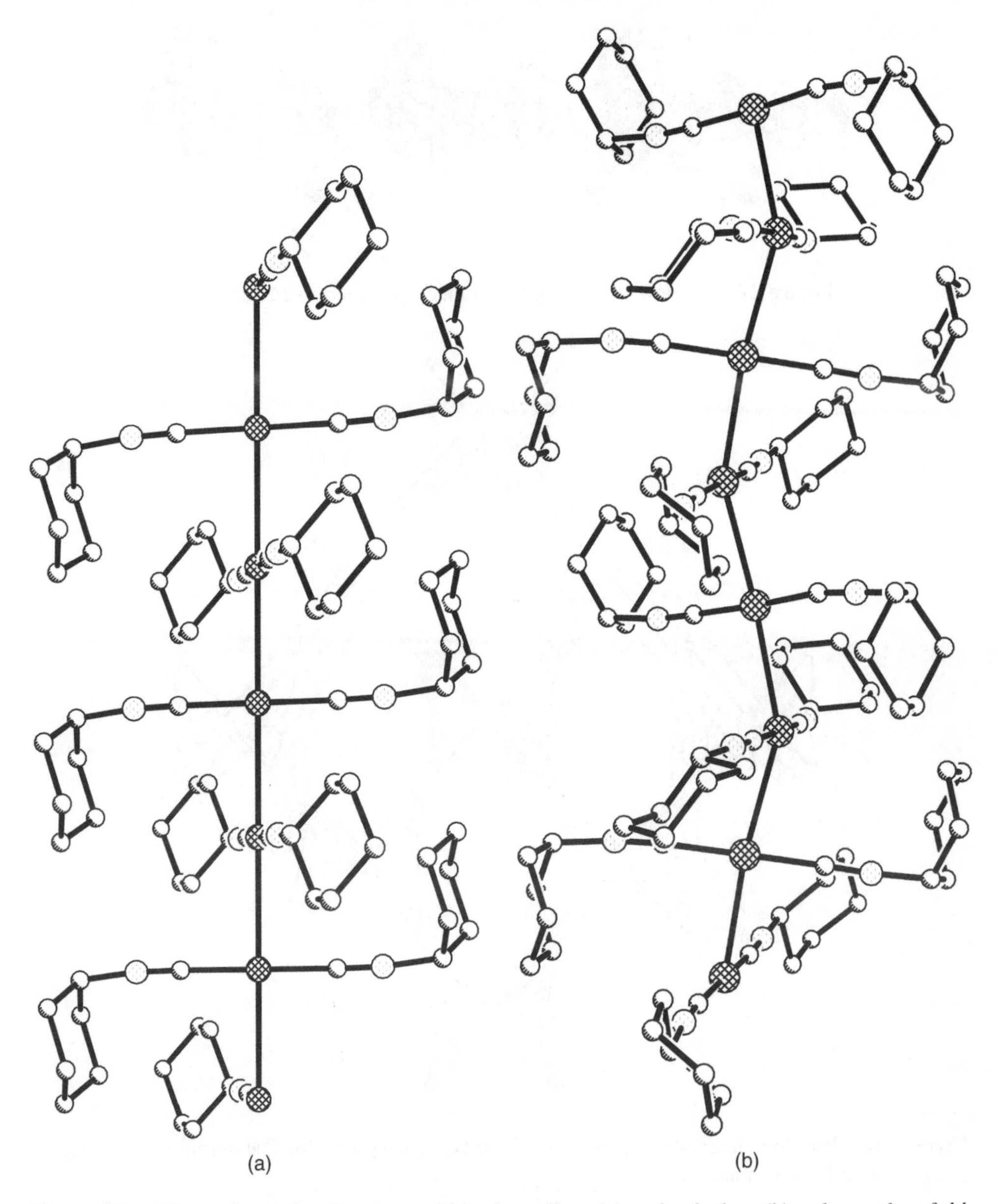

Figure 26. View of a chain of cations within the yellow (a) and colorless (b) polymorphs of **44**.

Similar to $[Ag(CN)_2]^-$, $[Au(CN)_2]^-$ has been used for the construction of an interesting triple interpenetrated 3-D framework, $[Fe(pmd)(H_2O)\{Au(CN)_2\}_2]\cdot H_2O$ (**46**) (pmd = pyrimidine), which shows thermally induced spin-crossover transitions with magnetic and chromatic bistability. The crystal structures of both the HS and LS forms have the monoclinic space group $P2_1/c$ [87]. The $[M(CN)_2]^-$ groups link the iron(II) centers to form $CdSO_4$ nets (Figure 28), which are triply interpenetrated to give the final structure. The independent nets are interconnected via weaker Au $\cdots$ Au interactions (average

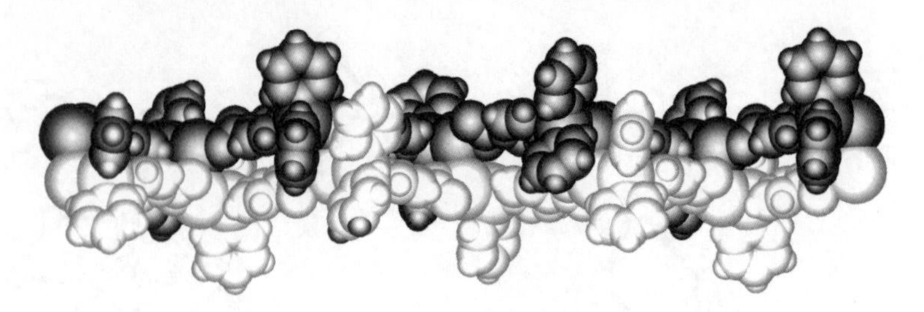

Figure 27. A view of the double-stranded polymeric structure of **45**.

Figure 28. The three interlocked networks of $CdSO_4$ topology in **46**. The aurophilic interactions are presented by thin lines.

3.2901(4) Å) as well as hydrogen bonds between the aqua ligands and uncoordinated pmd-nitrogen atoms. The spin-crossover phenomenon was verified by a magnetochemical study.

5. CONCLUDING REMARKS AND OUTLOOK

We have presented an up-to-date but not comprehensive review on the structural features of coordination molecular architectures featuring hydrogen-bonding, $\pi \cdots \pi$ and metallophilic interactions. Hitherto, investigations in this field have illustrated not only a large number of beautiful and interesting crystal structures, but also interesting physicochemical properties, such as guest storage/inclusion, guest exchange, photoluminescence, magnetism and electrochemistry. Some of them even have biological relevance.

Although the most fundamental factors controlling the structures of coordination molecular architectures are the coordination behavior of the metal ions and the structures of the ligands, supramolecular interactions, including hydrogen-bonding, $\pi \cdots \pi$, metallophilic interactions, and even van der Waals forces, can also demonstrate very significant influence on the crystal structures. Therefore, the design of crystal structure is not easy at all. Actually, many of the crystal structures reviewed in this article may not be predicted. Nevertheless, upon rational combinations of appropriate metal ions with organic ligands, it should be, in principle, possible to generate interesting and useful molecular architectures based on coordination bonding and supramolecular interactions.

It is also notable that supramolecular interactions are usually weaker in strength and directionality, and hence they can have higher reversibility and flexibility. On the other hand, investigations have shown that because of the corporative effect of the bulk supramolecular interactions, such molecular architectures or materials can be rather stable. As such, coordination molecular architectures may be holding a lot of promise in the development of future materials, not just for their intriguing structural diversity, but also as important materials of flexibility, as evidenced by the prototypes of flexible and dynamic microporous materials, as well as the prototype of electrochemical artificial muscle materials. Therefore, it can be expected that flexible and smart materials could be generated on the basis of coordination bonds and supramolecular interactions in the future.

6. ACKNOWLEDGMENTS

This work was supported by the National Natural Science Foundation of China (No. 20131020), the Ministry of Education of China (No. 01134, 20020558024, 2000122), and Guangdong Provincial Science and Technology Bureau (No. 04205405).

REFERENCES

1. (a) B. F. Hoskins and R. Robson, *J. Am. Chem. Soc.*, **112**, 1545–1554 (1990); (b) K. R. Seddon and M. Zaworotko (Ed.), *Crystal Engineering: The Design and Application of Functional Solids*, Kluwer Academic Publishers, Dordrecht, 1996.
2. (a) S. R. Batten and R. Robson, *Angew. Chem., Int. Ed. Engl.*, **37**, 1460–1494 (1998); (b) S. Leininger, B. Olenyuk and P. J. Stang, *Chem. Rev.*, **100**, 853–908 (2000); (c) G. F. Swiegers and T. J. Malefetse, *Chem. Rev.*, **100**, 3483–3538 (2000); (d) B. Moulton and M. J. Zaworotko, *Chem. Rev.*, **101**, 1629–1658 (2001); (e) B. J. Holliday and C. A. Mirkin, *Angew. Chem., Int. Ed. Engl.*, **40**, 2022–2043 (2001); (f) M. Eddaoudi, D. B. Moler, H. Li, B. Chen, T. M. Reineke, M. O'Keeffe and O. M. Yaghi, *Acc. Chem. Res.*, **34**, 319–330 (2001); (g) D. L. Caulder, K. N. Raymond, *Acc. Chem. Res.*, **32**, 975–982 (1999); (h) Y. Rodíguez-Martín, M. Hernández-Molina, F. S. Delgado, J. Pasán, C. Ruiz-Pérez, J. Sanchiz, F. Lloret and M. Julve, *CrystEngComm*, **4**, 522–535 (2002).
3. J. W. Steed and J. L. Atwood, *Supramolecular Chemistry*, Wiley, Chichester, 2000.
4. S. Subramanian and M. J. Zaworotko, *Coord. Chem. Rev.*, **137**, 357–401 (1994).
5. G. R. Desiraju, *Angew. Chem., Int. Ed. Engl.*, **34**, 2311–2327 (1995).

6. C. Janiak, *Dalton Trans.*, 3885–3896 (2000).
7. (a) H. Schmidbaur, W. Graf and G. Müller, *Angew. Chem., Int. Ed. Engl.*, **27**, 417–419 (1988); (b) D. E. Harwell, M. D. Mortimer, C. B. Knobler, F. A. L. Anet and M. F. Hawthorne, *J. Am. Chem. Soc.*, **118**, 2679–2685 (1996); (c) P. Pyykkö and Y. F. Zhao, *Angew. Chem., Int. Ed. Engl.*, **30**, 604–605 (1991); (d) P. Pyykkö, J. Li and N. Runeberg, *Chem. Phys. Lett.*, **218**, 133–138 (1994); (e) R. E. Bachman, M. S. Fioritto, S. K. Fetics and T. M. Cocker, *J. Am. Chem. Soc.*, **123**, 5376–5377 (2001); (f) A. Codina, E. J. Fernández, P. G. Jones, A. Laguna, J. M. López-de-Luzuriaga, M. Monge, M. E. Olmos, J. Pérez and M. A. Rodríguez, *J. Am. Chem. Soc.*, **124**, 5791–5795 (2002).
8. (a) H. L. Hermann, G. Boche and P. Schwerdtfeger, *Chem. Eur. J.*, **7**, 5333–5342 (2001); (b) L. Magnko, M. Schweizer, G. Rauhut, M. Schütz, H. Stoll and H.-J. Werner, *Phys. Chem. Chem. Phys.*, **4**, 1006–1013 (2002).
9. (a) A. M. Beatty, *CrystEngComm*, **51**, 1–13 (2001); (b) A. M. Beatty, *Coord. Chem. Rev.*, **246**, 131–143 (2003); (c) D. Braga, L. Maini, M. Polito, E. Tagliavini and F. Grepioni, *Coord. Chem. Rev.*, **246**, 53–71 (2003).
10. (a) M.-X. Li, G.-Y. Xie, Y.-D. Gu, J. Chen and P.-J. Zheng, *Polyhedron*, **14**, 1235–1239 (1995); (b) X.-M. Chen, M.-L. Tong, Y.-J. Luo and Z.-N. Chen, *Aust. J. Chem.*, **49**, 835–838 (1996); (c) A. J. Blake, S. J. Hill, P. Hubberstay and W. S. Li, *J. Chem. Soc., Dalton Trans.*, 913–914 (1997); (d) L. Carlucci, G. Ciano, D. M. Proserpio and A. Sironi, *J. Chem. Soc., Dalton Trans.*, 1801–1904 (1997); (e) S.-D. Huang and R.-G. Xiong, *Polyhedron* **16**, 3929–3939 (1997).
11. B. Q. Ma, H. L. Sun, S. Gao and G. X. Xu, *Inorg. Chem.*, **40**, 6247–6253 (2001).
12. H. W. Roesky and M. Andruh, *Coord. Chem. Rev.*, **236**, 91–119 (2003).
13. M.-L. Tong, H. K. Lee, X.-M. Chen, R.-B. Huang and T. C. W. Mak, *J. Chem. Soc., Dalton Trans.*, 3657–3659 (1999).
14. (a) M.-L. Tong, S.-L. Zheng and X.-M. Chen, *Polyhedron* **19**, 1809–1814 (2000); (b) M.-L. Tong, X.-L. Yu, X.-M. Chen and T. C. W. Mak, *J. Chem. Soc., Dalton Trans.*, 5–6 (1998).
15. M.-L. Tong, B.-H. Ye, J.-W. Cai, X.-M. Chen and S. W. Ng, *Inorg. Chem.*, **37**, 2645–2650 (1998).
16. H.-J. Chen, M.-L. Tong and X.-M. Chen, *Inorg. Chem. Commun.*, **4**, 76–78 (2001).
17. (a) C. V. K. Sharma and M. J. Zaworotko, *J. Chem. Soc., Chem. Commun.*, 2655–2656 (1996); (b) K. A. Hirsch, D. Venkataraman, S. R. Wilson, J. S. Moore and S. Lee, *J. Chem. Soc., Chem. Commun.*, 2199–2200 (1995).
18. C. B. Aakeröy, A. M. Beatty and D. S. Leinen, *Angew. Chem., Int. Ed. Engl.*, **38**, 1815–1819 (1999).
19. C. J. Kepert, D. Hesek, P. D. Beer and M. J. Rosseinsky, *Angew. Chem., Int. Ed. Engl.*, **37**, 3158–3160 (1998).
20. G. Yang, H.-G. Zhu, B.-H. Liang and X.-M. Chen, *J. Chem. Soc., Dalton Trans.*, 580–585 (2001).
21. R. Sekiya and S.-I. Nishikiori, *Chem. Eur. J.*, **8**, 4803–4810 (2002).
22. Y. Diskin-Posner and I. Goldberg, *Chem. Commun.*, 1961–1962 (1999).
23. (a) K. Kobayashi, M. Koyanagi, K. Endo, H. Masuda and Y. Aoyama, *Chem. Eur. J.*, **4**, 417–424 (1998); (b) M. Shmilovits, M. Vinodu and I. Goldberg, *New J. Chem.*, **28**, 223–227 (2004), and references cited therein.
24. (a) P. G. Desmartin, A. F. Williams and G. Bernardinelli, *New J. Chem.*, **19**, 1109–1112 (1995); (b) C. J. Matthews, M. R. J. Elsegood, G. Bernardinelli, W. Clegg and A. F. Williams, *Dalton Trans.*, 492–497 (2004).
25. D. W. Christianson and R. S. Alexander, *J. Am. Chem. Soc.*, **111**, 6412–6419 (1989).
26. J.-H. Yang, S.-L. Zheng, J. Tao, G.-F. Liu and X.-M. Chen, *Aust. J. Chem.*, **55**, 741–744 (2002).
27. (a) I. Katsuki, Y. Motoda, Y. Sunatsuki, N. Matsumoto, T. Nakashima and M. Kojima, *J. Am. Chem. Soc.*, **124**, 629–640 (2002); (b) Y. Ikuta, M. Ooidemizu, Y. Yamahata, M. Yamada,

S. Osa, N. Matsumoto, S. Iijima, Y. Sunatsuki, M. Kojima, F. Dahan and J.-P. Tuchagues, *Inorg. Chem.*, **42**, 7007–7017 (2003); (c) M. Yamada, M. Ooidemizu, Y. Ikuta, S. Osa, N. Matsumoto, S. Iijima, M. Kojima, F. Dahan and J.-P. Tuchagues, *Inorg. Chem.*, **42**, 8406–8416 (2003).

28. (a) C.-Y. Su, B.-S. Kang, H.-Q. Liu, Q.-G. Wang and T. C. W. Mak, *Chem. Commun.*, 1551–1552 (1998); (b) C.-Y. Su, B.-S. Kang, Q.-C. Yang and T. C. W. Mak, *J. Chem. Soc., Dalton Trans.*, 1857–1862 (2000).
29. M. Tadokoro, K. Isobe, H. Uekusa, Y. Ohashi, J. Toyoda, K. Tashiro and K. Nakasuji, *Angew. Chem., Int. Ed. Engl.*, **38**, 95–98 (1999).
30. (a) M. Tadokoro and K. Nakasuji, *Coord. Chem. Rev.*, **198**, 205–218 (2000); (b) R. Atencio, M. Chacón, T. González, Alexander. Briceño, G. Agrifoglio and A. Sierraalta, *Dalton Trans.*, 505–513 (2004); (c) K. Larsson and L. Öhrström, *CrystEngComm*, **5**, 222–225 (2003).
31. G. R. Desiraju, *Acc. Chem. Res.*, **29**, 441–449 (1996).
32. X.-C. Huang, S.-L. Zheng, J.-P. Zhang and X.-M. Chen, *Eur. J. Inorg. Chem*, 1024–1029 (2004).
33. (a) T. Steiner, *Angew. Chem., Int. Ed. Engl.*, **41**, 48–76 (2002); (b) M. Ohkita, M. Kavano, T. Suzuki and T. Tsuji, *Chem. Commun.*, 3054–3055 (2002); (c) R. Thaimattam, F. Xue, J. A. R. P. Sharma, T. C. W. Mak and G. R. Desiraju, *J. Am. Chem. Soc.*, **123**, 4432–4445 (2001).
34. X.-M. Zhang, M.-L. Tong, M.-L. Gong and X.-M. Chen, *Eur. J. Inorg. Chem.*, 138–142 (2003).
35. Y. B. Go, X. Wang, E. V. Anokhina and A. J. Jacobson, *Inorg. Chem.*, **43**, 5360–5367 (2004).
36. M.-H. Zeng, X.-L. Feng and X.-M. Chen, *Dalton Trans.*, 2217–2223 (2004).
37. (a) C. Piguet, G. Bernardinelli and G. Hopfgartner, *Chem. Rev.*, **97**, 2005–2062 (1997); (b) M. Albrecht, *Chem. Rev.*, **101**, 3457–3498 (2001).
38. S. G. Telfer, T. Sato and R. Kuroda, *Angew. Chem., Int. Ed. Engl.*, **43**, 581–584 (2004).
39. D. Dobrzyńska and L. B. Jerzykiewicz, *J. Am. Chem. Soc.*, **126**, 11118–11119 (2004).
40. H. Baruah, C. S. Day, M. W. Wright and U. Bierbach, *J. Am. Chem. Soc.*, **126**, 4492–4493 (2004).
41. L. J. Childs, N. W. Alcock and M. J. Hannon, *Angew. Chem., Int. Ed. Engl.*, **40**, 1079–1081 (2001).
42. G.-F. Liu, B.-H. Ye, Y.-H. Ling and X.-M. Chen, *Chem. Commun.*, 1442–1443 (2002).
43. H. Abourahma, B. Moulton, V. Kravtsov and M. J. Zaworotko, *J. Am. Chem. Soc.*, **124**, 9990–9991 (2002).
44. X. Wang, Y. Guo, Y. Li, E. Wang, C. Hu and N. Hu, *Inorg. Chem.*, **42**, 4135–4140 (2003).
45. S.-L. Zheng, M.-L. Tong, R.-W. Fu, X.-M. Chen and S. W. Ng, *Inorg. Chem.*, **40**, 3562–3569 (2001).
46. H. Mo, C.-J. Fang, C.-Y. Duan, Y.-T. Li and Q.-J. Meng, *Dalton Trans.*, 1229–1234 (2003).
47. B.-H. Ye, M.-L. Tong and X.-M. Chen, *Coord. Chem. Rev.*, **249**, 545–565 (2005).
48. X.-M. Chen and G.-F. Liu, *Chem. Eur. J.*, **8**, 4811–4817 (2002).
49. Q. Zhang, B.-H. Ye, C.-X. Ren and X.-M. Chen, *Z. Anorg. Allg. Chem.*, **629**, 2053–2057 (2003).
50. L.-Y. Zhang, G.-F. Liu, S.-L. Zheng, B.-H. Ye, X.-M. Zhang and X.-M. Chen, *Eur. J. Inorg. Chem.*, 2965–2971 (2003).
51. O. Evans and W. Lin, *Acc. Chem. Res.*, **35**, 511–522 (2002).
52. Y. Cui, H. L. Ngo and W. Lin, *Chem. Commun.*, 1388–1389 (2003).
53. M.-L. Tong, Y.-M. Wu, S.-L. Zheng, X.-M. Chen, T. Yuen, C. L. Lin, X. Huang and J. Li, *New J. Chem.*, **25**, 1482–1485 (2001).
54. (a) M.-L. Tong and X.-M. Chen, *CrystEngComm*, 1–5 (2000); (b) M. A. Withersby, A. J. Blake, N. R. Champness, P. Hubberstey, W.-S. Li and M. Schröder, *Angew. Chem., Int. Ed. Engl.*, **36**, 2327–2329 (1997); (c) D. Hagrman, R. P. Hammond, R. Haushalter and J. Zubieta, *Chem. Mater.*, **10**, 2091–2100 (1998); (d) J. Lu, C. Yu, T. Niu, T. Paliwala, G. Crisci, F. Somosa and A. J. Jacobson, *Inorg. Chem.*, **37**, 4637–4640 (1998).

55. Y. Cui, S. J. Lee and W. Lin, *J. Am. Chem. Soc.*, **125**, 6014–6015 (2003).
56. (a) O. M. Yaghi, H. Li and T. L. Groy, *Inorg. Chem.*, **36**, 4292–4293 (1997); (b) L. Carlucci, G. Ciani and D. M. Proserpio, *Chem. Commun.*, 449–450 (1999).
57. M.-L. Tong, H.-J. Chen and X.-M. Chen, *Inorg. Chem.*, **39**, 2235–2238 (2000).
58. C.-L. Chen, C.-Y. Su, Y.-P. Cai, H.-X. Zhang, A.-W. Xu, B.-S. Kang and H.-C. zur Loye, *Inorg. Chem.*, **42**, 3738–3750 (2003).
59. K.-J. Lin, S.-J. Fu, C.-Y. Cheng, W.-H. Chen and H.-M. Kao, *Angew. Chem., Int. Ed. Engl.*, **43**, 4186–4189 (2004).
60. X.-M. Zhang, M.-L. Tong, S.-H. Feng and X.-M. Chen, *J. Chem. Soc., Dalton Trans.*, 2069–070 (2001).
61. (a) Z. Dai, Z. Shi, G. Li, D. Zhang, W. Fu, H. Jin, W. Xu and S. Feng, *Inorg. Chem.*, **40**, 7396–7402 (2001); (b) D. Xiao, Y. Li, E. Wang, S. Wang, Y. Hou, G. De and C. Hu, *Inorg. Chem.*, **42**, 7652–7657 (2003); (c) J.-C. Dai, S.-M. Hu, X.-T. Wu, Z.-Y. Fu, W.-X. Du, H.-H. Zhang and R.-Q. Sun, *New J. Chem.*, **27**, 914–918 (2003).
62. S.-i. Noro, R. Kitaura, M. Kondo, S. Kitagawa, T. Ishii, H. Matsuzaka and M. Yamashita, *J. Am. Chem. Soc.*, **124**, 2568–2583 (2002).
63. R. Kitaura, K. Seki, G. Akiyama and S. Kitagawa, *Angew. Chem., Int. Ed. Engl.*, **42**, 428–431 (2003).
64. (a) P. Pyykkö, *Chem. Rev.*, **97**, 597–636 (1997); (b) P. Pyykkö, *Angew. Chem., Int. Ed. Engl.*, **40**, 3573–3578 (2001).
65. (a) C. M. Che, M. C. Tse, M. C. W. Chan, K. K. Cheung, D. L. Phillips and K. H. Leung, *J. Am. Chem. Soc.*, **122**, 2464–2468 (2000); (b) C. M. Che, Z. Mao, V. M. Miskowski, M. C. Tse, C. K. Chan, K. K. Cheung, D. L. Phillips and K. H. Leung, *Angew. Chem., Int. Ed. Engl.*, **39**, 4084–4088 (2000).
66. K. Jin, X. Huang, L. Pang, J. Li, A. Appel and S. Wherland, *Chem. Commun.*, 2872–2873 (2002).
67. A. Bondi, *J. Phys. Chem.*, **68**, 441–451 (1964).
68. X.-C. Huang, J.-P. Zhang and X.-M. Chen, *J. Am. Chem. Soc.*, **126**, 13218–13219 (2004).
69. X.-C. Huang, J.-P. Zhang, Y.-Y. Lin, X.-L. Yu and X.-M. Chen, *Chem. Commun.*, 1100–1101 (2004).
70. Y.-Q. Tian, H.-J. Xu, L.-H. Weng, Z.-X. Chen, D.-Y. Zhao and X.-Z. You, *Eur. J. Inorg. Chem.*, 1813–1816 (2004).
71. X.-M. Chen and T. C. W. Mak, *J. Chem. Soc., Dalton Trans.*, 3253–3257 (1991).
72. (a) A. N. Khlobystov, A. J. Blake, N. R. Champness, D. A. Lemenovskii, A. G. Majouga, N. V. Zyk and M. Schröder, *Coord. Chem. Rev.*, **222**, 155–192 (2001); (b) A. J. Blake, N. R. Champness, P. Hubberstey, W. S. Li, M. A. Withersby and M. Schröder, *Coord. Chem. Rev.*, **183**, 117–138 (1999).
73. (a) F. Robinson and M. J. Zaworotko, *Chem. Commun.*, 2413–2413 (1995); (b) O. M. Yaghi and H. Li, *J. Am. Chem. Soc.*, **118**, 295–296 (1996).
74. M.-L. Tong, X.-M. Chen and S. W. Ng, *Inorg. Chem. Commun.*, **3**, 436–441 (2000).
75. A. J. Blake, N. R. Champness, M. Crew and S. Parsons, *New J. Chem.*, **23**, 13–16 (1999).
76. A. J. Blake, G. Baum, N. R. Champness, S. S. M. Chung, P. A. Cooke, D. Fenske, A. N. Khlobystov, D. A. Lemenovskii, W. S. Li and M. Schröder, *J. Chem. Soc., Dalton Trans.*, 4285–4291 (2000).
77. M.-L. Tong, X.-M. Chen, B.-H. Ye and S. W. Ng, *Inorg. Chem.*, **37**, 5278–5281 (1998).
78. L. Carlucci, G. Ciani, D. W. Gudenberg and D. M. Proserpio, *Inorg. Chem.*, **36**, 3812–3813 (1997).
79. M.-L. Tong, Y.-M. Wu, J. Ru, X.-M. Chen, H.-C. Chang and S. Kitagawa, *Inorg. Chem.*, **41**, 4846–4848 (2002).
80. M.-L. Tong, X.-M. Chen, B.-H. Ye and L.-N. Ji, *Angew. Chem., Int. Ed. Engl.*, **38**, 2237–2240 (1999).
81. L. Carlucci, G. Ciani and D. M. Proserpio, *CrystEngComm*, **5**, 269–279 (2003).

82. S.-P. Yang, H.-L. Zhu, X.-H. Yin, X.-M. Chen and L.-N. Ji, *Polyhedron*, **19**, 2237–2242 (2000).
83. M. J. Hannon, C. L. Painting, E. A. Plummer, L. J. Childs and N. W. Alcock, *Chem. Eur. J.*, **8**, 2226–2238 (2002).
84. A. Galet, V. Niel, M. C. Muñoz and J. A. Real, *J. Am. Chem. Soc.*, **125**, 14224–14225 (2003).
85. R. L. White-Morris, M. M. Olmstead and A. L. Balch, *J. Am. Chem. Soc.*, **125**, 1033–1040 (2003).
86. F. Mohr, M. C. Jennings and R. J. Puddephatt, *Angew. Chem., Int. Ed. Engl.*, **43**, 969–971 (2004).
87. V. Niel, A. L. Thompson, M. C. Muñoz, A. Galet, A. E. Goeta and J. A. Real, *Angew. Chem., Int. Ed. Engl.*, **42**, 3760–3760 (2003).

11

The Structure-directing Influence of Hydrogen Bonding in Coordination Polymers*†

BRENDAN F. ABRAHAMS
School of Chemistry, University of Melbourne, Victoria 3010, Australia.

1. INTRODUCTION

Inspection of the chemical literature that has appeared since the late 1980s reveals that the area of coordination polymers has undergone rapid growth [1]. With scores of fascinating structures and topologies having been identified, exciting possibilities have begun to emerge in regard to the generation of coordination polymers that will possess novel and useful properties [2].

Workers in the area have employed rational design principles in attempts to create materials that may possess desirable properties [1, 3]. To a large extent, this has involved bringing together bridging ligands and metal centers of appropriate geometry in the hope that they will self-assemble into a network with a desired topology. This approach is not always successful as metal geometry and ligand behavior cannot always be predicted. Even when the desired coordination geometry is achieved and the ligand bridges metal centers in the anticipated manner, the desired network is far from guaranteed, mainly because of difficulties in controlling the relative orientation of connected nodes.

* *This chapter is dedicated to the memory of Dr Bernard Hoskins – a pioneer in the field of coordination polymers.*
† *The work described in this chapter was undertaken within Professor Richard Robson's research group in the School of Chemistry at the University of Melbourne. The author is most grateful to Professor Robson for helpful discussions and assistance in the preparation of this chapter.*

Frontiers in Crystal Engineering. Edited by Edward R.T. Tiekink and Jagadese J. Vittal

Consider, for example, networks constructed from perfect trigonal nodes. If all nodes are coplanar, that is, 0° twist along the nodes, then a 2-D (6,3) net results (Figure 1(a)). However, if twists of 109.5° occur along each connection in a manner that preserves threefold symmetry, then the cubic, 3-D (10,3)-*a* net results (Figure 1(b)). Twists of various angles along the connections between the nodes can provide access to a variety of other three-connected nets. Similarly, with four-connected tetrahedral nodes, the twist angle along the connection between nodes is critical in determining the topology of the resulting network. If the tetrahedral nodes are all perfectly staggered (twist angles of 60°) then the diamond net will be the inevitable result, as represented in Figure 2(a). However, if one of the four connections (in a unique direction) has a 0° twist, resulting in the tetrahedral nodes linked by this connection being eclipsed, then the result is a hexagonal network known as the *Lonsdaleite net* [4] (Figure 2(b)). The combination of very minor distortions of the tetrahedral geometry with variation in twist angles can give rise to a wonderful assortment of four-connected nets such as those adopted by the gas hydrates [5].

It is important to recognize in this discussion that nets like diamond and (10,3)-*a* can result when the twist angles and the geometry of nodes deviate from the ideal values. There are many examples in the literature of coordination polymers that have diamond or (10,3)-*a* topologies even though the nodes have twist angles and geometries that are a long way from the ideal [6]. The geometrical conditions described in the preceding paragraph represent what is required to *guarantee* each of the particular type of nets described.

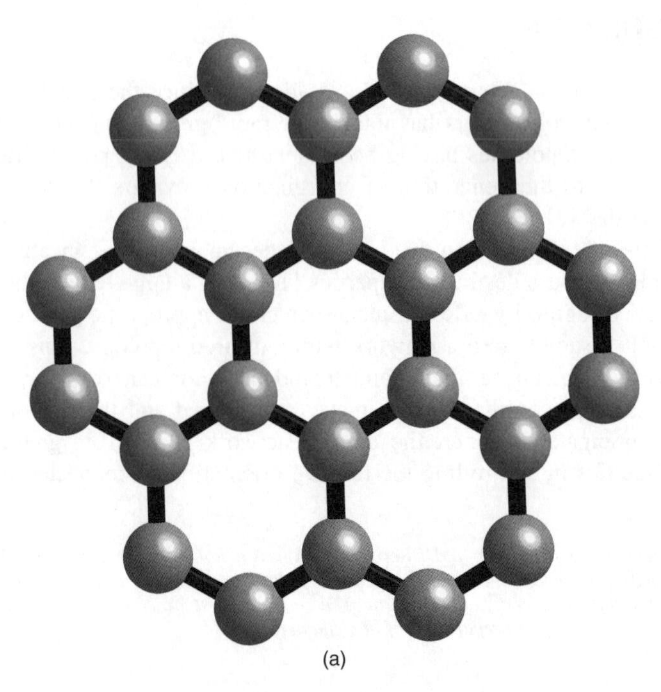

(a)

Figure 1. Nets constructed from trigonal nodes: (a) the 2-D (6,3) (honeycomb) net and (b) the chiral 3-D (10,3)-*a* net.

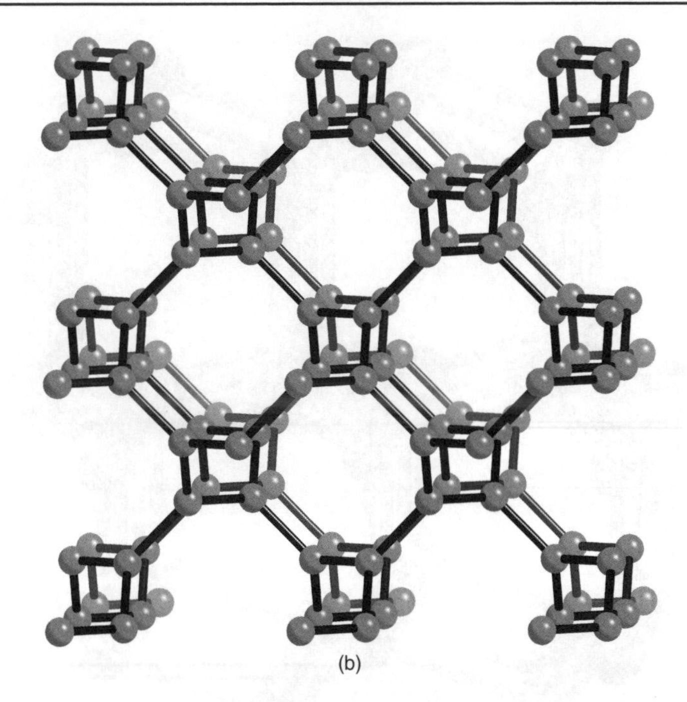

(b)

Figure 1. (*continued*)

(a)

Figure 2. Four-connected nets based on tetrahedral nodes: (a) diamond and (b) Lonsdaleite. Note that in diamond each tetrahedral node is staggered relative to the nodes to which it is connected, while in Lonsdaleite the nodes linked by vertical connections are eclipsed.

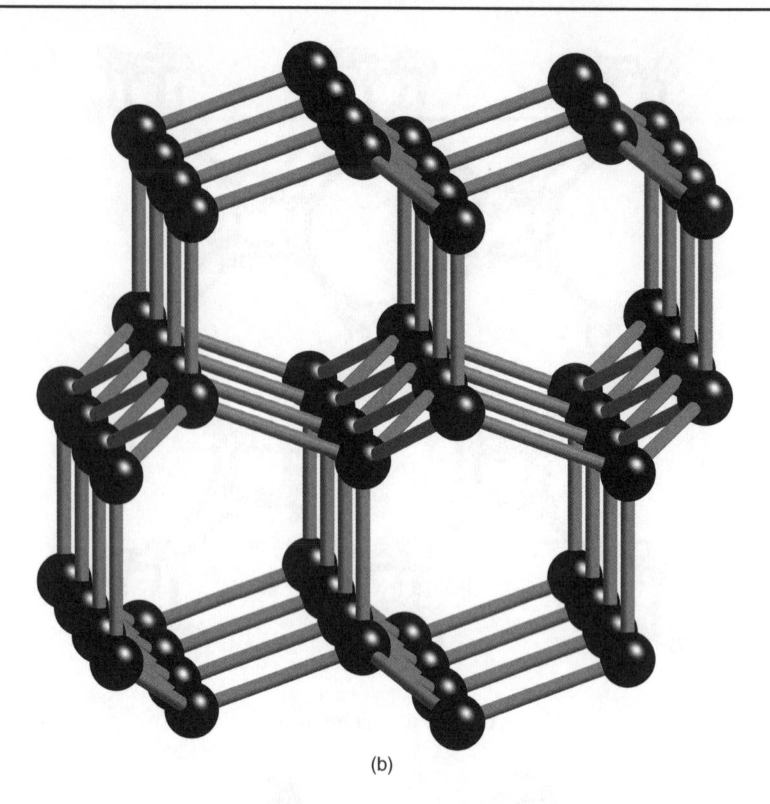

(b)

Figure 2. *(continued)*

Thus, the crystal engineer intending to "design" a coordination polymer faces a particularly challenging task. Even if ligand behavior and metal geometries can be correctly predicted, there is considerable difficulty in controlling the twist angles between the nodes. Until control can be exerted over such factors, the generation of desired coordination polymers will involve an element of "mix and hope".

Over the period that we have been interested in coordination polymers, it has become increasingly clear to us that secondary-bonding interactions play a critical role in directing the assembly of the network. This situation is not dissimilar to that observed with biological macromolecules, where interactions such as hydrogen bonds have a profound influence on the overall structure (and function) of the molecule [7]. Rather than seeing such interactions as unpredictable factors that compound the difficulties associated with structure design, we believe that it may be possible to exert greater control over the types of coordination polymer formed by exploiting such interactions.

In this chapter, four diverse systems that illustrate how hydrogen bonding can have a dramatic structural influence will be examined. The examples considered follow the development from situations in which the role of hydrogen bonding is unintentional toward cases where a hydrogen-bonded cation has an intended effect in directing the assembly of anionic coordination networks. Examination of the broad literature in this area reveals that many systems could be used to demonstrate the points made in this chapter. However, it was decided to limit the discussion to systems that our group has examined, as it illustrates a progression in the way we have approached this area of crystal engineering.

2. A NOVEL CADMIUM CYANIDE NETWORK

Cadmium cyanide [$Cd(CN)_2$] comprises two interpenetrating diamond networks formed by linking tetrahedral cadmium centers by bridging linear cyanide units as depicted in Figure 3. Although it is crystallized from water, the solid is not hydrated. When $Cd(CN)_2$ is recrystallized from aqueous *t*-butanol, water molecules serve as terminal ligands on some of the cadmium centers in a compound of formula: $Cd(CN)_2(H_2O)_{2/3}\cdot t$-BuOH [8–10].

There are two types of cadmium centers in $Cd(CN)_2(H_2O)_{2/3}\cdot t$-BuOH; tetrahedral centers much like that found in $Cd(CN)_2$, with the cadmium coordinated by four bridging cyanide units and octahedral centers with trans water molecules and four bridging cyanide units. Since the aqua ligands are nonbridging, the resulting network is made up of four-connecting centers, tetrahedral and square-planar nodes in a 2:1 ratio. A representation of the network is given in Figure 4. The hexagonal channels are occupied by the coordinated water molecules and highly disordered *t*-butanol molecules.

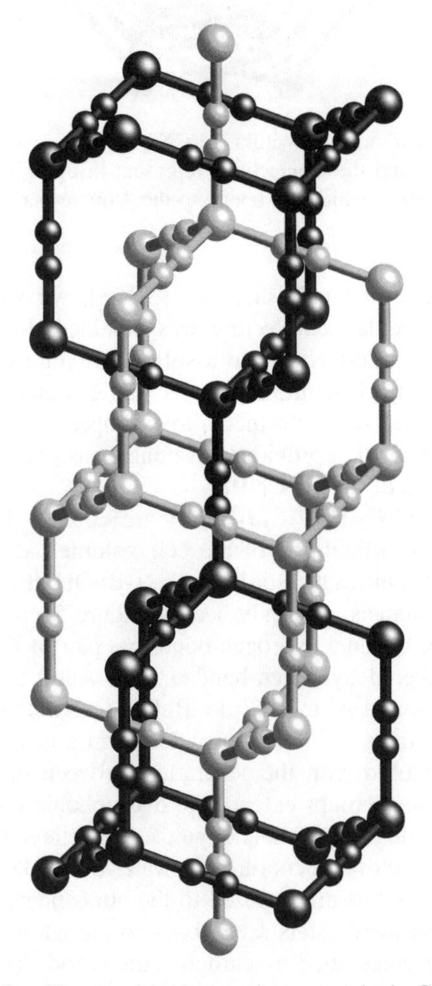

Figure 3. The two interpenetrating networks in $Cd(CN)_2$.

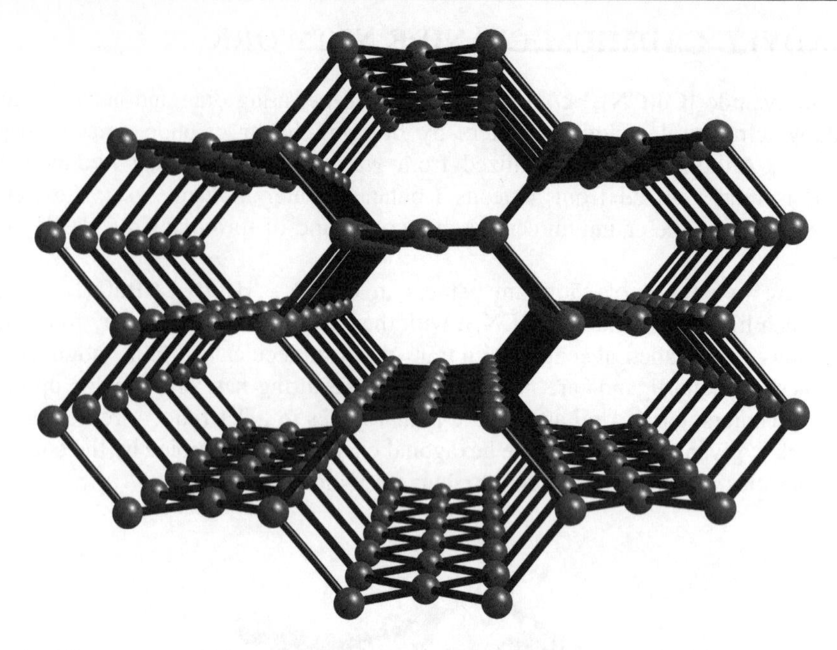

Figure 4. The honeycomblike network within $Cd(CN)_2{\cdot}(H_2O)_{2/3}{\cdot}t$-BuOH. The spheres represent the two types of Cd centers and the connections represent bridging cyanide ligands. For clarity, the coordinated water molecules (which are bound to the "square-planar" nodes) and the t-butanol molecules have been omitted.

When the structure was first published in 1990 [8], we were uncertain as to why such a very different molecular architecture was obtained. It seemed most peculiar to us that when we crystallize $Cd(CN)_2$ from a solvent mixture that has only 50% water, water acts as a ligand, yet when the solvent is 100% water, water is excluded from the solid product. Years later, we performed a low-temperature crystallographic study on $Cd(CN)_2(H_2O)_{2/3}{\cdot}t$-BuOH, which provided a valuable insight into the factors leading to such a vastly different type of network [10].

When crystals of Cd $(CN)_2(H_2O)_{2/3}{\cdot}t$-BuOH are cooled, they undergo a reversible phase change that leads to a doubling of the cell volume. At a temperature of 230 K, even though there is still significant disorder in the t-BuOH, it is possible to model the t-BuOH molecules in the channels. As can be seen in Figure 5, coordinated water molecules from opposite sides of the channel hydrogen bond to a pair of t-BuOH molecules (A and A′) to form a four-membered hydrogen-bonded ring. Another pair of hydrogen bonds extends from molecules A and A′ to a third t-BuOH molecule (B). Thus, the two water molecules along with the three t-BuOH molecules form a neat hydrogen-bonded cluster with a size that is compatible with the separation between opposing cadmium centers within a channel. The t-butyl groups extend into hydrophobic regions within the network and this allows efficient filling of the channel voids. We believe that the hydrogen-bonded interactions involving t-butanol and coordinated water play a key role in the stabilization of the crystal. When the crystals are exposed to the atmosphere, t-butanol and water are lost from the channels and the crystals revert back to the parent $Cd(CN)_2$ structure [11].

When tetrahedral molecules such as carbon tetrachloride (which does not have the potential for hydrogen bonding) are present during the crystallization of $Cd(CN)_2$, a

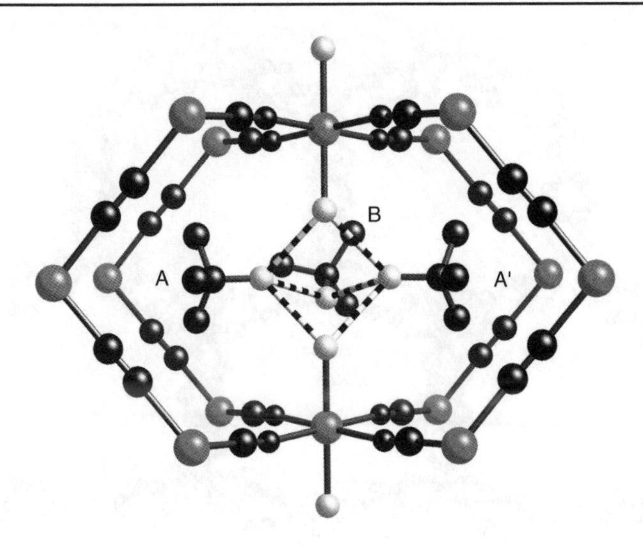

Figure 5. The hydrogen-bonded "cluster" involving the coordinated water molecules and the two types of *t*-butanol molecules (A and B) within $Cd(CN)_2{\cdot}(H_2O)_{2/3}{\cdot}t$-BuOH.

single diamond network is formed with the carbon tetrachloride molecules lying at the centers of the adamantane-type units [12].

The $Cd(CN)_2(H_2O)_{2/3}{\cdot}t$-BuOH structure provides a very clear indication of how secondary-bonding interactions can exert a dramatic influence on metal coordination and network topology. In the next example, we consider again how coordination of the metal and the overall network topology are influenced by hydrogen bonding.

3. DIHYDROXYBENZOQUINONE AND CHLORANILIC ACID DERIVATIVES OF LANTHANIDES

The reaction between cerium(III) nitrate and dihydroxybenzoquinone (H_2dhbq) (**1**) in aqueous solution yields crystals of formula $Ce_2(dhbq)_3{\cdot}24H_2O$ [13]. In this, the dhbq anion serves as a chelating ligand bridge between Ce(III) centers. Each Ce(III) center, which lies on a threefold axis, is linked to three other equivalent Ce(III) ions through these bridges. The coordination environment of the Ce(III) centers is completed by three water molecules. The resulting coordination polymer is an undulating 2-D (6,3) network (Figure 6) with the six-membered $Ce_6(dhbq)_6$ rings having a "chair" conformation. While such a network is perhaps unexceptional, an interesting structural feature results from the hydrogen bonding between coordinated water and lattice water molecules. The coordinated water molecules belonging to two Ce(III) centers from two different sheets combine with 12 lattice water molecules to form a $Ce_2(H_2O)_{18}$ cage, which resembles a pentagonal dodecahedron (Figure 7). Similar cages involving 20 H_2O molecules are found in some gas hydrates [5]. In addition to the intracage hydrogen bonding, hydrogen bonds extend to a $Ce_6(dhbq)_6$ ring, which encompasses the cage (Figure 8). The $Ce_6(dhbq)_6$ ring belongs to a sheet that lies half way between the opposing cerium centers of the cage. If the pair of Ce(III) ions belonging to each cage are considered as linked, then the structure may be considered as two interpenetrating diamond networks as shown in Figure 9. Remarkably,

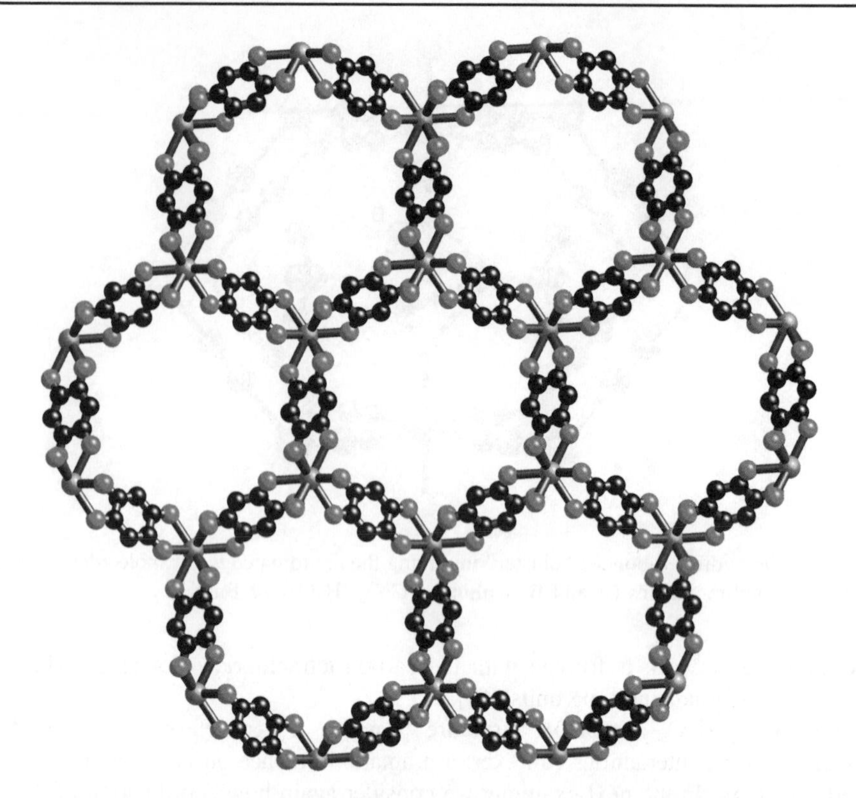

Figure 6. The 2-D $Ce_2(dhbq)_3$ network in $Ce_2(dhbq)_3 \cdot 24H_2O$.

the same structure is found for lanthanum, gadolinium, yttrium, ytterbium and lutetium despite the difference in the radii of the metal ions [13, 14]. For classes of lanthanide compounds, the normal trend is for a reduction in coordination number from lanthanum to lutetium. However, in the case of the hydrated dhbq complexes, the same nine-coordinate tricapped trigonal prismatic geometry persists from lanthanum to lutetium and includes yttrium.

O OH
HO O
1

The dianion of chloranilic acid (**2**, H_2can) is closely related to the dhbq dianion and forms compounds of formula $M_2(can)_3 \cdot xH_2O$ (M = La, Ce, Pr, Nd, Tb, Y, Eu, Lu and Yb). Structural investigations of this series reveal considerable variation in coordination number and network connectivity. In the case of $La_2can_3 \cdot 13H_2O$, a 2-D sheet is formed containing 10-coordinate La centers and bridging water molecules (Figure 10) [14]. For the M = Ce, Pr, Nd, Tb, Y and Eu complexes, 2D (6,3) nets are formed (Figure 11), but they are of two distinct types and though the compounds share the same topology as the dhbq series, the $Ln_2can_3 \cdot xH_2O$ (Ln = Ce, Pr, Nd, Tb, Y or Eu) structures are

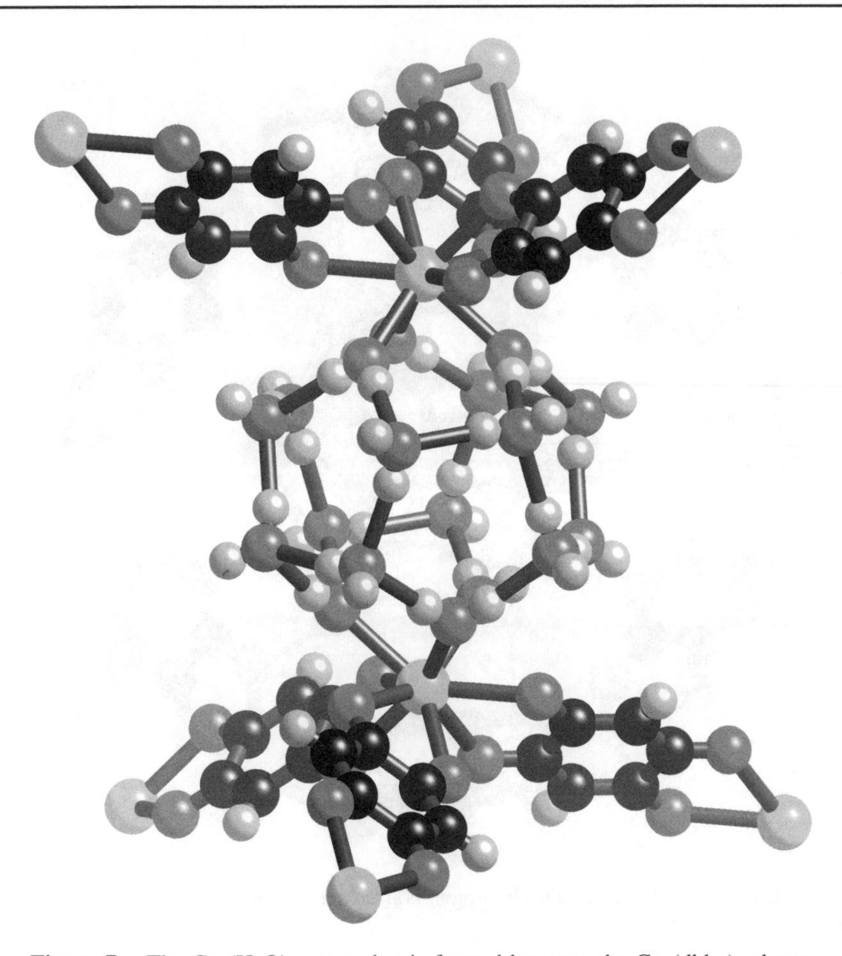

Figure 7. The $Ce_2(H_2O)_{18}$ cage that is formed between the $Ce_2(dhbq)_3$ sheets.

very different [14]. For the heavier members of this series, 1-D cationic chains of formula, $[Ln(can)(H_2O)_4{}^+]_n$, are produced, with discrete mononuclear complexes of formula $Ln(can)_2(H_2O)_4{}^-$ serving as counteranions (Ln = Lu or Yb) [14]. The metal centers in both the cation and anion are eight-coordinate (Figure 12).

Cl
O
OH
HO
O
Cl
2

The structural variation observed for the $Ln_2can_3{\cdot}xH_2O$ series provides a stark contrast with the $Ln_2(dhbq)_3{\cdot}24H_2O$ series in which the same structure is obtained across the

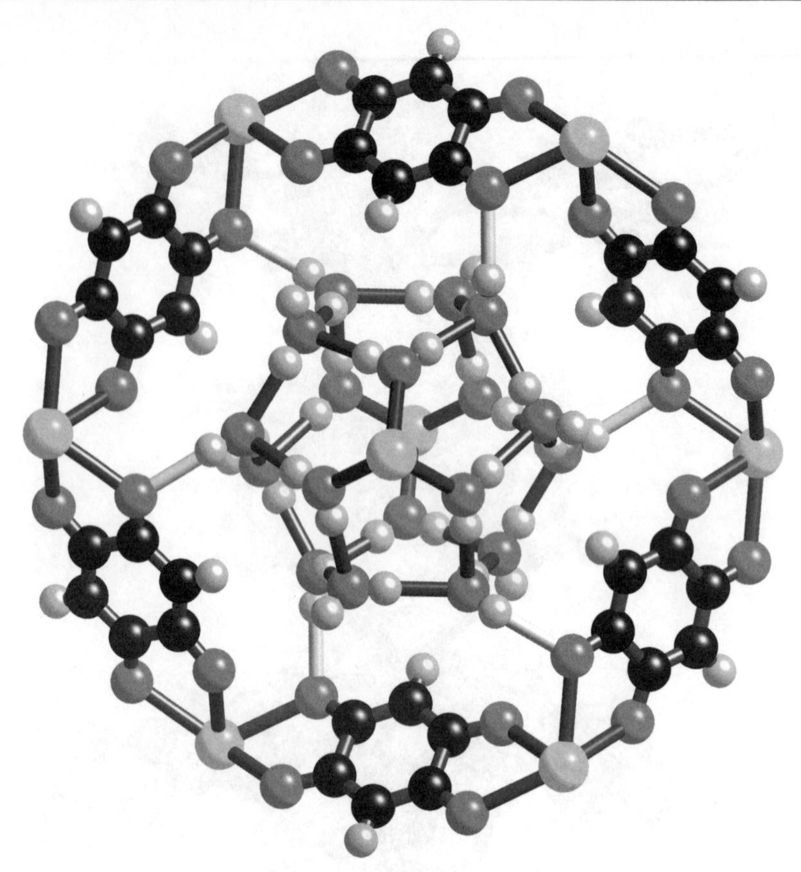

Figure 8. The $Ce_2(H_2O)_{18}$ cage with hydrogen bonds to a $Ce_6(dhbq)_6$ ring belonging to an intermediate $Ce_2(dhbq)_3$ sheet.

lanthanide period despite the lanthanide contraction. The persistence of the nine-coordinate geometry in the $Ln_2(dhbq)_3{\cdot}24H_2O$ series is remarkable and reflects the stability provided by the extensive hydrogen bonding within the network. In particular, the ability of the $Ln_6(dhbq)_6$ rings to neatly encompass the water cage through multiple hydrogen bonds results in the 2-D metal-ligand framework being locked into a highly symmetric and very stable, hydrogen bonded, 3-D network in which nine-coordination is imposed on the metal center.

The absence of an analogous isostructural series for the chloranilate complexes may be attributed to the chloride atoms that would protrude into potential hexagonal holes of an $Ln_6(can)_6$ ring and prevent the formation of a $Ln_2(H_2O)_{18}$ water cage. With the absence of a dominating structural influence, the size of the metal center becomes a more important factor in the type of structure that is formed.

The work with the $Ln_2(dhbq)_3{\cdot}24H_2O$ series demonstrated again the important role multiple weak interactions can play in influencing the type of structure that is formed. The work has also suggested that if a dominant structural influence can be exerted in the assembly of a crystalline material then it is possible to tinker with the nature of the components and still obtain essentially the same structure.

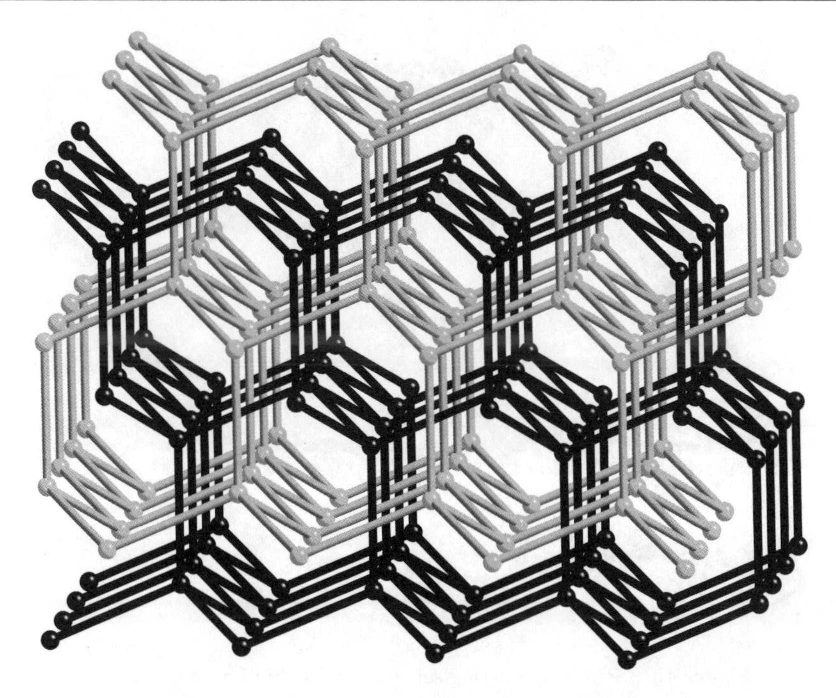

Figure 9. A representation of the connectivity of $Ce_2(dhbq)_3 \cdot 24H_2O$ in terms of two interpenetrating diamond-type networks. Spheres represent Ce atoms; nonvertical connections represent bridging dhbq anions and vertical connections represent the water cages.

4. A STABLE ZINC SACCHARATE NETWORK

The dicarboxylates, saccharate (**3**) and mucate (**4**) are derived from the oxidation of the aldohexoses, glucose and galactose, respectively. From the point of view of the crystal engineer who is attempting to exert control in the assembly of coordination polymers, the use of these dianions as ligands may seem very ambitious. The variability in coordination modes of the carboxylate unit makes it very difficult to predict the geometry at metal centers. Furthermore, the presence of eight oxygen atoms per ligand, all of which have the potential to be involved in hydrogen bonding means that the ligand's behavior in a

O=C–O⁻ | H–C–OH | HO–C–H | H–C–OH | H–C–OH | O=C–O⁻
3

O=C–O⁻ | H–C–OH | HO–C–H | HO–C–H | H–C–OH | O=C–O⁻
4

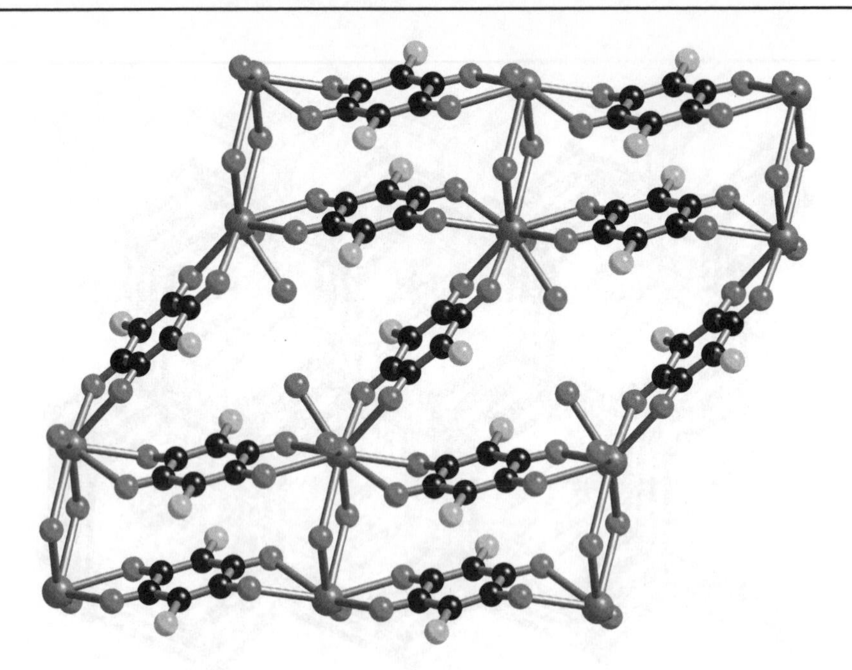

Figure 10. Part of the 2-D sheet structure within $La_2can_3{\cdot}13H_2O$.

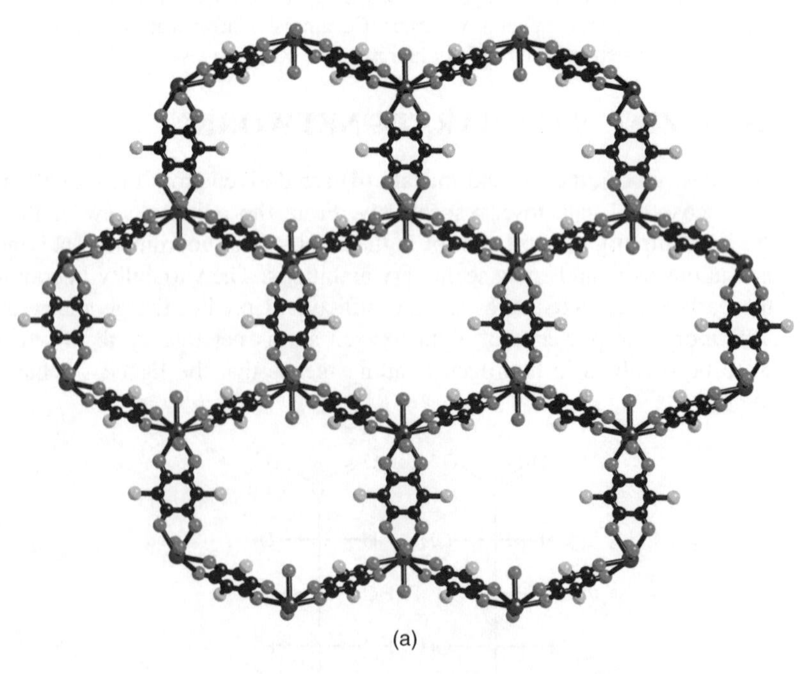

(a)

Figure 11. The 2-D (6,3) networks adopted by (a) $Ln_2can_3{\cdot}18H_2O$ (Ln = Ce, Pr, Nd and Tb) and (b) $Ln_2can_3{\cdot}16H_2O$ (Ln = Y, Gd, Eu).

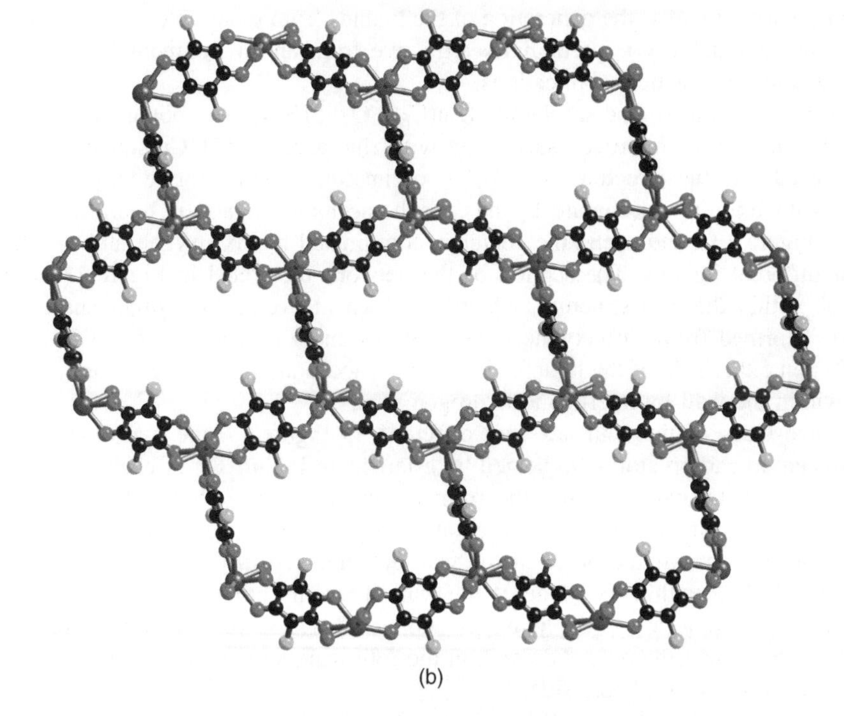

Figure 11. *(continued)*

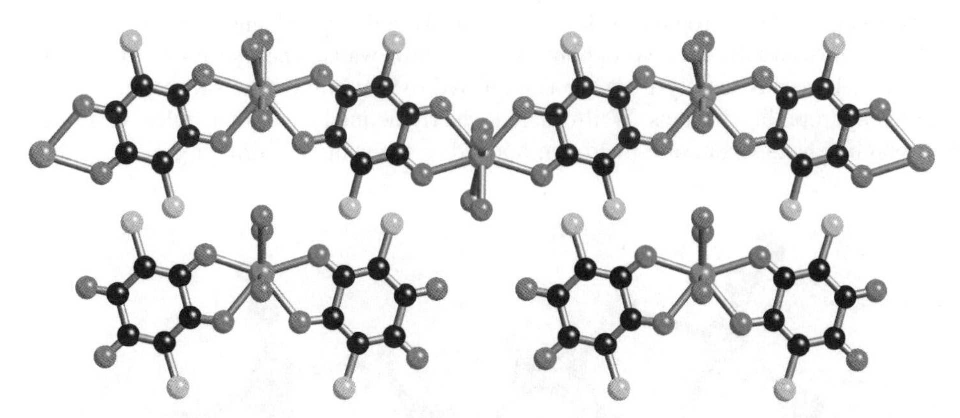

Figure 12. Part of the ionic structure of $Ln_2(can)_3 \cdot 12H_2O$ (Ln = Yb, Lu). The structure is composed of polymeric cationic chains of composition $[Ln(can)(H_2O)_4{}^+]_n$ and discrete $[Ln(can)_2(H_2O)_4{}^-]$ anions.

coordination polymer will be almost impossible to anticipate. Despite these disadvantages, the successful incorporation of these anions into some sort of network has many positive aspects. Firstly, the likelihood of extensive hydrogen bonding will potentially lead to a structure with significant reinforcement. Secondly, the prospect of an open network structure (although far from guaranteed) with "dangling" hydroxyl groups is appealing as it may allow discrimination between the types of guests that may be incorporated on the basis of polarity. Finally, the generation of the ligands from renewable sources in addition to their biodegradability is particularly attractive from an environmental perspective in the eventuality of practical applications.

Crystals of hydrated zinc saccharate [$Zn(C_6H_8O_8) \cdot 2H_2O$] were obtained by the direct reaction of potassium hydrogen saccharate with zinc acetate [15]. Crystal structure analysis showed that the structure is a 3-D coordination polymer formed by linking zinc centers with bridging saccharate ligands. Each saccharate ligand is bound to four zinc centers (Figure 13), and each zinc center is coordinated by six oxygen atoms belonging to four different ligands. The section of the network illustrated in Figure 13 is part of a "wall" within the 3-D structure, which is shown in Figure 14. Corner shared Zn_2O_2 "squares" formed from carboxylate oxygen atoms and zinc centers extend in a vertical direction on either side of the ligands. Individual saccharate dianions are "cemented" into place within the wall by interligand hydrogen bonds.

The two sides of the wall are very different. In Figure 14, the hydroxyl groups on the two central carbon atoms (indicated by asterisks in Figure 13) of each saccharate are directed out of the page, making the front of the wall a very hydrophilic surface. By contrast, the opposite side has no oxygen atoms pointing out and even the oxygen atoms that are in the approximate plane of the wall are involved in metal coordination and/or hydrogen bonding within the plane of the wall. The only atoms that extend away from the wall on this side are C–H hydrogen atoms. This side of the wall is certainly far less hydrophilic than the front side and, in the following structural description, it will be referred to as the hydrophobic side.

In the extended structure, walls are joined together to form infinite parallel channels (Figure 15). Two types of channels are produced: small channels where the interior surface is provided by the hydrophilic sides of four walls and large channels where the interior surface is provided by the hydrophobic sides of four walls. The difference in the size of the channels results largely from the central hydroxyl group on the saccharate extending into the hydrophilic channels. Well-ordered, clearly defined water molecules lie inside the hydrophilic channels and are hydrogen bonded to the central hydroxyl groups. Although

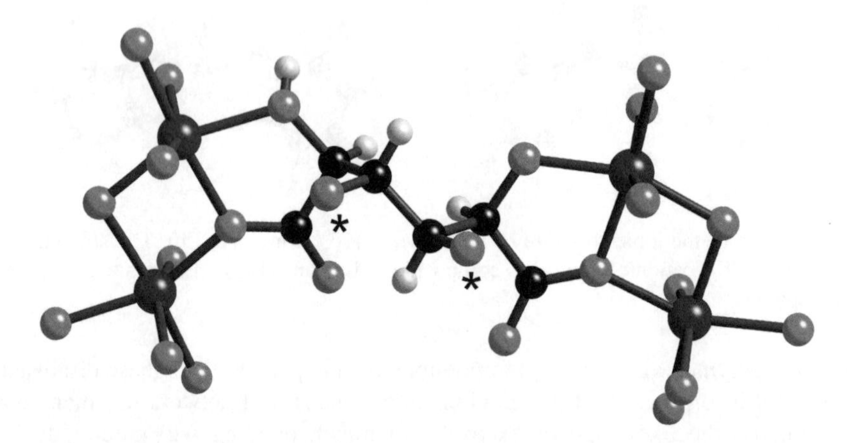

Figure 13. The saccharate ligand ($C_6O_8H_8^{2-}$) and the coordination environments of the four Zn centers to which it is bound within the structure of $Zn(C_6O_8H_8) \cdot 2H_2O$. The asterisks indicate the two central hydroxyl groups, which are directed out of the page.

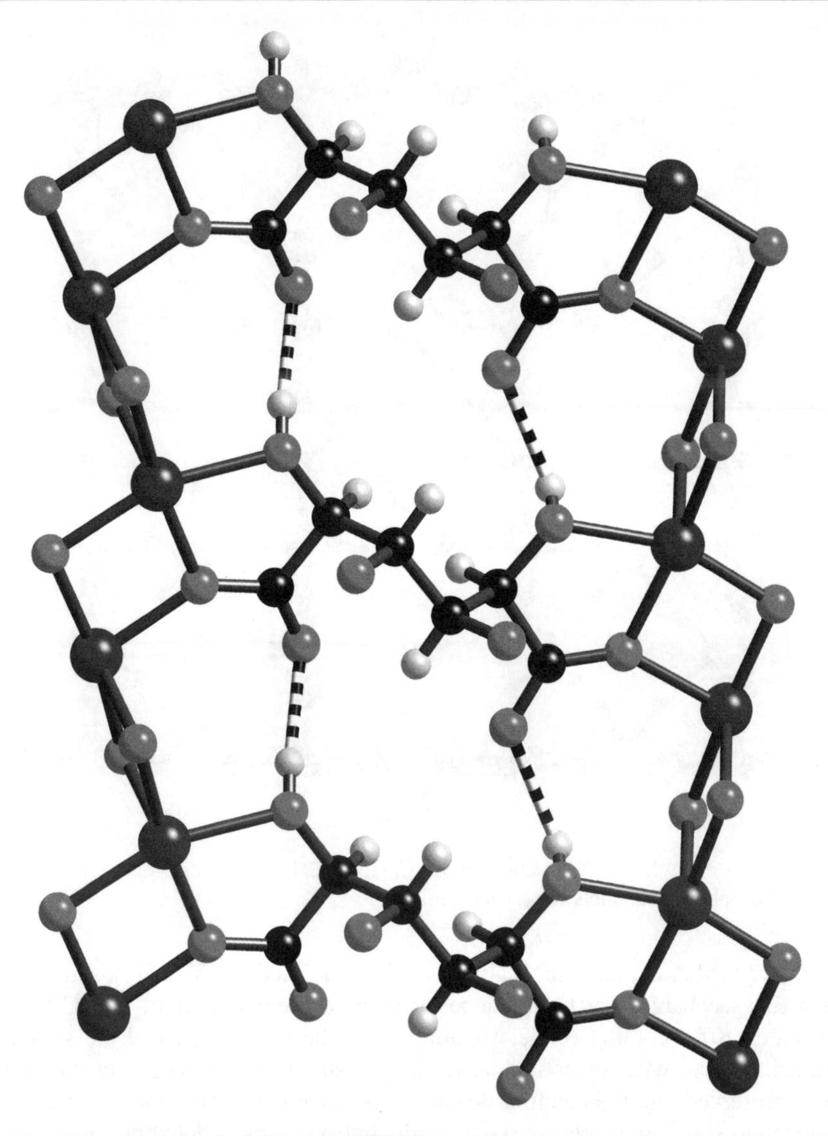

Figure 14. The structure of $Zn(C_6O_8H_8){\cdot}2H_2O$ showing part of an "impenetrable wall" with hydrogen bonds (striped connections) between saccharate ligands. Note that all the central hydroxyl groups are all on the front face of the wall, making the front surface hydrophilic.

the "hydrophobic" channels also contain water (in fact, three times as much), the molecules are highly disordered and as a consequence could not be modeled. Thermogravimetric analysis of the hydrated crystals indicated the loss of water in two steps. The first step begins at room temperature and ceases abruptly at 60 °C, while the second step begins at ∼100 °C and concludes at ∼150 °C. The first step may be attributed to the loss of loosely held water in the hydrophobic channels while the second step is consistent with the loss of the tightly bound water in the hydrophilic channels. Major decomposition of the network begins at 270 °C.

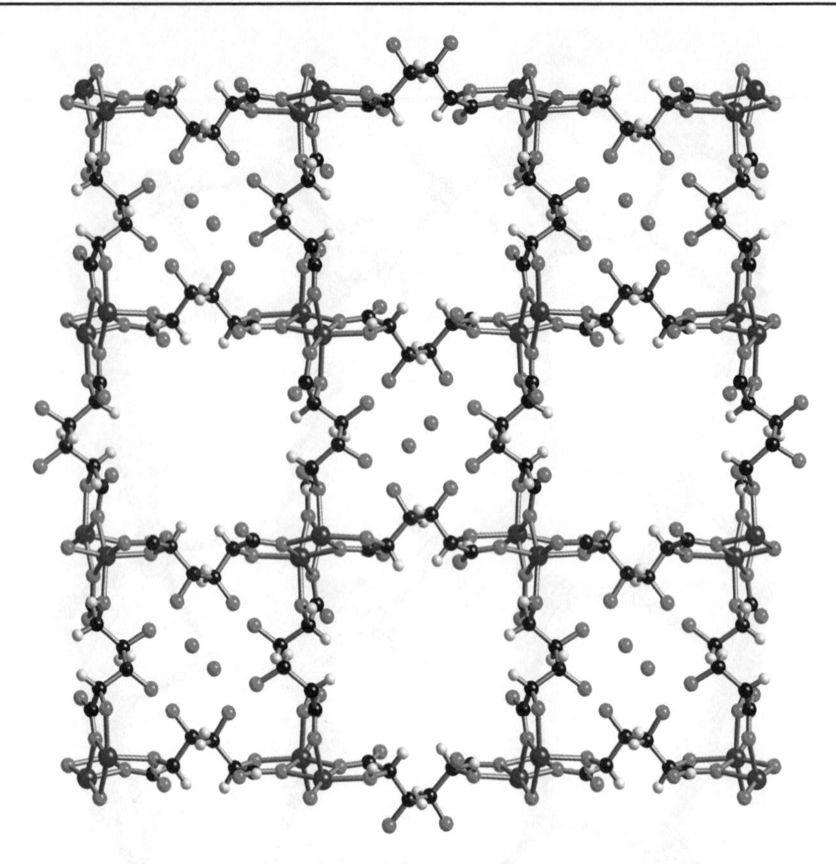

Figure 15. The extended 3-D structure of $Zn(C_6H_8O_8){\cdot}2H_2O$. The disordered water molecules within the hydrophobic channels have been omitted.

In order to further examine the stability of the coordination polymer, a single crystal of hydrated zinc saccharate was heated under vacuum to a temperature of 130 °C for a period of two hours. After cooling (under vacuum), the tube was sealed and the structure was determined. The x-ray analysis indicated that the single-crystal nature of the compound had been preserved and the electron density in the channels where the water was located had dropped to very low levels consistent with almost complete dehydration of the crystal. Without the water in the hydrophilic channels, the ligand atoms show much greater thermal motion. When the same crystal is reexposed to the atmosphere, the crystal becomes hydrated again and further x-ray analysis indicated that the crystal had returned to its original condition.

Along with the thermogravimetric analysis, the above diffraction experiments provide an indication of the stability of this network. We consider that the interligand hydrogen bonding, which supports the channel walls, plays an important role, not only in the stability of the network but also in the assembly of the coordination polymer.

The presence of what appeared to be very open channels prompted an investigation of the ability of the crystals to absorb guest molecules. After exposure to iodine vapor at room temperature, the crystals are able to absorb 2.6 molecules of I_2 per unit cell. X-ray analysis reveals that the single-crystal character is retained during the absorption of

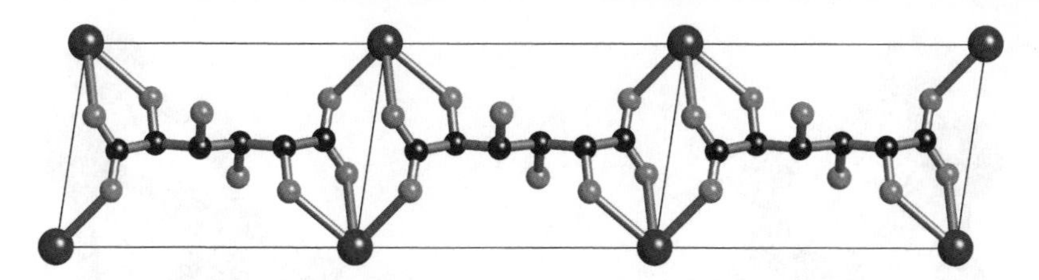

Figure 16. Three edge-sharing "panels" that form part of a wall that separates hexagonal channels in $Ln_2(muc)_3 \cdot 8H_2O$ (Ln = La, Ce, Pr and Nd).

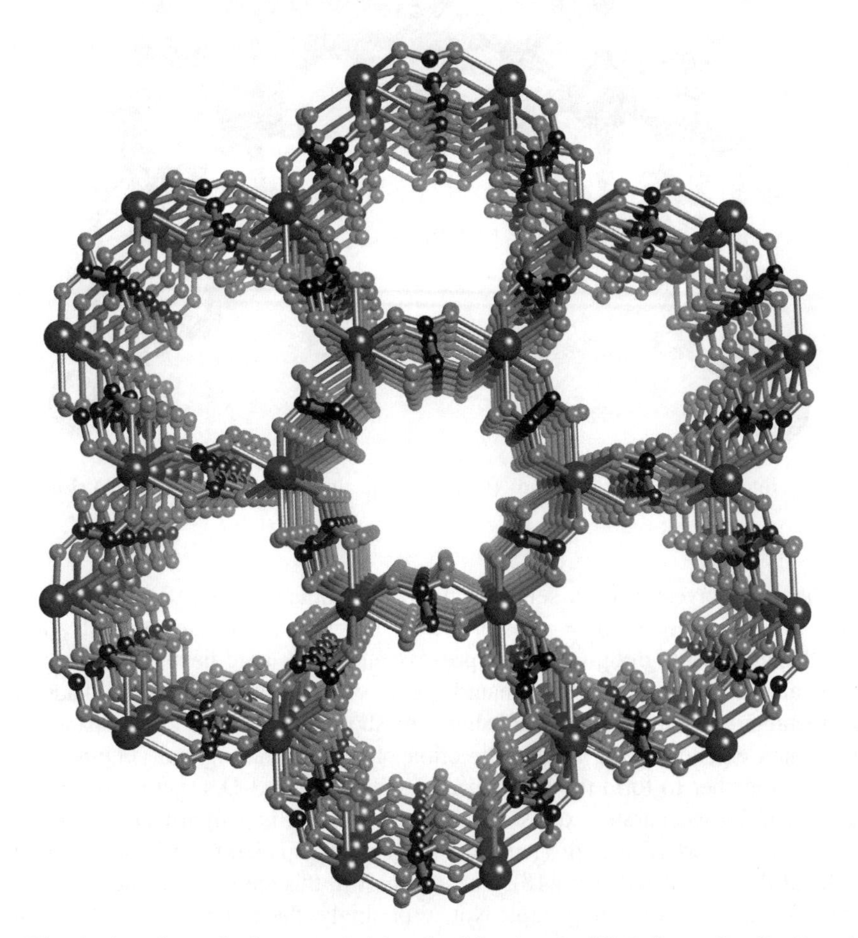

Figure 17. A view down the hexagonal channels of $Ln_2(muc)_3 \cdot 8H_2O$ (Ln = La, Ce, Pr and Nd).

the iodine into the hydrophobic channels. Preliminary investigations indicate that a range of guests, including hydrocarbons, elemental sulfur, CCl_4 and CI_4 may be successfully absorbed into the hydrophobic channels with retention of the single-crystal character.

As part of an investigation of other metal saccharate species, it was found that hydrated cobalt saccharate is isostructural with the zinc compound.

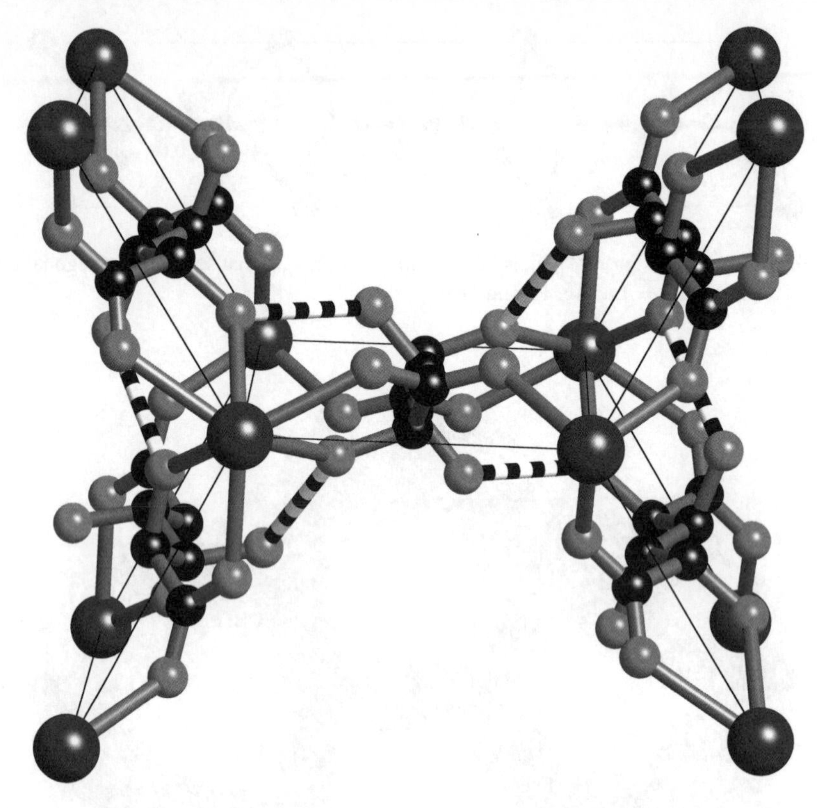

Figure 18. Interligand hydrogen bonds (striped connections) in $Ln_2(muc)_3{\cdot}8H_2O$ (Ln = La, Ce, Pr and Nd).

We have also investigated a series of hydrated lanthanide mucate structures of composition, $Ln_2muc_3{\cdot}xH_2O$ (Ln = La, Ce, Pr and Nd; muc = mucate (**4**), $C_6H_8O_8{}^{2-}$) that adopt a trigonal space group ($P\overline{3}$) and possess relatively large hexagonal channels [16]. Each mucate ligand links to four lanthanide centers to form what may be considered as an almost rectangular "panel". The panel shares its short edges with coplanar panels to form a wall, which extends in the unique direction of the trigonal crystal (Figure 16). These walls come together to form the hexagonal channels of the 3-D structure (Figure 17). As in the case of the saccharate, extensive hydrogen bonding is a significant feature of these structures, with interligand hydrogen-bonding sealing gaps between ligands, which form the walls of the channels (Figure 18). The channels in this case are identical and are lined with hydroxy and carboxylate groups. Not surprisingly, the channels are filled with water molecules, which are disordered.

5. ANIONIC METAL-CARBONATE NETWORKS

In recent times, we have become interested in forming coordination polymers by linking metal centers with relatively simple anions such as carbonate [17, 18]. We anticipated that, by bringing metal centers into relatively close contact within a 3-D network, it may be possible to produce materials with unusual electronic and magnetic properties. Metal-carbonate structures have a long history and so the generation of new classes of carbonate structures represents a significant challenge. Our experience with coordination polymers in which hydrogen bonding is an important feature provides encouragement that it may be possible to exploit hydrogen-bonding interactions in order to generate new types of materials.

We considered the guanidinium cation (**5**), with its trigonal symmetry and sextet of equivalent N–H protons, to have the potential to act as a powerful structure-directing cation for anionic metal-carbonate networks. The addition of $Cu(NO_3)_2$ to an aqueous solution of $KHCO_3$ and K_2CO_3 yields a deep-blue solution. The treatment of this solution with excess guanidinium results in the separation of royal-blue crystals of formula $[C(NH_2)_3]_2[Cu(CO_3)_2]$ [17]. In this structure, carbonate anions bridge four-coordinate copper centers to produce a 3-D network that has the same topology as diamond. An adamantane-type unit within this coordination network is represented in Figure 19. Guanidinium cations lie within $Cu_6(CO_3)_6$ cyclohexane-type rings, secured by hydrogen bonds to carbonate oxygen atoms (Figure 20).

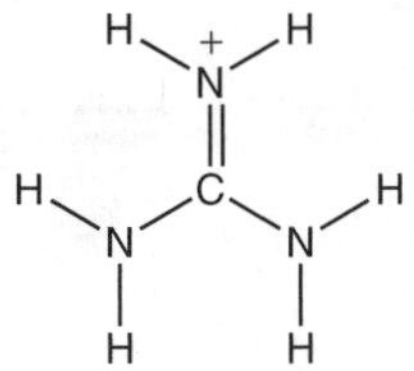

5

In the synthesis of $[C(NH_2)_3]_2[Cu(CO_3)_2]$, some attempts (involving slightly less guanidinium) yielded small quantities of a pale-blue crystalline material. A crystal structure analysis revealed a copper carbonate coordination polymer, but with a very different topology to the diamond-type net. The connectivity in this case was the same as that for the sodalite net, which is represented in Figure 21. The sodalite net, which may be represented by the Schlafli symbol, 4^26^4, is composed of four-connected nodes, which link together in a way that forms four-membered and six-membered rings. In the anionic copper carbonate network, the four-connecting nodes are represented by Cu(II) ions and the carbonates act as symmetrical two-connecting bridging ligands between Cu(II) centers that are in square-planar coordination environments. The overall composition of the crystal is $K_4[C(NH_2)_3]_8[Cu_6(CO_3)_{12}]$. On first inspection, it is somewhat surprising to discover that the replacement of only one-third of the guanidinium ions in $[C(NH_2)_3]_2[Cu(CO_3)_2]$ with potassium ions yields a very different type of copper carbonate network. This is even more surprising, given that the anionic copper carbonate network of $K_2[Cu(CO_3)_2]$ [19] is also a diamond network, although geometrically very different to that found for $[C(NH_2)_3]_2[Cu(CO_3)_2]$.

The crystal structure reveals that both the potassium and guanidinium ions play important roles in influencing the network topology. Guanidinium ions, located on either side of a $\bar{3}m$ site, lie within $Cu_6(CO_3)_6$ rings that form the "hexagonal" windows of the sodalite cage. As indicated in Figure 22, the size and shape of the pair of guanidinium ions neatly match the conformation of the $Cu_6(CO_3)_6$ rings, with 12 equivalent N–H$\cdots$O hydrogen bonds extending from the cations to the coordinated oxygen atoms of the bridging carbonates. The square windows of the sodalite net, which represent $Cu_4(CO_3)_4$ rings, are

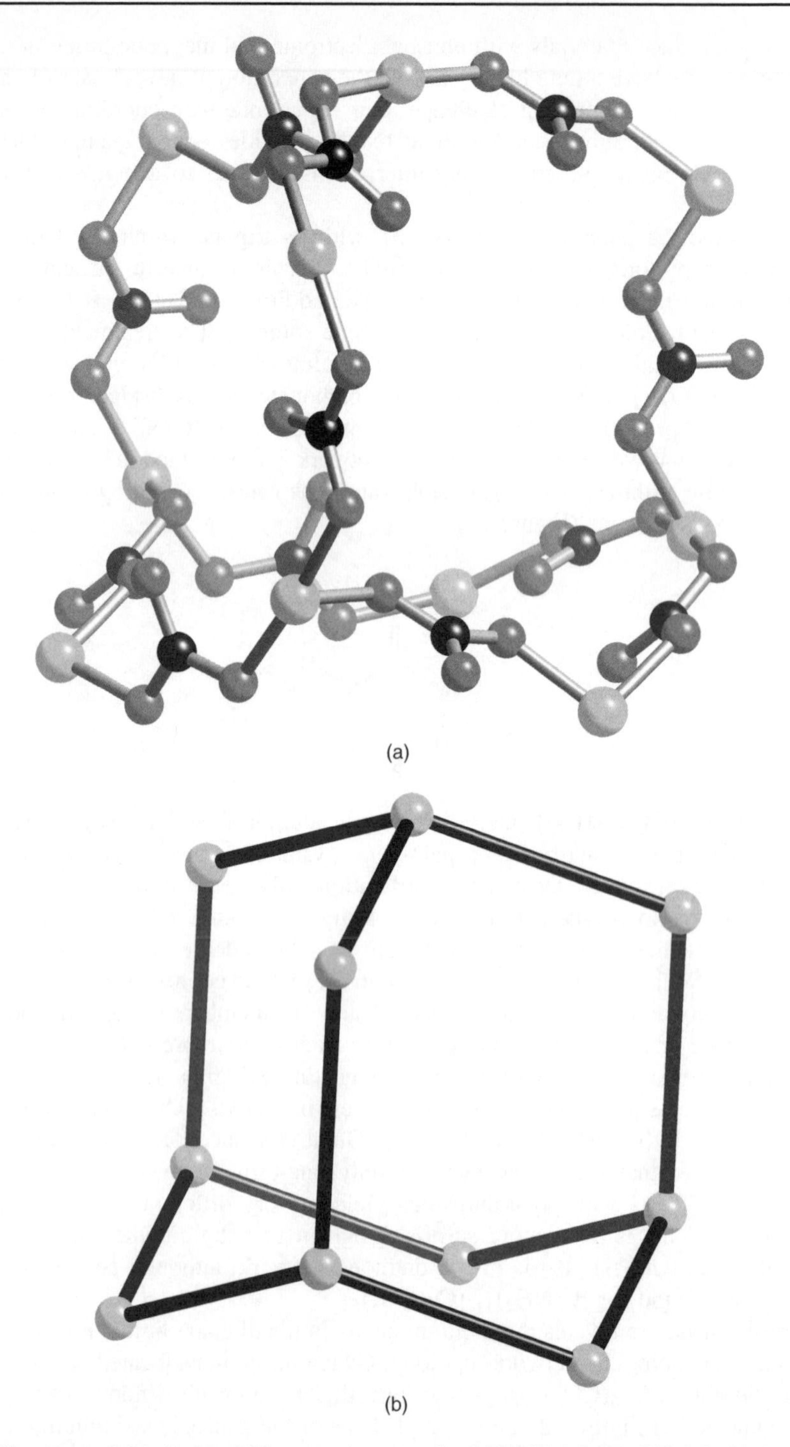

Figure 19. Part of the diamond-like anionic network of $[C(NH_2)_3]_2[Cu(CO_3)_2]$. (a) An adamantane-type unit with bridging carbonate ions included and (b) the adamantane-type unit with the carbonate bridge represented by a single rod.

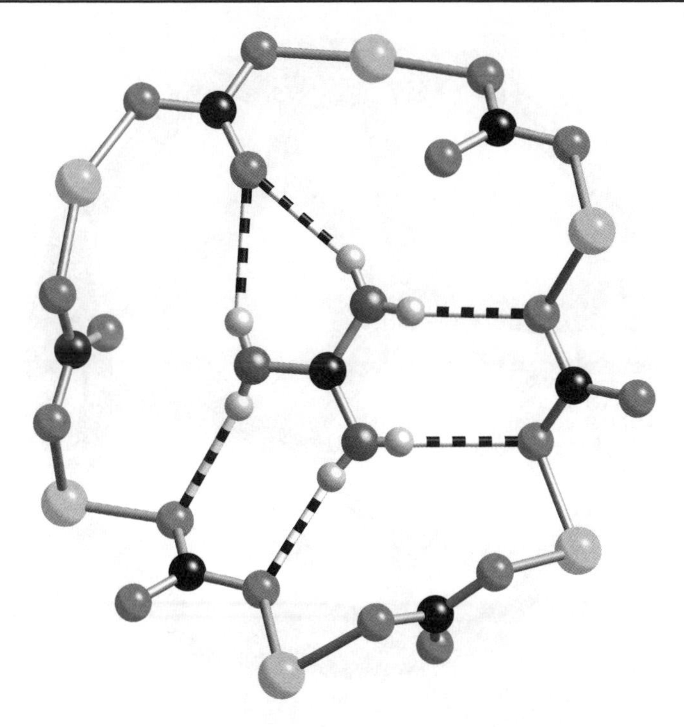

Figure 20. A guanidinium cation hydrogen bonded to a $Cu_6(CO_3)_6$ ring within the diamond-like network of $[C(NH_2)_3]_2[Cu(CO_3)_2]$.

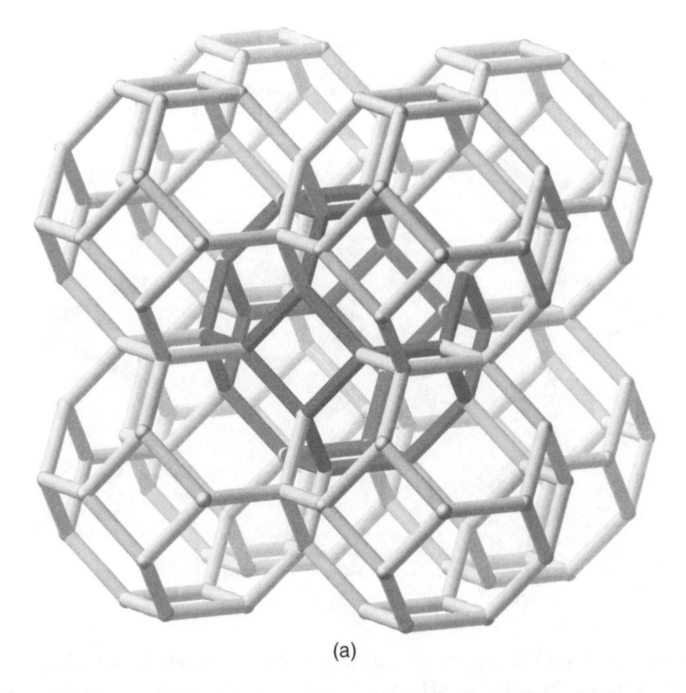

Figure 21. The sodalite network. (a) A representation of nine cages within the sodalite net; the central cage is highlighted and (b) a single sodalite cage.

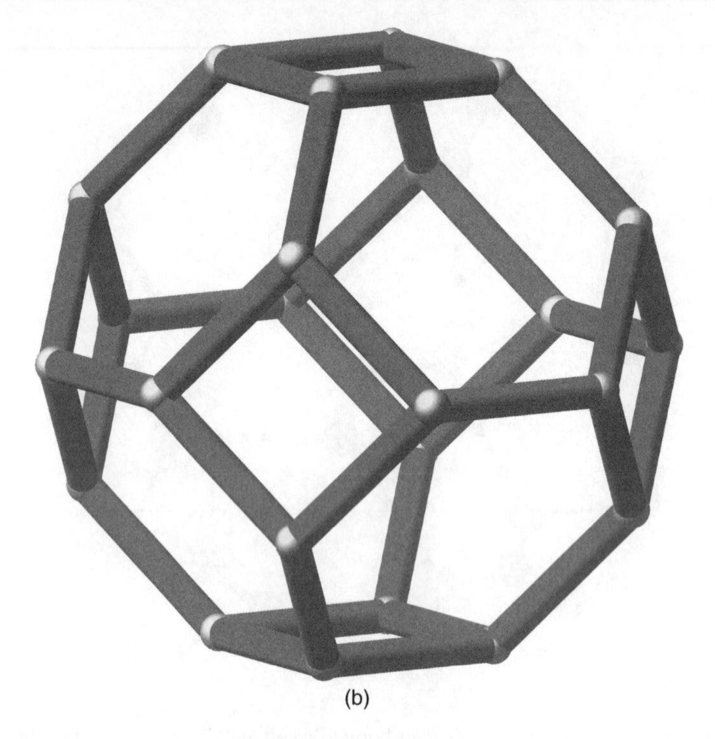

Figure 21. *(continued)*

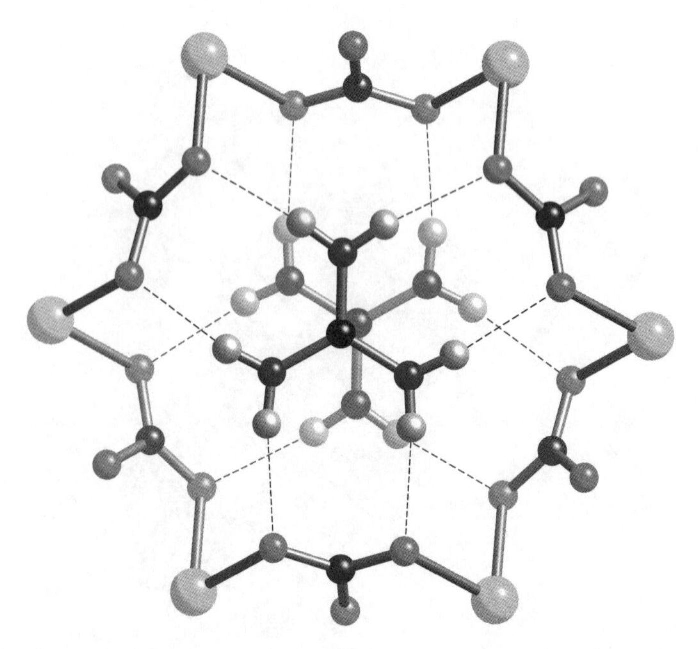

Figure 22. A pair of guanidinium cations forming hydrogen bonds (dashed lines) to a $Cu_6(CO_3)_6$ ring within $K_4[C(NH_2)_3]_8Cu_6(CO_3)_{12}\cdot 8H_2O$.

considerably smaller. The noncoordinated oxygen atoms of the carbonates are directed toward the center of this window and form a square with a diagonal, O···O separation of 4.16 Å. The K^+ ions are associated with these O_4 "holes" but are too large to lie within the plane. The K^+ ions are located away from the O_4 plane and sit equidistant from all four oxygen atoms (K···O 2.7 Å). Two-thirds of the O_4 holes are associated with one K^+ ion, which is disordered across both sides of the window. The remaining third have K^+ ions located on either side. From a crystallographic perspective, all windows are identical, and thus, if one considers a single sodalite cage, the four internal K^+ ions are disordered over six positions. Figure 23 shows four of the six positions occupied by K^+ ions. The sodalite-type cages, which are considerably larger, and in fact more symmetrical than their aluminosilicate counterparts, are able to accommodate eight disordered water molecules.

The mismatch between the size of the O_4 holes and the K^+ ion prompted an investigation of whether smaller cations such as Li^+ may be able to sit neatly within the O_4 planes. If every square window is able to accommodate one Li^+ ion, then, an extra cation per cage would be required. We considered the symmetrical tetramethylammonium ion to be of an appropriate size and shape to sit at the heart of each sodalite cage and provide the balance of the positive charge for the anionic network. The combination of aqueous solutions of NMe_4HCO_3, $(NMe_4)_2CO_3$ (in excess), $Cu(NO_3)_2$, $LiNO_3$ and $C(NH_2)_3NO_3$ at room temperature yielded the anticipated 3-D structure, $Li_3[N(CH_3)_4][C(NH_2)_3]_8[Cu_6(CO_3)_{12}]\cdot 5H_2O$, in a yield of

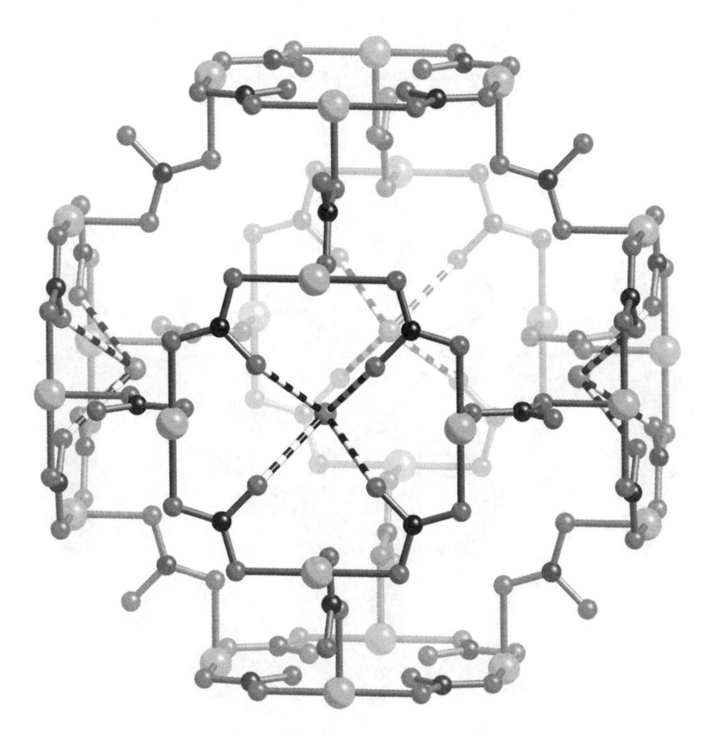

Figure 23. A sodalite-type unit in the structure of $K_4[C(NH_2)_3]_8Cu_6(CO_3)_{12}\cdot 8H_2O$. Four potassium ions are disordered over six sites within the cage; only one arrangement is shown. The striped connections represent O–K interactions. For clarity, the water molecules and the guanidinium ions have been omitted.

38% – a value much greater than the almost insignificant amount generated from the reaction that produced $K_4[C(NH_2)_3]_8[Cu_6(CO_3)_{12}]\cdot 8H_2O$. The five water molecules in $Li_3[N(CH_3)_4][C(NH_2)_3]_8[Cu_6(CO_3)_{12}]\cdot 5H_2O$ are disordered over six sites, with each water molecule located at 2.01 Å from the LiO_4 plane. The failure to incorporate six water molecules in each cage is a consequence of the $N(CH_3)_4^+$ pointing one of its methyl groups directly toward only one of the LiO_4 units and, in so doing, leaving insufficient room to accommodate a water molecule between the lithium and the methyl group. We consider that the synthesis of the intended compound from five components under nonforcing conditions represents an example of true crystal engineering.

Subsequently, we found that it is also possible to generate $Li_3[N(CH_3)_4][C(NH_2)_3]_8$ $[Cu_6(CO_3)_{12}]\cdot 5H_2O$ using potassium salts of carbonate and bicarbonate, even though the concentration of K^+ in the aqueous solution is considerably greater than that of Li^+. This indicates a clear preference for the incorporation of Li^+. All attempts to generate a sodalite cage using a $K^+/N(CH_3)_4^+$ mixture were unsuccessful, suggesting that there is not enough room to simultaneously accommodate the out-of-plane K^+ and the $N(CH_3)_4^+$ ions.

The successful inclusion of Li^+ into the O_4 holes raised the possibility of using Na^+ as the "square-window" cation. The intended compound, of composition $Na_3[N(CH_3)_4]$ $[C(NH_2)_3]_8[Cu_6(CO_3)_{12}]\cdot$hydrate, was formed in good yield. The structure is presented in Figure 24. As was the case with lithium, it was also possible to form the compound from solutions that contained K^+ in large excess.

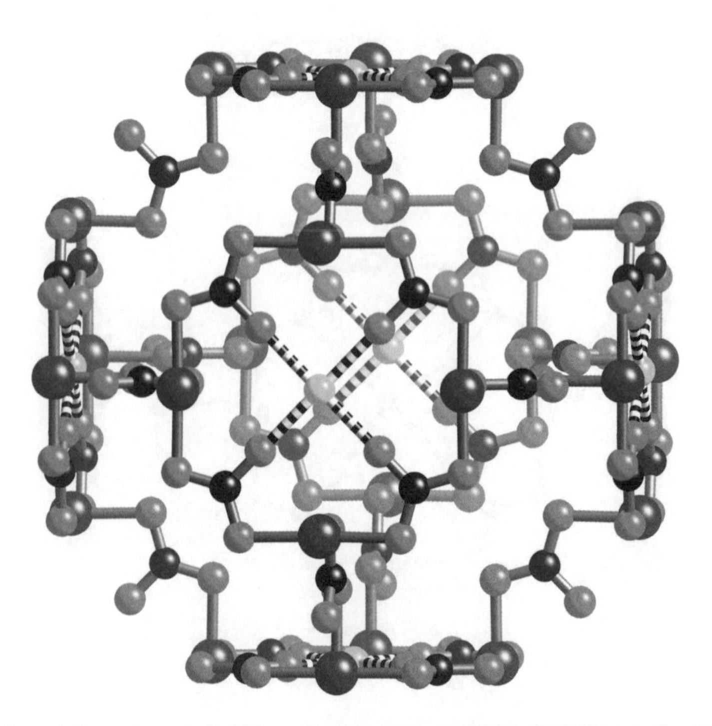

Figure 24. A sodalite-type unit in the structure of $Na_3[N(CH_3)_4][C(NH_2)_3]_8Cu_6(CO_3)_{12}\cdot 3H_2O$. Each "square window" is occupied by a Na^+ ion. The striped connections represent O–Na interactions. For clarity, the water molecules, the guanidinium ions and the tetramethylammonium ion have been omitted.

This work was extended to investigate whether other divalent metal centers could fulfill the role of Cu(II) in the sodalite network [18]. Anionic metal-carbonate networks were generated using the following divalent metals: magnesium, calcium, manganese, iron, cobalt, nickel, zinc and cadmium. Although all the compounds share the same topology for the anionic network, from a geometrical perspective, the compounds fall into two distinct classes. The division into these classes is related to the coordination preferences of the divalent metal, which influences whether the crystal adopts a body-centered cubic structure or a face-centered cubic structure.

5.1. Body-centered Cubic Structures

The compounds $Na_3[N(CH_3)_4][C(NH_2)_3]_8[Mn_6(CO_3)_{12}]\cdot 3H_2O$ and $Na_3[N(CH_3)_4][C(NH_2)_3]_8[Cd_6(CO_3)_{12}]\cdot 5H_2O$, have very similar structures to $Na_3[N(CH_3)_4][C(NH_2)_3]_8[Cu_6(CO_3)_{12}]\cdot H_2O$. On first inspection, this is somewhat surprising, given that in the case of the copper structure, the Cu(II) center is in a square-planar environment (Figure 25(a)) – a coordination geometry that one does not normally associate with Mn(II) and Cd(II) centers. As indicated in Figure 25(b), the divalent metal ions in the manganese and cadmium

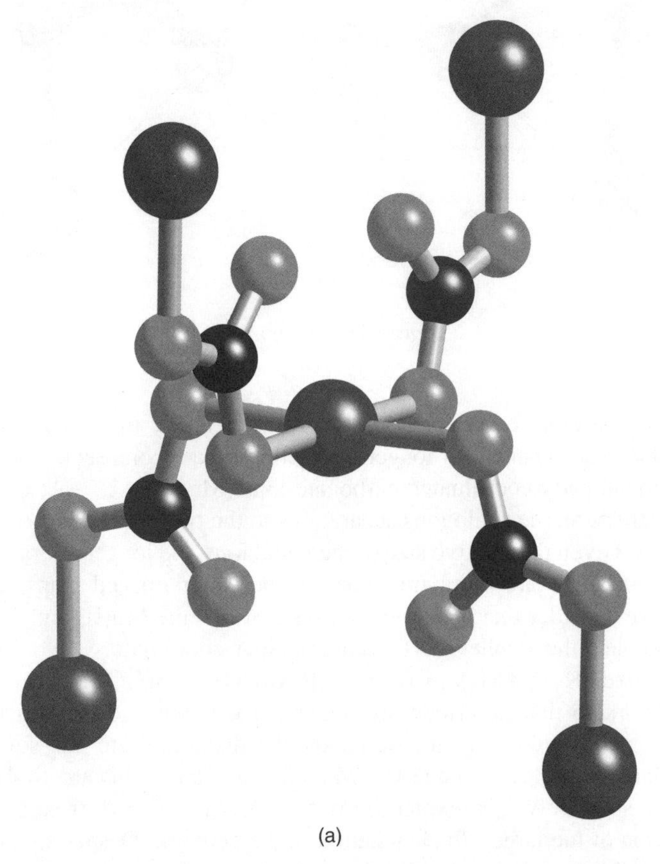

(a)

Figure 25. Coordination environments of (a) copper and (b) manganese metal centers in $Na_3[N(CH_3)_4][C(NH_2)_3]_8Cu_6(CO_3)_{12}\cdot 3H_2O$ and $Na_3[N(CH_3)_4][C(NH_2)_3]_8Mn_6(CO_3)_{12}\cdot 3H_2O$ respectively. The thin lines in (b) represent long Mn–O interactions.

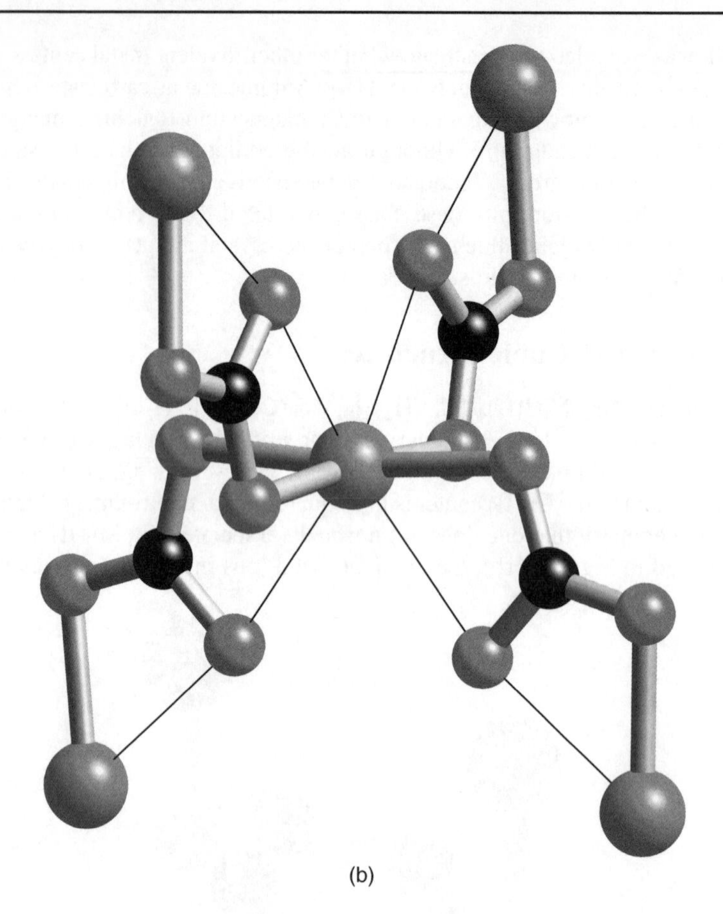

(b)

Figure 25. *(continued)*

structures form four close coordinate bonds to oxygen atoms in a square plane (Mn–O = 2.18 Å; Cd–O = 2.28 Å) and four longer interactions to a second set of four oxygen atoms belonging to the already coordinated carbonate ions (Mn–O = 2.59 Å; Cd–O = 2.62 Å). In the copper structure, the analogous separations to the two sets of four oxygen atoms are 2.04 and 2.58 Å. Given the relative size of the metal ions, the longer interactions at ∼2.6 Å found in each structure are more important in the cadmium and manganese structures. Accordingly, we consider that it may be best to describe the Mn(II) and Cd(II) centers as eight-coordinate and the smaller Cu(II) center as four-coordinate.

The compound $K_3[N(CH_3)_4][C(NH_2)_3]_8[Ca_6(CO_3)_{12}]\cdot 3H_2O$ has a similar metal-carbonate network to that described above for the Cu, Mn and Cd structures. The unit cell lengths (which are equal to the perpendicular distance from one square window to an opposing square window) for the Cu, Mn, Cd and Ca crystals are 14.41, 14.57, 14.75 and 14.82 Å, respectively. The expansion of the sodalite network from Cu to Ca reflects the incorporation of the larger divalent cation in the network. Despite numerous attempts, it was not possible to obtain the Ca framework with Na^+ ions in the O_4 sites. The preferred incorporation of the K^+ ion reflects the generation of larger O_4 holes ($O\cdots O$ diagonal = 5.02 Å) in the Ca^{2+} network. Even though the O_4 holes are larger, the K^+

ions are still unable to sit in the O_4 plane and are disordered over sites 0.93 Å from the O_4 plane. The intrusion of the K^+ ion into the sodalite cage reduces the space available for water molecules and as a consequence there are fewer water molecules in the unit cell of the Ca/K crystal compared to the smaller Cd/Na unit cell.

5.2. Face-centered Cubic Structures

Compounds of formula, $Na_3[N(CH_3)_4][C(NH_2)_3]_8[M_6(CO_3)_{12}]\cdot 2H_2O$ (M = Mg, Fe, Co, Ni and Zn) represent a second class of sodalite-type networks, in which the divalent metal is unambiguously in a six-coordinate environment. The unit cells for these compounds are face-centered cubic with cell lengths twice as long as that found for the body-centered structures. The reason for the increase in cell volume has its origins in the fact that for these compounds the metal has a clear preference for six-coordination over the four-coordinate and eight-coordinate environments found in the first class. The six-coordinate environment is provided by six oxygen atoms from four carbonate anions, each of which has become an asymmetric bridge (Figure 26). Two of the carbonates now chelate the metal center, while the other two are bound through a single oxygen atom.

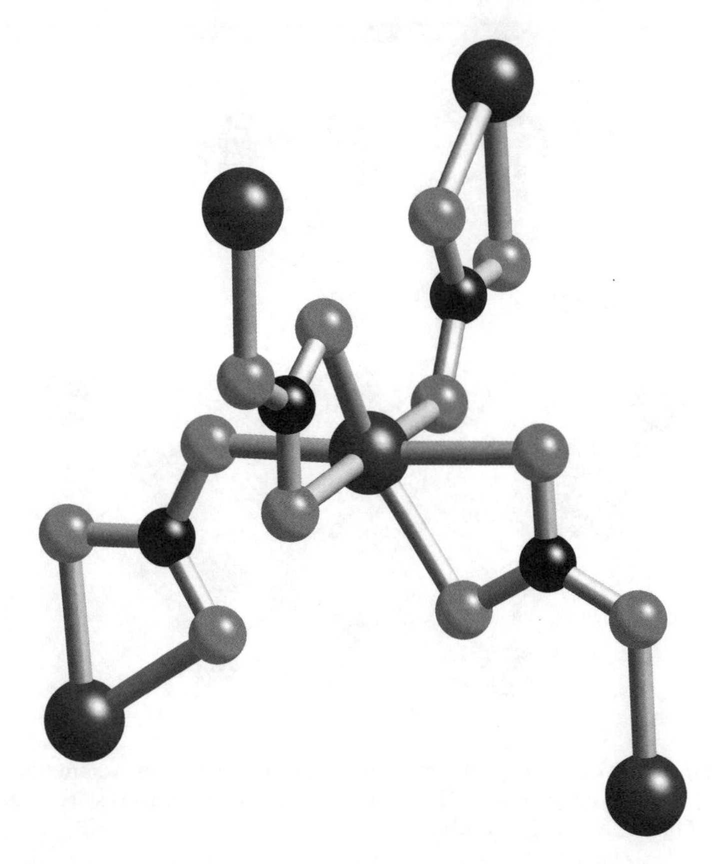

Figure 26. The coordination environment of the cobalt(II) metal center in $Na_3[N(CH_3)_4][C(NH_2)_3]_8Co_6(CO_3)_{12}\cdot 2H_2O$. Similar coordination environments are found for the corresponding Mg, Fe, Ni and Zn structures.

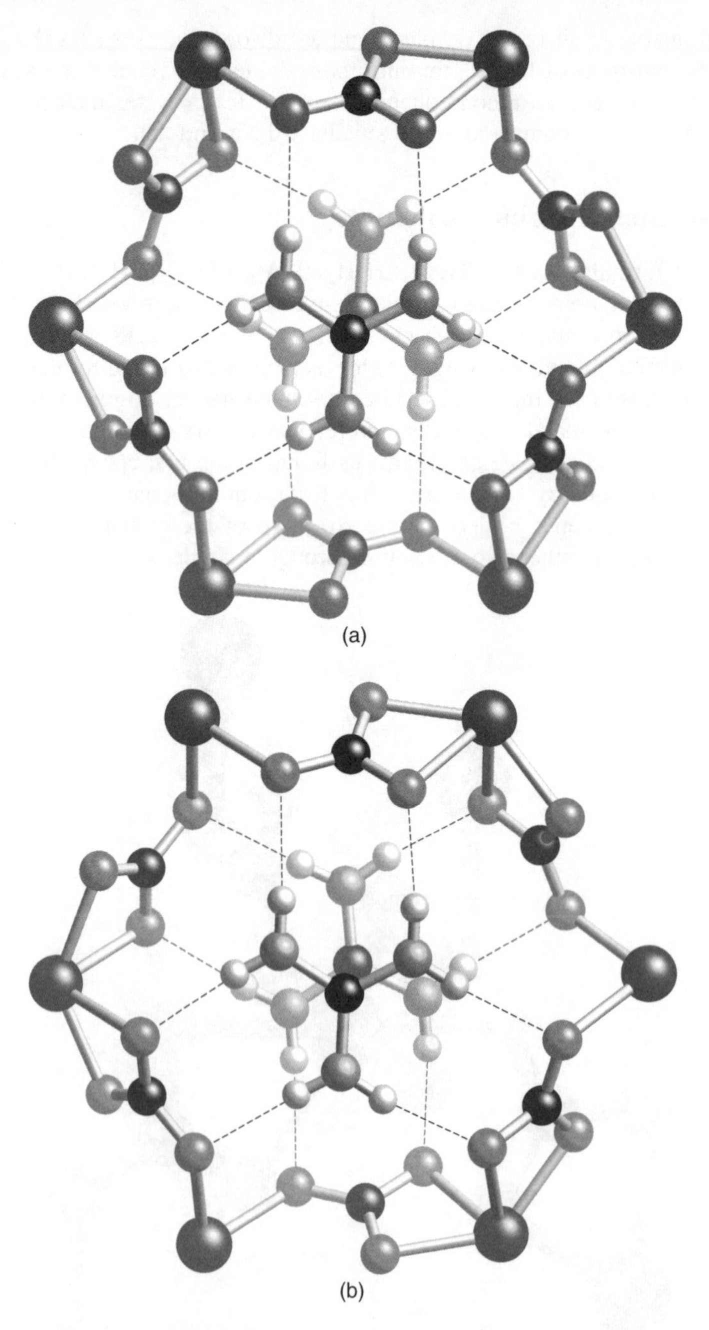

Figure 27. (a) and (b) are the two types of $Co_6(CO_3)_6$ windows within the structure of $Na_3[N(CH_3)_4][C(NH_2)_3]_8Co_6(CO_3)_{12}{\cdot}2H_2O$. Similar $M_6(CO_3)_6$ windows are found for the corresponding Mg, Fe, Ni and Zn structures.

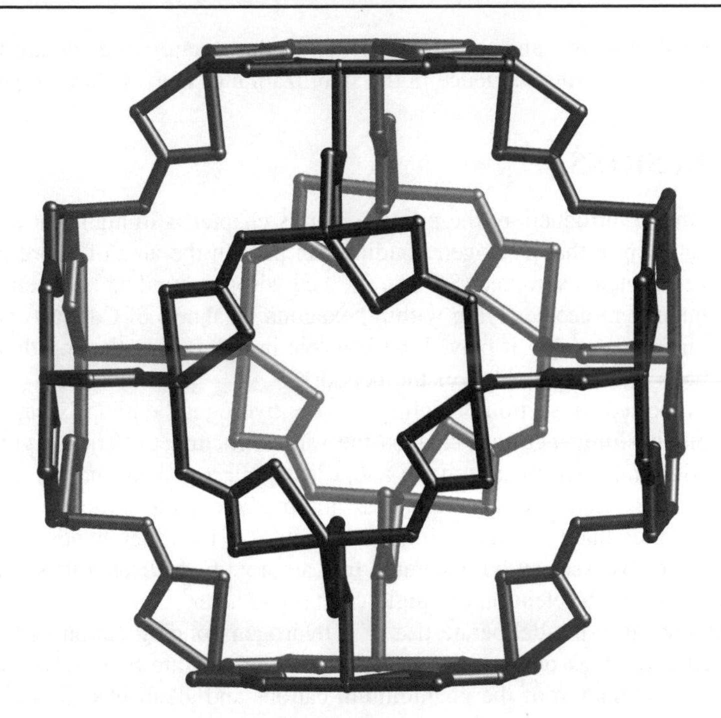

Figure 28. A chiral sodalite-type cage within the structure of $Na_3[N(CH_3)_4][C(NH_2)_3]_8Co_6(CO_3)_{12}\cdot 2H_2O$.

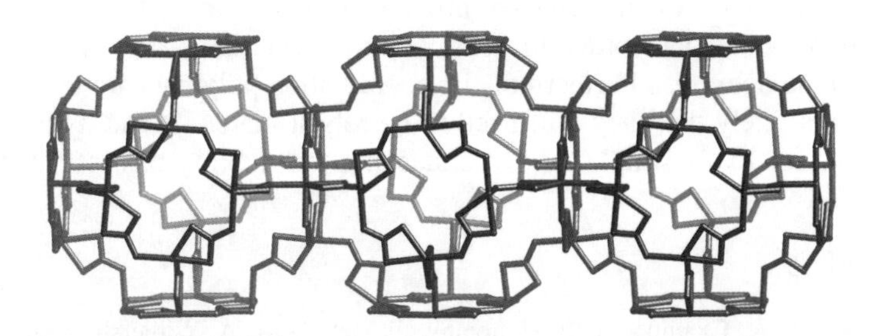

Figure 29. Neighboring sodalite-type cages within the structure of $Na_3[N(CH_3)_4][C(NH_2)_3]_8Co_6(CO_3)_{12}\cdot 2H_2O$. The handedness of the cages alternates within the structure.

In contrast to the body-centered structures, there are two distinct types of hexagonal windows in the sodalite cages of the face-centered structures (Figure 27). Despite the differences in the two types of hexagonal rings and the asymmetric bridging of the carbonates, pairs of guanidinium cations are still able to act as powerful structure-directing cations. In the face-centered structures, the sodalite-type cages are chiral (Figure 28). However, each cage shares a square window with a cage of the opposite hand (Figure 29), which means that the crystals themselves are not enantiomeric.

We consider it quite remarkable that the same structure persists for such a large number of divalent metal centers that have a distinctly different preference for coordination environments. The work provides us with a clear indication that the guanidinium cations in the hexagonal windows and the alkali metals in the square windows are providing a powerful structure-directing influence in the stabilization of these sodalite-type structures.

6. CONCLUSIONS

As indicated in the introduction, the purpose of this chapter is to highlight the important structure-directing role that hydrogen bonding can play in the area of coordination polymers. In the first system examined (Section 2), the hydrogen bonding between coordinated water and *t*-butanol molecules lying within hexagonal channels of $Cd(CN)_2$ network was completely unanticipated, yet it played the key role in influencing the coordination of the cadmium centers and the topology of the network.

The work reported in Section 3 highlights how hydrogen-bonding is capable of providing a stabilizing influence that enables the same structure to persist right across the lanthanide series and force metal ions to adopt less preferred coordination numbers.

While the zinc saccharate structure was not predicted, the structure determination clearly demonstrated that by encouraging the formation of hydrogen bonds, novel and particularly robust networks may be generated that are able to withstand loss and exchange of guest molecules with retention of single-crystal character.

In the final section, the deliberate use of a hydrogen-bonding cation in the assembly of anionic networks was demonstrated. Although, the sodalite-type networks were not predicted, the combination of the guanidinium cations and alkali metals provided access to previously unknown metal-carbonate networks. By recognizing the role each of these cations was playing in the structure, it was possible to "engineer" new materials in superior yields to those originally obtained in a serendipitous manner.

The results described in this chapter provide encouragement that hydrogen-bonded interactions may be used in order to exert control over the assembly of new generations of crystalline networks. Furthermore, if coordination polymers are to find real-world applications, it is likely that they will need to be robust materials and hydrogen bonding may represent a relatively simple way of enhancing the stability of useful networks.

REFERENCES

1. (a) A. Blake, N. R. Champness, P. Huttersby, W.-S. Li, M. A. Withersby and M. Schroder, *Coord. Chem. Rev.*, **183**, 117–138 (1999); (b) M. Eddaoudi, D. B. Moler, H. Li, B. Chen, T. M. Reineke, M. O'Keeffe and O. M. Yaghi, *Acc. Chem. Res.*, **34**, 319–330 (2001); (c) O. R. Evans, and W. Lin, *Acc. Chem. Res.*, **35**, 511–522 (2002); (d) P. J. Hagrman, D. Hagrman and J. Zubieta, *Angew. Chem., Int. Ed. Engl.*, **38**, 2638–2684 (1999); *Angew. Chem.*, **111**, 2798–2848 (1999); (e) A. N. Khlobystov, A. J. Blake, N. R. Champness, D. A. Lemenovskii, A. G. Majouga, N. V. Zyk and M. Schroder, *Coord. Chem. Rev.*, **222**, 155–192 (2001); (f) S. Kitagawa and S. Kawata, *Coord. Chem. Rev.*, **224**, 11–34 (2002); (g) R. Robson, *J. Chem. Soc., Dalton Trans.*, 3735–3744 (2000); (h) K. Kim, *Chem. Soc. Rev.*, **31**, 96–107 (2002); (i) C. Janiak, *J. Chem. Soc., Dalton Trans.*, 2781–2804 (2003); (j) B. Moulton and M. J. Zaworotko, *Chem. Soc. Rev.*, **101**, 1629–1658, (2001).
2. S. Kitagawa, R. Kitaura and S. Noro, *Angew. Chem., Int. Ed. Engl.*, **43**, 2334–2375 (2004).
3. (a) B. F. Hoskins and R. Robson, *J. Am. Chem. Soc.*, **111**, 5962–5964, (1989); (b) B. F. Hoskins and R. Robson, *J. Am. Chem. Soc.*, **112**, 1546–1554, (1990); (c) R. Robson, B. F. Abrahams, S. R. Batten, R. W. Gable, B. F. Hoskins and J. Liu, In *Supramolecular Architecture, ACS Symposium Series Vol. 499* (Ed. T. Bein), American Chemical Society, Washington,

1992; (d) N. L. Rosi, J. Kim, M. Eddaoudi, B. Chen, M. O'Keeffe and O. M. Yaghi, *J. Am. Chem. Soc.*, **127**, 1504–1518 (2005).

4. A. F. Wells, *Structural Inorganic Chemistry*, 5th ed., Oxford University Press, Oxford, 121, 913, 1984, In this book Wells refers to Lonsdaleite as the hexagonal form of diamond.
5. A. F. Wells, *Structural Inorganic Chemistry*, 5th ed., Oxford University Press, Oxford, 660, 1984.
6. S. R. Batten and R. Robson, *Angew. Chem., Int. Ed. Engl.*, **37**, 1460–1494 (1998).
7. J. D. Watson and F. H. C. Crick, *Nature*, **171**, 737–738 (1953).
8. B. F. Abrahams, B. F. Hoskins and R. Robson, *J. Chem. Soc., Chem. Commun.*, 60–61 (1990).
9. B. F. Abrahams, B. F. Hoskins, J. Liu and R. Robson, *J. Am. Chem. Soc.*, **113**, 3045–3051 (1991).
10. B. F. Abrahams, B. F. Hoskins, R. Robson, Y.-H. Lam, F. Separovic and P. Woodberry, *J. Solid State Chem.*, **156**, 51–56 (2001).
11. B. F. Abrahams, M. J. Hardie, B. F. Hoskins, R. Robson and G. A. Williams, *J. Am. Chem. Soc*, **114**, 10641–10643 (1992).
12. T. Kitazawa, S. Nishikiori, R. Kuroda and T. Iwamoto, *Chem. Lett.*, 1729–1732 (1988).
13. B. F. Abrahams, J. Coleiro, B. F. Hoskins and R. Robson, *J. Chem. Soc., Chem. Commun.*, 603–604 (1996).
14. B. F. Abrahams, J. Coleiro, K. Ha, B. F. Hoskins, S. D. Orchard and R. Robson, *J. Chem. Soc., Dalton Trans.*, 1586–1594 (2002).
15. B. F. Abrahams, M. Moylan, S. D. Orchard and R. Robson, *Angew Chem., Int. Ed. Engl.*, **42**, 1848–1851 (2003).
16. B. F. Abrahams, M. D. Moylan, S. D. Orchard and R. Robson, *CrystEngComm*, 313–317 (2003).
17. B. F. Abrahams, M. G. Haywood, R. Robson and D. A. Slizys, *Angew. Chem., Int. Ed. Engl.*, **42**, 1112–1115 (2003).
18. B. F. Abrahams, A. Hawley, M. G. Haywood, T. A. Hudson, R. Robson and D. A. Slizys, *J. Am. Chem. Soc.*, **126**, 2894–2904 (2004).
19. A. Farrand, A. K. Gregson, B. W. Skelton and A. H. White, *Aust. J. Chem.*, **33**, 431–434 (1980).

12
Hydrogen-bonded Coordination Polymeric Structures

JAGADESE J. VITTAL
Department of Chemistry, National University of Singapore, Singapore.

1. INTRODUCTION

Assembly of molecules is mainly achieved through non-covalent interactions in organic crystal engineering [1] while inorganic crystal engineering has an additional tool to assemble molecules, namely, through coordination bonds [2]. Despite the fact that these coordinate bonds are much stronger than non-covalent interactions, they can easily be broken and reassembled in solution, if the metal–ligand combination used is highly labile [3]. Further optical, magnetic and catalytic properties can be accessible by incorporating metals in the molecular assembly [4]. However, a major disadvantage is the lack of predictability of the polymeric network connectivity [5] since topology has been found to depend on several factors like coordination geometry and oxidation state of the metal, metal to ligand ratio, nature of the ligands used, presence of solvents and counter ions, and the experimental conditions employed for the crystal growth [6].

Hydrogen-bonded metal complexes or coordination polymers have the advantages of both coordination bonds and, perhaps, more predictable hydrogen bonds in constructing supramolecular structures [7]. Furthermore, the coordination polymers can be aligned in a particular fashion to form interesting network structures by incorporating hydrogen donors and acceptors at the backbone of the ligands. Lattice water and aqua ligands can also be used for this purpose. It may be noted that considerable success has been accomplished in aligning reactive functional groups, such as double bonds, in small molecules using the directional properties of the hydrogen bonds [8]. In this context, we have used flexible multidentate ligands as well as spacer ligands in the construction of hydrogen-bonded metal-coordination polymeric structures. Further, we have investigated the role played by different donors and acceptors present in the functional groups as well as the aqua

Frontiers in Crystal Engineering. Edited by Edward R.T. Tiekink and Jagadese J. Vittal

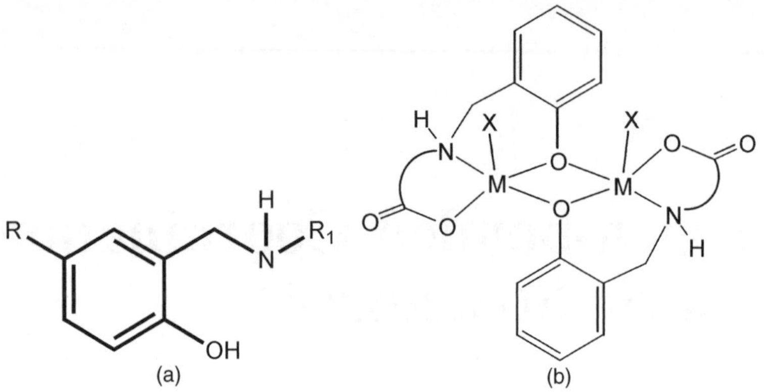

Figure 1. (a) Schematic diagram of the ligands used in this study. R = H, Cl, Me, R_1 = $-CH_2-CO_2H$ (H_2Sgly), $-CH(Me)-CO_2H$ (H_2Sala), $-CH(^iPr)-CO_2H$ (H_2Sval), -cyclopentane-1-carboxylic acid (H_2Scp11), $-CH(CO_2H)-CH_2CH_2CO_2H$ (H_3Sglu) and (b) the structure of a typical dimeric compound, M = Zn(II) or Cu(II), where X is solvent or the oxygen atoms of the neighboring carboxylate group of the ligand.

ligand in directing the formation of supramolecular structures in the solid state. A variety of tridentate ligands, *N*-(2-hydroxybenzyl)-aminoacid, as shown in Figure 1, have been employed to form complexes with Cu(II), Zn(II) and Ni(II) ions, for this purpose. Except for glycine, these aminoacid derivatives have chiral carbon centers. These ligands can easily be obtained by reducing the C=N bond in the Schiff base formed between 2-salicylaldehyde and various aminoacids. These ligands, also known as *Mannich bases*, have the ability to form conformationally flexible six- and five-membered rings on complexation (Figure 1(b)). In addition to donor sites, they have both hydrogen-donor and -acceptor functionalities capable of forming hydrogen bonds.

In the next few sections, the importance of strong N–H···O and O–H···O hydrogen bonds involving aqua ligands, N–H and carboxylate groups in the formation of 3-D network structures and transformation of network structures will be highlighted.

2. SOLID-STATE SUPRAMOLECULAR TRANSFORMATION OF HYDROGEN-BONDED 3-D NETWORK TO 3-D COORDINATION POLYMERIC NETWORK STRUCTURES BY THERMAL DEHYDRATION

The dimer, $[Zn_2(Sala)_2(H_2O)_2]\cdot 2H_2O$, can be synthesized by reacting deprotonated *N*-(2-hydroxybenzyl)-L-alanine (H_2Sala) with $Zn(ClO_4)_2\cdot 6H_2O$ in the presence NaO_2CCH_3 in an equimolar ratio [9]. In the solid-state structure of the Zn(II) complex (shown in Figure 2(a)), two Sala ligands coordinate in the plane of the square pyramidal Zn(II) center through phenolato, carboxylate-O and a secondary amine-N giving a phenolato bridged dimer. The two apical coordination sites are occupied by water. Coordination of the amine-N creates a new chiral center, which means the Sala ligand, when complexed, is a diastereomer with the hydrogen atoms on the amine and Cα adopting a trans arrangement about the C–N bond axis owing to steric interactions. These dimers are further sustained

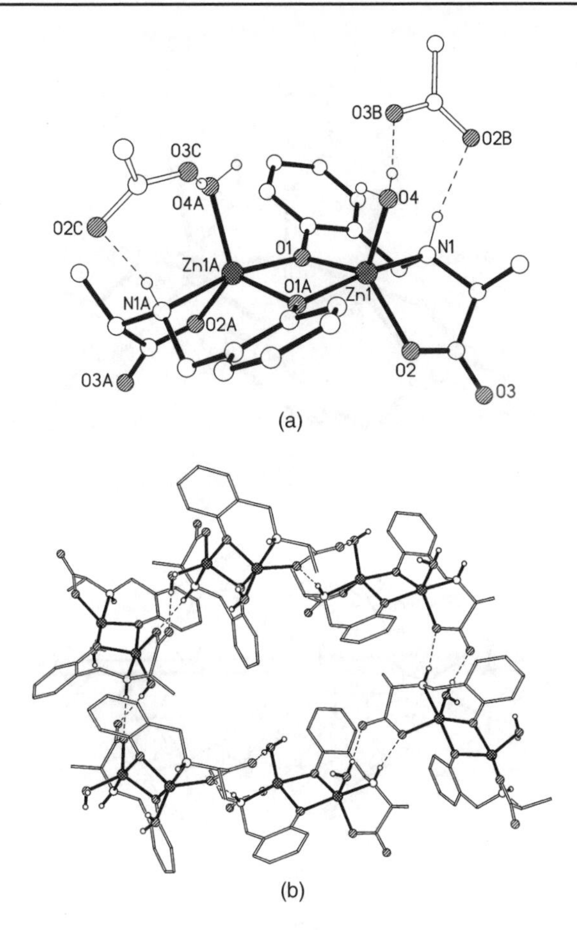

Figure 2. (a) Hydrogen-bonding interactions around the building block, $[Zn_2(Sala)_2(H_2O)_2]$ and (b) a portion of the connectivity in the 3-D coordination network structure of $[Zn_2(Sala)_2(H_2O)_2]$. The C–H hydrogen atoms have been omitted for clarity.

by O–H$\cdots$O and N–H$\cdots$O hydrogen bonding as illustrated in Figure 2(a). The water molecules attached to Zn(II) centers are on the same side, and a twofold crystallographic symmetry is present in the dimer.

The dimers self-assemble to form hydrogen-bonded honeycomblike 3-D network structure with chiral channels all aligned in one direction as shown in Figure 2(b). The close distance between the Zn(II) and oxygen atoms of the neighboring carboxylate group (3.74 Å) through a Zn–$OH_2\cdots$O(carboxylate) interaction suggests that the removal of the aqua ligands by thermal dehydration in the solid state might lead to the formation of new Zn(II)–O(carboxylate) bonds and hence a 3-D coordination network structure. Indeed, these Zn(II) complexes lose aqua ligands and lattice water molecules below 110 °C and convert to $[Zn_2(Sala)_2]$, having a coordination polymeric 3-D network structure as confirmed by x-ray crystallography (Figure 3). It should be noted that the formation of the Zn–O(carboxylate) bond from a Zn–OH_2 bond is supported by the presence of N–H$\cdots$O hydrogen bonds in the lattice. The anhydrous structure is also porous, containing chiral channels as shown in Figure 3c. Such materials are being sought as potential

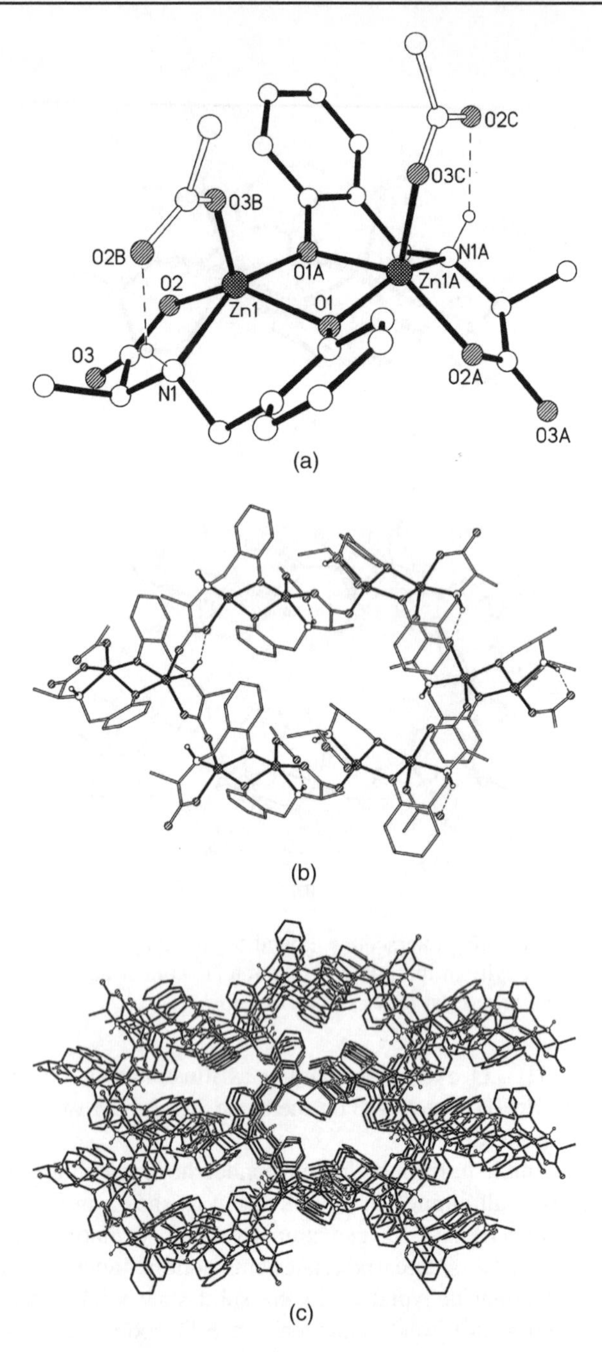

Figure 3. (a) A slice of the honeycomblike cavity in $[Zn_2(Sala)_2]$, and (b) and (c) a portion of the three-dimensional coordination polymeric network structure of $[Zn_2(Sala)_2]$. The C–H hydrogen bonds are omitted for clarity.

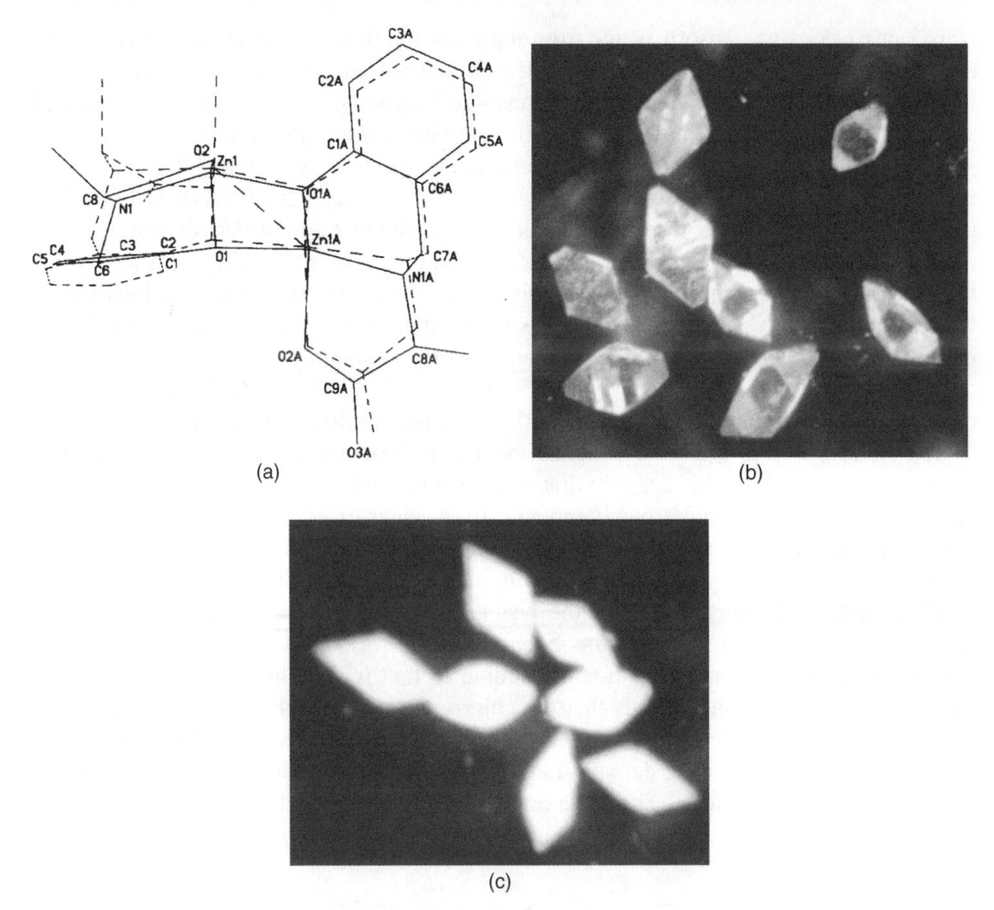

Figure 4. (a) Superimposition of the nonhydrogen atomic coordinates of the two dimers before and after dehydration. Photographs of single crystals of $[Zn_2(Sala)_2(H_2O)_2]\cdot 2H_2O$ before (b) and after (c) thermal dehydration.

chiral catalysts and molecular sieves. There is great potential to investigate such novel supramolecular structures for controlling their cavity size [10].

Both the Zn(II)-Sala compounds are crystallized in the tetragonal space group $P4_32_12$, but they are not isostructural, that is, their atomic coordinates are not superimposable, as shown in Figure 4(a). Hence, when the single crystals of $[Zn_2(Sala)_2(H_2O)_2]\cdot 2H_2O$ were carefully dehydrated thermally, they became opaque, but their single-crystal morphology has been retained, as illustrated in Figures 4(a) and (b). This is due to change of conformation of the dimeric building block in the lattice (opacity) but not their position in the lattice (retention of morphology because of being in the same space group based on structure determination) after the dehydration.

3. INTERCONVERTIBLE SOLID-STATE SUPRAMOLECULAR TRANSFORMATION

The solid-state structural transformation of the hydrogen-bonded 3-D network to a 3-D coordination polymeric network by a thermal dehydration reaction is irreversible and

$[Zn_2(Sala)_2]$ does not absorb water from aqueous solution. However, it is possible that the two supramolecular structures of Zn(II) compounds can be interconverted by substituting the backbone of the ligand by nonreactive functional groups [11]. Such reversible structural transformations may find potential applications as molecular switches.

The methyl or chloro substitution at the backbone of the H_2Sala ligand does not change the hydrogen-bonded 3-D architecture significantly. Hence, $[Zn_2(RSala)_2(H_2O)_2]\cdot 2H_2O$ (R = Cl, Me) compounds were also expected to undergo thermal dehydration reactions accompanied by the formation of new Zn–O bonds from neighboring RSala ligands, which would subsequently lead to the formation of the expected 3-D coordination network structure. However, the DTG curves reveal that only ~90% of the expected water molecules have been removed at 110 °C and that the remaining water has been removed only above ~180 °C, as shown in Figure 5. On the other hand, the Zn(II)-Sala compound could be completely dehydrated below 110 °C and the resultant dehydrated product could be recrystallized from water or aqueous methanol to obtain single crystals suitable for x-ray crystallography. However, hydrated Zn-RSala compounds are regenerated upon recrystallization of the anhydrous compound from water or aqueous MeOH.

Lattice water will normally be lost in the temperature range 50–100 °C and coordinated water molecules can only be removed at a higher temperature, depending on the $M–OH_2$ bond strength [12]. Since ~90% of the water loss occurred at approximately 110 °C, it appears that the driving force for this behavior is the formation of new Zn–OCO bonds leading to another 3-D network structure similar to that formed by the Sala derivative. A closer look at the packing reveals that the chloro and methyl substituents indeed occupy the channels in the lattice. The water molecules are naturally expected to use these chiral channels to escape during dehydration, which are partially blocked by the chloro and methyl substituents and hence complete removal of water molecules is hindered at approximately 100 °C. The reason for the dehydrated Zn(II)-RSala compounds to take back water

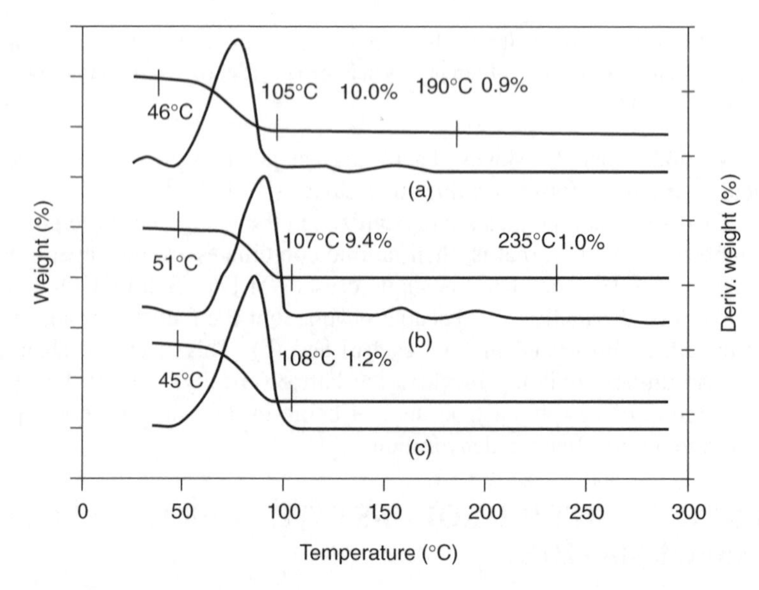

Figure 5. The TG and DTG curves for (a) $[Zn_2(ClSala)_2(H_2O)_2]\cdot 2H_2O$, and (b) $[Zn_2(MeSala)_2(H_2O)_2]\cdot 2H_2O$ and (c) $[Zn_2(Sala)_2(H_2O)_2]\cdot 2H_2O$.

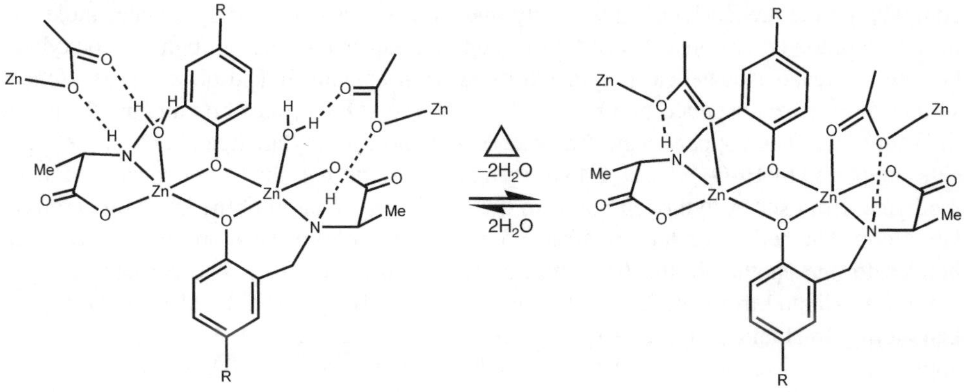

Figure 6. Schematic diagram depicting the interconversion between the two supramolecular structures.

during crystallization from aqueous solution is revealed by analyzing the packing of the close interactions in "$[Zn_2(RSala)_2]$" obtained by artificially introducing the chloro and methyl groups into the crystal structure of $[Zn_2(Sala)_2]$, since the structures of anhydrous Zn(II)-RSala are not available. The analyses revealed the presence of some repulsive interactions between these substituents and atoms in the other parts of the 3-D coordination network. It is likely that these interactions are the driving force for the observed interconvertibility, as depicted in Figure 6.

As mentioned above, the crystal structures of the dehydrated Zn(II)-RSala compounds could not be determined. However, when the single crystals of $[Zn_2(RSala)_2(H_2O)_2]\cdot 2H_2O$ were carefully dehydrated thermally, they became opaque while maintaining their single-crystal shape and morphology, similar to that observed for the Zn(II)-Sala complex. This observation supports the formation of new crystal structure with the point group 432 and space group, $P4_32_12$ in the Zn(II)-RSala compounds. Hence, it may be concluded that the Zn(II)-RSala compounds also undergo supramolecular transformation from hydrogen-bonded 3-D network to 3-D coordination polymeric network structures in the solid state in the thermal dehydration process.

The dream of crystal engineers encompasses the rational design and control of the crystal structures, and the fine-tuning of structures [1]. The knowledge gained in this work may provide an opportunity to predict the transformations of already reported compounds in the Cambridge Structural Database as well as to fine-tune the structures of the existing supramolecular architectures, and hence this is a step forward in the rationalization of the solid-state supramolecular reactivity.

4. SOLID-STATE TRANSFORMATION OF A HELICAL COORDINATION POLYMERIC STRUCTURE TO A 3-D COORDINATION NETWORK STRUCTURE BY THERMAL DEHYDRATION

Dark-green monoclinic crystals of $[Cu_2(Sala)_2(H_2O)]$ are formed when $Cu(OAc)_2\cdot H_2O$ is reacted with equimolar equivalents of LiOH and H_2Sala. In this neutral dicopper(II) complex, unlike the Zn(II) complex, only one axial site is occupied by a water molecule (site B) and the other (site A) forms a bridge to an adjacent dimer through a carboxylate-O. The net result is a helical polymer formed from apically linked dimers [13]. Helical strands are further sustained by O–H···O and N–H···O hydrogen bonding as illustrated in Figure 7(a). The copper-bound water molecule and carboxylate ligand are on the same side, supported by H–O–H···O–C hydrogen bonding. Further, the N–H hydrogen atom of the ligand from site A hydrogen bonds to the C=O oxygen atom of the bridging carboxylate group. The helical strands are arranged parallel to the *a*-axis and are further hydrogen bonded to one another in the *bc*-planes, as shown in Figure 7(b). The carbonyl-oxygen atom of the ligand from site B is further intermolecularly hydrogen bonded to two hydrogen atoms, N–H and O–H. As a result, chiral channels are created along the *a*-axis with approximate dimensions 4×9 Å (based on the nonhydrogen contacts), which is similar to the honeycombed hydrogen-bonded chiral network observed in $[Zn_2(Sala)_2(H_2O)_2]\cdot 2H_2O$. The geometric proximity of the carboxylate groups and the Cu(II) atoms of the B sites appear to be favorable for topochemical reactions to take place. The close Cu···O distance, 3.70 Å, between Cu1 and the neighboring carboxylate-O atom, is maintained by the hydrogen-bonded network in the *bc*-planes. As expected, removal of the aqua ligand by thermal dehydration yields anhydrous $[Cu_2(Sala)_2]$, where the basic dimeric structure has remained intact but now a porous, 3-D network coordination polymer has been established, similar to the Zn(II) analog.

This structural change is accompanied by minimum movement of atoms and conformational changes in the dimeric building block. However, the space group changes from $P2_12_12_1$ to $P4_32_12$. For this reason, single crystals of $[Cu_2(Sala)_2(H_2O)]$ become powder during thermal dehydration. Further, the anhydrous Cu(II) compound could not be rehydrated. It may also be noted that no Cu(II) analog similar to $[Zn_2(Sala)_2(H_2O)_2]\cdot 2H_2O$ has been isolated. Solid-state supramolecular reactions involving transformation of different structures are very rare since they involve breaking and making of bonds in more than one direction, compared to solid-state organic photochemical reactions where one or two new bonds are formed, usually involving two molecules [14]. Recently, several solid-state structural transformations have been reported [15].

Over and above having interesting structural chemistry, these systems, at least the Cu(II) complexes, possess interesting magnetochemical properties. Molecule-based magnetic materials have attracted attention in recent years as they are expected to exhibit technologically important properties [16]. In order to achieve magnetic ordering, it is essential to have unpaired electrons. The nature of coupling and their number will decide the resultant magnetic behavior. The electronic couplings, in turn, are a consequence of the structure motif in the solid state. In other words, a clear understanding of the structure-magnetic property relationship, in order to fine-tune magnetic properties, is necessary for the technological advancement in the field of magnetic materials. Most of the molecular magnets today possess magnetic ordering temperatures below room temperature. This limits their possible applications, since commercial materials usually require a magnetic ordering well above the room temperature. An exciting development is that, recently, molecular-based magnetic materials have been shown to have magnetic ordering temperatures well above room temperature. Ferlay and coworkers have reported a Curie temperature of 350 K in $V^{II/III}[Cr^{III}(CN)_6]\cdot 2.8H_2O$ [17]. A Curie temperature of 372 K was obtained by Miller *et al.* in $K_{0.058}V^{II/III}[Cr^{III}(CN)_6]_{0.79}\cdot(SO_4)_{0.058}\cdot 0.93H_2O$; vanadium has a mixed valence state in these complexes [18]. Interestingly, $[Cu_2(Sala)_2(H_2O)]$

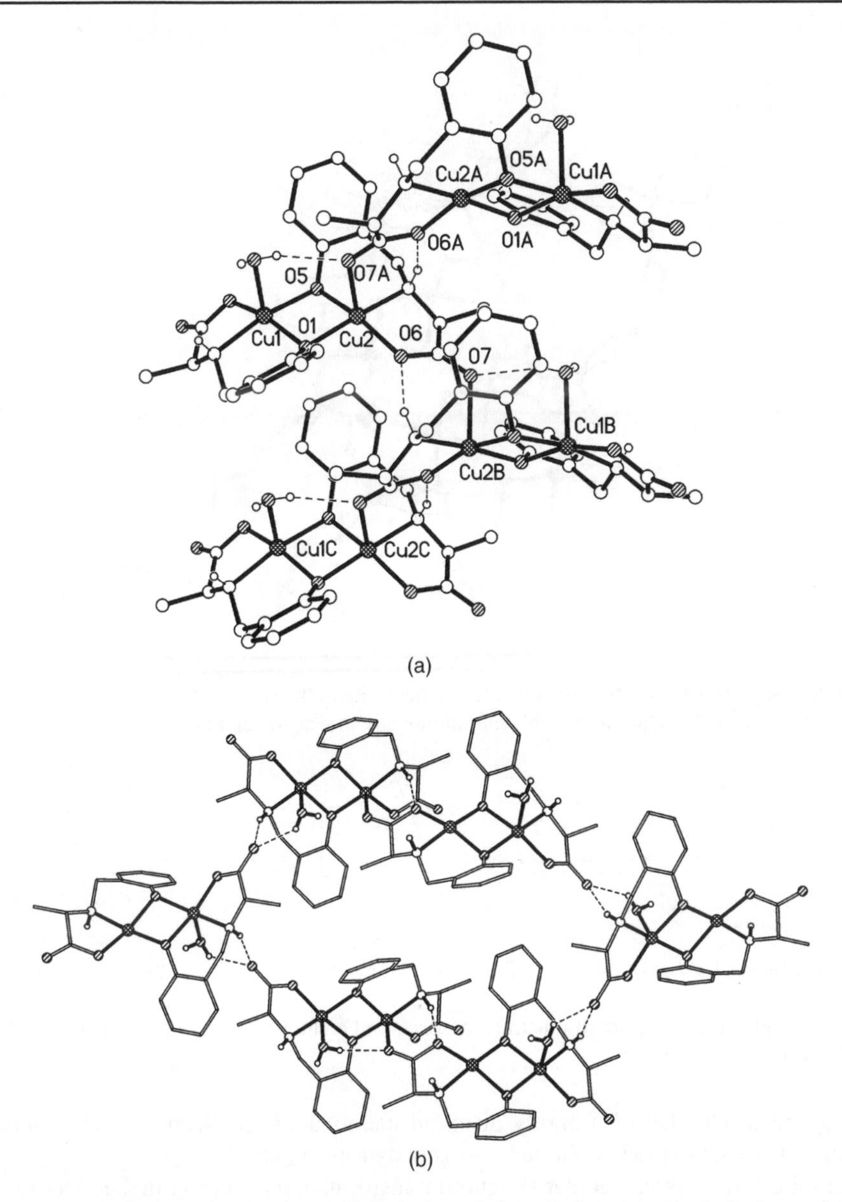

Figure 7. The hydrogen-bonding interactions in $[Cu_2(Sala)_2(H_2O)]$ along the helical strand (a) and (b) a slice of the honeycomblike structure viewed along the a-axis. The C–H hydrogen atoms have been omitted for clarity.

and $[Cu_2(Sala)_2]$ complexes exhibit canted antiferromagnetism at room temperature. The hydrated sample has a magnetic ordering temperature of 400 K, while the anhydrous structure has a magnetic ordering temperature as high as 435 K, which is the highest magnetic ordering temperature of molecular magnets to the best of our knowledge [19].

The connectivities operating in $[Zn_2(RSala)_2(H_2O)_2]\cdot 2H_2O$, and $[M_2(Sala)_2]$ (M = Zn and Cu) are worth mentioning here. In all of these 3-D network structures, the network

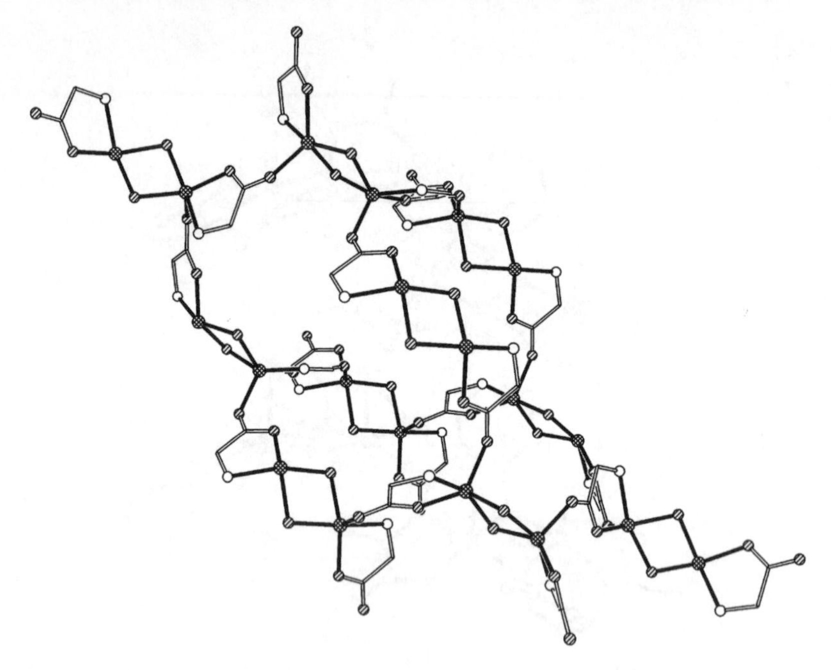

Figure 8. A portion of the diamondoid-like connectivity in the solid-state structures of $[M(Sala)_2]$, M = Cu(II) and Zn(II). Only a few relevant atoms are shown for clarity.

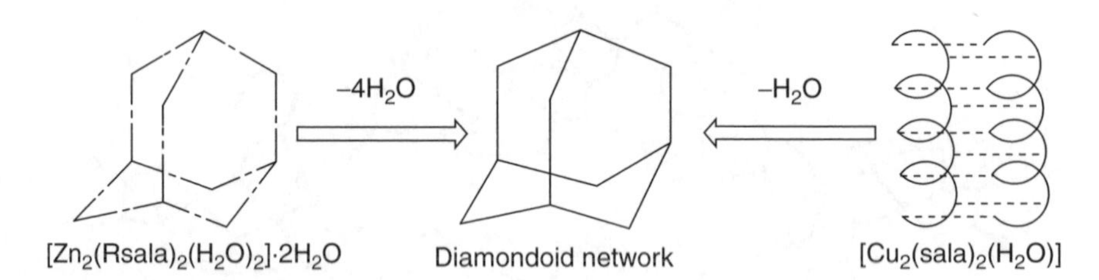

Figure 9. Schematic diagram to illustrate the structural connectivity and transformations of supramolecules in the solid state.

topology present is that of a highly distorted diamondoid architecture. The connectivity present in $[M_2(Sala)_2]$ (M = Zn and Cu) is shown in Figure 8.

The solid-state supramolecular structural transformation occurring in these coordination compounds may be summarized as shown in Figure 9.

5. INFLUENCE OF CHIRAL CENTERS ON THE HELICITY OF THE COORDINATION POLYMERS

Δ (or P) and Λ (or M), respectively, are the accepted descriptors for the right- and left-handed chirality of the helical structures [20]. The handedness of the helix can be predetermined through the use of chiral ligands [20]. The helicity in $[Cu_2(L\text{-}Sala)_2(H_2O)]$ has a Δ conformation with C_SN_R stereochemistry. The absolute stereochemistry determination of three additional, randomly selected, single crystals from different syntheses also

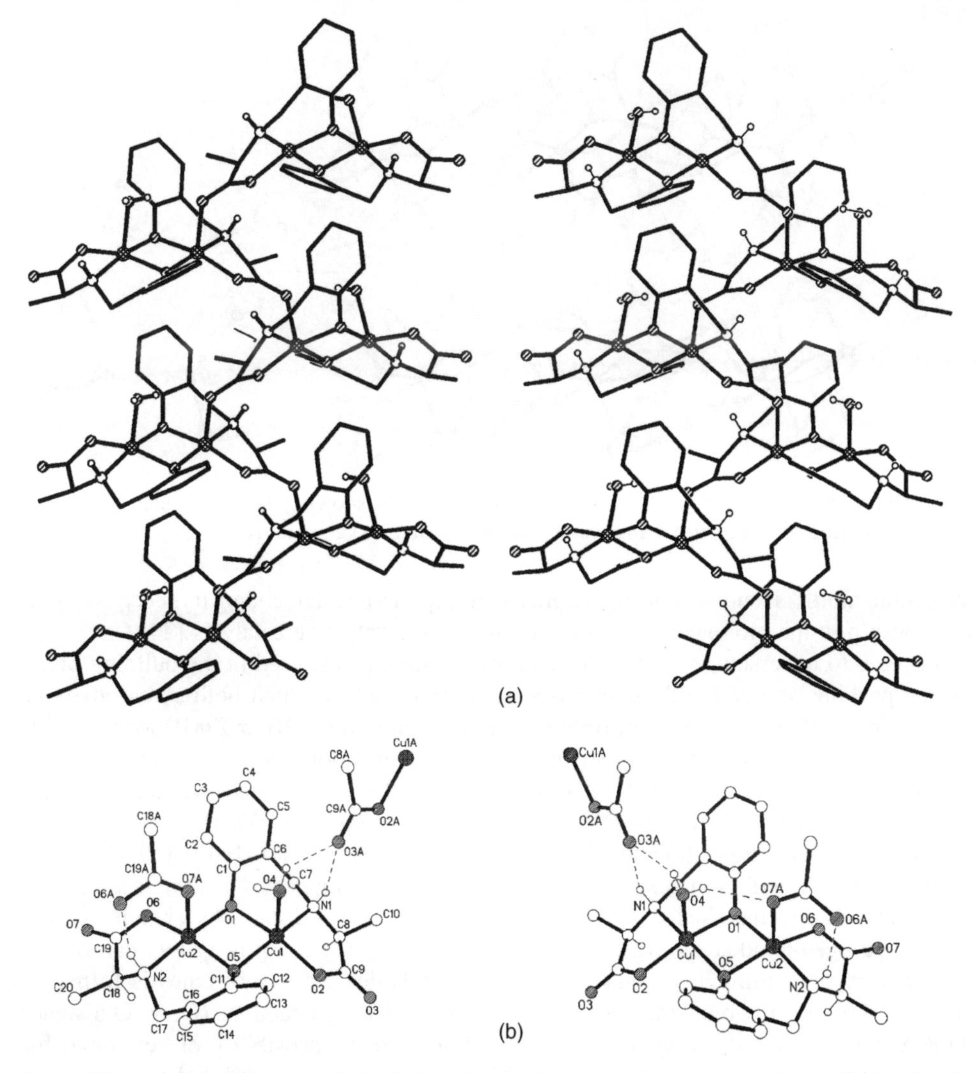

Figure 10. A comparison of the helical coils (a) $[Cu_2(L\text{-Sala})_2(H_2O)]$ and (b) $[Cu_2(D\text{-Sala})_2(H_2O)]$, as viewed along the *a*-axis.

proved to have a Δ helical conformation only. If L-H_2Sala gives rise to a Δ helix, the D-H_2Sala is expected to furnish a Λ conformation. Indeed, x-ray crystallographic studies of $[Cu_2(D\text{-Sala})_2(H_2O)]$ have shown that this is indeed the case [21]. The superimposition of the coordinates of these two dimeric building block structures shows they are exactly mirror images. The major difference is the absolute configuration at C8, C18 and N1, N2 atoms (Figure 10). As expected, the stereochemistry of the *D*-Sala complex is found to be C_RN_S. Furthermore, the helicity of the chain has changed from right-handed to left-handed (Λ) as shown in Figure 10. It appears that it might be possible to control the helicity by changing the chiral centers of the ligand. However, some literature indicated that chiral ligands can give rise to racemic crystals [22, 23]. The control of the helical

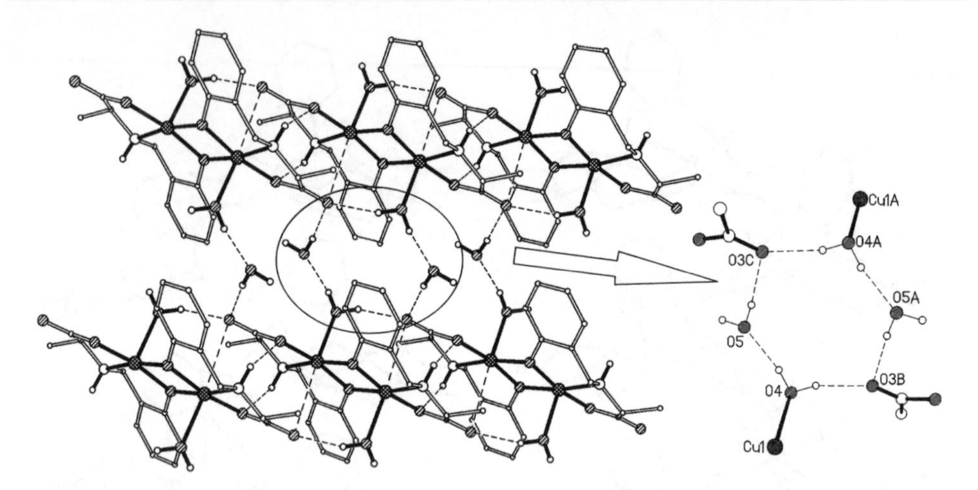

Figure 11. Packing diagram of $[D,L\text{-Sala})_2(H_2O)_2]\cdot 2H_2O$ showing intermolecular interactions and a cyclic ring, $R^6_6(12)$, involving water and carboxylate oxygen atoms.

structural motif is one of current interests in supramolecular chemistry because of its relationship to the biological systems and enantiomer selective catalysis [24].

In order to understand the effect of chirality of the ligands used in the building blocks on the packing, the *D*,*L*-Sala ligand has been employed in which both enantiomers are present in equal amounts. Complexation of *D*,*L*-Sala with Cu(II) or Zn(II) ions can give rise to either centrosymmetric dimer or a mixture of enantiomeric dimers. In the case of Cu(II), centrosymmetric $[Cu_2(D,L\text{-Sala})_2(H_2O)_2]\cdot 2H_2O$ was obtained, while a mixture of $[Zn_2(L\text{-Sala})_2(H_2O)_2]$ and $[Zn_2(D\text{-Sala})_2(H_2O)_2]$ was formed for Zn(II) [25].

The molecular composition and the structure of $[Cu_2(D,L\text{-Sala})_2(H_2O)_2]\cdot 2H_2O$ are different from the Cu(II) complexes containing optically pure ligands. The dimers are packed through O–H$\cdots$O and N–H$\cdots$O hydrogen bonding in the solid state. As a result, an interesting hydrogen-bonded ring (graphic notation $R^6_6(12)$[26]) comprising two carboxylate carbonyl-oxygen atoms, two aqua ligands and two lattice water molecules is formed as shown in Figure 11. Complementary Cu$\cdots$O(carboxylate) interactions (Cu$\cdots$O distance, 3.08 Å) between the centrosymmetric dimers indicated the possibility of new bond formation during thermal dehydration. The TG of this compound exhibited weight loss in the range 50–120 °C owing to dehydration. The observed weight loss matches well for the loss of all four water molecules. The loss of two metal bound aqua ligands along with two lattice water molecules below 120 °C is very similar to that observed during solid-state supramolecular transformations of Zn(II)-Sala complexes. The formation of new bonds between copper and carbonyl-oxygen atoms after dehydration is presumed to be the driving force for this behavior, ultimately leading to form a 1-D coordination polymer based on the packing in the solid state. Once formed, $[Cu_2(D,L\text{-Sala})_2]$ can easily be rehydrated, even on standing in air.

6. CONSEQUENCES OF C=O$\cdots\pi$ INTERACTIONS

The H_2Sgly (*N*-(2-hydroxybenzyl)-glycine) ligand is a prochiral molecule. But the complex $[Cu_2(Sgly)(H_2O)]\cdot H_2O$ crystallizes in the chiral space group $P2_12_12_1$. In the dimeric structure, the nitrogen atoms become chiral upon bonding to Cu(II) atoms and have the same chirality. The hydrogen bonding present in $[Cu_2(Sgly)(H_2O)]\cdot H_2O$ [27, 28] has been found to be very similar to that observed in $[Cu_2(Sala)_2(H_2O)]$, which undergoes a solid-state transformation after losing the water molecule below 110 °C. However, the Sgly complex fails to undergo such a transformation when thermally dehydrated and behaves like a normal coordination complex. In other words, the coordinated water molecule can only be removed when heated above 120 °C. The main difference between Cu(II)-Sgly and Cu(II)-Sala is the presence of one more water molecule in the lattice and the absence of a methyl group, which is replaced by a hydrogen atom in Sgly ligand. Furthermore, the O3A···Cu distance, 3.91 Å in $[Cu_2(Sgly)(H_2O)]\cdot H_2O$, is significantly longer than 3.70 Å found in the hydrated Cu(II)-Sala compound. The hydrogen-bond parameters are also considerably weaker in the Cu(II)-Sgly complex. Indeed, a comparison of hydrogen-bond parameters indicates that H1···O3A, 2.42 Å and H4A···O3A, 2.28 Å are significantly longer than the corresponding distances in Cu(II)-Sala compound, H1···O3A, 2.00 Å and H4A···O3A, 2.10 Å [13]. In addition, the plane containing carboxylate group of the neighboring dimers namely, C8A, C9A, O2A and O3A, is more parallel to the phenyl ring in Cu(II)-Sgly compound (interplanar angle = 8.9°) than in Cu(II)-Sala compound, (interplanar angle = 26°) owing to the absence of sterically bulky methyl groups as shown in Figure 12.

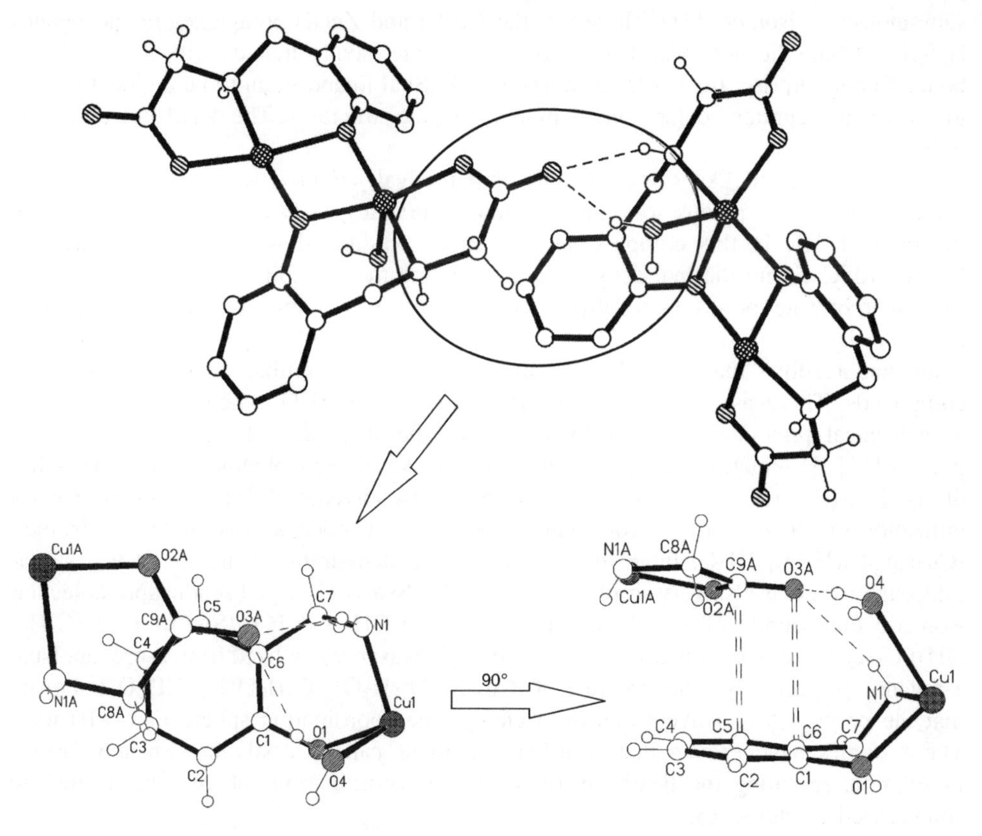

Figure 12. The C=O···π interaction present in $[Cu_2(Sgly)(H_2O)]$; the H_2O molecule is highlighted. Only relevant atoms are included for clarity. Adapted from Ref. [28].

A detailed analysis of the intermolecular interactions present in the Cu(II)-Sgly complex showed that the presence of $C{=}O\cdots\pi$ attractive intermolecular interactions between a carboxylate carbonyl and a phenyl group in the neighboring helices (Figure 12) appears to be responsible for the weakening of the hydrogen bonds, and thereby the thermal dehydration behavior by preventing the formation of a new Cu–O bond and the 3-D coordination network structure. This is a direct evidence for the influence of the weak interaction such as $C{=}O\cdots\pi$ on the solid-state structure and hence, as a consequence, upon the thermal dehydration properties. Such an influence in the Cu(II)-Sgly complex may not have been noticed in the absence of similar studies on the Cu(II)-Sala compounds. The $C{=}O\cdots\pi$ interaction has been well documented as early as 1979 by Addadi and Lahav [29] and later, by Moorthy and coworkers in 1994 [30]. This interaction can be used not only to control the reactivity in the solid state but also as an additional tool for the design and construction of novel multidimensional structures.

7. SUPRAMOLECULAR ISOMERISM

Conformationally flexible molecules are expected to exhibit polymorphism as well as supramolecular isomerism since the energies required to rotate the single bonds are comparable in magnitude to the lattice energy differences observed between polymorphs and supramolecular isomers [31]. However, the Cu(II) and Zn(II) complexes of the ligands H_2Sgly, H_2Sala did not exhibit this property, despite having ligands with flexible backbone. To our surprise, the Cu(II) complexes of H_2Sval ligand seem to be an ideal system to investigate supramolecular isomerism in solid-state structures. The details are described in this section.

Light-green needlelike single crystals of $[Cu_2(Sval)_2(H_2O)_3]$ were obtained by slow evaporation of an aqueous solution containing $Cu(OAc)_2{\cdot}H_2O$ and the Sval dianion in the ratio of 1:1. In this complex, all the three coordinated water molecules are lost below 110 °C during thermal dehydration, indicating the possibility of formation of new Cu–O(carboxylate) bonds upon dehydration. Unfortunately, single crystals of $[Cu_2(Sval)_2]$ could not be grown to confirm its structure by x-ray crystallography as the dehydrated compound readily "reacts" with the solvent used for crystallization to form solvated compounds. This was also true for $[Zn_2(RSala)_2(H_2O)_2]{\cdot}2H_2O$, where we have utilized topochemical principles successfully to deduce the structures of the anhydrous compounds [11]. Fascinating supramolecular isomers have been obtained when the anhydrous "$[Cu_2(Sval)_2]$" was recrystallized from various solvents [32]. For example, recrystallization of "$[Cu_2(Sval)_2]$" from aqueous solution afforded a supramolecular isomer, $[Cu_2(Sval)_2(H_2O)]{\cdot}2H_2O$. Interestingly, the thermal dehydration behavior of this supramolecular isomer is also very similar to that of $[Cu_2(Sval)_2(H_2O)_3]$. Pseudosupramolecular isomeric compounds of $[Cu_2(Sval)_2(H_2O)]{\cdot}0.5C_3H_7OH$ and $[Cu_4(Sval)_4(H_2O)_2(C_4H_9OH)]{\cdot}C_4H_9OH$ were obtained when "$[Cu_2(Sval)_2]$" was recrystallized from 2-propanol and 1-butanol, respectively. The structure of $[Cu_4(Sval)_4(H_2O)_2(C_4H_9OH)]{\cdot}C_4H_9OH$ showed that the properties of solvent can even change the coordination sphere of Cu(II) ions. These studies have proven that coordination bonds can be easily broken and formed in solution, revealing the flexibility of the Cu(II) coordination sphere. The results are summarized in Figure 13.

The main differences between the supramolecular isomers are as follows. In the crystal structure of $[Cu_2(Sval)_2(H_2O)_3]$, the dimeric building blocks are assembled almost

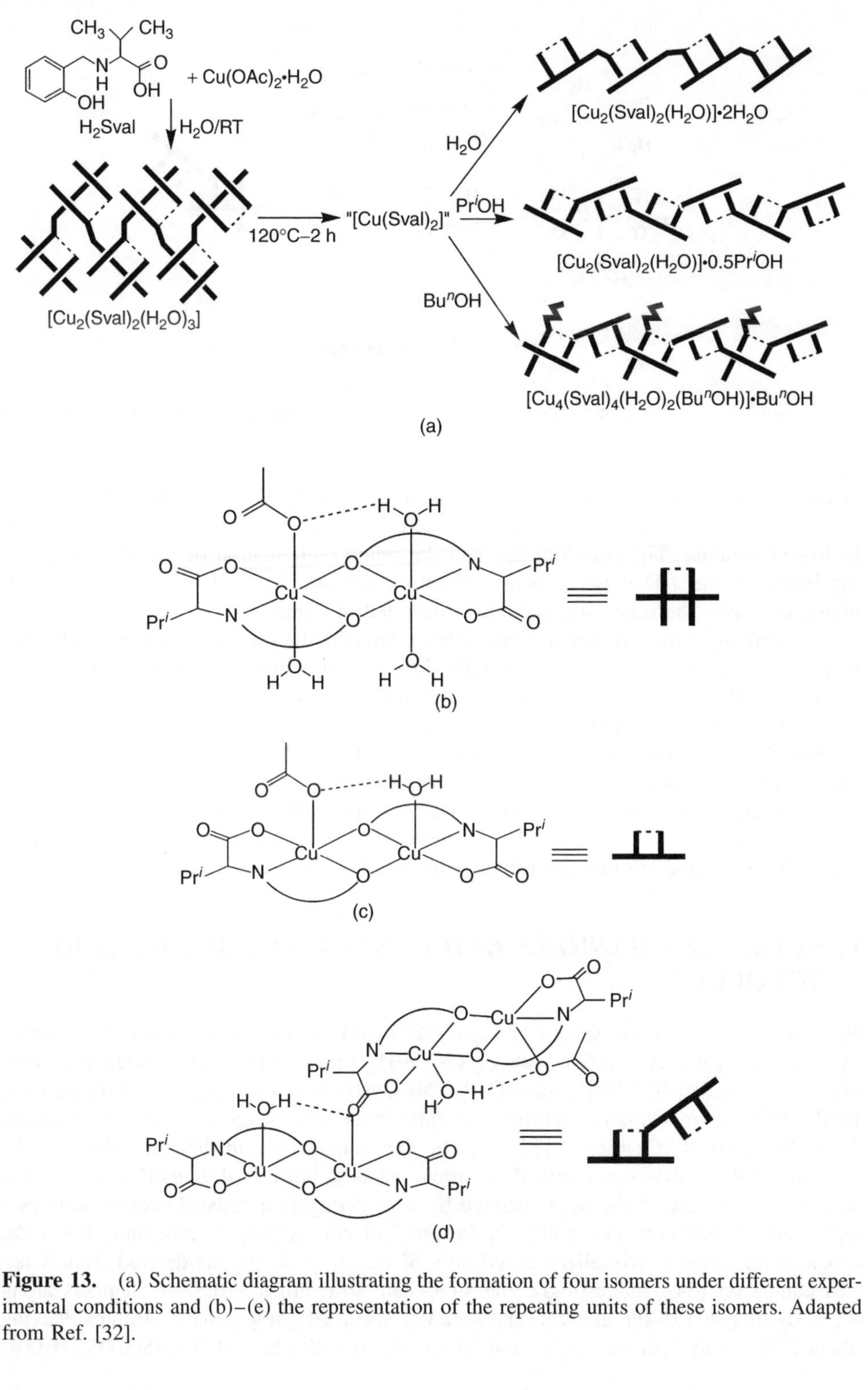

CH_3 CH_3
N H O OH OH
H_2Sval
+ $Cu(OAc)_2{\cdot}H_2O$
H_2O/RT
$[Cu_2(Sval)_2(H_2O)_3]$
120°C–2 h
"$[Cu(Sval)_2]$"
H_2O
$[Cu_2(Sval)_2(H_2O)]{\cdot}2H_2O$
Pr^iOH
$[Cu_2(Sval)_2(H_2O)]{\cdot}0.5Pr^iOH$
Bu^nOH
$[Cu_4(Sval)_4(H_2O)_2(Bu^nOH)]{\cdot}Bu^nOH$
(a)

O O H H O O O N Pr^i Cu Cu Pr^i N O O O H H H H ≡
(b)

O O H H O O O N Pr^i Cu Cu Pr^i N O O O ≡
(c)

O O O Cu N Pr^i N Cu O O Pr^i O O H H O H H O Pr^i N Cu O Cu O O O N Pr^i O O ≡
(d)

Figure 13. (a) Schematic diagram illustrating the formation of four isomers under different experimental conditions and (b)–(e) the representation of the repeating units of these isomers. Adapted from Ref. [32].

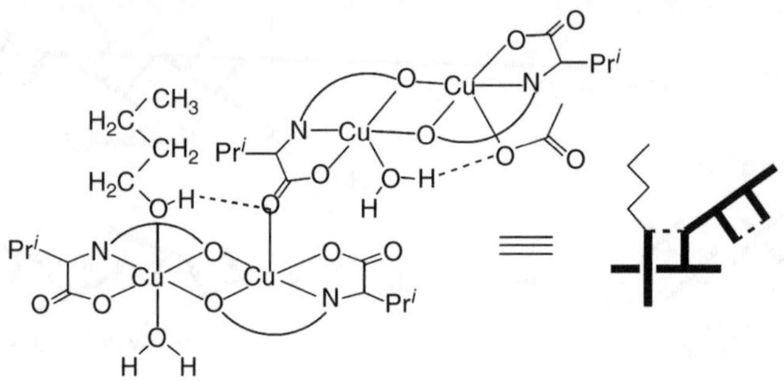

(e)

Figure 13. *(continued)*

perpendicular to each other about 2_1 screw axis. On the other hand, the dimers in $[Cu_2(Sval)_2(H_2O)]\cdot 2H_2O$ are parallel to each other and propagated by translational symmetry as depicted in Figure 13(a). Further, the packing efficiency of the latter isomer is better than the former isomer based on the calculated density. Owing to the presence of the aqua ligands in the latter, the N–H protons are not fully utilized in forming N–H···O=C hydrogen bonding. The conformation and the relative orientation of the dimeric building blocks in the 1-D polymer obtained from 2-propanol and 1-butanol may be better understood from the schematic diagrams illustrated in Figure 13.

One striking feature is that it is possible to "attach" the dimeric building blocks in a number of ways to form conformationally different one-dimensional coordination polymers. Secondly, coordination number five is predominant in these complexes. Thirdly, there is no correlation found between Cu–O–C angle and the nature of polymers obtained. Further, there is no correlation between the nature of the substituents on the phenolic ring and the nature of polymers obtained. This behavior may be due to the fact that all these polymers crystallized in the solid state are the kinetically favored products under the crystallization conditions employed. The coordination of water molecules to Cu(II) ions should be prevented in order to promote Cu–O(carboxylate) bond formation.

8. STARLIKE CHANNELS AND HEXAGONAL DIAMONDOID TOPOLOGY

Reaction of the disodium salt of H_2Scp11 (H_2Scp11 = *N*-(2-hydroxybenzyl)-1-amino-cyclopentyl-1-carboxylic acid), with $Cu(OAc)_2\cdot H_2O$ in an equimolar ratio gave a green crystalline powder, $[Cu_2(Scp11)_2(H_2O)_2]$. This Cu(II) complex when recrystallized from DMF/MeCN solvent mixture furnished mainly green cubic single crystals (space group, $Ia\bar{3}d$, No. 230) of $[Cu_2(Scp11)_2]\cdot H_2O$ [33]. The connectivity in this three-dimensional network structure may be described in simple terms as hexagonal diamondoid (or Lonsdaleite) architecture (if the net is defined by considering four fused dinuclear units as a single node). However, depending on the point of connectivity or redefining the node, this network could be visualized based on a S* ($6.6.6^2.6^2.6^2.6^2$) net derived from S lattice complexes [34]. Interestingly, the solid-state structure has starlike channels along the body diagonal of the unit cell and partially occupied guest water molecules, leaving about 27% empty space in the crystal lattice. Recrystallization of $[Cu_2(Scp11)_2(H_2O)_2]$

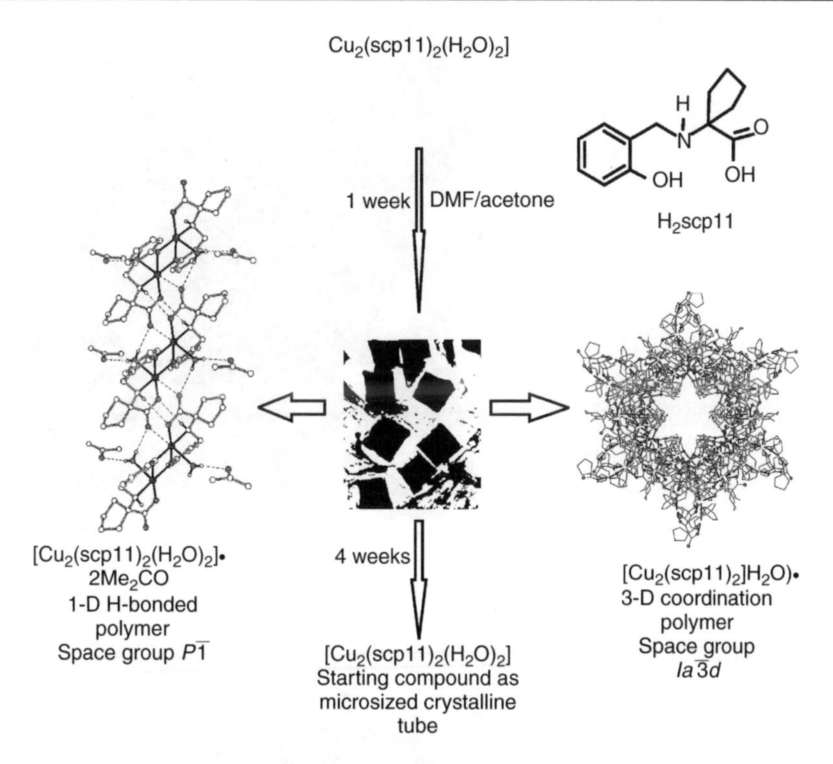

Figure 14. A schematic diagram representing the formation of three different structures of Cu(II) complexes of H_2Scp11 ligands. Adapted from Ref. [33].

from a mixture of DMF/acetone afforded a mixture of green and cubic crystals of $[Cu_2(Scp11)_2]{\cdot}H_2O$ and green-blue and long plates of $[Cu_2(Scp11)_2(H_2O)_2]{\cdot}2Me_2CO$, which has a 1-D hydrogen-bonded polymeric structure in the solid state (triclinic space group $P\overline{1}$, No. 2). When left in air at room temperature, the precipitating solvent is evaporated from the solution, both the above crystals slowly converted to microtubular crystals of $[Cu_2(Scp11)_2(H_2O)_2]$, which appears to be a thermodynamically more stable product. The reactions are summarized in Figure 14.

9. HYDROGEN-BONDED HELICAL WATER MOLECULES INSIDE A STAIRCASE 1-D COORDINATION POLYMER

Self-assembly of Ni^{II} and *N*-(2-hydroxybenzyl)-L-glutamic acid (H_3Sglu; Figure 15(a)) from aqueous solution lead to the formation of a staircaselike helical coordination polymer $[(H_2O)_2 \subset \{Ni(HSglu)\ (H_2O)_2\}]_n{\cdot}nH_2O$ [35]. The helical channels formed by coordination polymers host a stream of 1-D helical hydrogen-bonded lattice water molecules similar to those prevalent in biological systems where the helical protein matrix plays a role in stabilizing the host. The $HSglu^{2-}$ anion is coordinated through the neutral phenolic oxygen, secondary amine-N and carboxylate-O atoms in a *facial* manner along with another carboxylate oxygen from the neighboring molecule and two aqua ligands to define the octahedral geometry for Ni(II). This intermolecular carboxylate bridging generates a

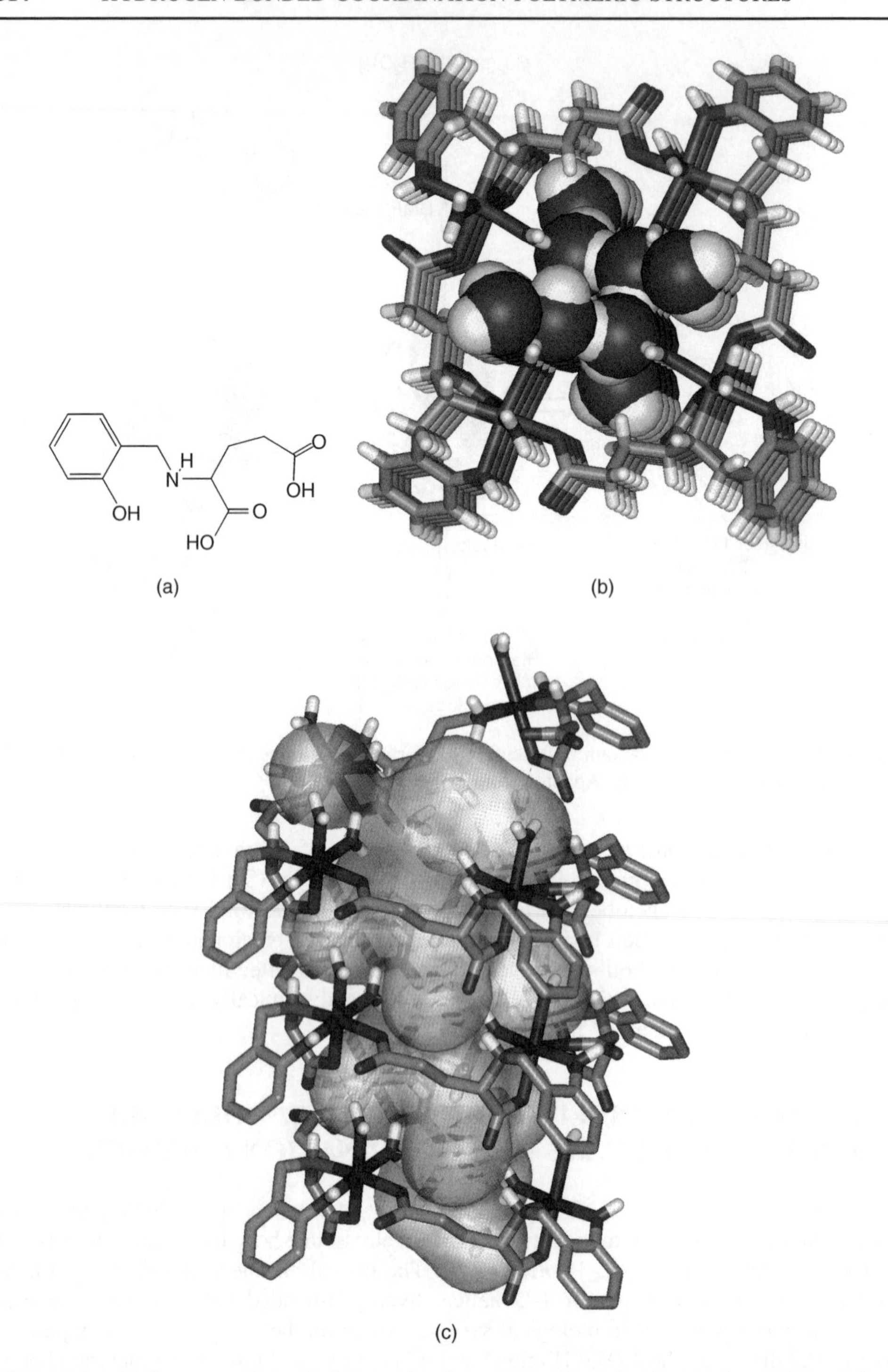

Figure 15. (a) Structure of the H_3Sglu molecule, and top (b) and side (c) views of the helical staircase coordination polymer containing a helical chain of water in the tube

left-handed staircaselike coordination polymer with a pseudo-4_1 screw axis. The H_2O and N–H donor groups form hydrogen bonds to the carboxylate-O atoms along the surface of the helical polymer, joining the coordination polymers together. Hence, this may be considered as hydrogen-bonded coordination polymeric tube. The dimensions of the approximately square-shaped cross section of the tube are 7.53 × 7.65 Å (closest distances between nickel atoms across the tube). One of the water molecules bonded to the Ni(II) atom is pointing into the tube and is normal to the axis of the helix. Such chiral channels are suitable for the guest water molecules to reside. This channel is indeed occupied by an array of helical hydrogen-bonded water stream as shown in Figures 15(b) and (c).

This novel display of lattice water assembled into hydrogen-bonded helical chain within a hollow and helical cavity defined by a metal-coordination polymer exemplifies, in the field of metalla-supramolecular chemistry, a nonbiological model for the water chains in membrane aquaporin proteins for the transport of water. Further, the intriguing structural feature as "helix inside the helix" illustrates yet another novel mode of the cooperative self-assembly of water molecules in the crystal host and these results may emulate the rationale that the degree of structuring imposed on water–water interactions by its diverse environments and vice versa can be profound. The 1-D zigzag

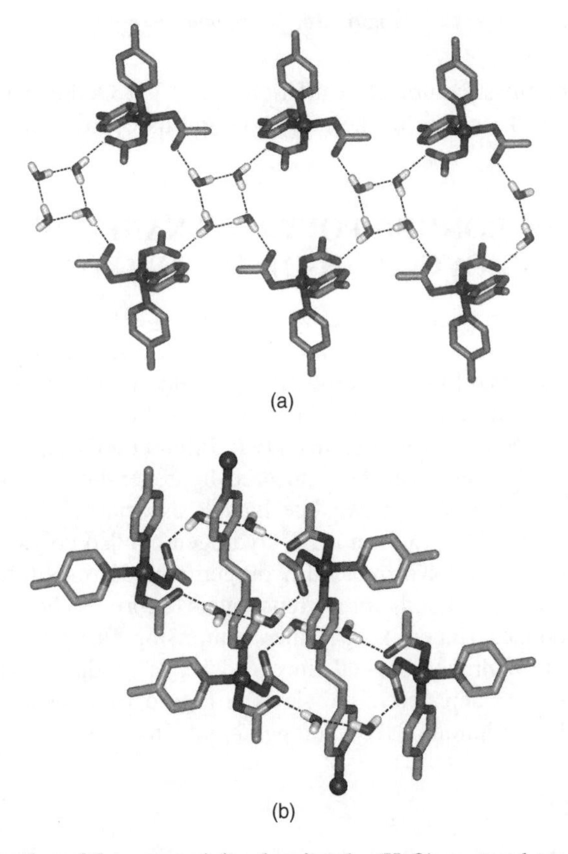

Figure 16. (a) A portion of the connectivity showing the $(H_2O)_4$ water clusters, (b) polyrotaxane-like interpenetration between the adjacent bpe ligands, and (c) a schematic diagram of the rotaxane-like structure.

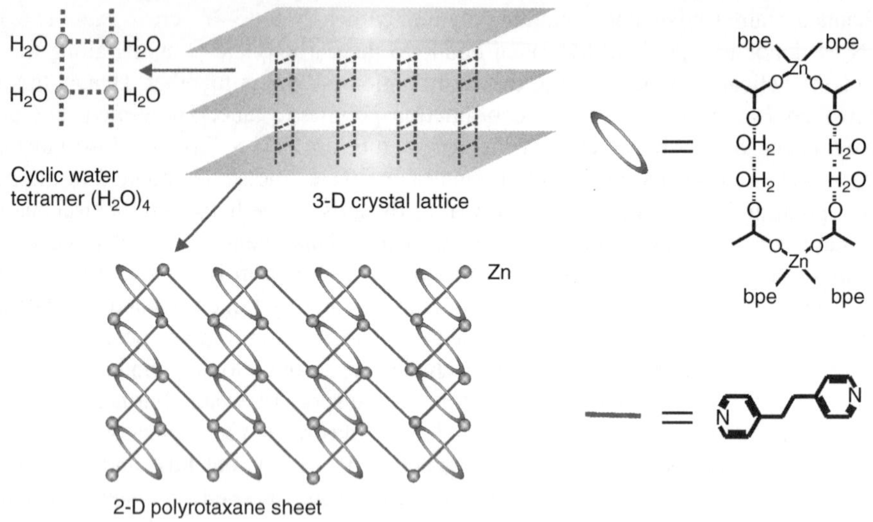

(c)

Figure 16. (*continued*)

coordination polymeric structure of $[Cu(HSglu)(H_2O)]\cdot H_2O$ illustrates that the overall topology depends on the nature of the metal ion and its coordination geometry in these systems [35].

10. HYDROGEN-BONDED POLYROTAXANE-LIKE STRUCTURE CONTAINING CYCLIC $(H_2O)_4$ IN $[ZN(OAC)_2(\mu\text{-}BPE)]\cdot 2H_2O$

The reaction of 4,4′-bipyridylethane (bpe) with $Zn(OAc)_2\cdot 2H_2O$ has led to the formation of a coordination polymer, $[Zn(OAc)_2(\mu\text{-}bpe)]\cdot 2H_2O$ with a zigzag coordination polymeric structure in the solid state [36]. However, the presence of two lattice water molecules makes the crystal structure particularly interesting. In the solid, the carboxylate C=O oxygen atoms of the $Zn(OAc)_2$ groups from two different adjacent zigzag polymers and four lattice water molecules form 24-membered hydrogen-bonded ring with graph set notation, $R^6_6(24)$ [26]. One of the two bpe ligands associated with each Zn(II) passes through the center of this ring to form a 2-D hydrogen-bonded polyrotaxane structure. In the solid state, the adjacent 24-membered hydrogen-bonded rings further fused together through O–H$\cdots$O hydrogen bonds among four waters to form cyclic $(H_2O)_4$. This resulted in 1-D hydrogen-bonded ribbonlike polymer comprising fused alternating 24- and 8-membered O–H$\cdots$O hydrogen-bonded rings and hence produce 3-D hydrogen-bonded polyrotaxane network, as depicted in the views of Figure 16. A neutron diffraction study provides a detailed description of the hydrogen bonds involved.

11. SUMMARY

The success in the construction of a variety of coordination polymeric structures is mainly attributed to the lability of metal–ligand bonds in solution. This property allows for the breaking and making of new bonds in solution. When a number of species are in equilibrium in solution, the crystallizing conditions determine the type of complexes formed in the solid state. The NH and C=O groups present in the ligands invariably form strong hydrogen bonds in all the coordination polymers studied. The water present in the lattice or bonded to the metal also participates in the hydrogen bonding to stabilize the crystal structures.

12. ACKNOWLEDGMENTS

The National University of Singapore is thanked for financial support. I would like to thank all my students and research assistants whose names appear in the reference list below.

REFERENCES

1. (a) G. R. Desiraju, *Crystal Engineering: The Design of Organic Solids*, Elsevier, Amsterdam, 1989; (b) G. R. Desiraju, *Acc. Chem. Res.*, **35**, 565–573 (2002); (c) G. R. Desiraju, in *Implications of Molecular and Materials Structure for New Technologies* (Eds. J. A. K. Howard, F. H. Allen, G. P. Shields), Kluwer Academic Publishers, Dordrecht, 1998; (d) G. R. Desiraju, *Angew. Chem., Int. Ed. Engl.*, **34**, 2311–2327 (1995); (e) G. R. Desiraju, *J. Mol. Struct.*, **374**, 191–198 (1996); (f) G. R. Desiraju, *Chem. Commun.*, 1475–1482 (1997).
2. (a) D. Braga, *J. Chem. Soc., Dalton Trans.*, 3705–3713 (2000); (b) A. J. Blake, N. R. Champness, P. Hubberstey, W.-S. Li, M. A. Withersby and M. Schröder, *Coord. Chem. Rev.*, **183**, 117–138 (1999); (c) K. Biradha and M. Fujita, in *Perspectives in Supramolecular Chemistry: Crystal Design: Structure and Function*, Vol. **7** (Ed. G. R. Desiraju), Wiley, England, 211–239, 2003; (d) A. Aoyama, *Top. Curr. Chem.*, **198**, 131–161 (1998); (e) R. Robson, *J. Chem. Soc., Dalton Trans.*, 3735–3744 (2000).
3. S. L. James, *Chem. Soc. Rev.*, **32**, 276–288 (2003).
4. (a) J. S. Miller and A. J. Epstein, in *Crystal Engineering: From Molecules and Crystals to Materials* (Eds. D. Braga, F. Grepioni and A. G. Orpen), Kluwer Academic Publishers, Dordrecht, 43–53, 1999; (b) O. Kahn, *Acc. Chem. Res.*, **33**, 647–657 (2000); (c) O. R. Evans and W. Lin, *Acc. Chem. Res.*, **35**, 511–522 (2002); (d) M. Eddaoudi, D. B. Moler, H. Li, B. Chen, T. M. Reineke, M. O'Keeffe and O. M. Yaghi, *Acc. Chem. Res.*, **34**, 319–330 (2001); (e) O. M. Yaghi, H. Li, C. Davis, D. Richardson and T. L. Groy, *Acc. Chem. Res.*, **31**, 474–484 (1998).
5. S. R. Batten and R. Robson, *Angew. Chem., Int. Ed. Engl.*, **37**, 1460–1494 (1998).
6. (a) L. Carlucci, G. Ciani, D. M. Proserpio and A. Sironi, *J. Chem. Soc., Chem. Commun.*, 2755–2756 (1994); (b) A. J. Blake, N. R. Champness, M. Crew and S. Parsons, *New J. Chem.*, **23**, 13–15 (1999); (c) O. M. Yaghi and H. Li, *J. Am. Chem. Soc.*, **118**, 295–296 (1996); (d) M. Kondo, T. Yoshitomi, K. Seki, M. Matsuzaka and S. Kitagawa, *Angew. Chem., Int. Ed. Engl.*, **36**, 1725–1727 (1997); (e) C. J. Kepert and M. J. Rosseinsky, *J. Chem. Soc. Chem. Commun.*, 375–376 (1999).
7. (a) A. D. Burrows, C.-W. Chan, M. M. Chowdry, J. E. McGrady and D. M. P. Mingos, *Chem. Soc. Rev.*, **24**, 329–339 (1995); (b) L. Brammer, in *Perspectives in Supramolecular Chemistry: Crystal Design: Structure and Function*, Vol. **7** (Ed. G. R. Desiraju), Wiley, England, 1–75, 2003; (c) C. B. Aakeröy, *Acta Crystallogr.*, **B53**, 569–586 (1997); (d) A. M. Beatty, *CrystEngComm*, **3**, 243–255 (2001); (e) C. B. Aakeröy and A. M. Beatty, *Aust. J. Chem.*, **54**, 409–421 (2001); (f) G. R. Desiraju, *J. Chem. Soc., Dalton Trans.*, 3745–3751 (2000); (g) S. Subramanian and M. J. Zaworotko, *Coord. Chem. Rev.*, **137**, 357–401 (1994).
8. (a) X. Gao, T. Friščić and L. R. MacGillivray, *Angew. Chem., Int. Ed. Engl.*, **43**, 232–236 (2004); (b) T. Friščić and L. R. MacGillivray, *Chem. Commun.*, 1306–1307 (2003); (c) B. Dush-

yant, G. Papaefstathiou and L. R. MacGillivray, *Chem. Commun.*, 1964–1965 (2002); (d) L. R. MacGillivray, *CrystEngComm*, **4**, 37–41 (2002); (e) see Chapter 3 of this book.

9. J. D. Ranford, J. J. Vittal and D. Wu, *Angew. Chem., Int. Ed. Engl.*, **37**, 1114–1116 (1998).
10. M. J. Zaworotko, *Angew. Chem., Int. Ed. Engl.*, **37**, 1211–1213 (1998).
11. J. J. Vittal and X.-D. Yang, *Cryst. Growth Des.*, **2**, 259–262 (2002).
12. (a) A. V. Nicolaev, Y. A. Logvinenko and L. I. Myachia, *Thermal Analysis*, Vol. 2, Academic Press, New York, 779–785, 1969; (b) L. G. Berg, in *Differential Thermal Analysis*, Vol. 1 (Ed. R. C. Mackenzie), Academic Press, London, 343–361, 1970; (c) F. Paulik, *Special Trends in Thermal Analysis*, John Wiley, England, 16, 1995.
13. J. D. Ranford, J. J. Vittal, D. Wu and X. Yang, *Angew. Chem., Int. Ed. Engl.*, **38**, 3498–3501 (1999).
14. (a) G. M. J. Schmidt, *Pure Appl. Chem.*, **27**, 647–678 (1971); (b) C. R. Theocharis and W. Jones, in *Organic Solid-State Chemistry* (Ed. G. R. Desiraju), Elsevier, Amsterdam, 47–68, 1987; (c) K. Tanaka and F. Toda, *Chem. Rev.*, **100**, 1025–1074 (2000); (d) G. W. Coates, A. R. Dunn, L. M. Henling, J. W. Ziller, E. B. Lobkovsky and R. H. Grubbs, *J. Am. Chem. Soc.*, **120**, 3641–3649 (1998).
15. (a) D. M. Shin, I. S. Lee, D. Cho and Y. K. Chung, *Inorg. Chem.*, **42**, 7722–7724 (2003); (b) X. Xue, X.-S. Wang, R.-G. Xiong, X.-Z. You, B. F. Abrahams, C.-M. Che and H.-X. Ju, *Angew. Chem., Int. Ed. Engl.*, **41**, 2944–2946 (2002); (c) B. Rather, B. Moulton, R. D. B. Walsh and M. J. Zaworotko, *Chem. Commun.*, 694–695 (2002); (d) O. S. Jung, S. H. Park, K. M. Kim and H. G. Jang, *Inorg. Chem.*, **37**, 5781–5785 (1998); (e) L. Iordanidis and M. G. Kanatzidis, *J. Am. Chem. Soc.*, **122**, 8319–8320 (2000); (f) L. Iordanidis and M. G. Kanatzidis, *Angew. Chem., Int. Ed. Engl.*, **39**, 1927–1930 (2000).
16. J. S. Miller, *Inorg. Chem.*, **39**, 4392–4408 (2000).
17. S. Ferlay, T. Mallah, R. Quahes, P. Veillet and M. Verdaguer, *Nature*, **378**, 701–703 (1995).
18. **Ø**. Hatlevik, W. E. Buschmann, J. Zhang, J. L. Manson and J. S. Miller, *Adv. Mater.*, **11**, 914–918 (1999).
19. X. Yang, L. Si, J. Ding, J. D. Ranford and J. J. Vittal, *Appl. Phys. Lett.*, **78**, 3502–3504 (2001).
20. (a) M. Albrecht, *Chem. Soc. Rev.*, **27**, 281–287 (1998); (b) C. Provent and A. F. Williams, in *Transition Metals in Supramolecular Chemistry* (Ed. J. P. Sauvage), John Wiley & Sons, London, 135–191, 1999; (c) C. Piguet, G. Bernadinelli and G. Hopfgartner, *Chem. Rev.*, **97**, 2005 (1997).
21. X. Yang, *Influence of Reduced Schiff Base Ligand Substituents on the Solid-State Structures of their Cu(II) and Zn(II) Complexes*, Ph.D. Thesis, National University of Singapore, 2003, 73–80.
22. M. Mizutani, N. Maejima, K. Jitsukwa, H. Masuda and H. Einaga, *Inorg. Chim. Acta*, **283**, 105–110 (1998).
23. E. Colacio, M. Ghazi, R. Kivekäs and J. M. Moreno, *Inorg. Chem.*, **39**, 2882–2890 (2000).
24. L. J. Prins, J. Huskens, F. Dejong, P. Timmerman and D. N. Reinhoudt, *Nature*, **398**, 498–502 (1999).
25. C.-T. Yang, M. Vetrichelvan, X. Yang, B. Moubaraki, K. S. Murray and J. J. Vittal, *Dalton Trans.*, 113–121 (2004).
26. (a) M. C. Etter, J. C. MacDonald and J. Bernstein, *Acta Crystallogr.*, **B46**, 256–262 (1990); (b) J. Bernstein, R. E. Davis, L. Shimoni, N.-L. Chang, *Angew. Chem., Int. Ed. Engl.*, **34**, 1555–1573 (1995).
27. (a) J. X. Xu, M. Q. Chen, Z. H. Wang and H. L. Zhang, *Acta Chimi. Sin.*, 434–437 (1989); (b) M.-Q. Chen, J.-X. Xu, H.-L. Zhang, *Jiegou Huaxue (J. Struct. Chem.)*, **10**, 1–9 (1991).
28. X. Yang, D. Wu, J. D. Ranford and J. J. Vittal, *Cryst. Growth Des.*, **5**, 41–43 (2005).
29. L. Addadi and M. Lahav, *Pure Appl. Chem.*, **51**, 1269–1284 (1979).
30. J. N. Moorthy, S. D. Samant and K. Venkatesan, *J. Chem. Soc., Perkin Trans.*, **2**, 1223–1228 (1994).

31. (a) B. Moulton and M. J. Zaworotko, *Chem. Rev.*, **101**, 1629–1658 (2001); (b) T. L. Hennigar, D. C. MacQuarie, P. Losier, R. D. Rogers and M. J. Zaworotko, *Angew. Chem., Int. Ed. Engl.*, **36**, 972–973 (1997); (c) M. J. Zaworotko, *Chem. Commun.*, 1–9 (2001); (d) J. A. R. P. Sarma and G. R. Desiraju, in *Crystal Engineering: The Design and Application of Functional Solids* (Eds. M. J. Zaworotko and K. R. Seddon), Kluwer Academic Publishers, Dordrecht, 325–356, 1999; (e) V. S. S. Kumar, A. Addlagatta, A. Nangia, W. T. Robinson, C. K. Broder, R. Mondal, I. R. Evans, J. A. K. Howard and F. H. Allen, *Angew. Chem., Int. Ed. Engl.*, **41**, 3848–3851 (2002); (f) G. R. Desiraju, *Nat. Mater.*, **1**, 77–788 (2002).
32. X. Yang, J. D. Ranford and J. J. Vittal, *Cryst. Growth Des.*, **4**, 781–788 (2004).
33. B. Sreenivasulu and J. J. Vittal, *Cryst. Growth Des.*, **3**, 635–637 (2003).
34. A. F. Wells, *Further Studies of Three-Dimensional Nets*, Vol. 8, ACA Monograph, Washington, 10, 1979.
35. B. Sreenivasulu and J. J. Vittal, *Angew. Chem., Int. Ed. Engl.*, **43**, 5769–5772 (2004).
36. M. T. Ng, T. C. Deivaraj, W. T. Klooster, G. J. McIntyre and J. J. Vittal, *Chem. Eur. J.*, **10**, 5853–5859 (2004).

Index

Figures are indicated by *italic numbers*

Frontiers in Crystal Engineering. Edited by Edward R.T. Tiekink and Jagadese J. Vittal

With kind thanks to Paul Nash for creation of this index.